EARTHQUAKE ENGINEERING IN EUROPE

GEOTECHNICAL, GEOLOGICAL AND EARTHQUAKE ENGINEERING

Volume 17

For further volumes:
http://www.springer.com/series/6011

Earthquake Engineering in Europe

edited by

MIHAIL GAREVSKI
Institute of Earthquake Engineering and Engineering Seismology (IZIIS), Skopje, R. Macedonia

ATILLA ANSAL
Kandilli Observatory and Earthquake Research Institute, Boğaziçi University, Istanbul, Turkey

Springer

Editors
Mihail Garevski
Institute of Earthquake
Engineering & Engineering Seismology
(IZIIS)
Skopje
Macedonia
garevski@pluto.iziis.ukim.edu.mk

Atilla Ansal
Kandilli Observatory and Earthquake
Research Institute
Boğaziçi University
Cengelkoy 34688
Istanbul, Turkey
ansal@boun.edu.tr

ISSN 1573-6059
ISBN 978-90-481-9543-5 e-ISBN 978-90-481-9544-2
DOI 10.1007/978-90-481-9544-2
Springer Dordrecht Heidelberg London New York

Library of Congress Control Number: 2010932608

Printed on acid-free paper

Springer is part of Springer Science+Business Media (www.springer.com)

Preface

The need for integrated books on earthquake engineering and management is increasing, particularly having in mind the rising trend of damage and loss of human lives due to earthquakes. The increase of the seismic risk is not because of the increase of the number of occurred earthquakes since their frequency does not change statistically, but it is rather due to the growing population in earthquake prone areas.

The book contains the invited papers to be presented at the 14th European Conference on Earthquake Engineering (14ECEE) to be held in Ohrid, R. Macedonia, during 30th August – 3rd September, 2010. This event takes place every 4 years and represents the most important forum for exchange of results from the latest investigations carried out in the field of earthquake engineering between two conferences. The present book containing the papers of the keynote and theme lecturers is published in addition to the Book of Abstracts and the DVD disc containing all the papers submitted to the Conference.

The unique character of this book is that the titles of the chapters are selected to enable a complete insight into the state-of-the-art in the field of earthquake engineering. The papers related to engineering seismology and seismic risk management, in addition to the papers on earthquake engineering in this book, add much to its value since they are concerned with problems related to high intensity earthquakes.

The book will be of a high benefit for the participants of the Conference and other scientists because the authors have presented much of the latest research done in this field. It can also be a useful tool for the engineers and students since the authors were allowed to present their investigations in much detail (up to 25 pages). All the illustrations (photos and figures) that are in full colour contribute to better visualization of the issues involved in this book.

The contributors to the book are among distinguished scientists from Europe. Submissions of renowned researchers from the USA are also included. The book starts with the contribution entitled "Seismic Engineering of Monuments" by T. Tassios, who is awarded to deliver the First Prof. Nicholas Ambraseys Distinguished Lecture. This presentation is followed by six sections on the following topics:

Engineering Seismology;
Geotechnical Earthquake Engineering;
Seismic Performance of Buildings;
Earthquake Resistant Engineering Structures;
New Techniques and Technologies;
Managing Risk in Seismic Regions.

Each of these sections starts with the contributions of the keynote speakers followed by those of the theme lecturers.

The first section begins with the presentation of A. Ansal et al. dealing with seismic microzoning and comparison of different microzoning maps elaborated on the basis of different parameters. The possibility of application of ground motion prediction equations (GMPEs) developed for a single region to other regions is considered in the subsequent presentation (J. Stewart).

The second section starts with soil response to earthquake ground motion. Reference is given to seismic response of structures including soil-structure interaction by linear and nonlinear analysis (A. Pecker and C.T. Chatzigogos). The second contribution is given by P. Bard et al. and it considers non-destructive techniques as the ambient vibration measurements for obtaining soil site amplifications. This section ends with the contribution of Cubrinvski et al. that investigates the decrease of liquefaction resistance with increased quantity of non-plastic fines.

The subsequent section referring to the behaviour of buildings under earthquake effect contains six chapters. All these chapters deal with themes that are very much of a current interest. F. Naeim analyses the behaviour (performance based seismic design) of tall buildings considering the growing popularity of construction of such buildings in seismically prone regions. The need for introducing nonlinear analysis of structures in engineering practice is pointed out (N. Aydinoglu and G. Onem). This part of the book also contains review of techniques used for retrofitting and strengthening of historic buildings (C. Oliveira and A. Costa). Further in this section, G. M. Calvi summarizes the historical background of implementation and improvement of engineering regulations contributing to safe seismic construction. In his paper, A. Kappos provides examples of different procedures for design of buildings with different number of storeys. At the end of the section, J. Wallace gives recommendations for performance based design of tall core wall structures.

In the fourth section of the book dedicated to earthquake resistance of engineering structures, P. Pinto and P. Franchin consider problems and methods used in analysis and design of bridges. Further in this section, in the chapter on development of health monitoring of structures (E. Safak et al.), consideration is given to a number of techniques for detection of damage to structures.

In the penultimate section, A. Pinto describes large scale pseudo-dynamic and hybrid (substructure) nonlinear tests. The next paper in this section written by R. Severn presents the great contribution of seismic shaking table tests of models to the progress of earthquake engineering in the period 1900–1980. This part of the book ends with a paper on manufacturing, testing and installation of low-cost rubber bearings (M. Garevski).

The last section of the book considers the possibilities for reduction and mitigation of earthquake consequences. S. Briceno presents the goals of the Hyogo Framework for Action. It is suggested that successful seismic risk reduction is only possible provided that risk management is incorporated in the planning policy of each country located in a seismically active region. This part also contains examples of necessary measures taken in the aftermath of the Abruzzi earthquake (M. Dolce). In the paper by C. Modena et al., special attention is paid to historic buildings damaged by the L'Aquila earthquake. The next paper (M. Erdik et al.) refers to almost real time damage assessment. The methodology and the software for rapid earthquake loss assessment is considered. The book ends with the contribution of H. Shah et al. that considers the possibility of renovation of dwellings damaged by earthquakes through development of a safety economic net in the form of micro insurance.

The editors wish to extend their gratitude to all the authors of the included papers for their cooperation in the creation of this book that provides a valuable contribution to the common goal of the earthquake engineering community, i.e., reduction of loss of human lives and damage to property due to earthquakes.

Skopje, Republic of Macedonia — Mihail Garevski
Istanbul, Turkey — Atilla Ansal

Contents

Contributors

J. Almeida Lisbon Fundaçao da Faculdade Ciencias da Universidade de Lisboa, IDL, 1749-016 Lisboa, Portugal, joanaribeiroalmeida@gmail.com

Atilla Ansal Kandilli Observatory and Earthquake Research Institute, Boğaziçi University, Çengelköy, Istanbul, Turkey, ansal@boun.edu.tr

M. Nuray Aydınoğlu Kandilli Observatory and Earthquake Research Institute, Boğaziçi University, 34684 Istanbul, Turkey, aydinogn@boun.edu.tr

P.-Y. Bard LGIT, Maison des Geosciences, Joseph Fourier University, 38041 Grenoble Cedex 9, France, bard@obs.ujf-grenoble.fr

Elisabeth Bowman Department of Civil and Natural Resources Engineering, University of Canterbury, Christchurch 0050 8140, New Zealand, elisabeth.bowman@canterbury.ac.nz

Sálvano Briceño UNISDR, Palais des Nations, Geneva, Switzerland, briceno@un.org

J. Burjanek Swiss Seismological Service, ETH Zürich, 8092 Zürich, Switzerland, j.burjanek@sed.ethz.ch

H. Cadet Institute of Engineering Seismology and Earthquake Engineering (ITSAK), 55102 Thessaloniki, Greece, kdhelo@gmail.com

Eser Çaktı Kandilli Observatory and Earthquake Research Institute, Bogaz ici University, 38684 Istanbul, Turkey, eser.cakti@boun.edu.tr

Gian Michele Calvi Department of Structural Mechanics, University of Pavia, 27100 Pavia, Italy, gm.calvi@eucentre.it

Filippo Casarin Department of Structural and Transportation Engineering, University of Padova, 35131 Padova, Italy, casarin@dic.unipd.it

Charisis T. Chatzigogos Géodynamique et Structure, 92220 Bagneux, France, charisis@lms.polytechnique.fr

C. Cornou LGIT, Maison des Geosciences, Joseph Fourier University, 38041 Grenoble Cedex 9, France, Cecile.Cornou@obs.ujf-grenoble.fr

Aníbal Costa Department of Civil Engineering Aveiro, Universidade de Aveiro, Aveiro, Portugal, agc@ua.pt

Misko Cubrinovski Department of Civil and Natural Resources Engineering, University of Canterbury, Christchurch 8140, New Zealand, misko.cubrinovski@canterbury.ac.nz

M. Betul Demircioglu Kandilli Observatory and Earthquake Research Institute, Bogazici University, 34684 Istanbul, Turkey, betul.demircioglu@boun.edu.tr

Mauro Dolce Seismic Risk Office, Italian Department of Civil Protection, 00189 Rome, Italy, mauro.dolce@protezionecivile.it

A.-M. Duval CETE Méditerranée, 06359 Nice Cedex 4, France, anne-marie.duval@developpement-durable.gouv.fr

B. Endrun Institute of Earth and Environmental Sciences, University of Potsdam, 14476 Potsdam OT Golm, Germany, endrun@geo.uni-potsdam.de

Mustafa Erdik Kandilli Observatory and Earthquake Research Institute, Bogazici University, 34684 Istanbul, Turkey, erdik@boun.edu.tr

D. Fäh Swiss Seismological Service, ETH Zürich, 8092 Zürich, Switzerland, faeh@seismo.ifg.ethz.ch

Paolo Franchin Department of Structural and Geotechnical Engineering, University of Roma "La Sapienza", 00197 Rome, Italy, paolo.franchin@uniroma1.it

Mihail Garevski Institute of Earthquake Engineering and Engineering Seismology, Ss. Cyril and Methodius University, 1000 Skopje, Republic of Macedonia, garevski@pluto.iziis.ukim.edu.mk

G. Gassner-Stamm Swiss Seismological Service, ETH Zürich, 8092 Zürich, Switzerland, stammg@student.ethz.ch

B. Guillier LGIT, Maison des Geosciences, Joseph Fourier University, 38041 Grenoble Cedex 9, France, bertrand.guillier@ird.fr

S. Hailemikael Ufficio Valutazione del Rischio Sismico, Dipartimento della Protezione Civile, 00189 Roma, Italy, salomon.hailemikael@protezionecivile.it

Ufuk Hancilar Kandilli Observatory and Earthquake Research Institute, Bogazici University, 34684 Istanbul, Turkey, hancilar@boun.edu.tr

H.B. Havenith Swiss Seismological Service, ETH Zürich, 8092 Zürich, Switzerland

M. Hobiger LGIT, Maison des Geosciences, Joseph Fourier University, 38041 Grenoble Cedex 9, France, manuel.hobiger@obs.ujf-grenoble.fr

Andreas J. Kappos Laboratory of Concrete and Masonry Structures, Department of Civil Engineering, Aristotle University of Thessaloniki, Thessaloniki 54124, Greece, ajkap@civil.auth.gr

Yavuz Kaya Kandilli Observatory and Earthquake Research Institute, Bogazici University, 38684 Istanbul, Turkey, kayaya@boun.edu.tr

A. Köhler Institute of Earth and Environmental Sciences, University of Potsdam, 14476 Potsdam OT Golm, Germany, akoehler@uni-potsdam.de

M. Kristekova Geophysical Institute, Academy of Sciences, 845 28 Bratislava, Slovak Republic, kristekova@savba.sk

Aslı Kurtuluş Kandilli Observatory and Earthquake Research Institute Boğaziçi University, Çengelköy, Istanbul, Turkey, Asli.kurtulus@boun.edu.tr

C. Lacave Résonance S.A., CH-1227 Carouge-Genève, Switzerland, corinne.lacave@resonance.ch

Claudio Modena Department of Structural and Transportation Engineering, University of Padova, 35131 Padova, Italy, modena@dic.unipd.it

Marco Munari Department of Structural and Transportation Engineering, University of Padova, 35121 Padova, Italy, marco.munari@unipd.it

Farzad Naeim Earthquake Engineering Research Institute, Oakland, CA, USA; John A. Martin & Associates, Inc., Los Angeles, CA, USA, FARZAD@johnmartin.com

M. Ohrnberger Institute of Earth and Environmental Sciences, University of Potsdam, 14476 Potsdam OT Golm, Germany, mao@geo.uni-potsdam.d

Carlos Sousa Oliveira Department of Civil Engineering and Architecture/ICIST Lisbon, Instituto Superior Técnico, Lisbon, Portugal, csoliv@civil.ist.utl.pt

Göktürk Önem Kandilli Observatory and Earthquake Research Institute, Boğaziçi University, 34684 Istanbul, Turkey, onemgokt@boun.edu.tr

Alain Pecker Géodynamique et Structure, 92220 Bagneux, France, alain.pecker@geodynamique.com

Artur Pinto ELSA, IPSC, Joint Research Centre, Ispra, 21020 VA, Italy, artur.pinto@jrc.ec.europa.eu

Paolo E. Pinto Department of Structural and Geotechnical Engineering, University of Roma “La Sapienza”, 00197 Rome, Italy, pinto@uniroma1.it

V. Poggi Swiss Seismological Service, ETH Zürich, 8092 Zürich, Switzerland

Francesca da Porto Department of Structural and Transportation Engineering, University of Padova, 35131 Padova, Italy, daporto@dic.unipd.it

Sean Rees Department of Civil and Natural Resources Engineering, University of Canterbury, Christchurch 8140, New Zealand, Sdr40@student.canterbury.ac.nz

F. Renalier LGIT, Maison des Geosciences, Joseph Fourier University, 38041 Grenoble Cedex 9, France, florence.renalier@obs.ujf-grenoble.fr

I. Rodrigues Lisbon Fundaçao da Faculdade Ciencias da Universidade de Lisboa, IDL, 1749-016 Lisboa, Portugal, isanovais@hotmail.com

F. Sabetta Ufficio Valutazione del Rischio Sismico, Dipartimento della Protezione Civile, 00189 Roma, Italy, fabio.sabetta@protezionecivile.it

Erdal Şafak Kandilli Observatory and Earthquake Research Institute, Boğaziçi University, 38684 Istanbul, Turkey, erdal.safak@boun.edu.tr

A. Savvaidis Institute of Engineering Seismology and Earthquake Engineering (ITSAK), 55102 Thessaloniki, Greece, alekos@itsak.gr

Karin Sesetyan Kandilli Observatory and Earthquake Research Institute, Bogazici University, 34684 Istanbul, Turkey, karin@boun.edu.tr

R.T. Severn Earthquake Engineering Research Centre, University of Bristol, Bristol, UK, R.T.Severn@bristol.ac.uk

Haresh C. Shah Department of Civil and Environmental Engineering, Stanford University, Stanford, CA 94305-4020, USA; RMS, Inc., Newark, CA 94560, USA; WSSI, Newark, CA, USA, Haresh.Shah@rms.com

Jonathan P. Stewart University of California, Los Angeles, CA, USA, jstewart@seas.ucla.edu

T.P. Tassios National Technical University of Athens, Athens, Greece tassiost@central.ntua.gr

P. Teves-Costa Lisbon Fundaçao da Faculdade Ciencias da Universidade de Lisboa, IDL, 1749-016 Lisboa, Portugal, ptcosta@fc.ul.pt

N. Theodulidis Institute of Engineering Seismology and Earthquake Engineering (ITSAK), 55102 Thessaloniki, Greece, ntheo@itsak.gr

S. Thomassin Résonance S.A., CH-1227 Carouge-Genève, Switzerland, sylvette.thomassin@resonance.ch

Gökçe Tönük Kandilli Observatory and Earthquake Research Institute, Boğaziçi University, Çengelköy, Istanbul, Turkey, gokce.tonuk@boun.edu.tr

I. Veludo Lisbon Fundaçao da Faculdade Ciencias da Universidade de Lisboa, IDL, 1749-016 Lisboa, Portugal, idalina.veludo@gmail.com

John W. Wallace NEES@UCLA Laboratory, Department of Civil and Environmental Engineering, University of California, Los Angeles, CA 90095-1593, USA, wallacej@ucla.edu

M. Wathelet LGIT, Maison des Geosciences, Joseph Fourier University, 38041 Grenoble Cedex 9, France, marc.wathelet@ujf-grenoble.fr

Can Zulfikar Kandilli Observatory and Earthquake Research Institute, Bogazici University, 34684 Istanbul, Turkey, can.zulfikar@boun.edu.tr

Chapter 1
Seismic Engineering of Monuments

T.P. Tassios

Abstract In this contribution the particularities of the seismic re-design of *Monuments* are discussed, related to their structural repair or strengthening. A Monument, besides its possible practical use and its economical value, requires a lot of other Values to be respected during its aseismic retrofitting, such as its aesthetic Form, the authenticity of its materials, etc. In order to respect these Values, the Engineer tends to minimize structural intervention – thus, violating social values such as the preservation of the Monument for future generations, the protection of human lifes, etc. An optimization is needed, and this contribution attempts to describe the necessary procedures to this end. On the other hand, emphasis is given to the particular difficulties in the determination of the resistance of masonry, as well as in the selection of suitable methods of Analysis, taking into account the specificity of each Monument. To this end, the contribution includes comments on monumentic Values and performance requirements, and emphasizes the need for an institutionalization of levels of importance, levels of visitability and of acceptable damage-levels for all monuments of each Country, as a basic input of aseismic design of monuments. Subsequently, criteria are given for the selection of methods of Analysis, and detailed comments are included about experimental investigations and strengths' determination. The final optimization procedure is then described, regarding the optimum seismic resistance level to be lent to a specific Monument.

1.1 The Significance of the Subject

It is broadly accepted that in seismicly prone regions, the seismic behaviour of Monuments is of a paramount importance.

First, because of the cultural need to maintain and transfer these Monuments to future generations (Figs. 1.1 and 1.2). To this end, more or less drastic structural

T.P. Tassios (✉)
National Technical University of Athens, Athens, Greece
e-mail: tassiost@central.ntua.gr

M. Garevski, A. Ansal (eds.), *Earthquake Engineering in Europe*, Geotechnical, Geological, and Earthquake Engineering 17, DOI 10.1007/978-90-481-9544-2_1,

Fig. 1.1 The Parthenon, Acropolis of Athens

Fig. 1.2 Historical muslim temple underneath Acropolis of Athens

interventions (repair or strengthening) are implemented,[1] with the lowest possible consequences on the "monumentic" values.

Second, our interest for the seismic resistance of Monuments is also encouraged by the legal obligation to protect human life (of the neighbours, curators, visitors or even inhabitants of the monumental building).

However, the preparation of design documents regarding structural interventions of Monuments is frequently facing several difficulties, not encountered in the case of non-monumental buildings, such as:

- Additional uncertainties related to the available resistances of building components
- Particularities in selecting the appropriate method of Analysis, suitable (i) to a given typology of the Monument and (ii) to the level of resistance uncertainties
- Difficulties in selecting
  - An appropriate design value of seismic action, such that the respective necessary intervention will not jeopardise the monumentic values of the Monument, and
  - Appropriate Techniques with an optimum level of reversibility/re-interventionality.

Because of these difficulties, the approval of submitted design-documents is frequently an occasion for controversial discussions between Engineers, on the one hand, and Architects and Archeologists, on the other.

1.2 Structural Interventions and the Conflict of Values

These controversies are but a reflection of the contrarieties between the following "Principles" (Values and Requirements) related to the structural interventions in Monuments:

(a) *Monumentic*[2] *Values*

a1: Form (aesthetic value).
a2: History (symbolic value).
a3: Preservation of ancient building-Techniques and Materials (technical value).

---

[1] In this respect, specialists do not anymore share the view that "since a Monument has withstood previous earthquakes, it will continue to resist any future seismic action".

[2] This neologism is a very useful term to express concepts and things related to Monuments, avoiding however the possible confusion with the secondary meanings of the term "monumental" (i.e. impressively large, outstanding, astounding).

(b) *Social Values*

b1: Preservation of the cultural Memory of a Monument (integrity, survival).
b2: Adequate safety against normal actions and Earthquakes (value of human life).
b3: Modern use of Monuments.
b4. Cost-reduction of the structural intervention.

(c) *Performance – requirements regarding the structural intervention* (Intervention Values)

c1: Reversibility level and or re-interventionality level
c2: Durability
c3: Technical reliability

Every intervention aiming at a structural repair or strengthening of a Monument entails some inevitable *harm* to several of these values and performance-requirements, depending on the actual condition of the Monument and the available technologies. It suffices perhaps to give a few typical *examples*:

- In order to preserve the Form and the integrity of a seismicly vulnerable Monument, a rather costly strengthening solution is adopted, consisting in (i) extensive grouting of masonry walls and (ii) change of the old (completely decayed) roof (Fig. 1.3). Thus, the following "principles" were violated: a3, b4 and c1.
- A second typical example may be the case of a monumental building made of precious historical materials to be completely preserved; the solution here was to offer seismic safety by means of external buttresses (Fig. 1.4). The violated "principles" here were: a1 and a2.
- In a third case, to avoid any harm to monumentic Values of a delicate Monument, some rather simple and provisional structural interventions were decided, offering a seismic resistance lower than the one required for modern important buildings. In this case, a remarkable violation of the social value of "human life safety" (principle b2) was accepted, together with a transgression of the durability requirement (c2).

Apparently, in all these cases, Authorities have sought an optimization[3] of Principles, and came to their final decision, knowingly of the partial violation of the "set of Principles". In this respect, it is reminded that such an optimization cannot be reached by means of just "scientific" judgments: The values entering the game are of different nature; they are *not* amenable to identical "units" – they are not quantitatively comparable between each other!

That is why, in the field of structural interventions of Monuments, only managerial (almost political) decisions are feasible; weighing factors for each of these

[3] See Sections 1.9 and 1.10.

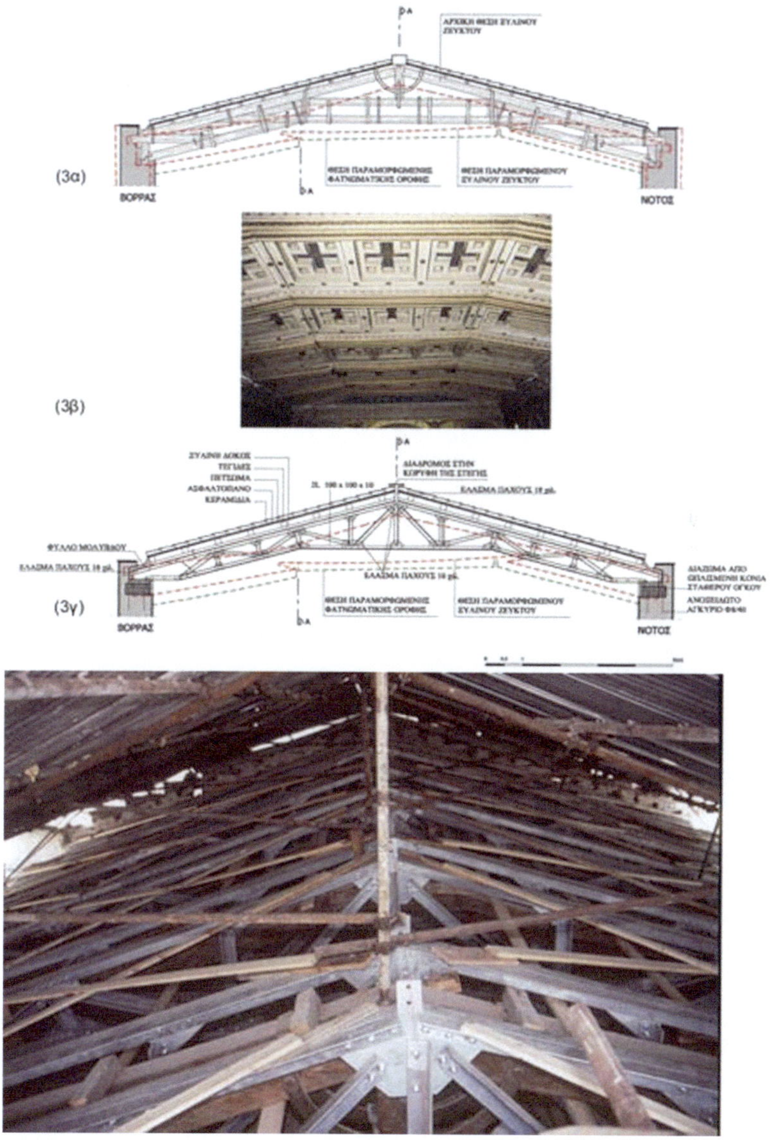

Fig. 1.3 Presentation of the historical ceiling of the church of St. Irene, Athens

Principles may be (directly or indirectly) discussed within an interdisciplinary group, and a final "optimal" decision be made.

Such an optimization process, directly affects some important technical issues related to the seismic (re)design of Monuments: The design value of seismic actions has to be decided taking (also) into account its eventual consequences on monumentic values, too, as well as on costs and technical performances. Thus, a sort of

Fig. 1.4 The Hosios Loukas monastery 11th cent. Church (Greece), strengthened by means of external buttresses (11th cent)

"negotiation"[4] of design seismic actions is initiated: Disproportionately high design-values, serve the "human life" and the "integrity" principles, but they may jeopardise some monumentic values and performance requirements. Therefore, a better overall intervention (an "optimal" solution) may be sought, based on possibly lower design-values of seismic actions, i.e. on higher exceedance probability. The same holds true for the selected intervention schemes and technologies; they should also be finally decided following a similar optimization process.

In order to facilitate such a decision making process, further rationalization of data is needed regarding the "Importance" of a given Monument, as well as its "Visitability" level.

[4]After all, design seismic actions regarding modern buildings are *also* negociated: The socially acceptable "probability of exceedance" of seismic actions imposed by actual Codes, depends on several variables, such as the actual economical level of the Country and the social importance of the building, i.e. on non-scientific data. The difference in Monuments is that the case of such a "negociation" is taking place within a broader *multiparametric* space, including many additional Values and Requirements.

1.3 Importance, Visitability and Acceptable Damage-Levels

- Seismic actions' values for the re-design of Monuments may also depend on acceptable damage-levels, which in their turn will be decided on the basis of the importance of each Monument. That is why in many Countries, a *categorisation* of Monuments is available to designers, as follow:
  - I1: Monuments of universal importance (Fig. 1.5)
  - I2: Monuments of national importance (Fig. 1.6)
  - I3: Monuments of local interest (Fig. 1.7)
- Another useful tool towards a rationalisation of decision making regarding structural interventions, is the categorization of the occupancy of Monuments: Higher occupancy means higher concern for human lives against earthquakes, and therefore higher seismic actions' design values. That is why engineering decisions would be facilitated if a "visitability" categorisation of Monuments would be made available, such as in the following list:

  V1: Almost continuous presence of public or frequent presence of large groups

  - Inhabited buildings in historical city centres
  - Monuments used as Museums
  - Monuments continuously used for worshipping

Fig. 1.5 The Hagia Sophia church in Istanbul, a monument of universal importance

Fig. 1.6 The Arta bridge (15th cent) in Greece, a monument of national importance

Fig. 1.7 A "neoclassical" building in Athens (19th cent), a monument of local importance

V2: Occasional habitation or intermittent presence of small groups

- Monuments visited only under specific conditions
- Remote and rarely visited Monuments

V3: Entrance allowed only to Service-Personnel. Visitors stand only outside the Monument.

- Combining the aforementioned Importance-levels and Visitability-levels, it is possible to decide *acceptable Damage-levels* ("I" for negligible damage, up to IV for serious damage, see Figs. 1.8, 1.9 and 1.10), under the re-design earthquake. Such a possible matrice is given here below (indicatively though):

Acceptable damage-levels (I to IV) under the re-design seismic actions		Prevailing values			
		Human life and monument's integrity			Form and history
Visitability		V1	V2	V3	
Importance level	I1	I	II	II	I
	I2	I	II	III	II
	I3	II	III	IV	III

To this end, a systematic description of each damage-level is needed, separately for traditional masonry buildings, arched structures or domes, and graeco–roman monuments.

Fig. 1.8 Local diagonal cracks only: level II damage (Anavatos, Chios Island, Greece)

Fig. 1.9 Compressive local failures; level III damage (Anavatos, Chios Island, Greece)

It is believed that such an approach may substantially facilitate rational decision making, related to structural interventions of Monuments against seismic actions.

1.4 Historical and Experimental Documentation, and Uncertainty Levels

1.4.1 Introduction

(a) Long experience shows that the structural design document regarding seismic strengthening of a Monument is an integral part of the broader study of the Monument; history and architecture of the Monument are indispensable prerequisites for the Structural Design, in order to account for all initial and consecutive construction *phases,* previous repairs etc.
(b) Description of existing and or repaired *damages* (visible or possibly hidden ones), together with their in-time evolution; monitoring, be it a short term one, may be helpful.

Fig. 1.10 Local collapses; level IV damage (Anavatos, Chios Island, Greece)

(c) Systematic description of the in-situ *materials*, including their interconnections – especially in the case of three leaf masonry walls. Connections of perpendicular walls are thoroughly investigated.

(d) Results of *experimental* investigations regarding:

- geometrical data,
- internal structure,
- in-situ strength of materials,
- structural properties of masonry walls,
- dynamic response of building elements,
- subterranean data,

as well as results of possible previous *monitoring* installations (displacements, settlements, internal forces, humidity, groundwater level, cracks' opening, seismic accelerations, environmental data etc).

(e) Description of the *structural system*

(f) Description of the soil and the foundation

1.4.2 Experimental Documentation

It is worth to inventorise first the categories of structurally useful data needed (thus making clear the scope of the experimental investigation), together with the particular methods used to this end. Thus, a better understanding of function, importance and interdependence of these methods will be achieved. Laboratory methods and monitoring are separately considered. Physical-chemical aspects are not examined in this contribution.

Instrumental In-Situ Methods

(a) *Geometrical data*

- Visual description of structural parts; Laser scanning, Photogrammetry
- Cracks
  - Opening: Lenses, Photogrammetry, Strain gauges
  - Depth: Ultrasonic tests
  - Length: Measuring tape, Photogrammetry
- Displacements: Phtogrammetry, Inclinometers, Penduli

(b) *Internal structure* (Van de Steen et al., 1997; Binda et al.; 1998, 2003; Maierhofer et al., 2004; Wenzel and Kahle, 1993; Silman and Ennis, 1993; Thomasen and Sears, 1993; Binda and Saisi, 2001)

- Hidden voids or discontinuities: Endoscopy, Thermography, Radar, Sonic tomography, Radiography
- Internal building details: Endoscopy (see also BIPS-method (Fig. 1.11)), Radar, Sonic tomography (Fig. 1.12).
- Hidden metals: Magnetometry, Radiography, Thermography, Radar

(c) *In-situ strength of constitutive materials*

- Stones: Rebound, Ultrasonic, Scratch width
- Infill material: Sonic cross-hole
- Mortars: Scratch width, Penetration test (Felicelti and Gattesco, 1998)
- Timber: Penetrometer (Giuriani and Gubana, 1991)
- Metals: Hardness test in-situ
- Bonding strength: Mortar pullout test

(d) *Structural properties of masonry*

Acting compressive stresses Compression resistance Shear resistance	} In situ jack-tests (Binda et al. 1997, 2003)

(e) *Effectiveness of Grouting:* Sonic tomography, Endoscopy, Radar (Berra et al., 1991; Côte et al., 2004)

(f) *Dynamic response of building elements*

- Microtremors
- Cable-release tests
- Vibrodyne

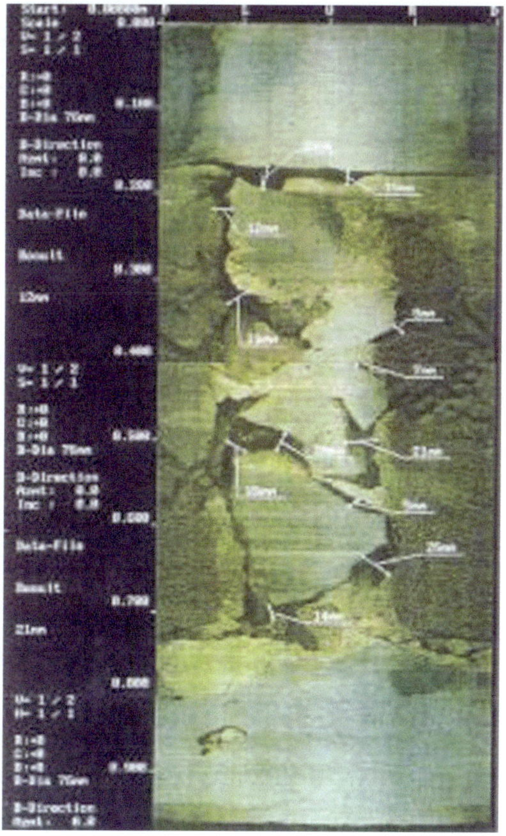

Fig. 1.11 Borehole Image Processing Systems (BIPS). The entire cylindrical surface of the boring is developed and examined in the case of an endoscopy in the Tower of Pisa, near the deficient area of the helicoidal stairs (Macchi and Ghelfi, 2006)

(g) *Subterranean data*

- Seismic tomography
- Ground radar

Laboratory Methods

(a) Core testing: Compression, Tension
(b) Tests of irregular mortar-fragments
(c) Testing on replicas: re-made masonry, subassemblages

Monitoring (in-Time)

(a) Displacements: Horizontal deformetric wires, Penduli, Laser measurements, Inclinometers
(b) Settlements: Leveling systems, Inclinometers

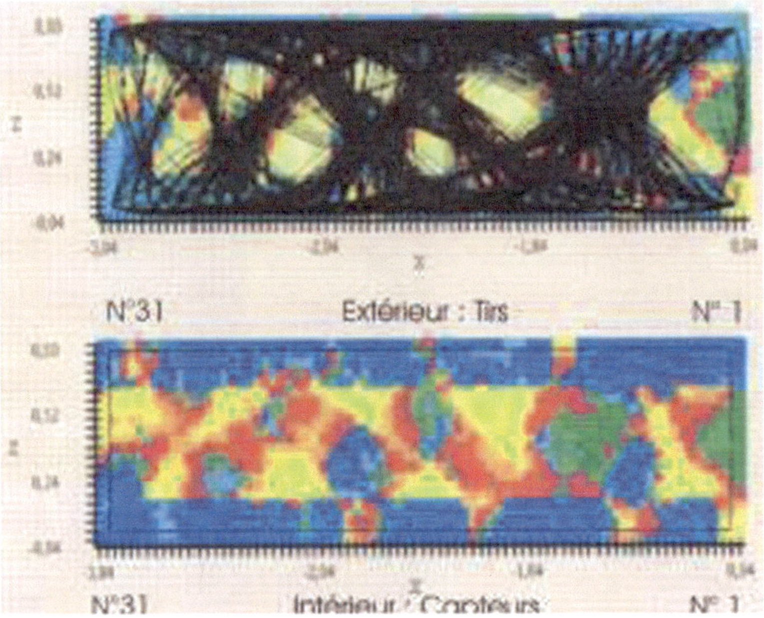

Fig. 1.12 Sonic tomography along a wall of Dafni Monastery, (Greece), after grouting (Côte et al., 2007)

(c) Internal forces: Inserted dynamometers, Flat jacks
(d) Humidity within masonry: Neutron probes, GPR, Thermograhy
(e) Ground-water level: Water pressure borings
(f) Cracks: Opening evolution and control
(g) Seismic actions: Seismometers, Accelerometers
(h) Environmental data: Temperature, Solar radiation, Wind.

For our purposes, it is good to know that such a rather impressive weaponry is made available to us, in order to "see" the interior of the (black and silent) "box" of the structural elements of a Monument. It has however to be reminded that this extensive inventory of experimental methods cannot be used in every case; not only because of economical and accessibility difficulties, but mainly because of the multiple limitations of applicability of each of these methods. Thus, an optimum use of these methods should be made, on a case-by-case basis, depending on:

– the importance of the Monument and its financing
– the inadequacies of available historical data
– the level of roughly estimated vulnerability of the Monument
– the technical characteristics of the Monument
– the time schedules, etc.

A lot of the aforementioned methods are well known to the Engineers; several books and papers describe them in detail (see i.a. Tassios and Mamillan, 1985).

1.4.3 Uncertainty Levels

The combination of historical and experimental documentation data, is an indispensable basis for the structural re-design of a Monument, since the structure of a Monument is a rather silent black-box; every effort is justified to make this box to talk. However, a lot of uncertainties will remain. And our scientific duty is to be conscious of these uncertainties and of the way they may affect Analysis (hidden discontinuities, behaviour of connections etc), as well as Resistance determination (hidden voids, weathered or inhomogenious materials, etc). Otherwise, our computational efforts may not be able to correctly evaluate the seismic resistance of the Monument and to appropriately design its best strengthening.

Long experience shows that for each primary structural member of the Monument, an appropriate level of reliability of documentation should be assigned, referring separately to basic data, such as:

- Dimensions, eveness and verticality
- Composition transversally to the element (e.g. three leaf masonry?)
- Connections with neighbouring elements.
- Strengths of constitutive materials, etc.

For each of these categories of data, a "Documentation-Reliability level" should be assessed in each particular case, (indicatively: "missing", "inadequate", "sufficient"). An important contribution to this end is offered in Section 1.4.2 (p. 35) of the Italian Guides, 2007, [D2]. In accordance with this characterization, Analysis and Resistance evaluation methodologies will be more accurately selected (see Sections 1.5 and 1.6).

1.5 Structural Analysis

1.5.1 Introduction

(a) The analytical procedure may include the following analysis, independently or in combination:

  - Analysis of the entire monument (occasionally without some possibly secondary elements)
  - Analysis of some selected structural sub-assemblages, in order to identify the most critical weaknesses of the monument.

(b) The *structural system* has to be clearly identified; the rich data included in the Documentation of the Monument are very helpful in this respect.

(c) *Selection criteria* of the method of seismic Analysis, appropriate for the Monument under consideration:

- For more important and or complicated Monuments, more sophisticated methods appear more justified.
- For each of the main typological categories of stone monuments, some particular methods may be more appropriate.
- Depending on acceptable *damage-level* (Section 1.3), a respective method of Analysis may be used. Indicatively:

  - In case of targeted negligible damage under the design earthquake, only linear Analysis is applicable.
  - In case of extensive accepted damages, a non-linear static Analysis seems to be more suitable.

- The available level of "Reliability of the Documentation" (Section 1.4) should also be taken into account when selecting the appropriate method of Analysis. Low RD-levels are not compatible with highly sophisticated methods.

(d) A warning may be useful, related to the use of data found by means of *dynamic experimental excitations*. These data, applicable for structural identification, should not be used for spectral response calculations; in fact, eigen frequencies determined by means of dynamic excitations, are considerably higher than real frequencies under actual displacement and damage conditions.

1.5.2 General Criteria for the Selection of the Methods of Analysis

(a) Depending on monument's *morphology:*

  - Depending on the extent and the complexity of the monument, non-linear analysis of all parts by means of three-dimensional models is not always possible.
  - Significant non-regularities necessitate linear dynamic analysis (independently, or in combination with static non-linear analysis)
  - Slender parts also necessitate the use of linear dynamic method.

(b) The final selection of an analytical method will be based on its capacity to "reproduce" (by computation) previous damage patterns, roughly though (Fig. 1.13).

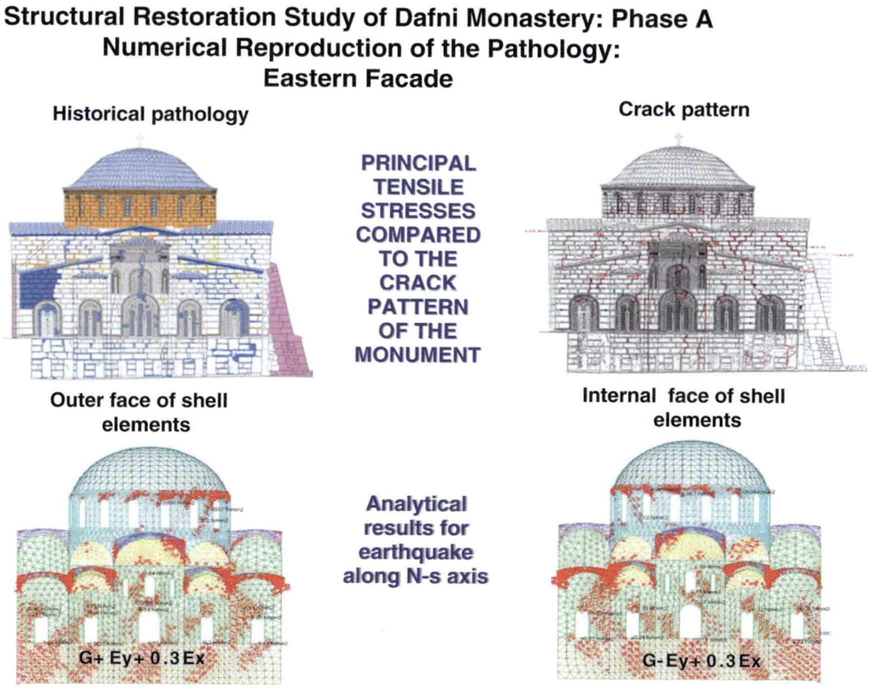

Fig. 1.13 An attempt to reproduce computationally previous damage patterns, roughly though (Dafni Monastery, Greece)

1.5.3 Reliable Discretisation for Each Method of Analysis

(a) Finite elements, in case of linear analysis: The reliability of results depend on the topology of the discretisation; sensitivity investigations are recommended.
(b) Equivalent framework (vertical pears and horizontal spandrels, connected by means of non-deformable regions): This approach may profitably be used for linear analysis (using however realistic stiffness values), as well as for non-linear static analysis (provided that ultimate angular deformations are correctly estimated).
(c) Macro-elements may also be used, their topology defined by the initial geometry of the structure alone, or/and on the basis of existing or expected cracks, predicted by means of a linear FEM.
(d) Kinematic mechanisms, usually to model specific parts of the monument (e.g. arched structures or out-of-plane failure of walls), analysed by means of linear or non-linear methods. Critical horizontal forces are estimated, mobilizing the selected mechanisms.

(e) Truss or struts-and-ties methods may be used in specific cases. Ties can be realized by steel or timber elements or, less accuratelly, making use of available tensile strength of masonry due to cohesion and friction resistance of its elements.

Note: As it is known, the results of Analysis in terms of forces (local stresses "σ", "τ", or global action effects "M", "N", "V") should be appropriately compared to corresponding available resistance values. To this end, too narrow (like in the case of local stresses) or too large resisting areas of masonry (like in the case of wide compressed regions) cannot reliably represent real conditions. The concept of "critical resisting volume" of masonry may be helpful in this respect, i.e. a volume containing approximately three blocks per each direction.

1.5.4 Analysis of "Sculptured Stones/Dry Joints" Monuments (Graeco–Roman)

Despite their apparent simplicity, seismic behaviour of such structures is extremely complicated and very sensitive to details. The main response mechanisms, friction-sliding that is and rocking (Fig. 1.14), are by nature non-linear. (Housner 1963, Konstantinidis et al. 2005, Makris et al. 2003, Papantonopoulos 1993, Psycharis 2000, Sinopoli 1989). Besides, neighbouring structural elements do not follow the rule of compatible deformations.

Fig. 1.14 Isolated columns of the Zeus temple in Nemea, Greece (N. Makris)

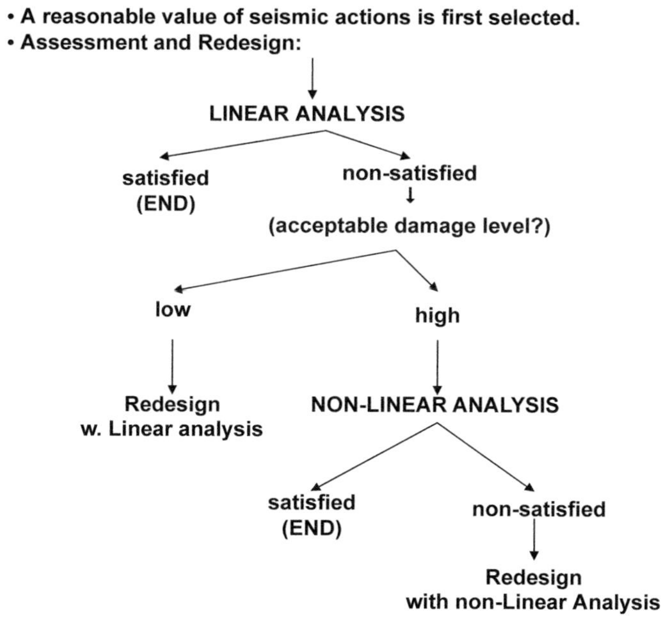

Fig. 1.15 Sequence of possible methods of seismic analysis of monuments

Thus, normally, only some main sub-systems is possible to be considered and analysed, taking into account all imperfections, such as weakness of foundation, large initial deformations, wearing of edges, existing cracks, initial rotations, etc.

Static analysis may be used only as a preliminary step, for a rapid evaluation of seismic risk – and only under rather small seismic loads.

Simplified spectral analysis may be useful if, by trial and error, the final rocking angle of each stone-element is compatible with its corresponding eigen-frequency. However, this approach cannot describe the behaviour of complex systems.

In fact, under substantial seismic actions such systems do not exhibit eigen-frequencies.

The directly "non-linear-dynamic" method is based on the integration of the equations of motion (based on a given accelerogram), step by step. Appropriate and reliable soft wares are needed.

Note: In Fig. 1.15 a schematic procedure is shown describing the sequence of use of possible methods of Analysis.

1.6 Evaluation of Resistances

One of the most remarkable particularities of the structural re-design of Monuments is our difficulty to determine the resistances of critical regions of stone building-elements, with an accuracy comparable to the precision of the determination of action-effects. Thus, the inequality of safety $E \not> R$ becomes a rather loose condition, leading to overconservative or to risky solutions.

This unpleasant situation is partly due to the fact that research financement and research glamour is normally offered to subjects related to Analysis, rather than Resistance determination...

In the field of structural assessment and re-design of Monuments, it is of a paramount importance to overcome this *scientific weakness*; thus, every effort is justified in order to better evaluate available Resistances of critical regions of "masonry" walls, or "sculptured stones/dry joints" building elements. To this end, empirical formulae or simple rules of thumb do not serve the purpose alone: Detailed preliminary inspection and in-situ experimental investigation will be the indispensable basis for any subsequent computational step of resistance determination.

1.6.1 Semi-analytic Strength Determination

(a) The following *data* are first needed in order to evaluate the basic resistance characteristic of a masonry, i.e. its compression strength perpendicularly to its layer:

- Photographic view of the face of each critical region of the wall or pillar.
- Calculation of mean values of length ($l_{\rm bm}$) and height ($h_{\rm bm}$) of the blocks, as well as the mean value of thickness of joints ($t_{\rm jm}$).
- By means of *vertical* sections, a nominal index (IL) of blocks' interlocking is calculated: Within a representative height (H) of masonry, such an index may be defined as follows.

  Some broken quasi-vertical lines are traced along consecutive vertical joints located underneath each-other; these lines are kept as close to vertical as possible. Measured are (Fig. 1.16) the lengths "U" of horizontal contacts of each block with its underlaid "supporting" block. For each broken quasi-vertical line, an "interlocking index" is calculated as follows

$$({\rm IL}) = \frac{2}{n}\sum_{1}^{n}(U_i : l_{bi}), \quad 0 < ({\rm IL}) < 1$$

  where $i = 1$ to n, denotes the order of block layers. Out of several such quasi-vertical lines (say 5–8 on each face of the wall), a characteristic average value $({\rm IL})_{\rm m}$ is taken as a representative estimator of the "interlocking capacity" of blocks of the wall in vertical direction.
- Now the *transversal* structural composition of the wall perpendicularly to its plane is examined. To this end, direct observation through temporarily opened holes, or televisioned endoscopy, or even georadar (and or sonic

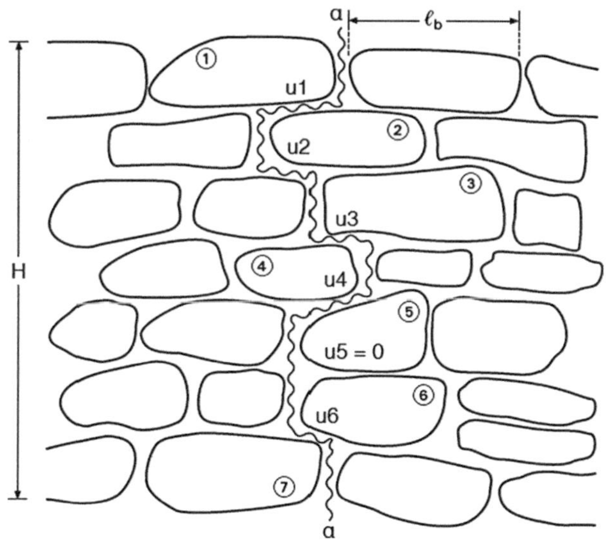

Fig. 1.16 Along a broken quasi-vertical line a-a, the contact lengths U_i are measured and normalised to the lengths l_{bi} of the "supported" block i

tomography) may be used, in order to find out if the wall may be considered as one-leaf or two (or three)-leaf masonry. In Monuments of lower importance, some transverse core-takings may also be helpful in this respect. The degree of interconnection of the two faces of the wall is of paramount importance for its structural behaviour. It would be therefore desirable to try to quantify, be it in a rudimentary way, this transversal interconnection index (TI) by estimating the sum of cross sections of large blocks along a representative area on a vertical plane section, situated in the middle between the two faces of the wall. In Fig. 1.17, this pseudo-quantitative index could be defined as

$$(\mathrm{TI}) = \sum A_i : \mathrm{LxB}$$

where

A_i = the cross-sections of sectioned large blocks on plane p-p
LxB = the vertical area of the entire region of the wall element under consideration

- Strength and deformability of blocks and mortars are measured in situ (by means of appropriate ND methods), or in-lab (on appropriate representative samples).

(b) Nowadays, it is well understood that an eminently discontinuum medium like historical masonries, cannot at all be handled as a pseudocontinuum (except in the rare case of one-leaf masonry made of well chiseled and well interlocked

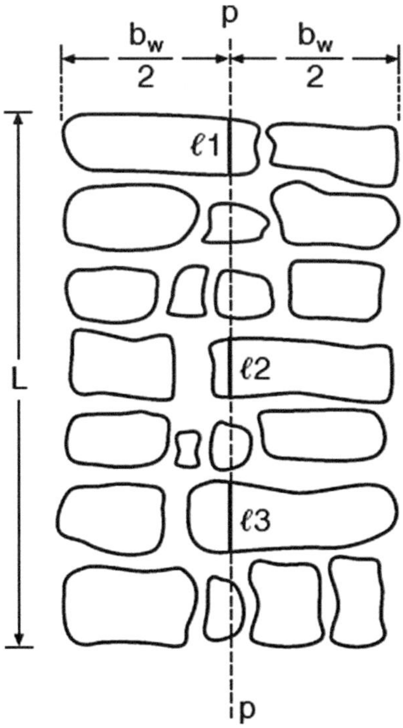

Fig. 1.17 Pseudo-quantitative assessment of the interconnection between the two external faces of a masonry wall: The length $l_1+l_2+l_3$ is reported to the height L

stones). Consequently, in normal cases, the designer needs to acquire a complete and 3-dimensional understanding of the apparent and the internal structure of the critical regions of masonry walls or pillars. Otherwise, any attempt to evaluate basic resistance properties, such as the compressive strength of masonry, will be superfluous. Besides, such compressive strengths may be completely different in various parts of a monument, or even in various areas of one wall. That is why it is now believed that the designer of structural interventions in a monument, should in most cases include in his/her design-documents detailed descriptions, pictures and measurements like those enumerated in the previous paragraph. The Italian school of thought had early enough contributed to this matter (see i.a. Giuffré, 1991), emphasising the significance of this 3-dimensional interconnections (Fig. 1.18). Following their views, instead of a quasi-quantitative approach via the indices (IL) and (TI), a more practical way could be followed, making use of rough typological classifications (like those indicatively illustrated in Fig. 1.18 for 1- and 2-leaf masonries).

Even if the state-of-the-Art on the subject is rather poor, experienced Engineers may eventually translate the characterizations of Table 1.1 into some numerical coefficients, modifying the empirical formulae which predict compressive stresses without taking into account interlocking and interconnection characteristics.

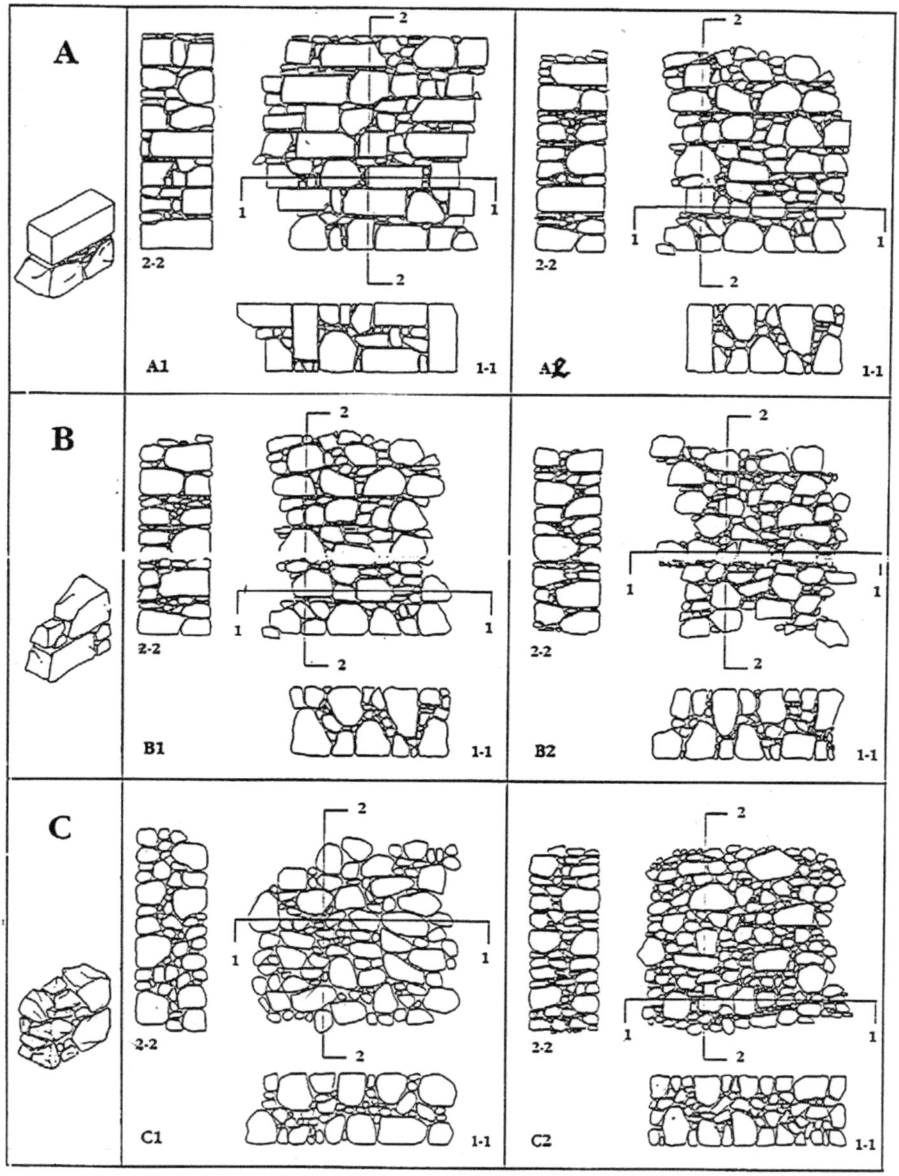

Fig. 1.18 A possible qualitative classification of efficient interconnections between stones and between external and internal faces of masonry walls (Giuffré, 1991)

Table 1.1 A practical typological classification of stones' interlocking on masonry faces and of transversal interconnection between the two faces of a wall

			Interlocking of stones on a face		
			G(ood)	M(edium)	L(ow)
1-leaf masonry	Transversal interconnection of faces	g(ood)	1 Gg	1 Mg	1 Lg
		m(edium)	1 Gm	1 Mm	2 Lm
		l(ow)	1 Gl	1 Ml	3 Ll
			2 Gl	2 Ml	2 Ll
2-leaf masonry		medium d(ense)	3 Gd	2 Md	3 Ld
3-leaf masonry	Infill material	p(orous)	3 Gp	3 Mp	3 Lp
		e(xtremely) porous	3 Ge	3 Me	3 Le

(c) It would now be possible to use a well established empirical expression predicting the vertical compression strength of masonry, under the condition that, as many as possible of the parameters shaping the overall compression strength are included in such an expression, like:

- nature of blocks (stones, bricks)
- roughness of blocks' surface
- strengths and deformability modulus of blocks and mortars
- average joint-thickness (and its normalized value $\alpha = t_{\mathrm{jm}}:h_{\mathrm{bm}}$)
- interlocking index (IL)
- transversal structural composition of the wall (one, two or three-leaf masonry), and transversal interconnection index (TI).

Under these basic conditions, such empirical formulae could be able to predict masonry compression strength (f_{wc}), separately for one-leaf and three-leaf masonry. However, the state-of-the-art on the subject, does not seem to be very satisfactory.

Indicatively some of these formulae (among quite a few of available ones) are reproduced here, only as an occasion to critisise their drawbacks. It is reminded that compressive strength f_{wc} of masonry as a *material* is dealt with here; system resistance such as buckling-effects or local-compression resistance are not considered here.

(i) Well built brick masonry (Tassios, 1988):

$$f_{\mathrm{wc}} = \left[f_{\mathrm{mc}} + 0.4(f_{\mathrm{bc}} - f_{\mathrm{mc}})\right] \cdot \left(1 - 0.8\sqrt[3]{a}\right), \quad f_{\mathrm{bc}} > f_{\mathrm{mc}}$$

$$f_{\mathrm{wc}} = f_{\mathrm{bc}} \cdot (1 - 0.8\sqrt[3]{a}), \quad f_{\mathrm{bc}} < f_{\mathrm{mc}}$$

where

f_{bc}, f_{mc}, compressive strengths of blocks and mortar, respectively
$\alpha = t_{jm}:h_{bm}$, the ratio between average (horizontal) joint thickness t_{jm}, and average block height h_{bm}.

In this particular case, empirical expressions do not need to account for interlocking and interconnection conditions. But they do not consider possible brittleness of bricks due to disproportionately low tensile strength of them.

(ii) Low-strength stone-masonry (Tassios and Chronopoulos, 1986):

$$f_{wc} = \xi \cdot \left[\left(\frac{2}{3}\sqrt{f_{bc}} - f_o \right) + \lambda f_{mc} \right] \quad \text{[in MPa]}$$

where

f_o, a reduction due to non-orthogonality of blocks, taken as

$$f_o = 0.5 \text{ for block stones} \\ 2.5 \text{ for rubble stones}$$

λ, mortar-to-stone bond factor, taken as

$$\lambda = 0.5 \text{ for rough stones} \\ 0.1 \text{ for very smooth-surface stones}$$

ξ, a factor expressing the adverse effect of thick mortar joints

$$\xi = 1 : [1 + 3.5(l - k_o)] \not> 1 \\ k = (\text{volume of mortar}) : (\text{volume of masonry}) \\ k_o = 0.3$$

for materials' strength values $f_{bc} = 25$ up to 75 MPa and $f_{mc} =$ 0.5 up to 2.5MPa

Among its drawbacks, this empirical formula completely disregards interlocking and interconnection conditions.

(iii) There is no space here to extent this discussion for the case of 3-leaf masonries; besides, empirical predictions of strength of such walls are rather inadequate (see i.a. Tassios, 2004).

(d) Deformational properties of masonry of Monuments should be known under two important situations.

First, before significant cracking, so that stiffness-values could be evaluated. To this end, the use of a modified modulus of elasticity may suffice, duly corrected to account for the loading level. Appropriate empirical expressions may be used on this purpose (see i.a. Psilla and Tassios, 2008)

$$E_w = \left[1 + \left(0.9\frac{\sigma_o}{f_{wc}} - 0.6 \right) \cdot \frac{V}{V_{cr}} \right] \cdot E_{wo} \qquad \text{for } \sigma_o \not> \frac{2}{3} f_{wc}$$

where

σ_o = normal compressive stress
f_{wc} = compressive strength
V = acting shear force
V_{cr} = cracking shear resistance
E_{wo} = the technical modulus of elasticity of the masonry

Second, post-elastic deformations are very important under seismic conditions, as a means to dissipate energy and to offer a certain ductility to the system – under the assumption, however, that local damages are acceptable for the monument in consideration. On this subject, the Annex may be of some assistance to the designers.

(e) *Tensile* strength (f_{wt}) should be carefully estimated on the basis of appropriate expressions, possibly accounting for the favourable contribution of friction resistance of blocks in case of weak mortar (see Fig. 1.19). Otherwise, tensile strength may be estimated as a variable percentage (1:3 up to 1:15) of compression strength, depending on the magnitude of compression strength. This is another source of important uncertainties in predicting cracking resistance of masonry walls.

(f) *Diagonal compression* strength (f_{ws}) may be estimated as a function of f_{wc}, f_{wt} and the transversal tensile stress (σ_t). However, direct reference to diagonal compression tests is advisable.

(g) *Shear resistances* are normally distinguished as follows:

- In-plane shear resistance: This however, is equivalent to a diagonal compression as per Section 1.6.1, step f.
- Sliding resistance between perpendicular walls, along their widths or their lengths. This very important strength should be evaluated considering (i) direct shear failure of interconnecting blocks, or (ii) deracination of these interconnecting blocks.

1.6.2 Additional In-Situ Strength Determination

In some Monuments, it may be allowed to complement the data mentioned in Section 1.6.1, by means of in-situ strength determinations. To this end, flat-jacks are inserted into appropriately deep horizontal slots, so that an in-situ masonry "specimen" to be shaped and tested. However, it has to be noted that, under these testing conditions, post-elastic deformational capacity of masonry may be clearly overestimated because of the surrounding confining conditions.

1.6.3 In-Lab Testing of Replicas

In the case of Monuments of higher importance, provided that the experimental investigations described in Section 1.6.1 (step a) have offered sufficient information,

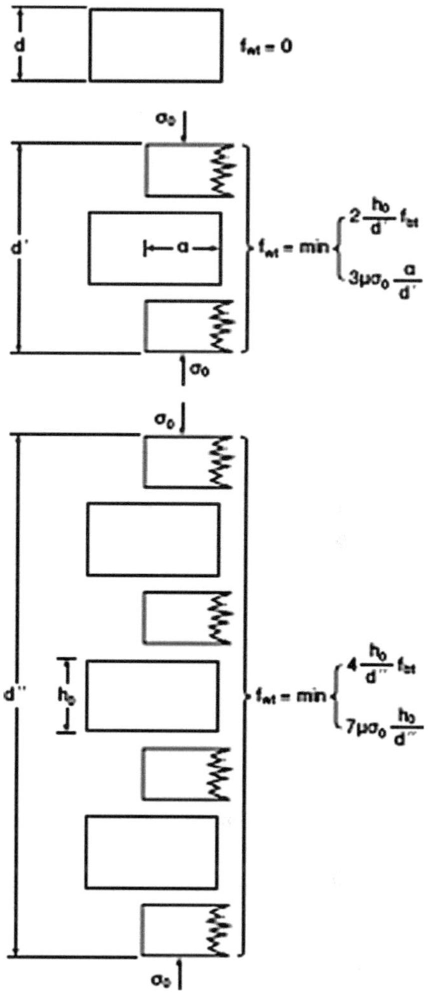

Fig. 1.19 Nominal tensile strength of masonry, as a function of the extent of the reference zone (d, d', d''); "σ_0" denotes the acting vertical compression stress, and "μ" is an appropriate minimal value of friction coefficient

it is suggested to reproduce in laboratory (piece by piece) a part of the critical region of the masonry under consideration. Thus, specimens will be available – replicas of the actual wall, to be subsequently tested under vertical and diagonal compression. The dimensions of these replicas should be large enough, for the failure mechanism to be freely developed.

1.6.4 Testing on Earthquake Simulators

There is also a possibility to gain an overall information on the seismic capacity of a (relatively simple in composition) Monument, by means of a scaled model tested on a shaking table. Several dynamic simulation problems have to be solved to this end; however, well conducted tests may offer precious information, as i.a. in the tests carried out in IZIIS, Skopje, on models of orthodox churches and adobe structures. If, instead of an entire monument, a sub-assemblage of it is tested on an earthquake simulator, scales may be more convenient and results may be more reliable (like in the case of the cross vaults of Dafni Monastery, tested in the Laboratory of Seismic Engineering, NTUA, Athens).

1.6.5 Resistance of Walls or Columns Made of Sculptured Stones/Dry Joints (Graeco–Roman)

Walls: Their compression strength under distributed load is influenced by possible internal and interface irregularities of some blocks, but it is generally of the same order as the strength of individual stones. Resistance against local load may be predicted via the provisions of Applied Mechanics.

Columns made of stone drums: Two resistance mechanisms should be documented in advance:

- Friction: (i) The relationship of friction coefficient and normal stress should be well known and a statistical dependence has to be established, for several humidity conditions of the joint. (ii) The constitutive law "imposed slip/mobilized friction resistance" should be experimentally found, for the specific conditions of the drums in consideration. To this end, in-lab tests on representative replicas are needed.
- Strength of a block or a drum in rocking position: Relevant in-lab tests on replicas may be needed for the determination of this strength (inclined compressive force, concentrated on a free edge), unless a reliable analytic model might to be satisfactory.

1.7 Assessment of the Actual Seismic Capacity of the Monument

(a) To this end, normally simple static linear models of Analysis are first used, up to a value of seismic actions revealing a local insufficiency. Subsequently, depending on the acceptable damage level of the Monument (Section 1.3), the assessment may end here, or for Monuments of relatively lower importance, further increase of seismic actions may be considered by using more sophisticated methods of Analysis, as described in Section 1.5, up to the level of acceptable damage.

Obviously, in the case of slender Monuments, the use of a simple dynamic method of Analysis is needed since the very beginning of the assessment.

(b) Besides the quantitative approach followed up to this point, the significance of a more qualitative approach should also be mentioned here towards a preliminary seismic assessment of Monuments. Reference is made to the method recommended by the Italian Guidelines, 2006, regarding the case of Churches. The methodology consists of the following steps.

- Several possible local failure mechanisms are identified and indicatively illustrated (e.g. Fig. 1.20).
- For each of them, an empirical vulnerability assessment is carried out, based on specific instructions.

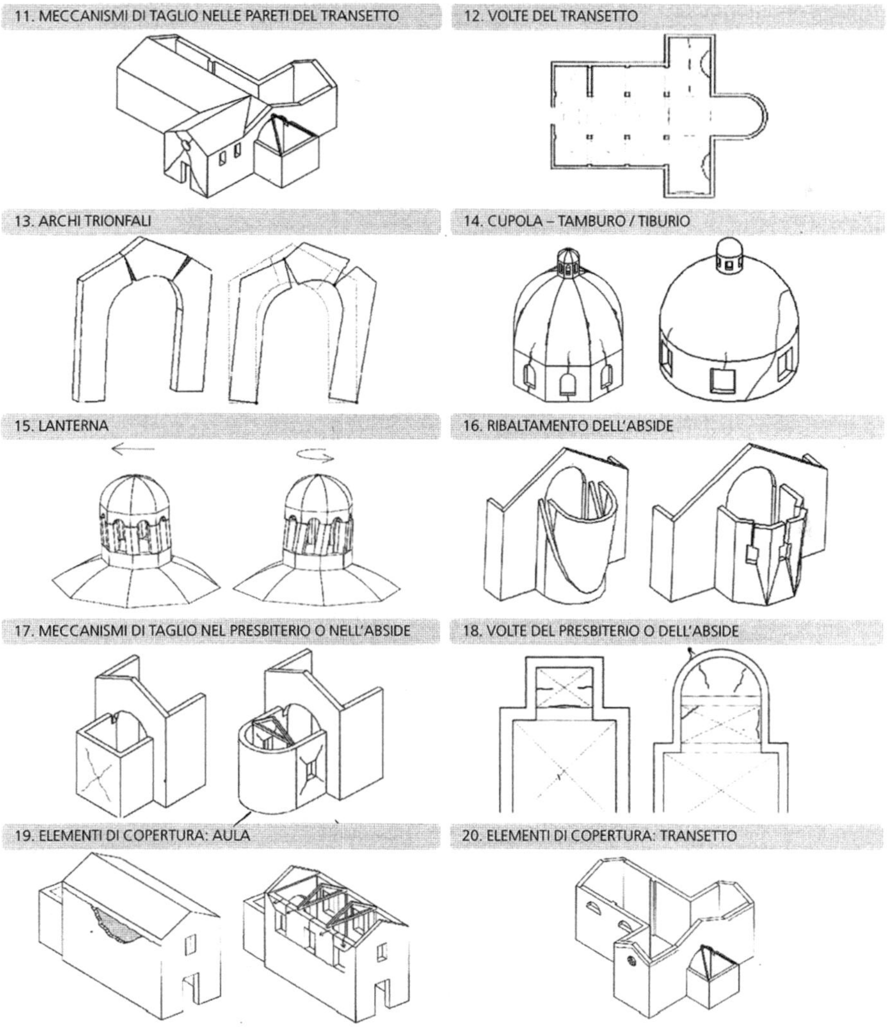

Fig. 1.20 Damage mechanisms of churches, for their vulnerability indices to be assessed, within a pseudo-quantitative procedure to evaluate seismic risk level of churches in Italy

- A weighed mean value of global vulnerability is calculated, and corresponding ground motion accelerations are found by means of empirical formulae, indicating a limited-damage state or a collapse limit state.
- The comparison of these accelerations with the provisions of the National Seismic Code at the location of the Monument, offers a statistical evaluation of priorities for structural interventions, and/or for civil protection strategies.

1.8 Selection of the "Re-design" Earthquake for the Structural Intervention

This issue is of fundamental importance. In Section 1.2 we faced multiple contrarieties: our wish to ensure a high seismic resistance for a Monument may necessitate structural interventions as "invasive" as to harm some monumentic or technical Values. Thus, we have recognised the need for an "optimization" or, in other words, the need to negociate the re-design earthquake to be finally used in our calculations. This chapter attempts to summarise the procedures to be followed in order to select an "optimal" value of seismic actions.

1.8.1 Historic Data

The behaviour of the Monument under specific earthquakes of the past, should be well studied and described in the Documentation (Section 1.4, steps a and b). Although it is not easy to quantify the intensity of those earthquakes, historical data do frequently offer helpful collateral information, such as: behaviour of normal buildings in adjacent areas, fall of objects etc. Archaeoseismology on the other hand may offer useful information regarding very old monuments. However, a certain reservedness is sometimes justified because of the incompleteness of data and the difficulties of interpretation of archeological, literary and geological data (Ambrazeys, 2005).

On the other hand, we may also derive quantitative information about the intensity of previous earthquakes, based on the behaviour of pre-existing repairs or strengthenings.

To the extent such methodologies are feasible, it will be possible to have a first estimate of the overall seismic capacity of the Monument as it stands. This knowledge may assist us in selecting a couple of seismic load values for the re-design of our planned structural interventions.

1.8.2 Actual Seismological Studies

This is the most "easier" source of information; but, as it is well known, the expected seismic load values are related (i) to several probabilities of exceedance, and (ii) to a conventional life-time of the building. Thus, this scientifically established knowledge has not an "absolute" validity: First, the life-time of a Monument cannot be clearly expressed; by definition, "monuments" should live "for ever". And second, the notion of an acceptable probability of exceedance is not too friendly to Curators of Monuments.

However, the designer may face these two issues as follows.

(i) Take into account first the probability actually accepted for existing old urban buildings, and second the probability specified by the Code in force for buildings of very high social importance.
(ii) In place of a life-time duration, one may consider a reasonable long period for radical repair and strengthening of the Monument in consideration.

Moreover, for monuments of higher importance, appropriate modifications or enrichments of the aforementioned data may be effectuated, by means of

- additional soil-dynamic consideration,
- geomorphological corrections
- an estimation of the expected duration of the quake, as an indication of the expected large amplitude cycles (one or two or three?).

In any event, the seismic load values dictated by actual seismological studies, are but one of the candidate-values for the final re-design: In fact, because of the contrariety of Monumental and Social values (Section 1.2), within the unavoidable "negociation" for optimisation, the a.m. seismological input loses its character of undeniability.

1.8.3 Pre-selected Seismic Loads

Based on these two sources of information, as well as on his/her own experience, the designer selects one (or two) temporary values of seismic load to be submitted to the necessary final optimisation (Section 1.10). The pre-selection of such a temporary value "E_0" may be facilitated by taking into account in advance the acceptable damage-level (Section 1.3) under the redesign earthquake: For higher damage-levels, higher probabilities of exceedence have a better chance to be satisfactory.

This "seismic load" may be expressed in appropriate terms (depending on the importance and the structural specificity of the Monument), such as ground acceleration, accelerograms and the like.

In any event, none of these seismic load values could be lower than a minimum level commonly acceptable in the region.

Note: Within the same context, the broader stability of foundations against earthquakes-induced displacements of slopes should also be considered (see i.a. Ambraseys and Srbulov, 1995).

1.9 Preliminary Designs

The fundamental process of the aforementioned optimisation, needs to be fed by at least two alternative solutions of structural intervention. Assuming that the repair or strengthening techniques were correctly chosen, it remains to check the favourable and adverse consequences of two seismic load design values. In what follows, the pertinent procedures are summarised.

(a) As it is well known, there are several available intervention philosophies, such as:

- Local repair or strengthening of a building member, (or its connections to adjacent elements), without external modification of its form.
- Installation of ties and confinements.
- Addition of new building elements (permanent or provisional) in order to ensure the wished global seismic resistance to the Monument. In this category of structural measures belongs also the case of adding or strengthening a diaphragm (for a better distribution of seismic forces), as well as the addition of hidden and completely reversible dumping devices.
- Possible construction of seismic isolation underneath foundations, (although such a technique may be better applicable in relatively recent Monuments on a "virgin" soil).
- Monolithic transportation of the Monument to a more favourable place – a solution suitable only in relatively small size Monuments.

Any intervention technique to be used, has to be appropriately checked against all Principles described in Section 1.2, especially those related to the specific performance requirements (Section 1.2, step c); in this connection too, there is space for another category of optimisation – which however is not further discussed in this contribution.

(b) Thus, assuming that the proposed intervention techniques were correctly chosen, at least two (E_{o1}, E_{o2}) seismic load values have to be used for the re-design, as they were pre-selected in Section 1.8.3. Corresponding seismic resistance levels (R_1 and R_2) are lent to the Monument, expressed e.g. in terms of ground acceleration or otherwise. For these seismic loads, the respective structural interventions are designed, and preferably illustrated in preliminary drawings.
The consequences of such interventions on all Principles enumerated in Section 1.2 will be now considered in details by the interdisciplinary decision-making Group, in order to deliberate as in the following paragraph.

(c) Strengthening resistance-models: The need for accurate Resistance-Models regarding the evaluation of the strengthening effects of various intervention methods cannot be overemphasised. In what follows, some of these models are enumerated and occasionally commented.

- *In low-tech walls*:
  - Adobe walls, strengthened by means of clay-grouting, need to be re-evaluated on the basis of the tensile resistance of grouts and the normalised volume of grout retained.
  - For brick-infilled timber trusses, strengthened with stronger diagonals or better built infills, there is no model predicting consequences on the ductility of the wall.
- *In plane masonry walls*:
  Not many of the "after-intervention" resistance models are established in literature, regarding the following repair or strengthening cases.
  - Replacing stone-blocks or addition of steel elements as shear connectors: along transversal walls intersections, in case of cracks' stitching, or in case of passing through connectors of three-leaf masonry.
  - Strengthening of the frames of doors and windows (increase of shear stiffness, increase of nominal strengths of pears and spandrels).
  - Hydraulic grouting: In this case strength predictions are somehow more developed (indicatively, see i.a. Vintzileou and Tassios, 1995; Valluzzi, 2000; Tassios, 2004).
  - Timber or metallic ties: The specific models needed are related to the interface resistance of transversal walls, to the increase of apparent tensile resistance of masonry mega-elements, etc.
  - Local re-building of critical regions: In this case, normal models apply.
- *Arches and domes*:
  Technical literature is more rich in this traditional field, but not necessarily under conceptually asymmetrical seismic conditions. One of the basic models needed is the quantification of structural consequences of chaining of domes, as a function of mobilised chain's response to deformations induced after initial seismic cracking of the dome.
- *Graeco–Roman antiquities*:
  Relevant resistance-models may be related to structural interventions like the following ones
  - Re-integration of pieces of epistyles or drums by means of titanium-connectors: Design criteria are sought regarding reinstatement of the initial strength of monoliths or, alternatively, of only the structurally required strength (Toumbakari, 2003)
  - Improvement of setting of irregular drums by means of extra-thin injections (Miltiadou-Fezans et al., 2005)

1.10 Selection of the Final Optimal Solution

In-view of the interdisciplinary nature of Principles (Section 1.2, steps a, b, and c) involved in the decision-making process, it is suggested that the final selection of the structural intervention scheme of seismic strengthening of the Monument, be effectuated by a representative Group – including the Owner of the Monument. However, depending on available local habits, decision may be taken by just one responsible agent of the Authorities, provided that he/she will follow the same decision-making process.

The process for the final judgement may be conceptually described as follows:

(a) For each of the Principles enumerated in Section 1.2, a scale of "level of satisfaction", graded $G = 1$ to $G = 5$ (from low to high degree of satisfaction) is established.
(b) For the specific Monument under consideration, a lowest acceptable level ($G_{i,\min}$) of each of these ten Values (i) is decided in advance.
(c) For the same Monument, weighing factors "f_i" are agreed, expressing the relative importance of each of the aforementioned ten Principles (with $\sum f_i = 1$). Obviously, it is expected that the highest values will be given to the factors regarding Form and Human life.
(d) An evaluation is needed of the way a given intervention solution affects the aforementioned Values; grades G_i ($i = 1$–10) are agreed respectively.
(e) The level of aseismic resistance (R) ensured to the Monument by the solution envisaged, was found in Section 1.9b. It is reminded however that $R \not< R_{\min}$, a commonly acceptable minimum level of seismic capacity – as for instance in the case of normal buildings of the Region.
(f) An estimator of overall efficiency of the proposed solution is calculated

$$e = \frac{R}{R_o} \cdot \sum_{i=1}^{10} f_i \cdot (G_i - G_{i,\min})$$

where R_o is an arbitrary, sufficiently high resistance-level (e.g. that wished for an extremely important modern building).

The solution with the higher score may be selected, unless (as it is very common) further *refinements* are requested by the Authorities.

It has to be noted that the whole procedure is also applicable without its quantifiable version: Experts may offer their opinion, based on their qualitative judgements after a thorough examination of the consequences of each solution. Afterall, the main interest of this procedure is that it may act as an open reminder of the multifold aspects of the design for the aseismic protection of Monuments.

Annex

Deformational Characteristics of Masonry in Monuments

1. *Basic* $\sigma_0 - \varepsilon_0$ *relationships*

   (a) It seems that our knowledge on this elementary matter is not very rich.
   First, because of practical causes like the experimental conditions (representative size of the specimen and in-lab boundary conditions) and the extreme variety of possible forms of blocks, as well as of the possible nature of blocks and mortars, and possible curing conditions.
   Second, because this kind of research does not seem to be "elegant" enough or attractive – although the variety of parameters constitutes a challenge to mathematical investigations. In what follows, only a minimum of information is included regarding the ultimate strains "ε_{cu}" of masonry under uniaxial or biaxial stress conditions.

   (b) The main parameters affecting the ε_{cu}-value may be identified as follows:

   - Strength of mortar ($\varepsilon_{cu} \searrow$)
   - Strength and brittleness of the blocks ($\varepsilon_{cu} \searrow$)
   - Interlocking between blocks ($\varepsilon_{cu} \searrow$)
   - Bond with mortar ($\varepsilon_{cu} \searrow$)
   - Completeness of joints ($\varepsilon_{cu} \searrow$)
   - Presence of transversal tensile stresses ($\varepsilon_{cu} \searrow$)

   - Pre-existing cracks ($\varepsilon_{cu} \nearrow$)
   - Number of cycles of repetitive compressive actions ($\varepsilon_{cu} \nearrow$)
   - Acceptable stress-response degradation or "damage level" ($\varepsilon_{cu} \nearrow$)
   - Reinforcement or confinement ($\varepsilon_{cu} \nearrow$)

   (c) Based on some experiments carried out in the Laboratory of R.C. of Nat. Tech. University of Athens, the following very rough values may be noted:

   - Normal compression:

     – Rubble stones masonry

     pick value (monotonic) $\varepsilon_{cu} = 2 \times 10^{-3}$ to 3×10^{-3}
     after 30% degradation $\varepsilon'_{cu} = 3 \times 10^{-3}$ to 5×10^{-3}

     – Full bricks' masonry ($f_{wc} \sim 3 - 8$ MPa)

     $$\varepsilon_{cu} \sim 4 \times 10^{-3}$$

     – Perforated bricks' masonry ($f_{wc} = 2 - 3$ MPa)

     $$\varepsilon_{cu} \sim 1.5 \times 10^{-3} \text{ to } 3.5 \times 10^{-3}$$

- Diagonal compression:
  - Perforated bricks' masonry ($f_{wc} = 3 - 4\,\text{MPa}$)

$$\varepsilon_{cu} \sim 1 \times 10^{-3} \text{ to } 2 \times 10^{-3}$$

- In case of timber-reinforced traditional stone masonry, ε_{cu} may be increased up to 300%, both under vertical and diagonal loading.

(d) Regarding 3-leaf walls, some indicative values are also reproduced here.

- Normal compression:
  - Well built external and internal leafs (Binda et al. 2006), $f_{wc} \sim 6$ MPa

$$\varepsilon_{cu} = 3 \times 10^{-3} \text{ to } 4 \times 10^{-3}$$

  - Byzantine semi-regular 3-leaf masonry (Vintzileou and Miltiadou, 2008), $f_{wc} \sim 2$ MPa

$$\varepsilon_{cu} = 1.5 \times 10^{-3}$$

  - Byzantine (semi-regular) *after* grouting (Vintzileou and Miltiadou, 2008), $f_{wc} \sim 3.5\,\text{MPa}$

$$\varepsilon_{cu} = 3.0 \times 10^{-3}$$

- Diagonal compression: (Vintzileou and Miltiadou, 2008)

$$\varepsilon_{cu} \sim 1 \times 10^{-3}$$

2. *Rotational capacity of masonry wall elements*

There is an understandable trend to apply non-linear static analysis in the case of masonry monuments (after all, the very concept of pushover analysis was first applied to masonry buildings). To this end, the analytical tools are well developed. However, the second term of the inequality of safety (resistance that is), expressed as "available rotational capacity θ_u", seems to be less well understood. That is why, an attempt towards a possibly better understanding of the post-elastic rotation capacity of orthogonal masonry elements is made in what follows.

(i) Vertical pier element
Obviously, the two components of "θ" should be recognized

$$\theta = \theta_{M+N} + \theta_{\nu}$$

where, approximately, $\theta_{\nu} = V : Gtl$, (t = width of the wall, l = length, and shear modulus $G \approx 100 f_c$ under cyclic conditions).

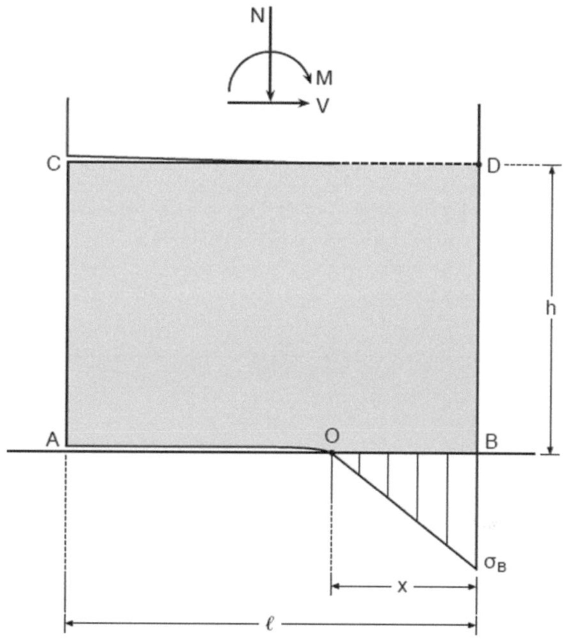

Fig. 1.21 Rectangular wall element under flexure (*M*, *N*) and shear (*V*)

Regarding the first term however, a basic correction should first be introduced in what is currently used. The vertical deformation at point *O* in Fig. 1.21 is NOT zero: This point during the gradual application of *M*, had undergone considerable compressive stresses, and after a maximum value, equal to $\sigma_{0,\max} = M{:}2\ tx$ (for approx. linear stress distribution), its stress comes rapidly down to zero (Fig. 1.22). Consequently, during this unloading stage, the residual deformation $\varepsilon_{0,\mathrm{res}}$ at point "*O*" is not equal to zero (Fig. 1.23).

Assuming approximately equal stresses acting on both ends (*AB* and *CD*) of the masonry element under consideration, the expression of available θ_{M+N} should be

$$\theta_{M+N} \approx (\varepsilon_{\mathrm{B}} - \varepsilon_{0,\mathrm{res}}) \cdot \frac{h}{x}$$

provided however that no shear failure has occurred.

The selection of the appropriate numerical ε-values is not an easy task: First, cyclic compression has to be taken into account, leading to increased ε_{c}-values valid under Monotonic loading. In doing so, however, the targeted *performance* level (i.e. acceptable damage level of the Monument) will be the basic criterion for the final ε_{c}-design values.

Besides, for those analytical methods using intermediate *spring*-elements (connecting discrete building elements), it is worth reminding that, because

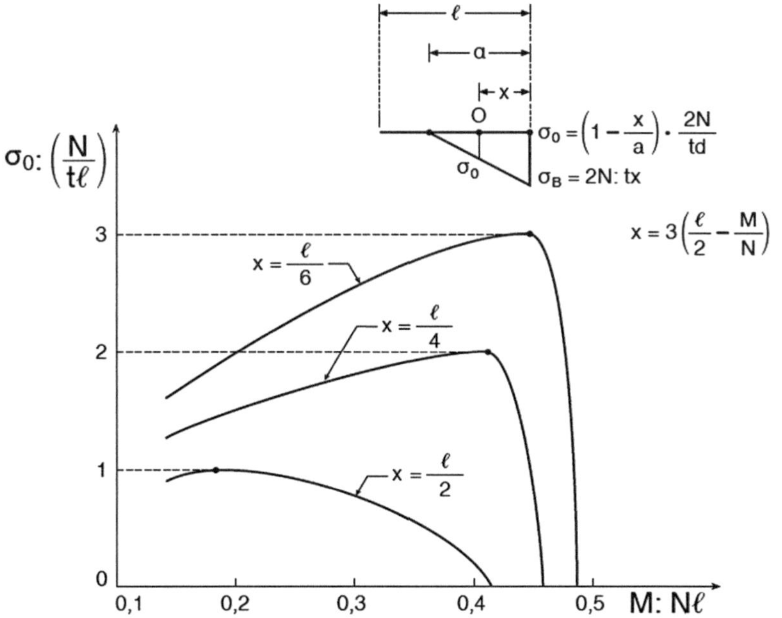

Fig. 1.22 During the gradual increase of acting moment "*M*", the actual zero-stress-point "*O*" was previously submitted to considerable stresses "σ_0"

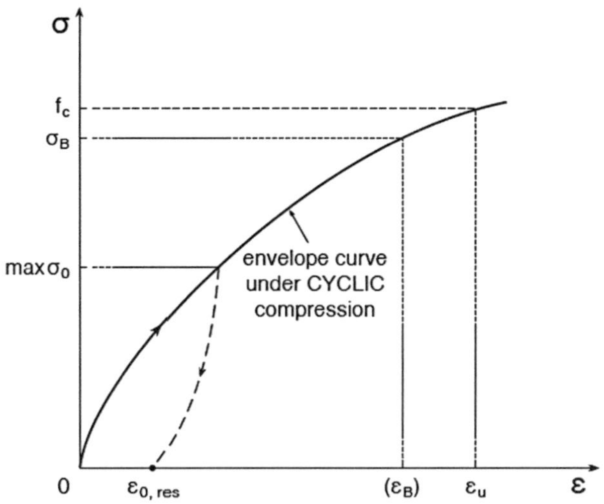

Fig. 1.23 Point "*O*" of Fig. 1.21, after unloading from its max σ_o (see Fig. 1.22), has a residual strain $\varepsilon_{0,\text{res}}$

of the presence of the axial force and the flexural cracking of the base cross section, rotations "θ" are *not* a linear function of moments "*M*". Thus a spring-"constant" $k = \theta{:}M$ can hardly be assessed. Roughly speaking, for (*M*: *Nl*)-values higher than 0.3, the average ratio ($\Delta\theta{:}\Delta M$) may increase up to 20 times as compared to previous values.

(ii) Spandrel element

In "frame"-like masonry, walls containing openings, ("perforated walls"), vertical "piers" are connected by "overlintel" two-dimensional elements; their available post-elastic rotational capacity is also of basic importance in applying a pushover method of analysis. Referring to Fig. 1.24, the compressive deformation of the "diagonal" equals ε'_{cu}, reduced value of ε_{cu} exhibited under vertical *cyclic* loading (see Fig. 1.25).

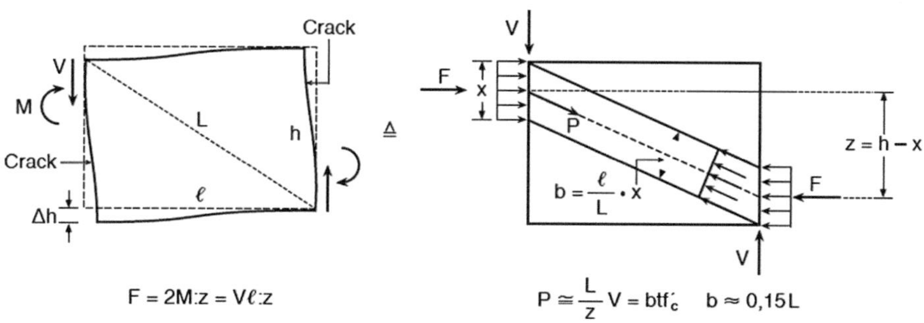

Fig. 1.24 Masonry spandrel element (of thickness "*t*") under ultimate shear-flexural loading

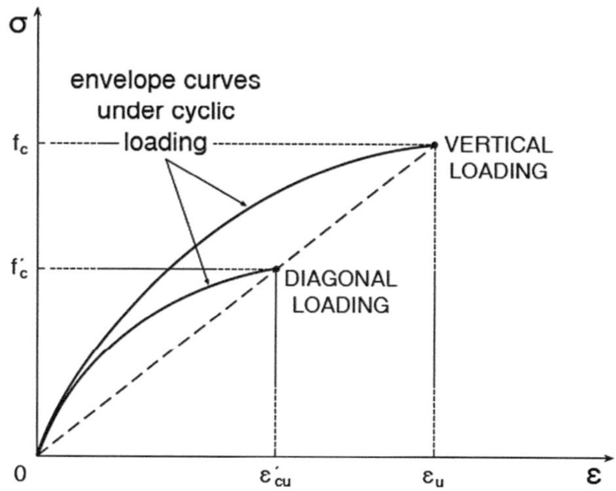

Fig. 1.25 Under diagonal compression (with simultaneous action of transversal tension stresses), it can be taken $f'_c : f_c \approx \varepsilon'_{cu} : \varepsilon_{cu} \approx 0.6$

Thus, finally, for the lowest possible performance level

$$\theta_u = \frac{\Delta h}{l} = \frac{h}{L} \cdot \varepsilon'_{\text{cu}}$$

provided however that shear sliding along the compressive zones does not occur.

As in the case of pier elements, increased ε'_{cu}-design values under cyclic loading will be highly dependent on the targeted performance level of the Monument.

References

References Related to Experimental Investigation

Berra M, Binda L, Anti L, Fatticcioni A (1991) Non destructive evaluation of the efficacy of masonry strengthening by grouting techniques. Proceedings of the 9th International Brick/Block Masonry Conference, vol 3. Berlin, Germany, pp 1457–1463

Binda L, Lenzi G, Saisi A (1998) NDE of masonry structures: use of radar tests for the characterization of stone masonries. NDT&E Int 31(6):411–419

Binda L, Modena C, Baronio G, Abbaneo S (1997) Repair and investigation techniques for stone masonry walls. Constr Build Mater 11(3):133–142

Binda L, Saisi A (2001) Non destructive testing applied to historic buildings: the case of some Sicilian Churches. In: Lourenço PB, Roca P (eds) Historical constructions. University of Minho, Guimarães

Binda L, Saisi A, Zanzi L (2003) Sonic tomography and flat-jack tests as complementary investigation procedures for the stone pillars of the temple of S. Nicolò l' Arena (Italy). NDT&E Int 36:215–227

Côte Ph, Dérobert X, Miltiadou-Fezans A, Delinikolas N, Durand O, Alexandre J, Kalagri A, Savvidou M, Chryssopoulos D, Anamaterou L (2007) Application of non-destructive techniques at the Katholikon of Dafni Monastery for mapping the mosaics substrata and grouting monitoring. In: Proceedings of the 6th international conference on stuctural analysis of historical construction, vol II. Bath, UK, Jul 2008

Côte Ph, Dérobert X, Miltiadou-Fezans A, Delinikolas N, Minos N (2004) Mosaic-grouting monitoring by ground-penetrating radar. SAHC2004: In: Proceedings of the 4th international seminar on structural analysis of historical constructions, Padova, Italy, Nov 2004

Felicelti R, Gattesco N (1998) A penetration test to study the mechanical response of mortar in ancient masonry buildings. Mater Struct 31:350–356

Giuriani E, Gubana A (1991) A penetration test to evaluate wood decay and its application to Laggia monument. Mater Struct 26:8–14

Macchi G, Ghelfi S (2006) Indagini strutturali. In: La Torse Restituita, vol III. Bolletino d' Arte, Ministry of Cultural Affairs, Rome, Italy

Maierhofer C, Hamann M, Hennen C, Knupfer B, Marchisio M, da Porto F, Binda L, Zanzi L (2004) Structural evaluation of historic walls and columns in the Altes museum in Berlin using non-destructive testing methods. In: Modena C, Lourenço PB, Roca P (eds) Proceedings of the 4th international seminar on structural analysis of historical constructions, vol 1. Taylor & Francis Group, London, pp 331–341

Maierhofer C, Köpp C, Wendrich A (2004) On-site investigation techniques for the structural evaluation of historic masonry buildings – a European research project. In: Modena C, Lourenço

PB, Roca P (eds) Proceedings of the 4th international seminar on structural analysis of historical constructions, vol 1. Taylor & Francis Group, London, pp 313–320
Tassios Th, Mamillan M (1985) Structural investigations of ancient monuments – Valutazione strutturale dei monumenti antichi. Ed. Kappa, Rome, Italy
Thomasen SE, Sears CL (1993) Historic preservation. High and low tech diagnostic technology. International Association for Bridge and Structural Engineering, Zurich, pp 91–98
Silman R, Ennis M (1993) Non-destructive evaluation to document historic structures. International Association for Bridge and Structural Engineering (IABSE), Zurich, pp 195–203
Van de Steen B, Van Balen K, Halleux L, Mertens R (1997) Nondestructive testing techniques applied for the investigation of a typical baroque church facade: the St-Michiels church in Leuven. Inspection and monitoring of the architectural heritage, international colloquium seriate, Italian group of IABSE, pp 177–185
Wenzel F, Kahle M (1993) Indirect methods of investigation for evaluating historic masonry. International Association for Bridge and Structural Engineering (IABSE), Zurich, pp75–90

References on Analysis and Resistance

Ambraseys N (2005, Dec) Earthquakes and archaeology. J Archaeol Sci
Ambraseys N, Srbulov M (1995) Earthquake induced displacements of slopes. Soil Dyn Earthquake Eng 14
Binda L, Pina-Henriques J, Anzani A, Fontana A, Lourenço PB (2006) A contribution for the understanding of load-transfer mechanisms in multi-leaf masonry walls: testing and modelling. Eng Struct 28
Giuffré A (1991) Letture sulla Meccanica delle Murature strutturali. Kappa, Rome, Italy
Miltiadou-Fezans A, Papakonstantinou E, Zambas K, Panou A, Frantzikinaki K (2005) Design and application of hydraulic grouts of high injectability for structural restoration of the column drums of the Parthenon Opisthodomos. In: Brebbia CA, Torpiano A (eds) Advances in Architecture 20, STREMHA IX "Structural studies, repairs and maintenance of architectural heritage". WIT Press, Malta, pp 461–471
Psilla N, Tassios TP (2008, Dec) Design models of reinforced masonry walls under monotonic and cyclic loading. Eng Struct
Tassios TP (1988) Meccanica delle Murature. Liguori Editore, Napoli
Tassios TP (2004) Rehabilitation of 3-leaf masonry. In: Evoluzione nella sperimentazione per le costruzioni, Seminario Internationale (26 Sept–03 Oct), Centro Scientifico Internazionale di Aggiornamento Sperimentale-Scientifico (CIAS)
Tassios TP, Chronopoulos M (1986) Aseismic dimensioning of interventions on low-strength masonry buildings. Middle East Mediterranean regional conference "Earthen and low strength masonry buildings", Ancara
Toumbakari EE (2003) Structural restoration of the architectural members of Parthenon Opisthodomos. In Proceedings of 5th international meeting for the restoration of the acropolis monuments, Athens, 4–6 Oct 2002, pp 149–160 (in greek)
Valluzzi MR (2000) Comportamento meccanico di murature storiche consolidate con materiali e tecniche a base di calce. Doctor Thesis, University of Trieste
Vintzileou E, Miltiadou A (2008) Mechanical properties of 3-leaf stone masonry grouted with ternary on hydraulic line-based grouts. Eng Struct 30
Vintzileou E, Tassios TP (1995) Three leaf masonry strengthened by injecting cement grouts. J Struct Eng ASCE 5
Vintzileou E (2009) The effect of timber ties on the behaviour of historic masonry. ASCE J Str Eng (accepted for publication)

References on Graeco–Roman Monuments

Housner GW (1963) The behavior of inverted pendulum structure during earthquakes. Bull Seism Soc Am 53(2):403–417

Konstantinidis D, Makris N (2005) Seismic Response analysis of multidrum classical columns. Earthquake Eng Struct Dyn 34:1243–1270

Makris, N. and D. Konstantinidis (2003) The rocking spectrum and the limitations of practical design methodologies. Earthquake Eng Struct Dyn 32:265–289

Papantonopoulos C (1993) The "Articulated" structural system: studying the earthquake response of a classical temple. In: Proceedings of the 3rd international conference on structural studies, repairs and maintenance of historical buildings, Bath, UK

Psycharis IN, Papastamatiou DY, Alexandris Ap (2000) Parametric investigation of the stability of classical columns under harmonic and earthquake excitations. Earthquake Eng Struct Dyn 29:1093–1109

Sinopoli A (1989) Dynamic analysis of a stone column excited by a sine wave ground motion. Appl Mech Rev 44(11):S246–S255

Guide Lines on Seismic Strengthening of Monuments

Aplicación del Código Técnico de la Edificación a las obras de restauración arquitectrónica, Spain 2008

Linee Guida per la valutazione e riduzione del rischio sismico del patrimonio culturale. Ministero per i Beni e le Attivitá Culturali, Gangemi Ed., Roma, 2007

Recommendations for the Analysis and Restoration of Historical Structures, ISCARSAH/Icomos, 2001

Part I
Engineering Seismology

Chapter 2
Microzonation for Earthquake Scenarios

Atilla Ansal, Gökçe Tönük, and Aslı Kurtuluş

Abstract Seismic microzonation involves generation of seismic hazard maps with respect to estimated ground motion characteristics on engineering bedrock outcrop based on a regional seismic hazard study compatible with the scale of the microzonation. A grid system is implemented dividing the investigation area into cells according to the availability of geological, geophysical and geotechnical data. Site characterizations are performed based on available borings and other relevant information by defining representative soil profiles for each cell with shear wave velocities extending down to the engineering bedrock. 1D site response analyses are conducted to estimate site specific earthquake ground motion characteristics on the ground surface for each representative soil profile to estimate elastic response spectrum based on calculated acceleration time histories. Average of spectral accelerations between 0.1 and 1 s periods of elastic acceleration response spectrum are calculated as one of the two parameters representing earthquake shaking intensity on the ground surface. Site specific peak spectral accelerations corresponding to 0.2 s period are also calculated as the second microzonation parameter using the empirical amplification relationships proposed by Borcherdt (1994) based on equivalent shear wave velocities for the top 30 m of the soil profiles. Superposition of these two parameters is assumed to represent overall effect of site conditions and is adopted as the criteria for the microzonation with respect to ground shaking intensity. Recently, an extensive site investigation study was carried out on the European side of Istanbul as the first phase of the large-scale microzonation project for the Istanbul Metropolitan Municipality. A detailed microzonation with respect to earthquake ground shaking intensity is carried out for the Zeytinburnu town in Istanbul using part of these recently compiled soil data and the regional probabilistic seismic hazard scenario proposed by Erdik et al. (2004). The microzonation maps

---

A. Ansal (✉)
Kandilli Observatory and Earthquake Research Institute, Boğaziçi University,
Çengelköy, Istanbul, Turkey
e-mail: ansal@boun.edu.tr

M. Garevski, A. Ansal (eds.), *Earthquake Engineering in Europe*, Geotechnical,
Geological, and Earthquake Engineering 17, DOI 10.1007/978-90-481-9544-2_2,

are compared with the previously generated Zeytinburnu microzonation maps for the European Union Framework FP6 LessLoss Project (Ansal et al., 2007a) and for the Zeytinburnu Pilot Microzonation Project (Ansal et al., 2005; Kılıç et al., 2006; Özaydın et al., 2004) where microzonation maps were produced with limited number of site investigations and site response analyses using more approximate microzonation procedures.

2.1 Introduction

Seismic microzonation can be considered as the methodology for estimating the response of soil layers under earthquake excitations and the relative variation of earthquake ground motion characteristics on the ground surface for a specific area. The purpose of microzonation is to provide input for urban planning and for the assessment of the vulnerability of the building stock for different hazard (performance) levels.

Site specific free field earthquake characteristics on the ground surface are the essential components for microzonation with respect to ground shaking intensity, liquefaction susceptibility and for the assessment of the seismic vulnerability of the urban environment. The adopted microzonation methodology is based on a grid system and is composed of three stages. In the first stage, regional seismic hazard analyses are conducted to estimate earthquake characteristics on rock outcrop for each cell. In the second stage, the representative site profiles are modelled based on available borings and in-situ tests. The third stage involves site response analyses for estimating the earthquake characteristics on the ground surface and the interpretation of the results for microzonation (Ansal et al., 2004a, b). In addition to the generation of base maps for urban planning, microzonation maps with respect to spectral accelerations, peak acceleration and peak velocity on the ground surface can be estimated to assess the vulnerability of the building stock and lifeline systems (Ansal et al., 2005, 2006b).

Recently, a very comprehensive site investigation study was carried out on the European side of Istanbul as part of the microzonation project for the Istanbul Metropolitan Municipality (OYO, 2007). 2,912 borings (mostly down to 30 m depth with approximately 250 m spacing) were conducted within an area of about 182 km^2 to investigate local soil conditions. Standard Penetration Test (SPT), Cone Penetration Test (CPT), PS-Logging, Refraction Microtremor (ReMi), seismic reflection and refraction measurements were carried out at each borehole location. Samples collected in the field were tested in the laboratory to determine index and engineering properties of local soils within the investigated area. A detailed microzonation study with respect to earthquake ground shaking parameters is carried out for Zeytinburnu using part of these recently compiled soil data and based on probabilistic seismic hazard study by Erdik et al. (2004) to demonstrate the applicability of the methodology proposed to generate microzonation maps for urban areas and to show the effects of detailed site investigation and more comprehensive microzonation procedure.

2.2 Seismic Hazard and Earthquake Motion

The regional earthquake hazard analysis may be probabilistic or deterministic. In the case of microzonation for urban planning, it is preferable to adopt a probabilistic earthquake hazard assessment but in the case of earthquake damage scenarios for estimating possible damage distribution, deterministic approach may be more suitable (Ansal et al., 2009; Erdik et al., 2004). In the case study conducted for Zeytinburnu town based on the study by Erdik et al. (2004) the variation of peak ground acceleration at the bedrock outcrop corresponding to 475 year return period used for site response analysis is shown in Fig. 2.1.

The results of the earthquake hazard analysis corresponding to 475 year return period were calculated in terms of peak ground (PGA) and spectral accelerations (SA) at $T = 0.2$ s and $T = 1$ s periods for each cell and used for microzonation

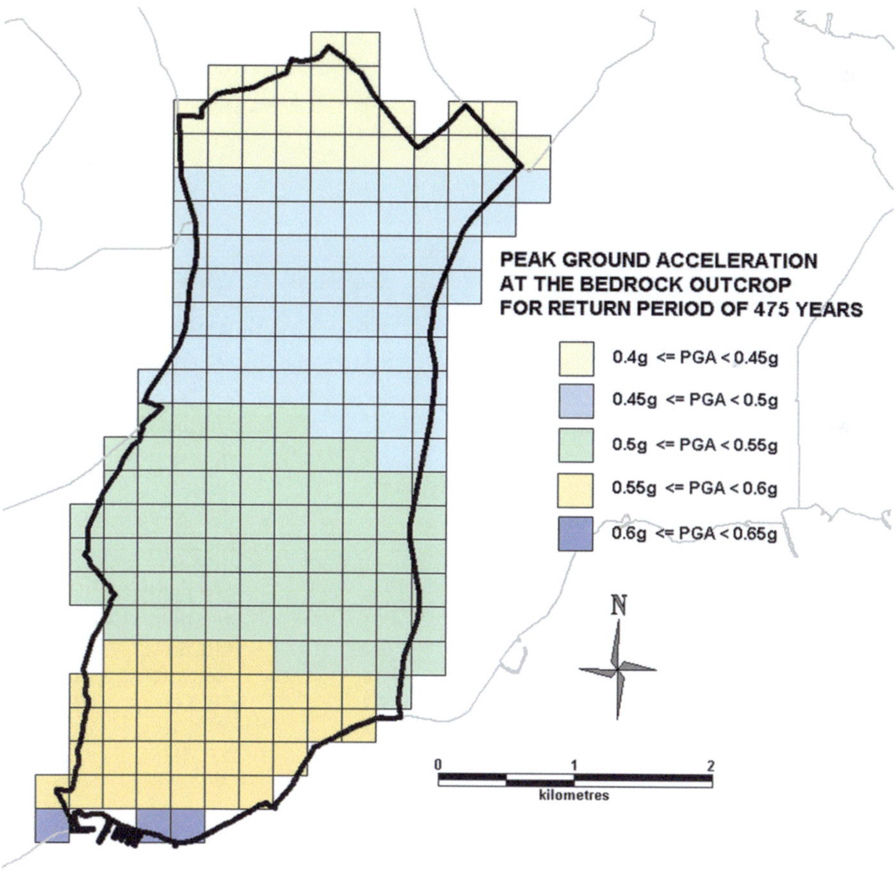

Fig. 2.1 Variation of peak ground acceleration at bedrock outcrop

of Zeytinburnu town. Independent of the methodology adopted for the earthquake hazard analysis, whether it is probabilistic or deterministic, realistic recorded or simulated acceleration time histories are needed to conduct site response analyses to determine earthquake characteristics on the ground surface. It was demonstrated by Ansal and Tönük (2007b) that if limited number of acceleration time histories (e.g. Three records as specified in some earthquake codes) is used even with scaling to the same PGA amplitudes for site response analysis, the results in terms of PGA and ground shaking intensity can be different for different sets of input acceleration time histories. Therefore, one option is to conduct site response analyses using large number of input acceleration time histories to eliminate the differences that are observed between different sets (Ansal and Tönük, 2007a) and also to take into account the variability due to the earthquake characteristics by adopting average response parameters on the ground surface for design and vulnerability assessment.

In the Zeytinburnu microzonation study, all available acceleration time histories compatible with the earthquake hazard analysis in terms of probable magnitude range ($M_W = 7.0 - 7.4$) and distance range (20–30 km) with strike slip fault mechanism that were recorded on sites with NEHRP (BSSC, 2001) site classification of B/C boundary were selected as input outcrop motion and were downloaded from PEER website (PEER, 2009) as listed in Table 2.1.

Table 2.1 List of earthquake acceleration records used for site response analysis

Earthquake	Station	Magnitude	Component	PGA (g)
Duzce 11/12/99	375 Lamont	7.1	375E	0.514
Duzce 11/12/99	375 Lamont	7.1	375 N	0.970
Duzce 11/12/99	531 Lamont	7.1	531E	0.118
Duzce 11/12/99	531 Lamont	7.1	531 N	0.159
Duzce 11/12/99	1059 Lamont	7.1	1059E	0.133
Duzce 11/12/99	1059 Lamont	7.1	1059 N	0.147
Duzce 11/12/99	1061 Lamont	7.1	1061E	0.134
Duzce 11/12/99	1061 Lamont	7.1	1061 N	0.107
Duzce 11/12/99	1062 Lamont	7.1	1062E	0.254
Duzce 11/12/99	1062 Lamont	7.1	1062 N	0.114
Duzce 11/12/99	Bolu	7.1	BOL000	0.728
Duzce 11/12/99	Bolu	7.1	BOL090	0.822
Kocaeli 08/17/99	Arçelik	7.4	ARC000	0.219
Kocaeli 08/17/99	Arçelik	7.4	ARC090	0.150
Kocaeli 08/17/99	Gebze	7.4	GBZ000	0.244
Kocaeli 08/17/99	Gebze	7.4	GBZ270	0.137
Kocaeli 08/17/99	Duzce	7.4	DZC180	0.312
Kocaeli 08/17/99	Duzce	7.4	DZC270	0.358
Imperial Valley 5/19/40	El Centro Array #9	7.0	I-ELC180	0.313
Imperial Valley 5/19/40	El Centro Array #9	7.0	I-ELC180	0.215
Landers 6/28/92	Morango Valley	7.3	MVH090	0.182
Landers 6/28/92	Morango Valley	7.3	MVH000	0.138
Landers 6/28/92	Joshua Tree	7.3	JOS000	0.274
Landers 6/28/92	Joshua Tree	7.3	JOS090	0.284

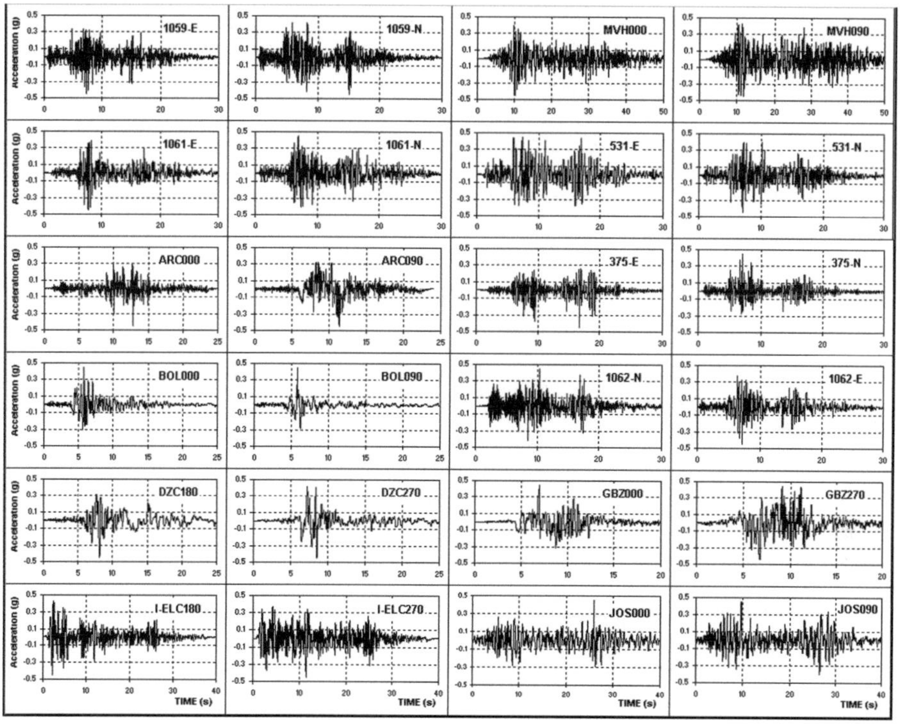

Fig. 2.2 Acceleration time histories used as input in the site response analysis

The input acceleration time histories are scaled with respect to the peak ground accelerations determined from regional seismic hazard study since this approach is observed to be practical and yielding consistent results as shown by Ansal et al. (2006a). For the Zeytinburnu case study, 24 scaled acceleration time histories were used as input motion for site response analyses by Shake91 (Idriss and Sun, 1992) and the average of the acceleration response spectra on the ground surface were determined to obtain the necessary parameters for microzonation. Selected time histories scaled with acceleration values at engineering bedrock level are listed in Table 2.1 and are shown in Fig. 2.2.

2.3 Site Characterizations

The investigated region is divided into cells by a grid system (preferably 250 m × 250 m) and site characterization is performed for each cell based on available borings and other relevant information by defining representative soil profiles. Shear wave velocity profiles are established down to the engineering bedrock with estimated shear wave velocity of 750 m/s. Typically, representative soil profiles for each cell where one or more borehole data are available are generated by considering the

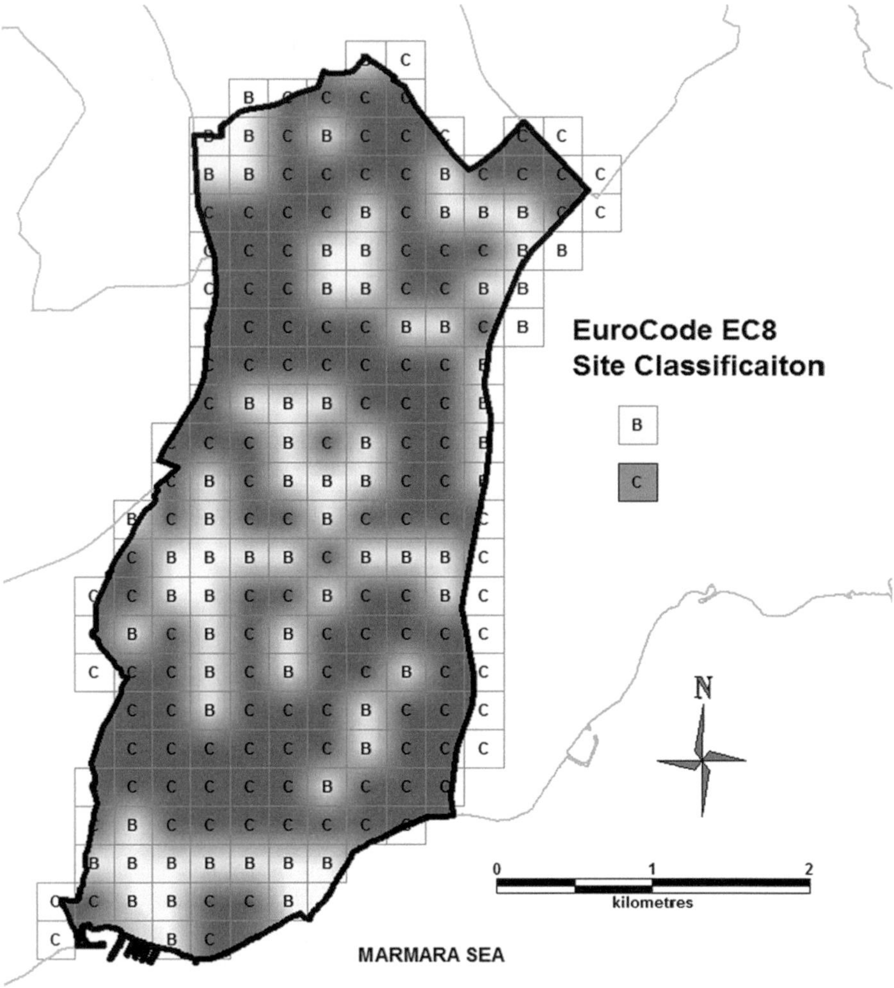

Fig. 2.3 Microzonation map with respect to EC8 site classification

most suitable borehole, and for the cells with no available borehole information, representative soil profiles are selected from the neighbouring cells by utilizing the available data. Interpolations between neighbouring boreholes may be performed taking into consideration the surface geology.

For the most recent Zeytinburnu case study there was at least one boring for each cell. Geotechnical data included a borehole with depth of at least 30 m for each cell where SPT, REMI and/or PS Logging measurements and laboratory index test results were available. Geological data together with seismic measurements provided engineering bedrock ($V_S > 750$ m/s) depths for all the cells. Variations of shear wave velocity with depth for the top 30 m of soil profiles were determined from SPT blow counts using empirical relationships proposed in the literature. Shear wave velocity profiles down to the engineering bedrock were estimated based on seismic wave velocity measurements. The calculated shear wave velocity profiles were compared with respect to shear wave velocities obtained from in-situ borehole seismic wave velocity measurements and were modified when necessary.

The preliminary output of any microzonation study may be the microzonation map with respect to site classification based on earthquake codes. The site classification in accordance with EuroCode 8 Part 5 (EC8, 2002) is given in Fig. 2.3. The microzonation with respect to site classification is useful in evaluating the effects of site conditions. However, it only reflects the characteristics of existing site conditions in relatively broad sense. It is obvious that in developing microzonation maps to assess earthquake hazard scenarios, probable earthquake characteristics is an essential input to achieve reliable results since site response as well as building vulnerability is directly related to the characteristics of the earthquake input. In addition, site classification with respect to earthquake codes (i.e. NEHRP, EC8) would be based on relatively large ranges of site parameters (e.g. average shear wave velocity) as shown in Fig. 2.3.

2.4 Microzonation with Respect to Ground Motion

In assessing the ground shaking intensity the purpose is to estimate relative effects of local site conditions on the level of ground motion characteristics. Therefore, all available data from site characterisation such as equivalent shear wave velocity (V_{s30}) as well as results of site response analyses conducted for each cell should be evaluated together to achieve a realistic and consistent result. The empirical amplification relationships such as the one proposed by Borcherdt (1994) enables the estimation of site-specific peak spectral accelerations based on equivalent (average) shear wave velocities (V_{s30}) measured or estimated for the top 30 m of soil profile. Site response analyses using Shake91 (Idriss and Sun, 1992) yields acceleration time histories to estimate peak ground acceleration as well as elastic acceleration response spectrum on the ground surface. Peak ground velocities on the ground surface are determined by integration of acceleration time histories. The results obtained are mapped using GIS techniques by applying linear interpolation among the grid points, thus enabling a smooth transition of the selected parameters. Soft transition boundaries are preferred to show the variation of the mapped parameter. More defined clear boundaries are not used due to the accuracy of the study. This allows some flexibility to the urban planners and avoids misinterpretation by the end users that may consider the clear boundaries as accurate estimations for the different zones.

The proposed methodology for microzonation is based on the division of the investigated urban area into three zones (as A, B, and C) with respect to frequency distribution of the selected ground shaking parameter corresponding to 33 and 67% percentiles (Ansal et al., 2004a, b). The site characterizations, as well as all the analyses performed, require various approximations and assumptions. Therefore, the absolute numerical values for the selected ground motion parameters may not be very accurate and besides may not be necessary for urban planning. Their relative values are more important then their absolute values.

Site response analysis, whether it is conducted by Shake91 (Idriss and Sun, 1992) or using similar programs can sometimes yield relatively high spectral amplifications or low peak ground acceleration values depending on the thickness of the deposit, estimated initial shear moduli, and on the characteristics of the input acceleration time histories. Even though the amplification relationships by Borcherdt (1994) are empirical, the spectral accelerations calculated using equivalent shear wave velocities are more consistent compared with the selected soil profiles.

The ground shaking intensity microzonation map that should reflect the estimated relative shaking intensity levels is based on the combination of two parameters: the cumulative average spectral acceleration between $T = 0.1$ s and $T = 1$ s periods of the average acceleration spectrum of all site response analyses conducted for each cell is adopted as the first microzonation parameter and peak spectral accelerations at short period range calculated from Borcherdt (1994) using V_{s30} is adopted as the second microzonation parameter.

The approach adopted to determine peak ground accelerations and elastic acceleration response spectra on the ground surface as first microzonation parameter was to conduct one dimensional site response analysis using Shake91 (Idriss and Sun, 1992). For each soil layer encountered in the soil profiles, total unit weight, thickness, shear wave velocity, and G/G_{max} and damping ratio relationships are provided as input. The strain dependent G/G_{max} and damping ratio relationships used in the site response analysis are summarized in Table 2.2.

For microzonation with respect to ground shaking intensity, the first microzonation parameter adopted was the average spectral accelerations between 0.1 and 1 s

Table 2.2 G/G_{max} and damping ratio versus shear strain relationships used in site response analysis

Material no.	Soil type	References
1	Clay (CH) PI = 60%	Vucetic and Dobry (1991)
2	Clay (CL) PI = 45%	Vucetic and Dobry (1991)
3	Clay (CH) PI = 30%	Vucetic and Dobry (1991)
4	Clay (CL) PI = 15%	Vucetic and Dobry (1991)
5	Silt	Darendeli (2001)
6	Sand (SC-SM)	Darendeli (2001)
7	Sand	Seed et al. (1984)
8	Gravel	Seed et al. (1984)
9	Gravel	Menq et al. (2003)
10	Rock 0–6 m	EPRI (1993)
11	Rock 6–16 m	EPRI (1993)
12	Rock 16–37 m	EPRI (1993)
13	Rock 37–76 m	EPRI (1993)

periods using the average acceleration spectra determined from the results of the all site response analyses conducted for each cell. The range of average spectral accelerations computed for the period interval of 0.1–1.0 s was between 0.885 and 1.283 g for Zeytinburnu case and since the difference between 33 and 67% percentiles was in the order of 45%, the area was divided into three zones with respect to spectral accelerations corresponding to 33 and 67% percentiles. In Fig. 2.4, A_{AVG} shows the most favourable regions with lower 33% percentile and C_{AVG} shows the most unsuitable regions with higher 33% percentile with respect to average spectral accelerations.

In the adopted methodology as the second microzonation parameter, the peak spectral accelerations for the short period ($T = 0.2$ s) were determined based on

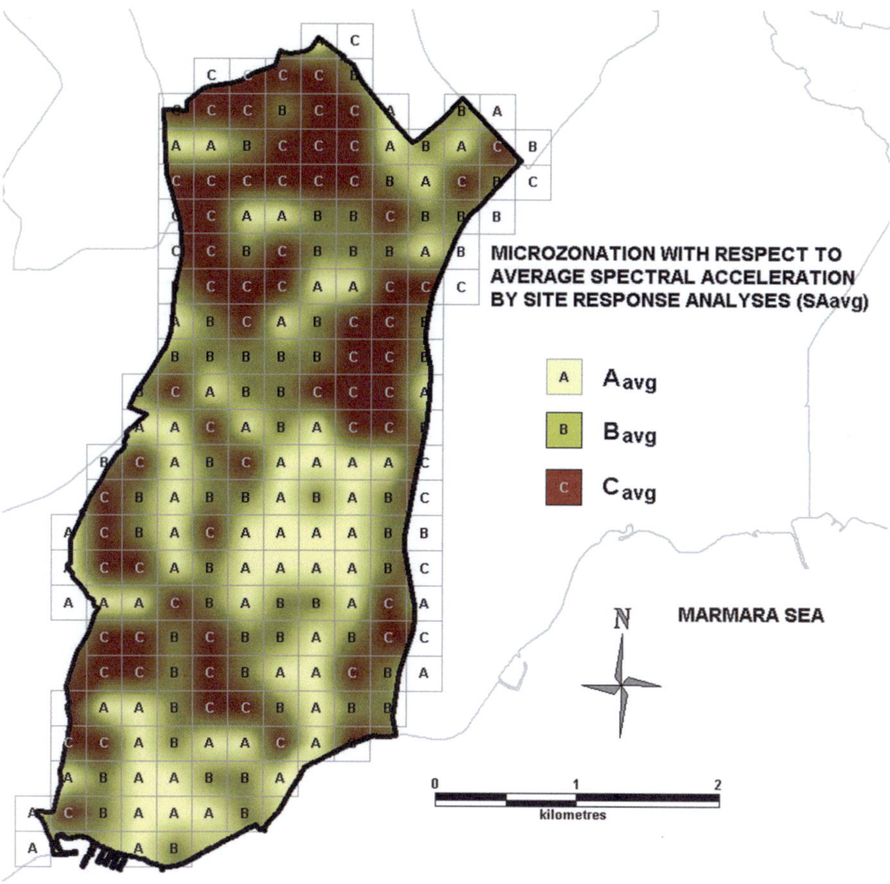

Fig. 2.4 Microzonation with respect to average spectral accelerations calculated by site response analyses

average (equivalent) shear wave velocity of each soil profile using the empirical relationships proposed by Borcherdt (1994);

$$S_a = F_a S_S \tag{1}$$

where S_S is the spectral acceleration at $T = 0.2$ s on the rock outcrop obtained from the seismic hazard analysis. The spectral amplification factor, F_a was defined based on the average shear wave velocity V_{s30}.

$$F_a = \left(760/V_{s30}\right)^{m_a} \tag{2}$$

where

$$m_a = \begin{cases} -\text{PGA} + 0.45 & : 0.1\ \text{g} < \text{PGA} \leq 0.2\ \text{g} \\ -1.5\ \text{PGA} + 0.55 & : 0.2\ \text{g} < \text{PGA} \leq 0.4\ \text{g} \\ -0.05 & : \text{PGA} > 0.4\ \text{g} \end{cases} \tag{3}$$

where PGA is the peak ground acceleration at the rock outcrop estimated based on the seismic hazard analysis.

For Zeytinburnu case, since the relative difference between peak spectral accelerations (at $T = 0.2$ s) calculated from Borcherdt (1994) relationships corresponding to 33 and 67% percentiles of the distribution (0.658 and 0.706 g) was smaller than 20%, the area was divided into two zones instead of three zones using 50% percentile (median) value of 0.678 g as recommended by Studer and Ansal (2004). Microzonation map was produced in accordance with the relative mapping as shown in Fig. 2.5, where A_{borch} shows the more favourable regions for lower 50% percentile where spectral accelerations are less than 0.678 g and C_{borch} shows the more unsuitable regions with higher 50% percentile with respect to peak spectral accelerations where the spectral accelerations are higher than 0.678 g.

As can be seen from these maps (Figs. 2.4 and 2.5), there are similarities and differences between the average spectral accelerations obtained by site response analyses with the spectral accelerations calculated using Borcherdt (1994) equation based on equivalent shear wave velocity. The most important difference is in the range of values for both parameters. In the case of site response analysis the range of average spectral accelerations was much larger allowing microzonation with respect to three zones.

The microzonation map with respect to ground shaking intensity was calculated by the superposition of these maps with respect to these two parameters. Superposition of empirically and analytically calculated spectral accelerations is assumed to provide a realistic assessment of the variation of site effects. The approach was developed and used for most of the seismic microzonation studies conducted in Turkey during the last decade (Ansal et al., 2007a, b, 2006b, 2005, 2004a, b; Kılıç et al., 2006).

The final microzonation map is superposition of microzonation maps with respect to average spectral accelerations obtained from site response analyses (A_{avg},

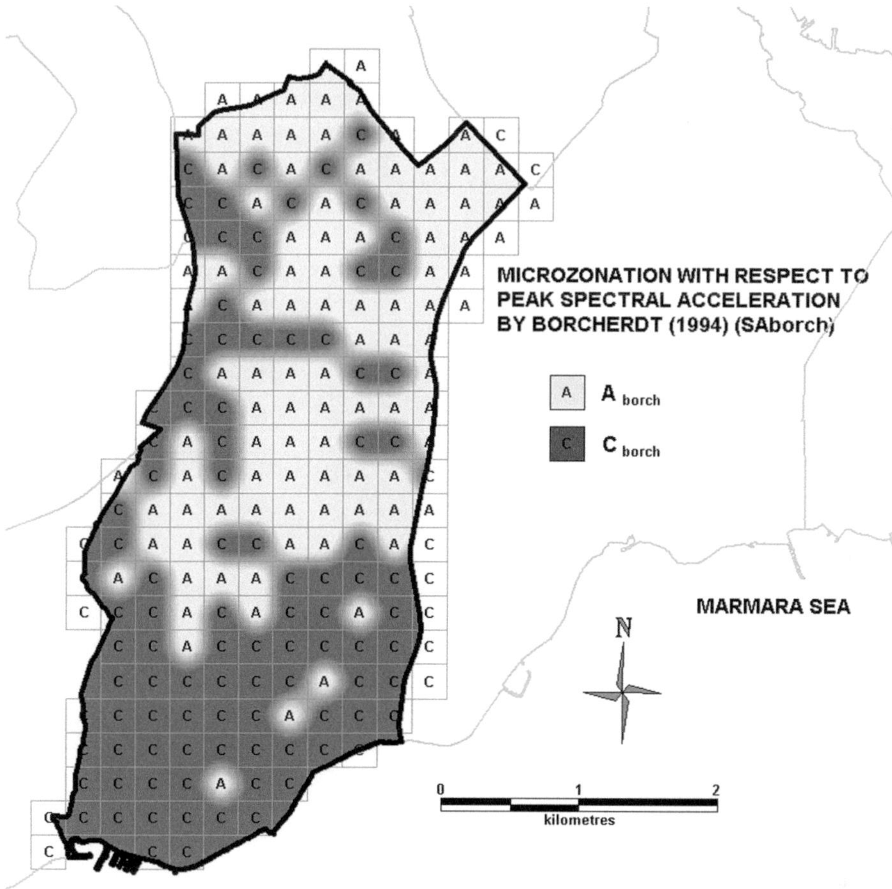

Fig. 2.5 Microzonation with respect to peak spectral accelerations based on Borcherdt (1994) formulations

B_{avg}, C_{avg}) and short period spectral accelerations calculated according to Borcherdt (1994) (A_{borch}, B_{borch}, C_{borch}). It is independent of the absolute value of the ground shaking intensity. The superposition of zones is achieved by applying following conditions:

$$A_{GS} \quad \text{if} \quad A_{avg} \text{ and } A_{borch} \text{ or } A_{avg} \text{ and } B_{borch} \text{ or } B_{avg} \text{ and } A_{borch},$$

$$B_{GS} \quad \text{if} \quad B_{avg} \text{ and } B_{borch} \text{ or } A_{avg} \text{ and } C_{borch} \text{ or } C_{avg} \text{ and } A_{borch},$$

$$C_{GS} \quad \text{if} \quad C_{avg} \text{ and } C_{borch} \text{ or } C_{avg} \text{ and } B_{borch} \text{ or } B_{avg} \text{ and } C_{borch}.$$

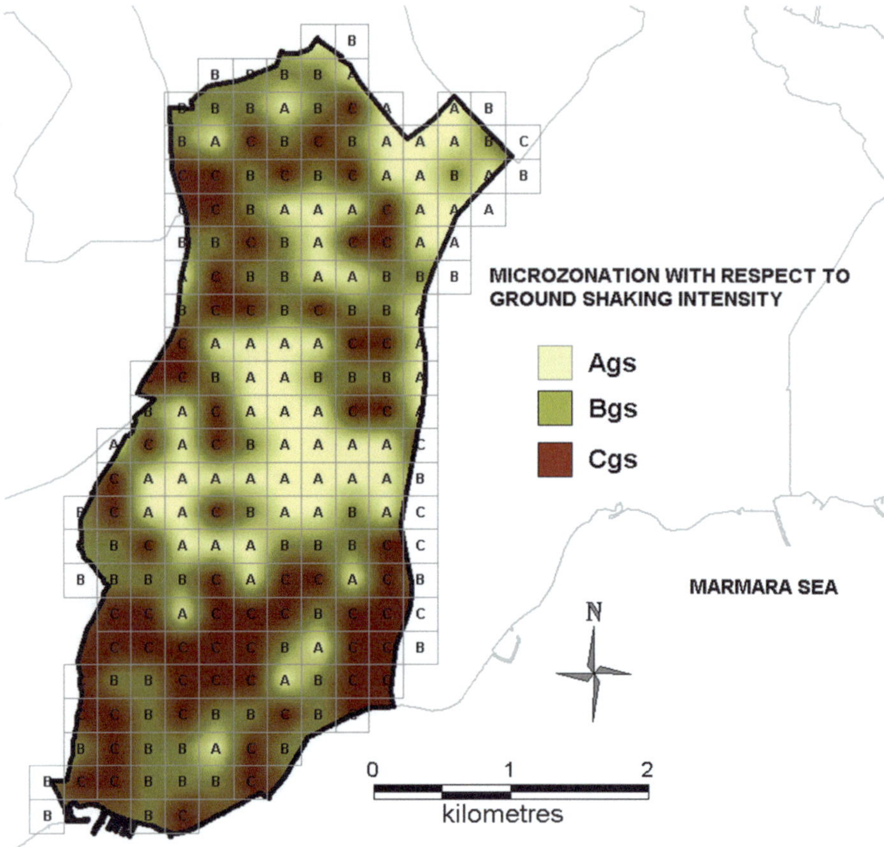

Fig. 2.6 Microzonation for ground shaking intensity

Hence, the superposed map is composed of three relative zones (A_{GS}, B_{GS}, C_{GS}) where A_{GS} shows the areas with lower ground shaking and C_{GS} shows the areas with higher ground shaking intensity as shown in Fig. 2.6.

2.5 Comparisons with Previous Microzonation Studies

Microzonation with respect to ground shaking intensity as given in Fig. 2.6 is compared with two previous microzonation studies conducted for Zeytinburnu. The first one was the pilot study conducted within the framework of Istanbul Earthquake Master Plan (EMPI, 2003, Ansal et al., 2005, Kılıç et al., 2006). This study was of preliminary nature and was carried out to demonstrate the applicability of the previously developed microzonation methodology (Ansal et al., 2004b; Studer and Ansal, 2004) utilising all the available boring data in the area from previous investigations. The grid size adopted was 250 m × 250 m. However, the number of borings was relatively limited such that there were borings only in 100 cells out of 230. Representative soil profiles for each cell with no available borehole information were estimated based on the borings in the neighbouring cells by utilizing the available data. Interpolations between neighbouring cells were performed taking into consideration the surface geology.

In this earlier version of the microzonation procedure for ground shaking intensity (Studer and Ansal, 2004), the approach adopted for the estimation of the peak spectral amplifications based on equivalent shear wave velocity was using the empirical relationship proposed by Midorikawa (1987).

$$A_{\mathrm{K}} = 68\, V_{\mathrm{S30}}^{-0.6} \tag{4}$$

where A_{K} is the spectral amplification and V_{S30} is the average shear wave velocity, in m/s.

The approach adopted for the estimation of the second microzonation parameter was to conduct one dimensional site response analysis using the Excel Subroutine EERA (Bardet et al., 2000) to determine elastic acceleration response spectra on the ground surface (Ansal et al., 2005). Site response analyses were conducted using three earthquake hazard spectra compatible simulated acceleration time histories (Papageorgiou et al., 2000). Microzonation with respect to ground shaking intensity from this first study is shown in Fig. 2.7.

The microzonation maps shown in Figs. 2.6 and 2.7 are significantly different from each other. Since the microzonation given in Fig. 2.6 is based on very detailed site investigation and based on large number of site response analyses it can be considered more reliable. However, the microzonation as given in Fig. 2.7 which was based on limited soil borings mostly based on surface geology and in addition to the use of a slightly different and more simplified approach yielded results that can be considered to be on the unsafe side in comparison to Fig. 2.6.

The second study conducted was part of the EU FP6 Project "LessLoss – Risk Mitigation for Earthquakes and Landslides" (Ansal et al., 2006b). In this study, site characterization was identical to the first study but site response analysis was performed for different sets of input acceleration time histories as well as for large number of earthquake hazard compatible real acceleration time histories (same used in the most recent study) that were scaled with respect to peak ground acceleration calculated for each cell at the bedrock outcrop again based on the earthquake hazard study (Erdik et al., 2004). Microzonation for ground shaking intensity was estimated based on the same approach as explained in detail in the previous section (Fig. 2.8).

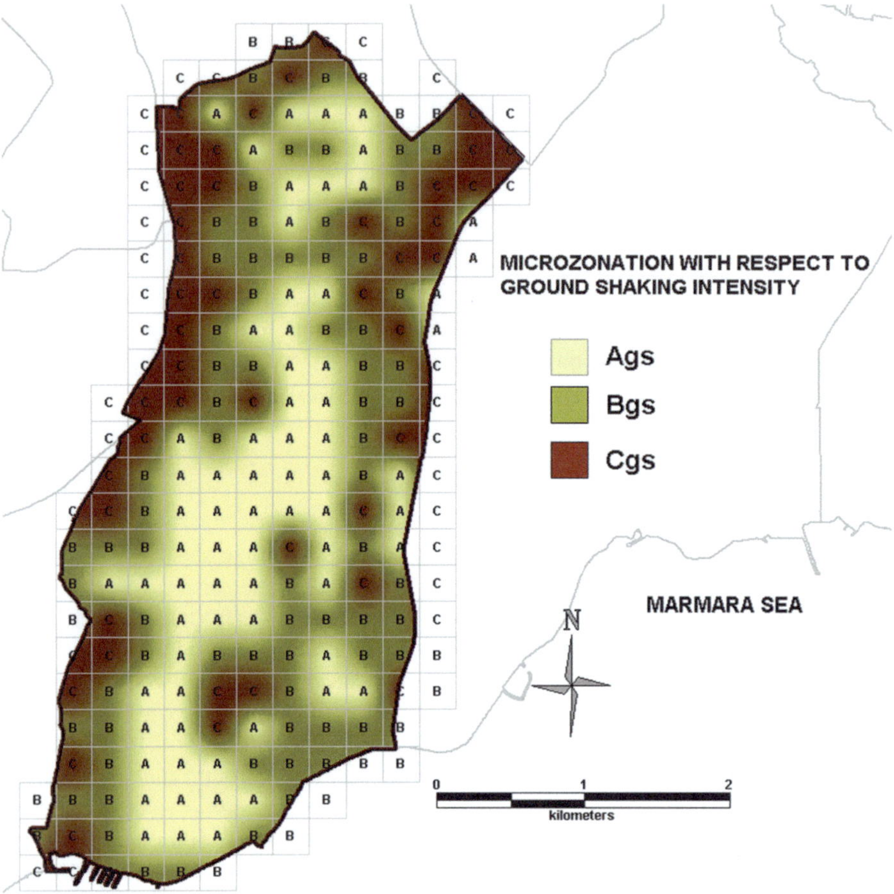

Fig. 2.7 Microzonation for ground shaking intensity based on limited site investigations and limited number of site response analysis (Ansal et al., 2005)

As can be observed from the comparison of Figs. 2.6 and 2.8, there are again significant differences between the ground shaking intensity microzonation maps and as in the previous case the results are on the unsafe side in comparison to the detailed microzonation shown in Fig. 2.6. In this case, since the methodology was almost identical and the only difference was the site characterisation data set, it is clearly evident that quantity and quality of site investigations and site characterisation are the main controlling factors in seismic microzonation. The differences between the three levels of microzonation with respect to ground shaking intensity is shown in Fig. 2.9, where it is apparent that there can be significant differences in the final microzonation maps if soil data is limited and limited number of site response are calculated.

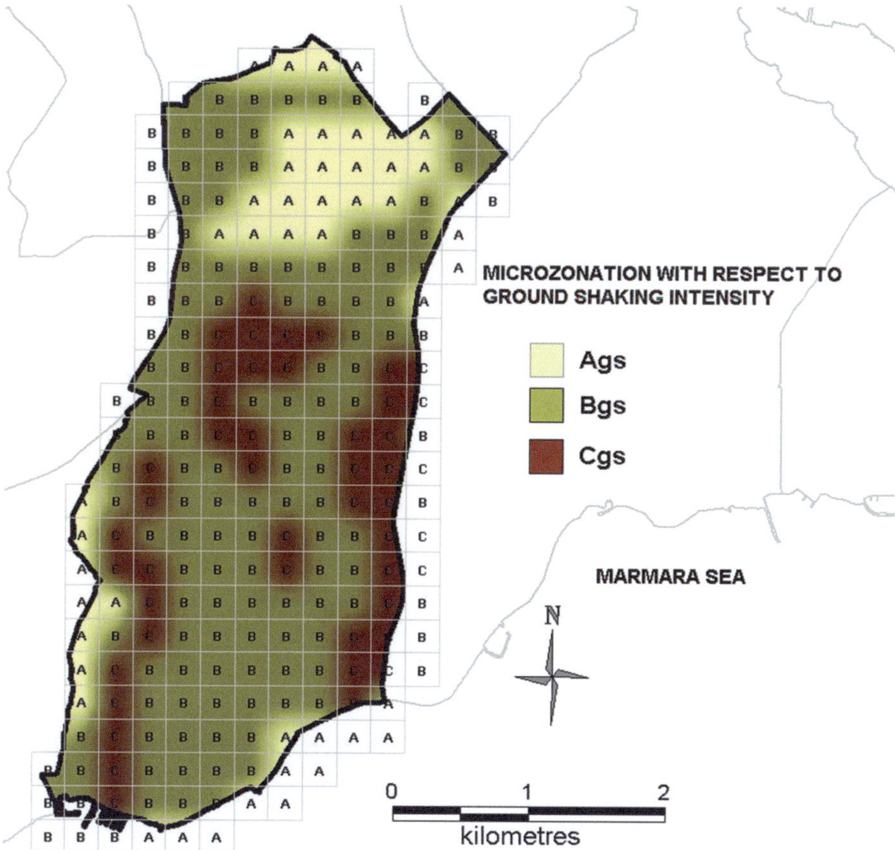

Fig. 2.8 Microzonation for ground shaking intensity based on limited site investigations and site response analyses using large number of PGA scaled hazard compatible acceleration time histories

2.6 Microzonation with Respect to PGA

Even though microzonation with respect to ground shaking intensity can be considered as a suitable criterion for land use and urban planning, it represents only the relative level of shaking intensity. Since detailed site characterisation and large number of site response analyses were performed, the results obtained in terms of average peak ground acceleration can also be used as additional microzonation maps with respect to ground shaking intensity that are relevant to liquefaction susceptibility and building vulnerability (Fig. 2.10).

The microzonation with respect to PGA based on detailed site investigation and large number of site response analyses as shown in Fig. 2.10 can be compared with

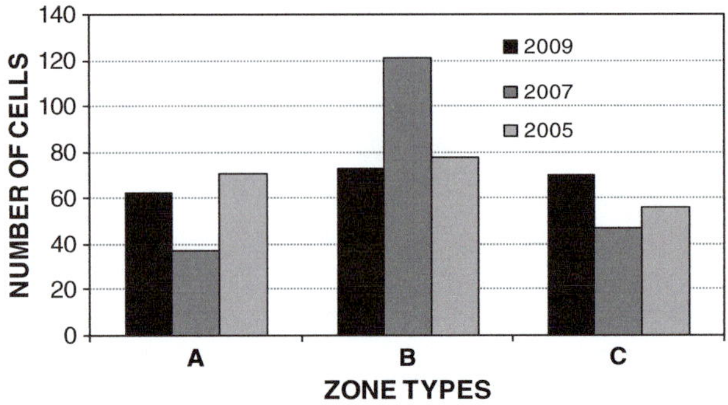

Fig. 2.9 Comparison of three levels of microzonation with respect to number of cells in each zone for three cases: very detailed site characterisation and site response analysis (2009), limited site data with detailed site response analysis (2007) and limited site data with limited number of site response analysis (2005)

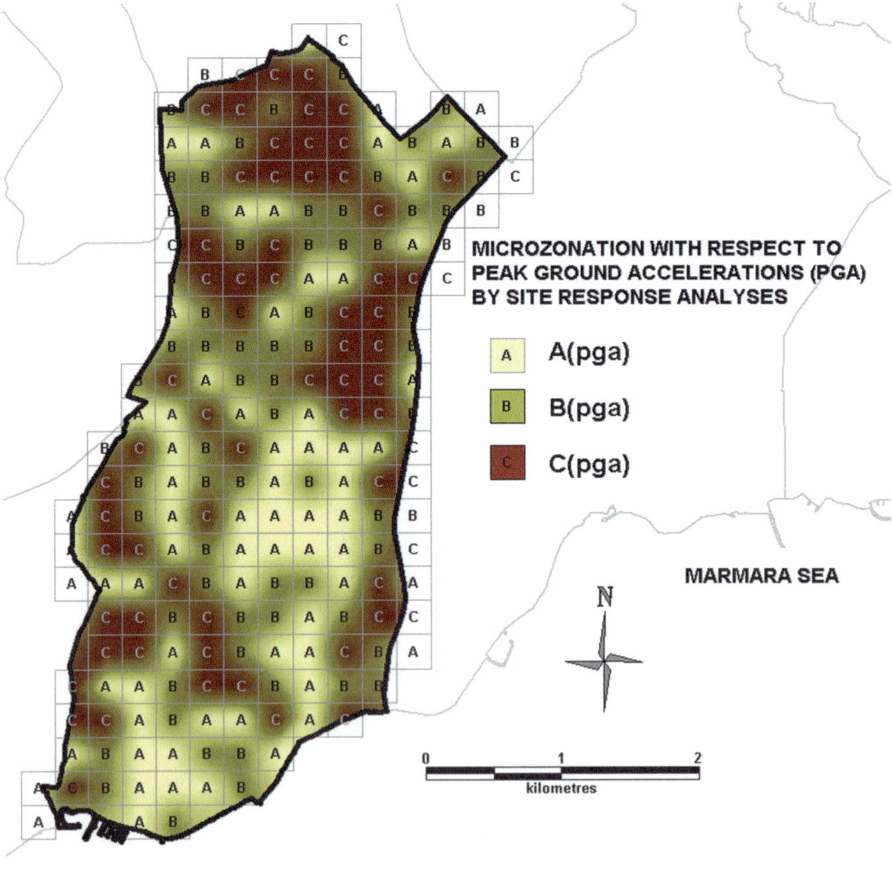

Fig. 2.10 Microzonation map with respect to peak ground acceleration (PGA) based on detailed site characterisation and site response analysis

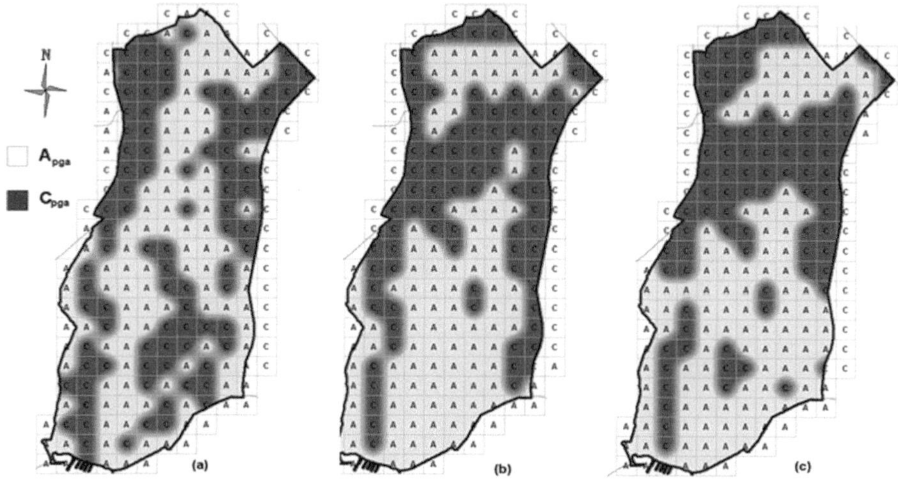

Fig. 2.11 Microzonation map with respect to peak ground acceleration calculated using (**a**) three earthquake hazard spectrum compatible simulated acceleration time histories; (**b**) first set of three real acceleration time histories; (**c**) second set of three real acceleration time histories scaled to the same PGAs estimated by the earthquake hazard study

the PGA microzonation maps obtained from the previous studies based on limited number of site investigations and using different sets of input acceleration time histories.

In Fig. 2.11, three sets of PGA microzonation maps are given to demonstrate the importance of the input motion characteristics in the site response analysis with respect to earthquake ground motion characteristics calculated on the ground surface. Even though the difference in the microzonation maps was not very significant between the two PGA microzonations calculated using different sets of real acceleration time histories, it is still important as pointed out by Ansal and Tönük (2007b). The other issue is the difference of all three PGA microzonation maps given in Fig. 2.11, with respect to the PGA microzonation based on detailed site characterisation given in Fig. 2.10. This difference again indicates the importance of the detailed site investigations, as well as the number of input motion used in site response analysis.

2.7 Microzonation with Respect to PGV

In addition to microzonation with respect to peak ground acceleration, microzonation maps can be generated with respect to peak ground velocity calculated by the integration of acceleration time histories calculated as output of site response analyses. The results obtained in terms of average peak ground velocity can also be used as additional microzonation maps with respect to ground shaking intensity that are relevant to building and lifeline vulnerabilities.

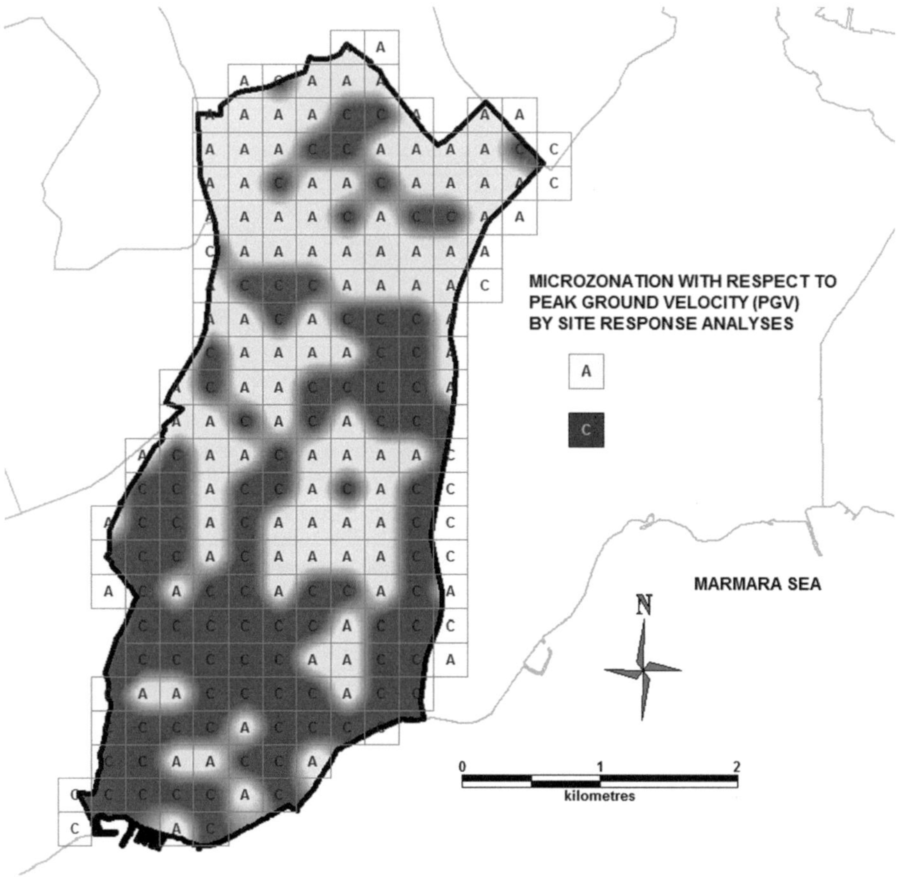

Fig. 2.12 Microzonation map with respect to peak ground velocity (PGV) based on detailed site characterisation and site response analysis using PGA scaled 24 seismic hazard compatible acceleration time histories

The peak ground velocity microzonation map as shown in Fig. 2.12 is determined by the integration of the acceleration time histories calculated on the ground surface using 24 PGA scaled real acceleration time histories for the detailed site characterisation as in the case of PGA microzonation given in Fig. 2.10. The comparison of PGA and PGV microzonation maps is significantly different indicating the importance of the selected microzonation parameter and the resulting earthquake damage scenario estimations.

2.8 Conclusions

Microzonation with respect to ground shaking intensity was based on two parameters: (1) average spectral accelerations calculated between 0.1 and 1 s periods using the average acceleration spectrum calculated from the results of 24 site response analysis conducted for each boring, (2) the peak spectral accelerations calculated from Borcherdt (1994) using equivalent shear wave velocities. The microzonation with respect to ground shaking intensity is produced with respect to three regions where zone A_{GS} shows the areas with very low ground shaking intensity, zone B_{GS} shows the areas with low to medium ground shaking intensity, and zone C_{GS} shows the areas with high ground shaking intensity.

Based on the microzonation studies conducted during the recent years, two conclusions may be drawn: (1) the detailed site investigation and related detailed site characterisation is very important and essential when performing site response analyses to have reliable and more accurate information on ground shaking characteristics for microzonation, and (2) the methodology followed and the type and number of acceleration time histories used for site response analysis to generate microzonation maps can have significant effect on the final microzonation.

The last issue is the selection of microzonation parameters. It was shown that microzonation with respect to different parameters such as PGA and PGV can give significantly different microzonation maps. Therefore, the selection of the microzonation parameter needs to be compatible with the main purpose of the microzonation project.

Acknowledgments The Authors would like to acknowledge the support and contributions of all their colleagues in the Earthquake Engineering Department of Kandilli Observatory and Earthquake Research Institute with special thanks to Prof. Mustafa Erdik, Dr. Mine Demircioğlu, and Dr. Karin Şeşetyan.

References

Ansal A, Akinci A, Cultrera G, Erdik M, Pessina V, Tönük G, Ameri G (2009) Loss estimation in Istanbul based on deterministic earthquake scenarios of the Marmara Sea region (Turkey). Soil Dyn Earthquake Eng 29(4):699–709

Ansal A, Durukal E, Tönük G (2006a) Selection and scaling of real acceleration time histories for site response analyses. In: Proceedings of the ISSMGE ETC12 workshop, Athens, Greece

Ansal A, Erdik M, Studer J, Springman S, Laue J, Buchheister J, Giardini D, Faeh D, Koksal D (2004a) Seismic microzonation for earthquake risk mitigation in Turkey. In: Proceedings of the 13th World Conference of Earthquake Engineering, Vancouver, BC, p 1428

Ansal A, Kurtuluş A, Tönük G (2007a) Earthquake damage scenario software for Urban areas. In: Papadrakakis M, Charmpis DC, Lagaros ND, Tsompanakis Y (eds) Keynote lecture, Computational Methods in Structural Dynamics and Earthquake Engineering, Rethymno, Crete.

Ansal A, Laue J, Buchheister J, Erdik M, Springman S, Studer J, Koksal D (2004b) Site characterization and site amplification for a seismic microzonation study in Turkey. In: Proceedings of the 11th International Conference on Soil Dynamics and Earthquake Engineering and 3rd Earthquake Geotechnical Engineering, San Francisco, CA

Ansal A, Özaydın K, Erdik M, Yıldırım Y, Kılıç H, Adatepe S, Özener PT, Tonaroglu M, Sesetyan K, Demircioglu M (2005) Seismic microzonation for urban planning and vulnerability assessment. In: Proceedings of the International Symposium of Earthquake Engineering (ISEE2005), Awaji Island, Kobe

Ansal A, Tönük G (2007a) Ground motion parameters for loss estimation. Keynote lecture, 4th International Conference on Urban Earthquake Engineering, Tokyo Institute of Technology, Tokyo, Japan, 7–14

Ansal A, Tönük G (2007b) Source and site effects for microzonation. In: Pitilakis K (ed) Theme lecture, 4th International Conference on Earthquake Geotechnical Engineering, Earthquake Geotechnical Engineering, Chapter 4, Springer, Berlin, pp 73–92

Ansal A, Tönük G, Bayraklı Y (2007b) Microzonation with respect to ground shaking intensity based on 1D site response analysis. In: Proceedings of the 14th European Conference on Soil Mechanics and Geotechnical Engineering, Madrid

Ansal A, Tönük G, Demircioglu M, Bayraklı Y, Sesetyan K, Erdik M (2006b) Ground motion parameters for vulnerability assessment. In: Proceedings of the 1st European Conference on Earthquake Engineering and Seismology, Geneva, Switzerland, p 1790

Bardet JP, Ichii K, Lin CH (2000) EERA. A computer program for equivalent linear earthquake site response analysis of layered soils deposits. University of Southern California, Los Angeles, CA

Borcherdt RD (1994) Estimates of site dependent response spectra for design (Methodology and Justification). Earthquake Spectra 10(4):617–654

BSSC-Building Seismic Safety Council (2001) NEHRP (National Earthquake Hazards Reduction Program) Recommended provisions for seismic regulations for new buildings and other structures, 2000 edn, Part 1: provisions (FEMA 368), Chapter 4, Washington, DC

Darendeli MB (2001) A new family of normalized modulus reduction and material damping curves. PhD Dissertation, University of Texas at Austin, p 362

EC8 (2002) Eurocode 8: design of structures for earthquake resistance, Part 5: foundations, retaining structures and geotechnical aspects, European committee for standardization, Central secretariat: rue de Stassart 36, B1050, Brussels

EMPI (2003) Earthquake master plan for Istanbul, Bogaziçi university, Istanbul Technical university, Middle East Technical University, and Yildiz Technical University, Metropolitan municipality of Istanbul, Planning and construction directorate geotechnical and earthquake investigation department, p 569

EPRI (1993) Guidelines for determining design basis ground motions. Electric power research institute, vol 1, EPRI TR-102293. Palo Alto, CA

Erdik M, Demircioglu M, Sesetyan K, Durukal E, Siyahi B (2004) Earthquake hazard in Marmara region. Soil Dyn Earthquake Eng 24:605–631

Idriss IM, Sun JI (1992) Shake91, A computer program for conducting equivalent linear seismic response analysis of horizontally layered soil deposits modified based on the original SHAKE program Published in December 1972 by Schnabel, Lysmer and Seed

Kılıç H, Özener PT, Ansal A, Yıldırım M, Özaydın K, Adatepe S (2006) Microzonation of Zeytinburnu region with respect to soil amplification: a case study. J Eng Geol 86: 238–255

Menq FY, Stokoe KH, Kavazanjian E (2003) Linear dynamic properties of sandy and gravelly soils from largescale resonant tests. International symposium IS Lyon 03, Deformation characteristics of geomaterials, Lyon, France, 22–24 Sept 2003

Midorikawa S (1987) Prediction of isoseismal map in the Kanto plain due to hypothetical earthquake. J Struct Eng 33B:43–48

OYO Inc., Japan (2007) Production of microzonation report and maps on European side (South). Final report to Istanbul Metropolitan Municipality

Özaydın K, Ansal A, Erdik M, Yıldırım M, Kılıç H, Adatepe Ş, Özener PT, Tonoroğlu M, Şeşetyan K, Demircioğlu M (2004) Earthquake Master Plan for Istanbul, Zeytinburnu Pilot Project. "Report on geological and geotechnical evaluation for seismic microzonation and seismic microzonation for ground shaking" Yıldız Technical University, Faculty of Civil Engineering Geotechnical Department, Boğaziçi University, Kandilli Observatory and Earthquake Research Institute (In Turkish)

Papageorgiou A, Halldorsson B, Dong G (2000) Target acceleration spectra compatible time histories, Department of civil, structural and environmental engineering, University of Buffalo, http://civil.eng.buffalo.edu/engseislab/

PEER (2009) Strong motion data bank. http://peer.berkeley.edu

Seed HB, Wong RT, Idriss IM, Tokimatsu K (1984) Moduli and damping factors for dynamic analyses of cohesionless soils. Earthquake engineering research center, Report no. UCB/EERC-84/14, University of California, Berkeley, CA, p 37

Studer J, Ansal A (2004) Manual for seismic microzonation for municipalities. Research report for republic of Turkey, Ministry of Public Works and Settlement, General Directorate of Disaster Affairs prepared by World Institute for Disaster Risk Management, Inc

Vucetic M, Dobry R (1991) Effect of soil plasticity on cyclic response. J Geotech Eng ASCE 117(1):89–107

Chapter 3
Analysis of Regional Ground Motion Variations for Engineering Application

Jonathan P. Stewart

Abstract An important question for many ground motion hazard analyses is the degree to which ground motion prediction equations (GMPEs) developed for one region may have bias for a different region. A closely related problem is the applicability of multi-regional GMPEs to a particular region, even if that region contributed some fraction of the database. It is well known that ground motions show distinct characteristics for stable continental regions, subduction zones, and active tectonic regions with shallow crustal earthquakes. Here I consider variations among active regions with shallow crustal earthquakes. For such regions having sufficient data that meaningful comparisons are possible, I review four approaches for evaluating regional variations: (1) direct comparisons of medians from GMPEs; (2) analysis of variance; (3) overall goodness of fit metrics; and (4) verification of specific GMPE attributes relative to regional data. For engineering application, the objective of the comparison should be to evaluate whether median predictions show statistically similar trends with respect to magnitude-scaling, distance-scaling, and site effects across the range of magnitudes and distances controlling the seismic hazard, as well as consistent standard deviation terms.

3.1 Introduction

The attributes of earthquake ground motion intensity measures (IMs) are predicted using ground motion prediction equations (GMPEs), which describe the variation of the median and standard deviation of an IM with respect to source, path, and site parameters. A review of the vast literature on GMPEs is beyond the scope of this article (see Douglas, 2003, 2006 for reviews).

Earthquakes from subduction zones, shallow sources in active regions, and stable continental regions produce ground motions with distinct attributes, and

J.P. Stewart (✉)
University of California, Los Angeles, CA, USA
e-mail: jstewart@seas.ucla.edu

M. Garevski, A. Ansal (eds.), *Earthquake Engineering in Europe*, Geotechnical, Geological, and Earthquake Engineering 17, DOI 10.1007/978-90-481-9544-2_3,

hence different GMPEs are needed. However, within active regions having crustal earthquakes, there remains a lack of consensus on the manner by which to manage regionalization. As described by Douglas (2007), many investigators in Italy, Turkey, France, Spain and elsewhere assume a high degree of regionalization and develop GMPEs from small datasets derived from local regions (sometimes only a few thousand km^2 in size or arbitrarily defined by political boundaries). Due to the small size of these datasets, the GMPEs are not useful for prediction of the effects of relatively large earthquakes that form the basis for engineering design.

Another approach is to develop GMPEs from a large database derived from multiple regions. This was the approach of the Next Generation Attenuation (NGA) project that resulted in 2008 GMPEs by Abrahamson and Silva (AS), Boore and Atkinson (BA), Campbell and Bozorgnia (CB), and Chiou and Youngs (CY). The database used in that project (Chiou et al., 2008) included world-wide shallow crustal earthquakes from active regions including California, Taiwan, Japan, Turkey, Greece, Italy, New Zealand, and elsewhere.

The arguments for and against regionalization are well described elsewhere (e.g., Bommer, 2006; Douglas, 2007). The objective of this paper is to review and critique four methods for comparing ground motions from different regions. The emphasis here is on engineering application; hence, I assume that the intent of regionalization studies is the development or verification of GMPEs that can be used in probabilistic analyses that often find the hazard to be controlled by earthquakes of moderate to large magnitude at modest to close distance.

The four methods I will describe are comparisons of GMPE attributes; analysis of variance (Douglas, 2004a, b, 2007); overall goodness-of-fit (Scherbaum et al., 2004; Stafford et al., 2008); and verification of specific GMPE attributes relative to regional data (Scasserra et al., 2009a).

3.2 Methods for Analyzing Regional IM Variations

3.2.1 Comparison of GMPEs

Ground motion prediction equations (GMPEs) provide estimates of the median and log-normal standard deviation of ground motion. The attribute of GMPEs that is most often compared is the median and its variation with distance and magnitude for a reference site condition. As described by Douglas (2007), such comparisons are often problematic when one or more of the GMPEs is derived from small datasets because the standard error of the median (i.e., the uncertainty in the location of the median) is high and is not considered in the comparison. The standard deviation of the GMPEs is also critical for ground motion hazard analysis, but is seldom compared.

Figure 3.1 compares medians from GMPEs derived from large data sets (hence relatively small uncertainty in medians). Median peak horizontal ground accelerations (PGA) and 5%-damped pseudo spectral acceleration from two European

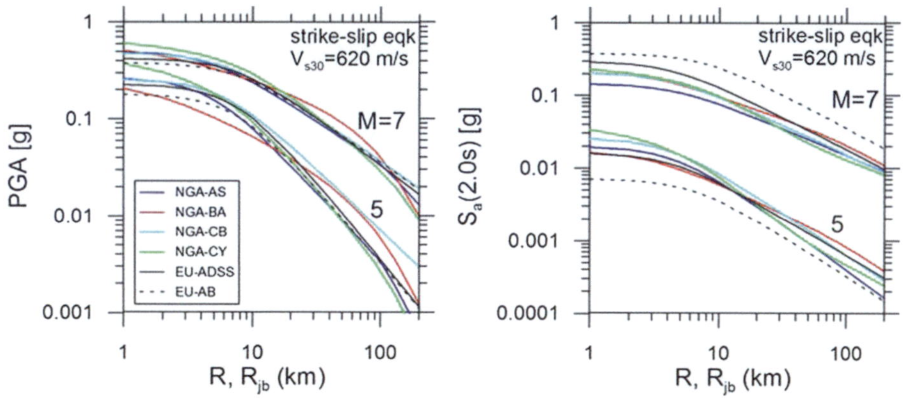

Fig. 3.1 Comparison of median predictions of PGA and 2.0 s pseudo spectral acceleration for strike slip earthquakes and soft rock site conditions from NGA and European GMPEs. AS = Abrahamson and Silva (2008); BA = Boore and Atkinson (2008); CB = Campbell and Bozorgnia (2008); CY = Chiou and Youngs (2008); ADSS = Ambraseys et al. (2005); AB = Akkar and Bommer (2007). Adapted from Scasserra et al. (2009a)

models (Akkar and Bommer, 2007; Ambraseys et al., 2005) are compared to those from NGA models. The European and NGA predicted medians generally compare well over the range of distances and magnitudes well constrained by the data. The bands of results for the two magnitudes generally show reasonably consistent vertical offsets from model-to-model (e.g., the difference between M7 and M5 PGA at R_{jb} = 30 km is reasonably consistent across models). This suggests generally consistent levels of magnitude scaling. The slopes of the median curves for a given magnitude are generally steeper for the European relations than the NGA relations for PGA, suggesting faster distance attenuation of this parameter.

In Fig. 3.2, I compare standard deviations from the AB European GMPE to a representative NGA GMPE (CY). Note that two standard deviation terms are shown. Standard deviation σ represents intra-event dispersion, which can be interpreted as the average level of dispersion from individual, well-recorded earthquakes. Term τ represents inter-event dispersion, or the standard deviation of event terms. Since event terms represent the average misfit of a GMPE to the data for a given event, τ represents event-to-event variability of the IM.

The comparison in Fig. 3.2 indicates relatively consistent τ terms and σ terms at large magnitude, but larger low-magnitude σ terms in the AB model relative to the CY model. This represents a potential example of regional variability, as both data sets apply for shallow crustal earthquakes in active regions.

3.2.2 Analysis of Variance

This approach was applied by Douglas (2004a) to compare ground motions for five local regions within Europe, Douglas (2004b) to compare ground motions from

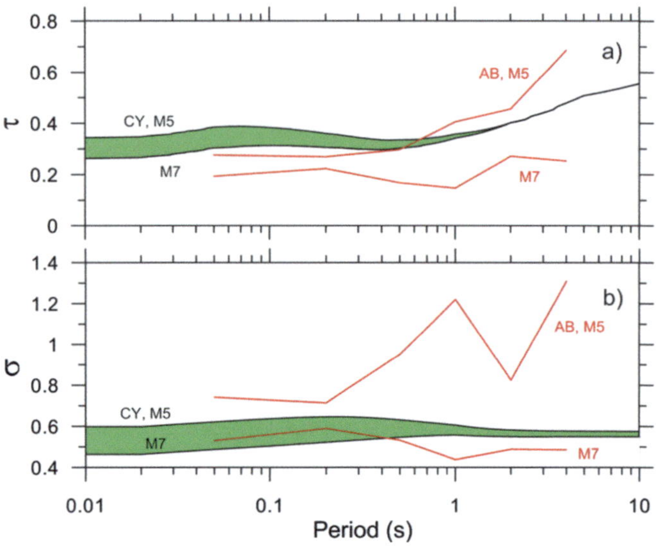

Fig. 3.2 Standard deviation terms σ and τ from European and NGA relation

Europe, New Zealand, and California, and Douglas (2007) for two local regions within Italy. The procedure involves calculating the mean (μ) and total variance (σ_T^2) of the log of data inside particular magnitude and distance bins (*M-R* bins) for two different regions (e.g., Europe and California) and combined data for those regions. The distance metric used by Douglas is the closest distance to the surface projection of the fault for $M > 6$ and epicentral distance otherwise. Individual data points are adjusted for a linear site factor before the calculation of mean and variance. These results are then used in two ways. First, for a given *M-R* bin and pair of regions, the variance of the combined data for both regions [termed $(\sigma_T^2)_{inter-region}$] is compared to the within-region variance [termed $(\sigma_T^2)_{intra-region}$] using statistical tests that evaluate whether the data sets are significantly distinct. If $(\sigma_T^2)_{inter-region} > (\sigma_T^2)_{intra-region}$ in a statistically significant way, there is likely to be significantly different medians between regions. The second use of the binned results is to plot medians for each *M-R* bin together for pairs of regions as shown for example in Fig. 3.3.

Using the above approach, Douglas (2004a) found similar variances for the various regions in Europe, indicating a lack of regional variations. Accordingly, Douglas (2004b) combined all of the European data into a single category for comparison to New Zealand and California data. The Europe-California comparisons indicate that approximately half of the *M-R* bins demonstrate significantly different inter- and intra-region variances. The distinction was towards larger ground motions in California (Douglas, 2004b). Figure 3.3 shows an example comparison of California and European medians from Douglas (2004b). The results indicate that the California and European medians for most *M-R* bins are similar at short

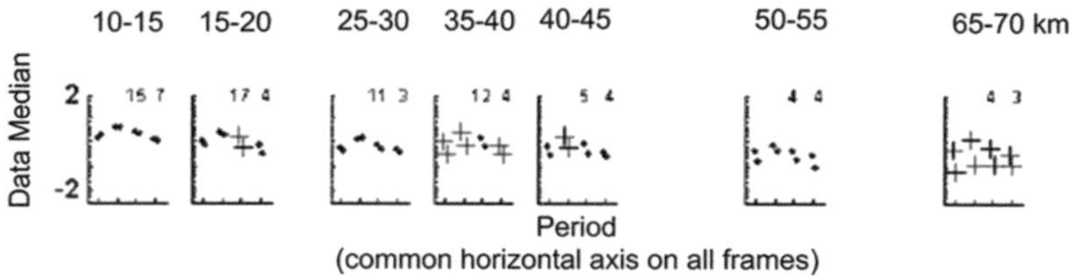

Fig. 3.3 Medians of data from $M = 6$–6.25 earthquakes within distance bins from California and Europe. Data from the two regions are shown side-by-side (California on left, Europe on right) with dots when the differences are not statistically significant and with crosses when significant at the 95% confidence level. Frames not shown for poorly populated bins. Modified from Douglas (2004b)

distance (<20 km), whereas California amplitudes are larger at larger distances (>30 km). Thus, Douglas' (2004b) finding of larger California ground motions could be expressed as more rapid distance attenuation in Europe.

3.2.3 Overall Goodness of Fit of GMPE to Data

This approach, developed by Scherbaum et al. (2004), provides an evaluation of overall goodness-of-fit of a GMPE to a dataset. A normalized residual is calculated for recording *j* from event *i* in a dataset as:

$$Z_{T,ij} = \frac{\ln\left(IM_{\mathrm{obs},ij}\right) - \ln\left(IM_{\mathrm{mod},ij}\right)}{\sigma_T} \tag{1}$$

where ln $(IM_{\mathrm{obs},ij})$ represents the IM value from the record, ln $(IM_{\mathrm{mod},ij})$ represents the median model prediction for the same magnitude, site-source distance, and site conditions of the record, and σ_T represents the total standard deviation of the model ($\sigma_T^2 = \sigma^2 + \tau^2$). If the data is unbiased with respect to the model and has the same dispersion, the normalized residuals (Z_T) should have zero mean and standard deviation of one – i.e., the properties of the standard normal variate. Accordingly, in simple terms, the procedure of Scherbaum et al. (2004) consists of comparing the actual Z_T distribution to that of the standard normal variate. Note that this procedure tests both misfit of the median and standard deviation.

Figure 3.4 shows an example application of this approach by Stafford et al. (2008), who also extended the method to consider both inter- and intra-event variability. They compared European data to the BA NGA relation and several European GMPEs. The specific example shown in Fig. 3.4 is intra-event normalized residuals for PGA relative to the BA relation. The BA relation was shown to match the median of the European data nearly as well as European GMPEs. The BA standard deviation, however, is lower than implied by the European data, resulting in the misfit of the histogram relative to the standard normal variate shown in Fig. 3.4.

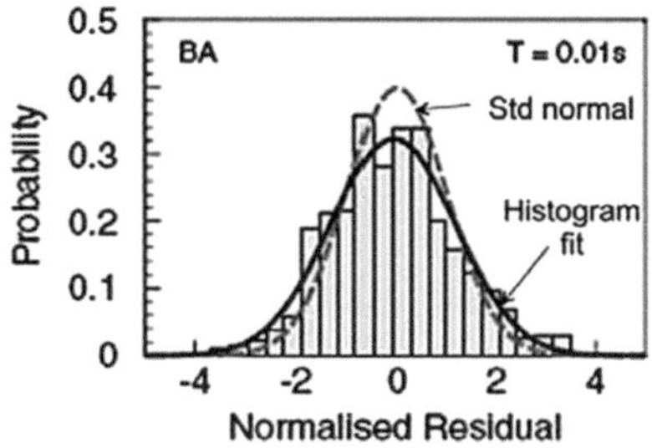

Fig. 3.4 Histogram of intra-event residuals of European data for PGA relative to BA GMPE. Note lack of bias indicated by median near zero. The histogram fit has a standard deviation larger than unity, indicating larger σ in the dataset than in the model from Stafford et al. (2008)

3.2.4 Verification of Specific GMPE Attributes Relative to Regional Data

This approach involves comparing regional data to GMPEs developed for a different region or for a diverse set of regions (e.g., shallow crustal earthquakes in active regions world-wide, as in NGA). As noted by Douglas (2007), this approach can provide misleading results if the available data lies outside the magnitude and distance range for which the GMPE is valid. However, when sufficient data is available to enable a valid comparison, it is possible to verify specific attributes of the GMPE relative to the data such as magnitude-scaling, distance-scaling, site effects, and standard deviation terms. This approach has been applied by Scasserra et al. (2009a) using the NGA GMPEs and data from pre-2009 Italian earthquakes.

To begin, residuals are evaluated between the data and a particular GMPE referred to with index *k*. Residuals are calculated as:

$$\left(R_{i,j}\right)_k = \ln\left(IM_{i,j}\right)_{\text{data}} - \ln\left(IM_{i,j}\right)_k \tag{2}$$

Index *i* refers to the earthquake event and index *j* refers to the recording within event *i*. Hence, $(R_{i,j})_k$ is the residual of data from recording *j* in event *i* as calculated using GMPE *k*. Term ln $(IM_{i,j})_{\text{data}}$ represents an IM computed from recording *j*. Term ln $(IM_{i,j})_k$ represents the median calculated using GMPE *k* in natural log units.

The analysis of residuals with respect to magnitude-, distance, and site-scaling requires that event-to-event variations be separated from variations of residuals within events. This is accomplished by performing a mixed effects regression (Abrahamson and Youngs, 1992) of residuals according to the following function:

$$\left(R_{i,j}\right)_k = c_k + (\eta_i)_k + \left(\varepsilon_{i,j}\right)_k \tag{3}$$

where c_k represents a mean offset (or bias) of the data relative to GMPE *k*, η_i represents the event term for event *i* (explained below), and $\varepsilon_{i,j}$ represents the intra-event residual for recording *j* in event *i*. Event term η_i represents approximately the mean offset of the data for event *i* from the predictions provided by the GMPE median (after adjusting for mean offset c_k, which is based on all events). Event terms provide

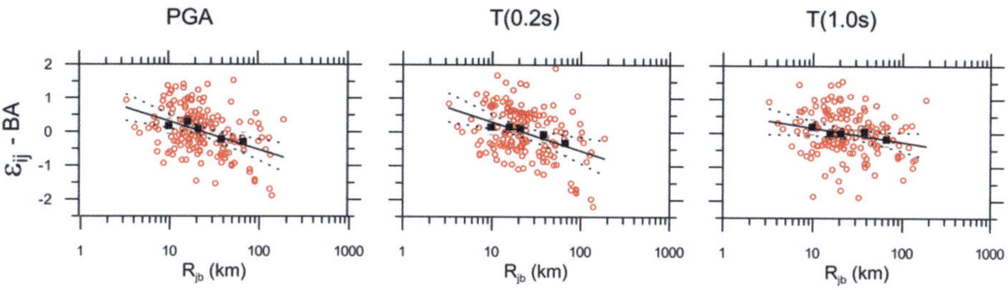

Fig. 3.5 Trend of intra-event residuals of Italian data for PGA relative to BA GMPE. Modified from Scasserra et al. (2009a)

a convenient mechanism for testing the ability of a GMPE to track the magnitude scaling of recordings in a dataset. Event terms are assumed to be normally distributed, and have zero mean and standard deviation $= \tau$ (in natural log units). Intra-event error ε is also assumed to be normally distributed with zero mean and standard deviation $= \sigma$.

Scasserra et al. (2009a) applied the above methodology to an Italian data set that had been carefully screened for record quality and for which the necessary metadata on site conditions, magnitudes, and distance parameters was available (Scasserra et al., 2009b). Careful and consistent screening of the data in that manner is necessary to obtain meaningful results from the analyses.

The analysis of Scasserra et al. (2009a) showed a general lack of significantly non-zero values of c, suggesting a lack of overall bias in the NGA GMPEs relative to the Italian data. As shown in Fig. 3.5, intra-event residuals (ε) demonstrated a negative trend with distance, indicating faster attenuation of the Italian data than in the NGA GMPEs for high-frequency spectral accelerations. That bias was not present for low-frequency spectral accelerations. After removing the distance attenuation bias from the GMPEs, no statistically significant bias of the event terms with respect to magnitude was found, indicating consistent magnitude-scaling between the NGA GMPEs and Italian data. As found by Stafford et al. (2008), intra-event standard deviation term σ from the Italian data was larger than in the NGA GMPEs, but inter-event standard deviation term τ was similar.

3.3 Discussion and Conclusions

The multi-regional database used to develop the NGA GMPEs is large (3,551 recordings from 173 earthquakes; subsets used for particular GMPEs). As noted by Stafford et al. (2008), because of the large size and high quality of the NGA database, certain effects are captured that could not be evaluated using only data from a single region. Examples include depth to top-of-rupture, magnitude- and/or site-dependent standard deviations, and nonlinear site response. The NGA data also provides the opportunity to constrain relatively complex functional forms for magnitude and distance scaling as compared to regional models.

Because of the relative sophistication of the NGA GMPEs, it is of interest to evaluate whether they can be applied to specific geographic regions. In this article,

I have briefly described four procedures by which this evaluation can be performed. I assume for the sake of this discussion that a strong motion database is available for the region under consideration, and that this database has an appropriate level of processing, screening, and available meta-data.

For such situations, the goodness-of-fit approach of Scherbaum et al. (2004), later modified by Stafford et al. (2008), assesses GMPE performance in an overall sense – i.e., all aspects of the model (magnitude-scaling, distance-scaling, site effects) are evaluated together. If one or more of these model components is in error, that effect could be obscured through compensating errors in the analysis of normalized residuals. Accordingly, while the results of Stafford et al. (2008) are certainly promising with respect to the application of NGA relations in Europe, they do not specifically address whether individual components of the NGA models are adequate with respect to European data.

When the required data can be assembled, a more complete picture of regional variations emerges when specific attributes of a GMPE are tested against the regional dataset in the manner described by Scasserra et al. (2009a). These comparisons require selection of data that lies within the range of applicability of the GMPE. For the Italian dataset, this approach enabled the finding of faster distance attenuation and higher intra-event standard deviation described above. Those specific misfits, in turn, can be corrected within the multi-region GMPE for application in a local region. This allows advantages of the relatively sophisticated GMPEs to be leveraged in the local region without the introduction of obvious, first-order bias.

The above approaches rely on the availability of a GMPE that can be used for comparison to regional data. If such a GMPE is unavailable or judged to be problematic, an alternative approach is needed. The analysis of variance approach of Douglas (2004a, b, 2007) avoids reliance on GMPEs, focusing instead on comparisons of binned data from two regions. This approach provides valuable insights into fundamental features of datasets from two regions, but does not directly result in a usable GMPE, which is needed for hazard analyses.

All of the approaches described in this article have advantages and limitations. My goal here is not to recommend a single approach, but to highlight available methodologies and the types of insights that can be gained from them. Ultimately, the goal of regionalization studies should be to support the development and use of reliable GMPEs in engineering practice. Those GMPEs should be constrained to the maximum extent possible by recordings/events that span the range of magnitudes and distances controlling seismic hazard at the return periods of engineering interest. GMPEs based solely on small, local databases are unlikely to meet this standard for the foreseeable future.

References

Abrahamson NA, Silva WJ (2008) Summary of the Abrahamson and Silva NGA ground motion relations. Earthquake Spectra 24(S1):67–97

Abrahamson NA, Youngs RR (1992) A stable algorithm for regression analyses using the random effects model. Bull Seism Soc Am 82:505–510

Akkar S, Bommer JJ (2007) Prediction of elastic displacement response spectra in Europe and the Middle East. Earthquake Eng Struct Dyn 36:1275–1301

Ambraseys NN, Douglas J, Smit P, Sarma SK (2005) Equations for the estimation of strong ground motions from shallow crustal earthquakes using data from Europe and the Middle East: horizontal peak ground acceleration and spectral acceleration. Bull Earthquake Eng 3(1):1–53

Bommer JJ (2006) Empirical estimation of ground motion: advances and issues. In: Proceedings of the 3rd international symposium on the effects of surface geology on seismic motion, vol 1. Grenoble, France, pp 115–135

Boore DM, Atkinson GM (2008) Ground motion prediction equations for the average horizontal component of PGA, PGV, and 5%-damped PSA at spectral periods between 0.01 and 10.0 s. Earthquake Spectra 24(S1):99–138

Campbell KW, Bozorgnia Y (2008) NGA ground motion model for the geometric mean horizontal component of PGA, PGV, PGD, and 5%-damped linear elastic response spectra for periods ranging from 0.01 to 10 s. Earthquake Spectra 24(S1):139–171

Chiou BS-J, Darragh R, Dregor D, Silva WJ (2008) NGA project strong-motion database. Earthquake Spectra 24(S1):23–44

Chiou BS-J, Youngs RR (2008) An NGA model for the average horizontal component of peak ground motion and response spectra. Earthquake Spectra 24(S1):173–215

Douglas J (2003) Earthquake ground motion estimation using strong-motion records: a review of equations for the estimation of peak ground acceleration and response spectra ordinates. Earth Sci Rev 61:43–104

Douglas J (2004a) An investigation of analysis of variance as a tool for exploring regional differences in strong ground motions. J Seism 8:485–496

Douglas J (2004b) Use of analysis of variance for the investigation of regional dependence of strong ground motion. In: Proceedings of the 13th world conference on earthquake engineering, Vancouver, BC, Paper 29 (electronic file)

Douglas J (2006) Errata of and additions to "Ground motion estimation equations 1964–2003". Intermediary Report BRGM/RP-54603-FR, Bureau de recherches géologiques et minières

Douglas J (2007) On the regional dependence of earthquake response spectra. ISET J Earthquake Technol 44(1):71–99

Scasserra G, Stewart JP, Bazzurro P, Lanzo G, Mollaioli F (2009a) A comparison of NGA ground-motion prediction equations to Italian data. Bull Seism Soc Am 99(5):2961–2978

Scasserra G, Stewart JP, Kayen RE, Lanzo G (2009b) Database for earthquake strong motion studies in Italy. J Earthquake Eng 13(6):852–881

Scherbaum F, Cotton F, Smit P (2004) On the use of response spectral reference data for the selection and ranking of ground motion models for seismic hazard analysis in regions of moderate seismicity: the case of rock motion. Bull Seism Soc Am 94(6):2164–2185

Stafford PJ, Strasser FO, Bommer JJ (2008) An evaluation of the applicability of the NGA models to ground motion prediction in the Euro-Mediterranean region. Bull Earthquake Eng 6:149–177

Part II
Geotechnical Earthquake Engineering

Chapter 4
Non Linear Soil Structure Interaction: Impact on the Seismic Response of Structures

Alain Pecker and Charisis T. Chatzigogos

Abstract The paper presents results of incremental dynamic analyses (IDA) of a simple structural system with consideration of non linear soil structure interaction. The analyses are facilitated using a non linear dynamic macroelement for the soil-foundation system. Three base conditions are examined, namely fixed base, linear foundation and non-linear foundation including uplift and soil plasticity. IDA curves are produced for a variety of intensity and damage parameters describing both the maximum and the residual response of the system. The results highlight the beneficial role of foundation non linearities in decreasing the ductility demand in the superstructure but point out the need to carefully assess the variability of the response when non linearity is allowed at the foundation design.

4.1 Introduction

The topic of soil structure interaction (SSI) has long been recognized as a major factor controlling the design of the structure. During an earthquake, the soil deforms under the influence of incident seismic waves and imposes its motions to the foundation and to the supported structure. In turn, the induced motion of the foundation creates inertial forces in the superstructure that are transmitted back to the foundation and to the underlying soil. Therefore, the induced deformations create additional waves that emanate from the soil-foundation interface. Both phenomena occur simultaneously and therefore are closely linked and dependent on one another. SSI increases in significance as the supporting soil becomes softer. Although recognized by recent building codes, like Eurocode 8 (EC8, 2000), which requires to take into consideration SSI for massive structures founded on soft deposits, most building codes ignore the effect of SSI: "For the majority of usual building structures, the effects of SSI tend to be beneficial, since they reduce the bending moments and

---

A. Pecker (✉)
Géodynamique et Structure, 92220 Bagneux, France
e-mail: alain.pecker@geodynamique.com

M. Garevski, A. Ansal (eds.), *Earthquake Engineering in Europe*, Geotechnical, Geological, and Earthquake Engineering 17, DOI 10.1007/978-90-481-9544-2_4,

shear forces acting in the various members of the superstructure" (EC8, 2000). As pointed in Gazetas (2006) this statement may hold for a large class of structures but may be misleading for others and mainly relies on the smooth shape of normalized code spectra. To the best, SSI is considered in the dynamic analysis assuming a linear behavior of the soil foundation interface. Even though, the results obtained by various authors on the beneficial or detrimental effect of SSI are controversial. Within the framework of performance based design, the question becomes essential to know how non linear soil structure interaction may affect the seismic demand in the superstructure. With the advance of efficient numerical tools to model the non linear behavior of foundations, it becomes possible to investigate this effect through the concept of incremental dynamic analysis (IDA) as defined by Vamvatsikos and Cornell (2002). The paper presents preliminary results obtained with this technique which may help to clarify this issue and provide guidelines for the seismic design of foundations. For that purpose, the studied structure is a reinforced concrete bridge pylon founded on the surface of a homogeneous cohesive soil by means of a circular footing.

4.2 Linear Soil Structure Interaction

Despite the fact that SSI is not very often considered in building codes, it has a long history which started back in 1936 with the work of Reissner. Since then, several improvements have been achieved and the present state of the art is well developed and understood. For the interested readers a comprehensive review of the early history of SSI is presented in Kausel (2009).

Several modeling techniques are available to account for SSI in the dynamic analysis. The most sophisticated ones are based on finite element analyses in which the supporting medium is explicitly modeled as a continuum. This technique is very demanding, both in computer time and manpower, and is not very efficient at early design stages of a project. Therefore, a substructure approach is often preferred in which all the degrees of freedom of the supporting medium are lumped at the soil-foundation interface. With the assumption of a rigid surface foundation subjected to the vertical propagation of body waves in a horizontally layered profile, the method breaks down to the calculation of the dynamic response of the structure subjected to the free field motion and connected to the so-called dynamic foundation impedances (Kausel and Roësset, 1974). The dynamic impedances can be viewed as frequency-dependent springs and dashpots; if those impedances are assumed frequency-independent or if simple rheological models are used (Wolf and Deeks, 2004), the analysis is rendered very attractive and efficient. Such simple models can therefore be implemented to analyze the impact of linear soil structure interaction on the seismic response of structures.

A recent very comprehensive study (Moghaddasi Kuchaksarai et al., 2010), has investigated the effects of soil-shallow foundation-structure interaction on the seismic response of structures using Monte Carlo simulations. The structure was modeled as a non linear one degree of freedom system and SSI was taken into account with conventional springs and dashpots; in other words, SSI was treated as a linear phenomenon even though soil non linearities were accounted for through a reduction of the soil secant shear modulus. Forty time histories recorded in recent earthquakes were used as input motions; the magnitudes ranged from 6.5 to 7.5 and the distances from 15 to 40 km; the original records were scaled to produce peak ground accelerations between 0.3 and 0.8 g. In addition the impact of the following parameters was investigated: fundamental period of the fixed-base structure, soil shear wave velocity, mass density and Poisson's ratio, constitutive non linear model for the superstructure. To summarize the findings, it appears that SSI effects on the median response of a structure exhibiting a non linear behavior is relatively small; however there is a 30–50% probability for an increase in the total structural displacement of more than 10% due to SSI. Therefore, based on these results, SSI does not seem to be a major issue, at least in terms of median response. However, one may wonder how much these conclusions are influenced by the initial modeling assumptions regarding SSI.

4.3 Non Linear Soil Structure Interaction

More than 30 years ago the earthquake engineering community realized that the increase of strength of a structural system does not necessarily enhance its safety. This recognition has lead to the development of new design principles, aiming at rationally controlling seismic damage and rendering the structure "fail-safe". This concept is embedded in the capacity design philosophy which is widely implemented in structural design, but is given less attention in geotechnical engineering. Even when foundation compliance is taken into account, little care is given to the nonlinearity of soil and foundation. Such an approach may lead to non conservative oversimplifications, especially in the case of strong geometric nonlinearities, such as foundation uplifting and sliding. Most importantly, neglecting such phenomena prohibits the exploitation of strongly non-linear energy dissipating mechanisms in case of occurrence of ground motions larger than design. Today, a growing body of evidence suggests that soil-foundation plastic yielding under seismic excitation is not only unavoidable, but may even be beneficial (Anastasopoulos et al., 2009; Paolucci, 1997; Pecker, 1998, 2003; Martin and Lam, 2000; Gazetas et al., 2003; Gajan and Kutter, 2008). Such evidences has even led some authors to make the proposal of totally reversing the foundation design philosophy by allowing significant yielding in the foundation to protect the structure (Anastasopoulos et al., 2009).

However, implementation of a design philosophy in which, even partial, yielding is allowed at the foundation level requires that efficient and reliable tools be available for design. Non linear structural analyses are very sensitive to small changes in the structural properties and in the input motion. Obviously, the situation is even worse in foundation engineering where the properties of the soil are never known with a great accuracy. A safe design will therefore require a large amount of analyses to be run and this can hardly be efficiently achieved with heavy, although rigorous, numerical models such as finite element models. The concept of dynamic macroelements, developed over the last decade, offers a unique opportunity to evaluate the effect of non linear soil structure interaction on the response of a yielding structure.

Advantage of macroelement modelling is used in this paper, to examine the effect of non linear soil structure interaction on the response of a yielding structure. This is performed with a series of Incremental Dynamic Analyses (IDA). The results are further compared to analyses with linear SSI and without SSI (fixed-base structure) to highlight the changes in behaviour of the structure when SSI is accounted for either with a linear assumption or with a non linear one.

4.4 Problem Description

The studied structure is depicted in Fig. 4.1; it represents a typical highway bridge pier under seismic excitation. The deck of mass m_d is monolithically connected to the reinforced concrete circular column of diameter d and height h. The pier is founded on a relatively stiff homogeneous clay stratum by means of a shallow circular foundation of height h_f and diameter D. Separation (uplift) and no sliding are allowed along the soil-footing interface. The system is subjected to seismic loading only along the transverse (with respect to the bridge axis) horizontal direction.

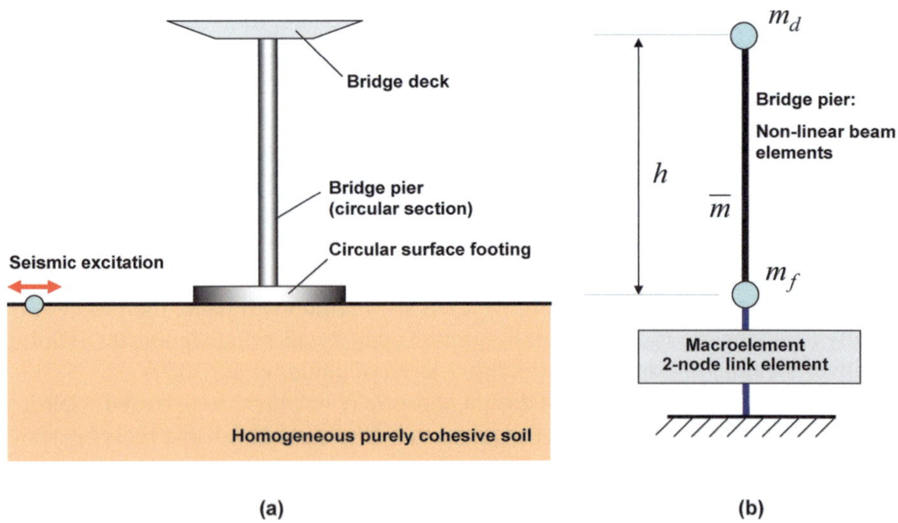

Fig. 4.1 Soil-foundation-structure system **a** physical, **b** model

A direct displacement-based design procedure (DDBD) (Priestley et al., 2007), appropriately modified to take into account soil-structure interaction effects, has been implemented for the pier design. The procedure is detailed in Figini (2010). The design of the bridge pylon has been performed considering a seismic input represented by the Eurocode 8 design spectrum, Type 1, with firm soil conditions and a peak ground acceleration $a_g = 0.5$ g. The following design performance criteria have been defined:

- System drift limit $\Delta_d = 0.03\,h$.
- Maximum foundation rotation $\theta_{\text{lim}} = 0.01$.
- Maximum structure ductility demand $\mu_{\text{lim}} = 3.2$.

The bridge pier is modeled with non-linear beam elements. The foundation and the soil are replaced by one unique 2-node link element, which is the non-linear dynamic macroelement for shallow foundations as developed in Chatzigogos et al. (2009a, b). The first node of the macroelement is attached to the superstructure. The mass of the foundation is lumped at this node; the input motion is applied at the second node. The constitutive behavior of the macroelement reproduces the non-linear phenomena arising at the soil-footing interface: elastoplastic soil behavior leading to irreversible foundation displacements, possibility for the footing to get detached from the soil (foundation uplift). Additionally, the macroelement is coupled with a viscous dashpot reproducing radiation damping.

The numerical parameters defining the problem are given in Table 4.1.

Table 4.1 Properties of the soil-structure system

Physical quantity	Symbol	Unit	Value
Mass of deck	m_d	kt	0.973
Column height	h	m	20.
Column diameter	d	m	2.5
Column mass	m_c	kt	0.245
Concrete compression strength	f_c	MPa	30.
Steel yield strength	f_y	MPa	400.
Number of longitudinal rebar	n	–	100
Diameter of longitudinal rebar	d_{long}	mm	26
Diameter of transverse rebar	d_{trans}	mm	12
Spacing of transverse rebar	s	mm	70
Foundation diameter	D	m	7.5
Foundation height	h_f	m	2.0
Foundation mass	m_f	kt	0.221
Total weight of structure	W_{tot}	MN	14.12
Soil undrained shear strength	c_u	MPa	0.15
Soil shear modulus	G_s	MPa	104
Soil shear wave velocity	V_s	m/s	255
Static bearing capacity factor	FS	–	2.84
Fixed base period of structure	T_0	s	1.379
Period for structure with SSI	T_{SSI}	s	1.650

4.5 Description of the Macroelement

Several macroelement models have been developed during the last decade to account for non linear soil structure interaction. A comprehensive review of the existing models is presented in Chatzigogos et al. (2007). For the sake of completeness the model used in the present study is briefly presented. More details are provided in Chatzigogos et al. (2009a, b).

4.5.1 Generalized Variables

The macroelement model implemented in this study is formulated with respect to the resultant forces and moments acting at the centre of the footing. Although the footing geometry is circular (implying 3D kinematics), the system is subjected to planar loading; therefore, the in-plane dimensionless force parameters are assembled in the following vector:

$$\mathbf{Q} = \begin{bmatrix} N/M_{\max} \\ V_x/N_{\max} \\ M_y/DN_{\max} \end{bmatrix} \tag{1}$$

N is the resultant vertical force on the footing centre, V_x is the resultant horizontal force, M_y the resultant rocking moment, D the footing diameter and $N_{\max}$ the maximum centred vertical force supported by the foundation.

The kinematics of the problem is simplified by considering that the footing is perfectly rigid and will, in all cases, undergo a planar rigid body motion. Its kinematics is thus described by three displacement parameters, identified with the in plane translations and rotation of the footing centre. The displacement parameters are normalized and assembled in a displacement vector as follows:

$$\mathbf{q} = \begin{bmatrix} u_z/D \\ u_x/D \\ \theta_y \end{bmatrix} \tag{2}$$

4.5.2 Elastic Range

In the *linear elastic* range the force and displacement vectors are linked through a complex frequency dependent symmetric matrix $\tilde{\mathbf{K}}$. The tilde ($\sim$) over $\mathbf{K}$ is used to denote the linear case. For shallow foundations, off-diagonal terms are negligible and $\tilde{\mathbf{K}}$ is diagonal. In the context of non-linear time domain analyses it is common practice to use constant stiffness terms that correspond to quasi-static loading or to some characteristic frequency of the soil-structure system. Similarly, radiation damping effects are accounted for with an equivalent viscous damping matrix with constant (frequency-independent) damping coefficients. The linear (visco-elastic) part of the response in the macroelement model is thus defined by 6 numerical parameters: $\tilde{K}_{NN}, \tilde{K}_{VV}, \tilde{K}_{MM}, \tilde{C}_{NN}, \tilde{C}_{VV}, \tilde{C}_{MM}$.

The possibility of the footing to get partially detached from the soil surface is introduced within the macroelement through a phenomenological *non-linear elastic* model. The adopted uplift model consists in writing the stiffness matrix **K** as a function of the displacement parameters as follows:

$$\mathbf{K} = \mathbf{K}(\mathbf{q}) \tag{3}$$

In Chatzigogos et al. (2009a, b), the following explicit relationships have been introduced:

$$K_{NN} = \tilde{K}_{NN} \quad , \quad K_{VV} = \tilde{K}_{VV} \tag{4}$$

$$K_{NM} = K_{MN} = \begin{cases} 0 & \text{if } \left|\theta_y\right| \leq \left|\theta_{y,0}\right| \\ \varepsilon \tilde{K}_{NN}\left(1 - \frac{\theta_{y,0}}{\theta_y}\right) & \text{if } \left|\theta_y\right| > \left|\theta_{y,0}\right| \end{cases} \tag{5}$$

$$K_{MM} = \begin{cases} \tilde{K}_{MM} & \text{if } \left|\theta_y\right| \leq \left|\theta_{y,0}\right| \\ \gamma\delta\tilde{K}_{MM}\left(\frac{\theta_{y,0}}{\theta_y}\right)^{\delta+1} + \varepsilon^2\tilde{K}_{NN}\left(1 - \frac{\theta_{y,0}}{\theta_y}\right)^2 & \text{if } \left|\theta_y\right| > \left|\theta_{y,0}\right| \end{cases} \tag{6}$$

$$\theta_{y,0} = \pm\frac{N}{\alpha K_{MM}} \tag{7}$$

In Eqs. (4), (5), (6) and (7), the quantity $\theta_{y,0}$ represents the foundation rotation angle that corresponds to uplift initiation. The equations indicate that uplift introduces a non-zero coupling term between the vertical force and the rocking moment and an appropriate modification of the rocking stiffness K_{MM}. The moment of uplift initiation is defined as follows:

$$M_0 = \pm\frac{N}{\alpha} \tag{8}$$

Parameters α, γ and δ are numerical constants that solely depend on the footing shape. The numerical parameter ε controls the coupling between the rocking and the vertical degree of freedom during uplift. Values of these parameters have been proposed in Chatzigogos et al. (2009a, b) for strip and circular footings.

4.5.3 Plasticity Model

The second non-linear mechanism introduced in the macroelement constitutive relationship is related to the irreversible soil behaviour. A bounding surface hypoplasticity model (Dafalias and Hermann, 1982), is developed independently from the uplift model presented in the previous paragraph. The yield surface of classical plasticity is replaced by a bounding surface denoted f_{BS}: in the interior of this surface a continuous plastic response is obtained as a function of the distance between the actual force state **Q** and an image point **I**(**Q**) on the bounding surface, defined through an appropriately chosen mapping rule. As the force state **Q** approaches the bounding surface, the plastic response becomes more and more pronounced with eventual plastic flow occurring when the force state reaches the bounding surface: this situation actually corresponds to a bearing capacity failure of the foundation. The bounding surface f_{BS} can therefore be identified with the *ultimate surface* of a footing resting on a cohesive soil with a perfectly bonded interface (neither uplift nor sliding allowed). A sufficient, for the scope of macroelement modelling, and extremely simple approximation is obtained by considering that the ultimate surface f_{BS} is an ellipsoid centred at the origin:

$$\left(\frac{N}{N_{\max}}\right)^2 + \left(\frac{V_x}{\psi N_{\max}}\right)^2 + \left(\frac{M_y}{\xi D N_{\max}}\right)^2 - 1 = 0 \tag{9}$$

The bounding surface is thus defined by three numerical parameters: the ultimate vertical force supported by the footing $N_{\max}$ and the parameters ψ, ξ which are used for the definition of the maximum horizontal force and the maximum moment supported by the footing.

In the present formulation a simple radial mapping rule is selected; such a mapping rule actually corresponds to the case of proportional loading of the footing up to bearing capacity failure. The image point **I**(**Q**) is thus defined by the following expression:

$$\mathbf{I}(\mathbf{Q}) = \{\lambda\,\mathbf{Q} | \mathbf{I} \in \partial f_{BS}, \lambda \geq 1\} \tag{10}$$

In (10), ∂f_{BS} represents the boundary of the bounding surface f_{BS}. The image point **I**(**Q**) is used to define the direction of plastic displacements, the magnitude of the plastic modulus and the situations of plastic loading, neutral loading and unloading for a given force increment $\dot{\mathbf{Q}}$, as in classical plasticity. For unloading and neutral loading the response is elastic, for plastic loading the plastic modulus is given by:

$$\dot{\mathbf{Q}} = \mathbf{H} \cdot \dot{\mathbf{q}}^{pl} \tag{11}$$

where $\dot{\mathbf{q}}^{pl}$ is the increment of plastic displacements. The inverse plastic modulus $\mathbf{H}^{-1}$ can in turn be written as:

$$\mathbf{H}^{-1} = \frac{1}{h}\mathbf{n} \otimes \mathbf{n}_g \tag{12}$$

n being the unit normal on **I**(**Q**), $\mathbf{n}_g$ defines the direction of the plastic incremental displacements which is in general different from **n** and h a scalar quantity, which expresses the extent of plastic response; in the context of bounding surface hypoplasticity h is a function of the distance between the force state **Q** and its image point **I(Q)**. A convenient measure of this distance is the scalar parameter λ defined in (10). For cyclic loading the functional dependence between h and λ is given by the following simple relationship:

$$h = h_0 \ln\left(\frac{\lambda^{p+1}}{\lambda_{\min}^{p}}\right) \tag{13}$$

In (13), h_0 and p are numerical constants and $\lambda_{\min}$ is the minimum value attained by the parameter λ during loading. The meaning of (13) is the following: in virgin loading $\lambda = \lambda_{\min}$; in reloading $\lambda > \lambda_{\min}$ and the response is less plastic since the ratio $\lambda/\lambda_{\min}$ is always greater than 1.

Finally, the unit vector $\mathbf{n}_g$ defining the direction of plastic displacements is typically defined as the normal vector to a plastic potential surface. In the context, of the proposed model we adopt a much simpler definition for $\mathbf{n}_g$ by simply relating its components to the unit normal vector to the bounding surface as follows:

$$\mathbf{n}_g = \left(p_g \frac{\partial f_{BS}}{\partial N}, \frac{\partial f_{BS}}{\partial V_x}, \frac{\partial f_{BS}}{\partial M_y}\right)^T \tag{14}$$

In other words, $\mathbf{n}_g$ is identical to $\mathbf{n}$ with the exception of the component parallel to the vertical force N on the footing, which is modified by the factor p_g. This parameter expresses the extent of vertical settlement of the foundation when subjected to load cycles under horizontal force or moment.

4.5.4 Uplift – Plasticity Coupling

The non-linear uplift and soil plasticity mechanisms presented in the previous paragraphs are defined independently from one another; they become coupled when they are assembled within the macroelement. For example, the moment of uplift initiation is no longer proportional to the applied vertical force but is approximated by the following relationship:

$$M_0 = \pm \frac{N}{\alpha} e^{-\zeta N} \tag{15}$$

The numerical parameter ζ generally varies between 1.5 and 2.5.

The uplift-plasticity coupling that occurs under dynamic loading leads in general to an effect of rounding of the soil-footing contact area and consequently to a reduction of the effective size of the footing and of the ultimate vertical force $N_{\max}$. In the present formulation this progressive "damage" effect has not been incorporated. This phenomenon has proven to be more pronounced in the case of dry cohesionless soils with very limited rebound capacity and less important in the case of cohesive soils examined in this paper.

4.5.5 Model Parameters

The model parameters are listed in Table 4.2. Derivation of those parameters is briefly commented below.

Table 4.2 Macroelement model parameters

Parameter description	Symbol	Unit	Value
Footing diameter	D	m	7.50
Ultimate vertical force	N_{max}	MN	40.09
Ultimate horizontal force	V_{max}	MN	6.63
Ultimate moment	M_{max}	MN.m	33.30
Bounding surface parameter	ψ	–	0.17
Bounding surface parameter	ξ	–	0.11
Vertical elastic stiffness	K_{NN}	MN/m	2225
Horizontal elastic stiffness	K_{VV}	MN/m	1833
Rocking elastic stiffness	K_{MM}	MN.m	20862
Vertical dashpot coefficient	C_{NN}	MN.s/m	27.8
Horizontal dashpot coefficient	C_{VV}	MN.s/m	18.0
Rocking dashpot coefficient	C_{NN}	MN.m.s	4.8
Plastic parameter (initial loading)	h_0/K_{NN}	–	4.0
Plastic parameter (reloading)	p	–	0.5
Non-associative parameter	p_g	–	5.0
Uplift initiation parameter	α	–	6.0
Uplift parameter	γ	–	2.0
Uplift parameter	δ	–	0.5
Uplift parameter	ε	–	0.2
Uplift plasticity coupling parameter	ζ	–	1.5

4.5.5.1 Viscoelastic Parameters

They are determined using the classical impedance functions for a circular footing on a half space (Gazetas, 1991).

4.5.5.2 Bounding Surface Parameters

The ultimate vertical force for a centred load is given by the ultimate bearing capacity of a circular footing on a cohesive soil $N_{\text{max}} = 6.05c_{\text{u}}A$ where c_{u} is the soil undrained shear strength and A the footing area. The parameters ψ and ξ are given by $\psi = V_{\text{max}}/N_{\text{max}}$ and $\xi = M_{\text{max}}/DN_{\text{max}}$. The ultimate shear force and overturning moment for a perfectly bonded footing are given by $V_{\text{max}} = c_{\text{u}}A$ and $M_{\text{max}} = 0.67c_{\text{u}}AD$ (Chatzigogos et al., 2007).

4.5.5.3 Plasticity Model Parameters

These are the only parameters (h_0, p, p_g) that require a calibration from a 3D static finite element model. The soil is modelled using the multi-yield elastoplastic model developed in Prevost (1985) and numerically implemented in the finite element code DYNAFLOW Prevost (2008). To this end, a quasi-static test analysis is performed in two stages:

- A vertical force, representing approximately the weight of the structure, is imposed to the foundation;
- The vertical force is kept fixed, while a cycle of loading under a horizontal force is applied. The imposed horizontal force can typically be selected to vary between $0.5V_{\max}$ and $0.8V_{\max}$.

The numerical parameters h_0, p are chosen to reproduce the soil hardening behavior in the diagrams of vertical force versus vertical displacement and the diagram of horizontal force versus horizontal displacement. The numerical parameter p_g is calibrated to fit the accumulated vertical settlement during the phase of loading under the horizontal force. This calibration procedure has been implemented for the bridge pylon under consideration. The results of the calibration procedure are presented in Figs. 4.2 and 4.3, which compare the vertical force versus vertical displacement and horizontal force versus horizontal displacement curves obtained with the finite element model and with the macroelement. The fit is satisfactory for the vertical force diagram for both loading phases and for the first-half cycle in the horizontal force diagram. The difficulty to fit both the vertical and the

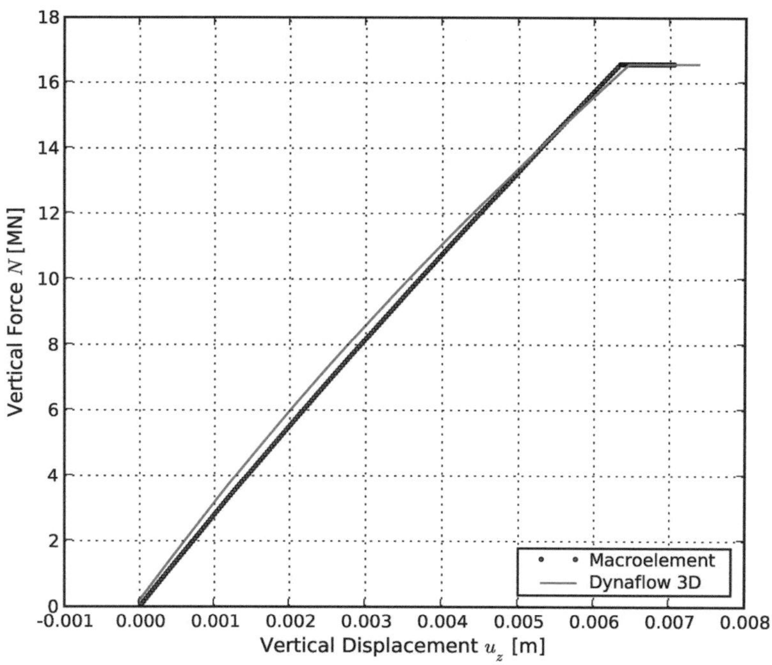

Fig. 4.2 Load-displacement curve for vertical loading

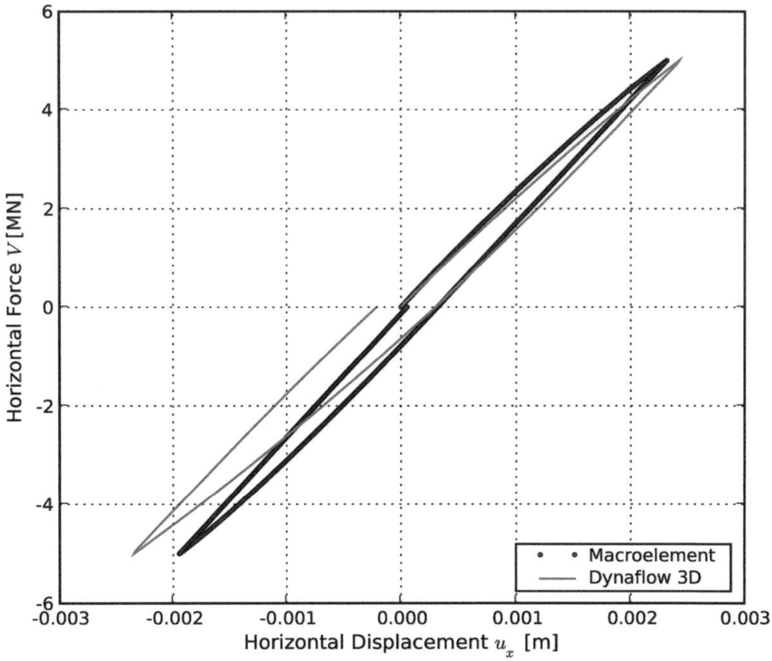

Fig. 4.3 Load-displacement curve for cyclic horizontal loading

horizontal force versus displacement curves stems from the fact that an "isotropic" formulation has been selected for the bounding surface hypoplastic model, in the sense that the hardening relationship (13) applies equally to all degrees of freedom. Possibilities of improving the model performance have been discussed in Chatzigogos et al. (2009) but they are counter-balanced by the complexity they would add to the adopted plasticity model. In general, the implemented bounding surface hypoplasticity model is deemed to provide a satisfactory compromise between model accuracy and calibration simplicity.

4.6 Superstructure Model

The bridge pier is modeled using small-displacement/small-rotation Timoshenko beam elements with an elastoplastic constitutive law. For simplicity, an elastoplastic bilinear model is adopted in the analyses. The moment-curvature diagram for the examined concrete column has been calculated in Priestley et al. (2007). The parameters used for the definition of the bilinear moment-curvature diagram are the yield moment M_y and the post yield stiffness of the beam in pure tension. The elastic stiffness of the beam elements is calculated from the geometric characteristics of the cross section and the elastic properties of reinforced concrete. The numerical

Table 4.3 Numerical parameters for structural model

Parameter	Symbol	Unit	Value
Yield moment for elastoplastic beam element	M_y	MN.m	37.5
Yield curvature	κ_y	–	6.52×10^{-4}
Post-yield stiffness for beam elements under traction	K_{post}	MPa	5.0

parameters used for the definition of the bilinear model for the beam elements are presented in Table 4.3.

4.7 Incremental Dynamic Analyses

An incremental dynamic analysis (IDA) consists in performing a series of non-linear time-history analyses, using as input motion the same acceleration record scaled to increasing amplitudes, and keeping track of some characteristic quantities of the response of the structure. Using the terminology introduced in Vamvatsikos and Cornell (2002), we refer to *intensity measures* (IMs), characterizing the severity of the input motion and to *damage measures* (DMs), characterizing the response of the structure.

The output of an IDA is an *IDA curve*, i.e. a plot of a selected IM versus a selected DM. Similarly, an *IDA curve set* is a collection of IDA curves of the same structural model under different records that have been parameterized on the same IM and DM.

4.7.1 Intensity Measures

Different options are available for the IM to be used in the IDA curves. In the following, we use three IMs and in particular:

- The PGA of the input motion.
- The cumulative absolute velocity (CAV) of the input motion.
- The spectral acceleration of the input motion at the natural period of vibration of the bridge pylon with consideration of soil flexibility, denoted as $S_A(T_{SSI})$.

The choice of the cumulative absolute velocity is guided by its cumulative character which may be a better proxy for the residual response parameters of the structure. Similarly, the spectral accelerations at characteristic periods of vibration reflect the intensity of the input motion but also the dynamic characteristics of the structure.

4.7.2 Damage Measures

For the DMs, the following quantities may be considered:

- The residual settlement of the foundation
- The maximum foundation rotation
- The maximum total horizontal drift of the bridge deck
- The residual horizontal drift of the bridge deck
- The maximum horizontal deck displacement due to structural drift
- The maximum structural ductility demand in the concrete column: $\max\{\mu_d\}$ defined in terms of curvature as:

Table 4.4 Selected records for incremental dynamic analyses

Record	Event	Year	Station	φ[a]	Soil[b]	M[c]	R[d]
1	Loma Prieta	1989	Agnews State Hospital	090	C,D	6.9	28.20
2	Northridge	1994	LA, Baldwin Hills	090	B,B	6.7	31.30
3	Imperial Valley	1979	Compuertas	285	C,D	6.5	32.60
4	Imperial Valley	1979	Plaster City	135	C,D	6.5	31.70
5	Loma Prieta	1989	Hollister Diff Array	255	C,D	6.9	25.80
6	San Fernando	1971	LA, Hollywood Stor. Lot	180	–,D	6.6	21.20
7	Loma Prieta	1989	Anderson Dam Downstrm	270	C,D	6.9	21.40
8	Loma Prieta	1989	Coyote Lake Dam Downstrm	285	B,D	6.9	22.30
9	Imperial Valley	1979	El Centro Array #12	140	B,D	6.5	18.20
10	Imperial Valley	1979	Cucapah	085	C,D	6.5	23.60
11	Northridge	1994	LA, Hollywood Storage FF	360	C,D	6.7	25.50
12	Loma Prieta	1989	Sunnyvale Colton Ave	270	C,D	6.9	28.80
13	Loma Prieta	1989	Anderson Dam Downstrm	360	C,D	6.9	21.40
14	Imperial Valley	1979	Chihuahua	012	B,D	6.5	28.70
15	Imperial Valley	1979	El Centro Array #13	140	C,D	6.5	21.90
16	Imperial Valley	1979	Westmoreland Fire Station	090	C,D	6.5	15.10
17	Loma Prieta	1989	Hollister South & Pine	000	C,D	6.9	28.80
18	Loma Prieta	1989	Sunnyvale Colton Ave	360	–,D	6.9	28.80
19	Superstition Hills	1987	Wildlife Liquefaction Array	090	C,D	6.7	24.40
20	Imperial Valley	1979	Chihuahua	282	C,D	6.5	28.70
21	Imperial Valley	1979	El Centro Array #13	230	C,D	6.5	21.90
22	Imperial Valley	1979	Westmoreland Fire Station	180	C,D	6.5	15.10
23	Loma Prieta	1989	Halls Valley	090	C,D	6.9	31.60
24	Loma Prieta	1989	WAHO	000	–,D	6.9	16.90
25	Superstition Hills	1987	Wildlife Liquefaction Array	360	C,D	6.7	24.40
26	Imperial Valley	1979	Compuertas	015	C,D	6.5	32.60
27	Imperial Valley	1979	Plaster City	045	C,D	6.5	31.70
28	Loma Prieta	1989	Hollister Diff Array	165	–,D	6.9	25.80
29	San Fernando	1971	LA, Hollywood Stor. Lot	090	C,D	6.6	21.20
30	Loma Prieta	1989	WAHO	090	–,D	6.9	16.90

[a]Component.
[b]USGS, Geomatrix soil classification.
[c]Moment magnitude.
[d]Closest distance to fault rupture (km).

$$\mu_d = \frac{\max\{\kappa\}}{\kappa_y} \tag{16}$$

In (16), $\max\{\kappa\}$ is the maximum curvature developed at the base of the column during the seismic excitation and κ_y is the yield curvature of the column given in Table 4.3.

4.7.3 Time Histories

A set of 30 acceleration records has been chosen for the incremental dynamic analyses. The compilation of the suite of records has been given in Vamvatsikos and Cornell (2002). The selected acceleration records are from relatively large-magnitude earthquakes ($M = 6.5$–6.9) with moderate distances and exhibiting no marks of directivity. Additionally, they have all been recorded on firm soil

Table 4.5 Characteristics of selected unscaled records

Record	PGA (g)	CAV (m/s)	$S_A(T_0)$ (g)	$S_A(T_{SSI})$ (g)
1	0.159	4.94	0.133	0.127
2	0.239	6.03	0.151	0.114
3	0.147	2.67	0.034	0.022
4	0.056	1.54	0.043	0.032
5	0.279	6.77	0.189	0.189
6	0.173	3.66	0.046	0.046
7	0.244	5.97	0.144	0.181
8	0.179	4.38	0.198	0.160
9	0.144	4.47	0.162	0.115
10	0.308	6.81	0.296	0.199
11	0.358	10.42	0.178	0.117
12	0.207	6.70	0.267	0.157
13	0.240	6.10	0.170	0.113
14	0.270	9.03	0.205	0.138
15	0.117	3.73	0.097	0.100
16	0.074	2.46	0.069	0.076
17	0.370	9.66	0.533	0.484
18	0.208	5.95	0.272	0.182
19	0.181	5.11	0.059	0.070
20	0.254	8.94	0.266	0.182
21	0.139	3.63	0.106	0.085
22	0.110	2.43	0.090	0.083
23	0.102	3.42	0.151	0.092
24	0.398	15.17	0.229	0.104
25	0.208	8.26	0.407	0.305
26	0.186	4.18	0.060	0.050
27	0.042	1.07	0.017	0.021
28	0.269	6.05	0.331	0.302
29	0.210	4.53	0.266	0.125
30	0.672	20.25	0.352	0.169

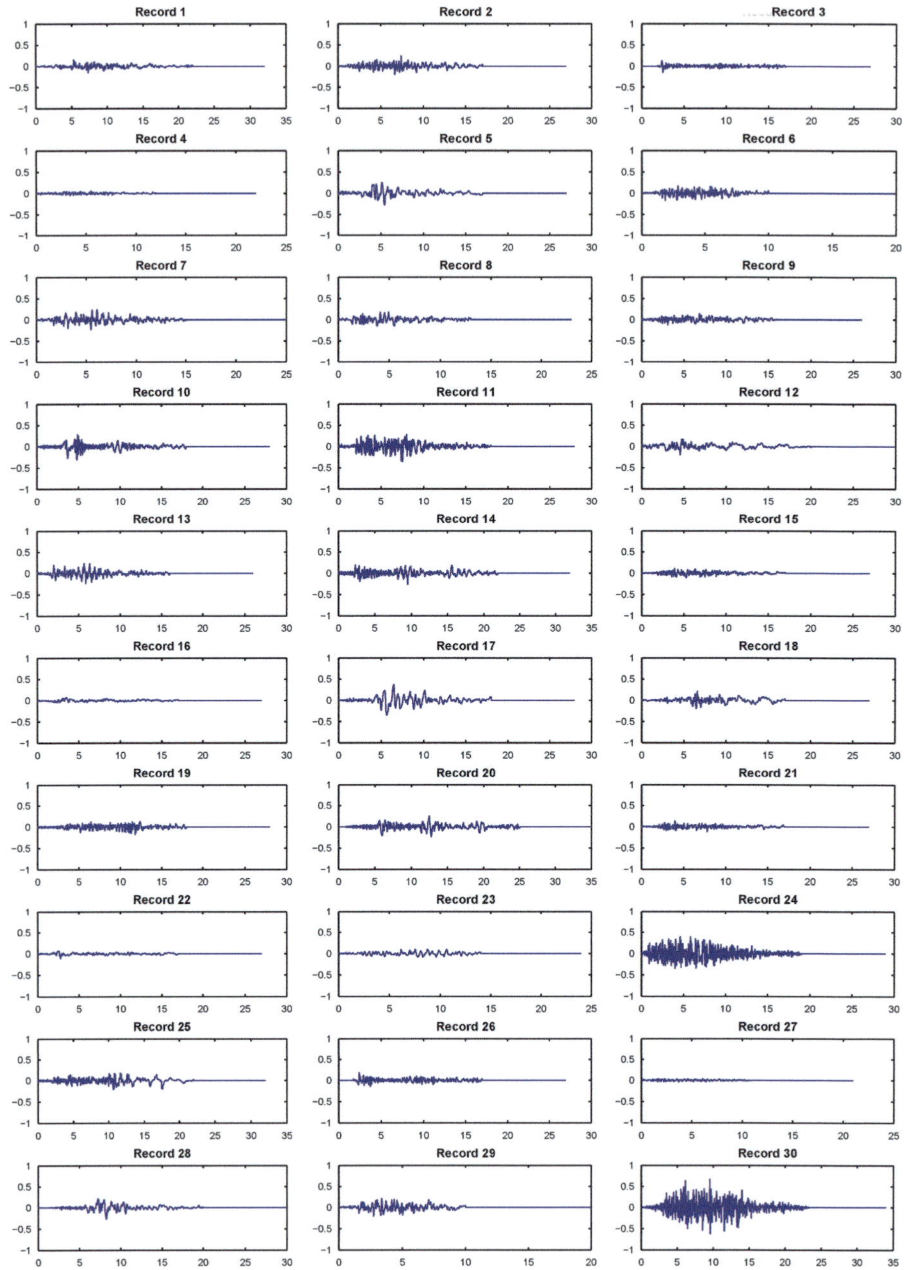

Fig. 4.4 Acceleration time histories of the 30 unscaled records

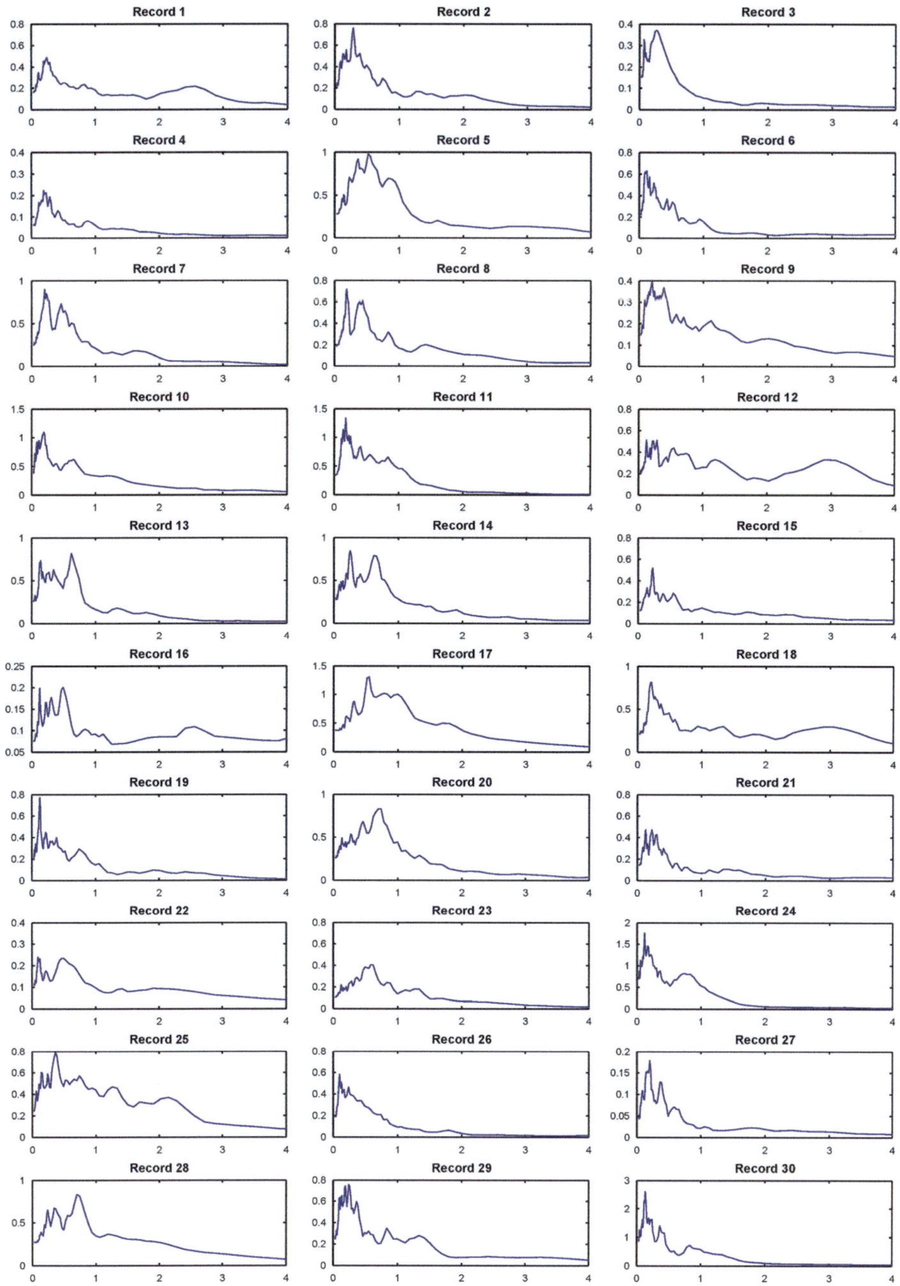

Fig. 4.5 Acceleration response spectra of the 30 unscaled records

conditions. They represent a realistic earthquake scenario for the examined soil-structure system.

Table 4.4 summarizes the suite of thirty ground motion records used in the analyses and Table 4.5 gives a list of characteristic quantities of the selected records: maximum recorded acceleration, cumulative absolute velocity and spectral acceleration at the fixed-base natural period of the structure ($S_A(T_0)$) and with consideration of soil flexibility ($S_A(T_{SSI})$). Note that the quantities presented in Table 4.5 refer to the *unscaled* records.

In addition to Table 4.4, Fig. 4.4 presents the acceleration time histories of the selected unscaled records and Fig. 4.5 the corresponding acceleration response spectra.

4.8 Structural Behavior in the Light of Incremental Dynamic Analyses

The incremental dynamic analyses have been performed for every possible combination of the IMs and DMs. Each IDA analysis is made of 30 curves corresponding to the 30 time histories. One such example is depicted in Fig. 4.6 showing the ductility demand versus CAV for the fixed-base structure. Each curve on the diagram corresponds to one of the 30 records scaled downward and upward to produce 11 IMs with CAVs spanning the range 0–35 m/s. A common feature to all IDA analyses is that some curves exhibit instabilities, possibly followed by regain at higher levels,

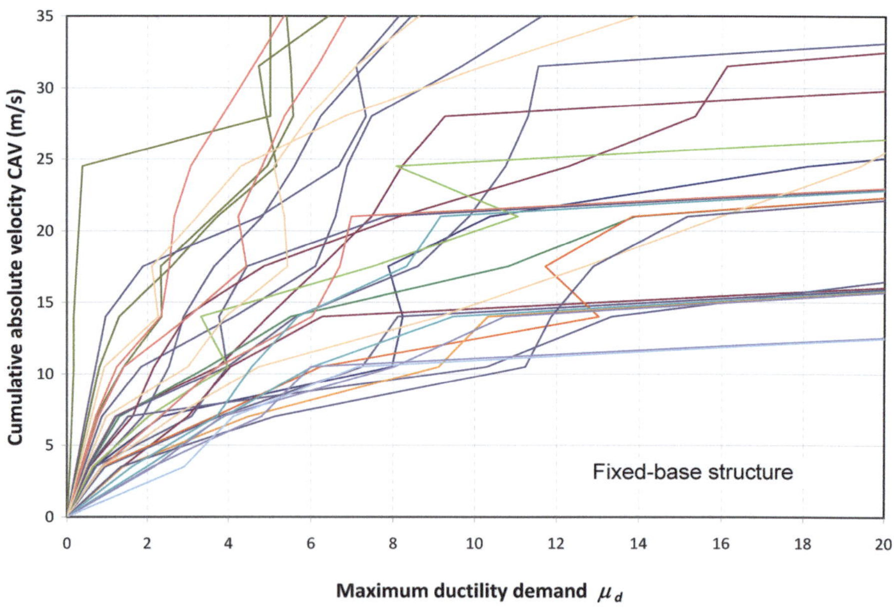

Fig. 4.6 IDA curves for a fixed-base structure

while other do not show any sign of instability, at least up to the highest tested IM. These kinds of curves are instructive because they clearly evidence the variability of the response as a function of the individual records, although all records are deemed to represent an almost unique earthquake scenario (see Table 4.4). These results can be used to derive statistical results (or probability distributions) of the response, which can be further incorporated in a PBEE framework (Vamvatsikos and Cornell, 2002).

The content of an IDA set is more compact and meaningful if, instead of individual curves, the median and some fractiles, for instance 16 and 84% fractiles, are presented (Vamvatsikos and Cornell, 2002). IDA curves have been constructed for the three cases involving (or not) soil-structure interaction, namely:

- Non linear fixed-base structure
- Non linear structure with linear soil-structure interaction
- Non linear structure with non linear soil-structure interaction

As mentioned previously, it is anticipated that IMs that reflect the cumulative damaging effect of the earthquake are good proxies for correlation with a DM related to residual states (permanent settlement, permanent foundation rotation). Therefore ductility demand and permanent settlements have been related to CAV. On the other hand, DMs related to peak responses, like the maximum deck displacement, should be better correlated to the spectral acceleration at the fundamental period of the system.

4.8.1 Typical Result of a Dynamic Analysis

A typical dynamic response produced with the macroelement is depicted in Fig. 4.7 showing the variation during excitation of typical quantities: pier curvature versus bending moment, horizontal displacement and deck drift versus time, foundation rotation versus rocking moment and foundation settlement versus rotation.

This figure illustrates the capability of the macroelement to produce permanent settlement under horizontal excitation and uplift of the foundation, evidenced by the S-shaped of the moment-rotation diagram. The deck develops significant horizontal displacements (of the order of 0.8 m) almost entirely due to foundation rotation. However, the structure remains elastic as revealed by the moment-structural curvature diagram: it is clear that uplift acts as an isolation mechanism for the superstructure. In this example, both the residual foundation settlement and the residual foundation rotation seem to remain at acceptable levels (less than 0.03 m and 0.01 rad respectively).

4.8.2 Statistical Results

For each of the IDA set of curves, similar to those of Fig. 4.6, statistical values corresponding to the median and to the 16 and 84% fractiles are computed. Comparisons

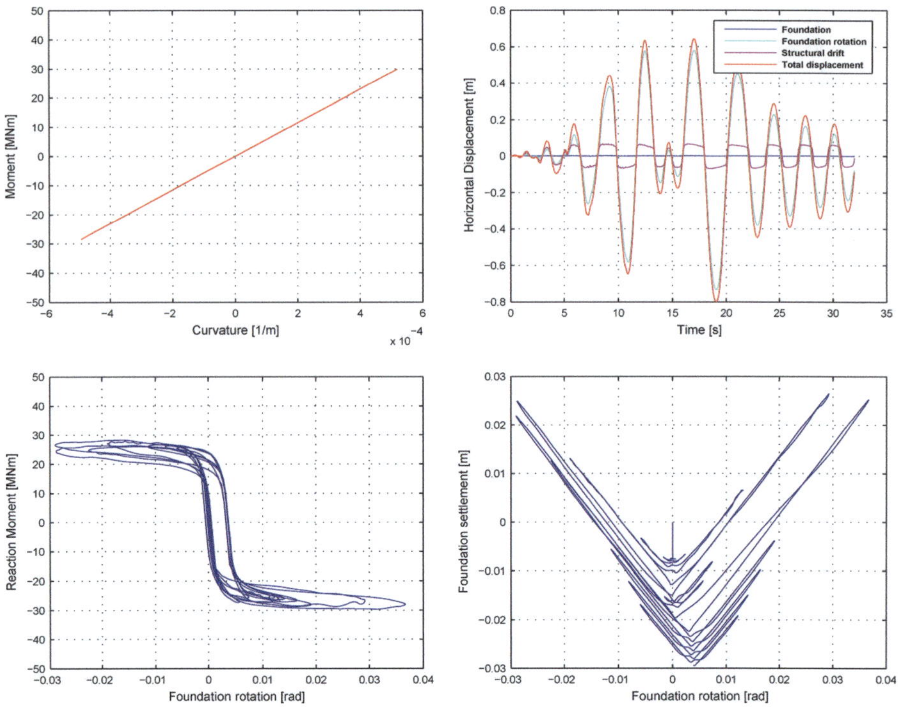

Fig. 4.7 Example of a dynamic analysis with the macroelement; non-linear foundation, record 1, PGA = 0.5 g

are made in terms of structural behavior for the three possible assumptions for the foundation behavior: fixed-base structure, elastic linear foundation (linear SSI) and non linear behavior. All results are obtained with the macroelement model, with the proper options activated, and the non linear structural model for the structure. Due to space limitations, only few significant results are presented. They correspond to the ductility demand in the bridge pier, the permanent foundation settlement (for the non linear foundation), and the maximum deck displacement. As mentioned previously the ductility demand and the residual settlements are related to the CAV, while the maximum deck displacement is related to the SSI period of vibration.

Figures 4.8, 4.9 and 4.10 present the statistical curves for the ductility demand for the three cases of foundation behavior. The overall behavior is not so different between the fixed-base structure and the linear elastic foundation: beyond a CAV of the order of 20, the ductility demand increases at a very rapid rate, denoting the onset of instability. It is interesting to note that up to a CAV of 10 m/s, both systems produce the same median curve; for larger CAV, the ductility demand is slightly larger, for a given IM value, for the linear elastic SSI system indicating that SSI may not be favorable. This result is in line with the more extensive study of Moghaddasi Kuchaksarai et al. (2010). For the non linear foundation system, the behavior is strikingly totally different: up to a CAV of 35 m/s, the ductility demand

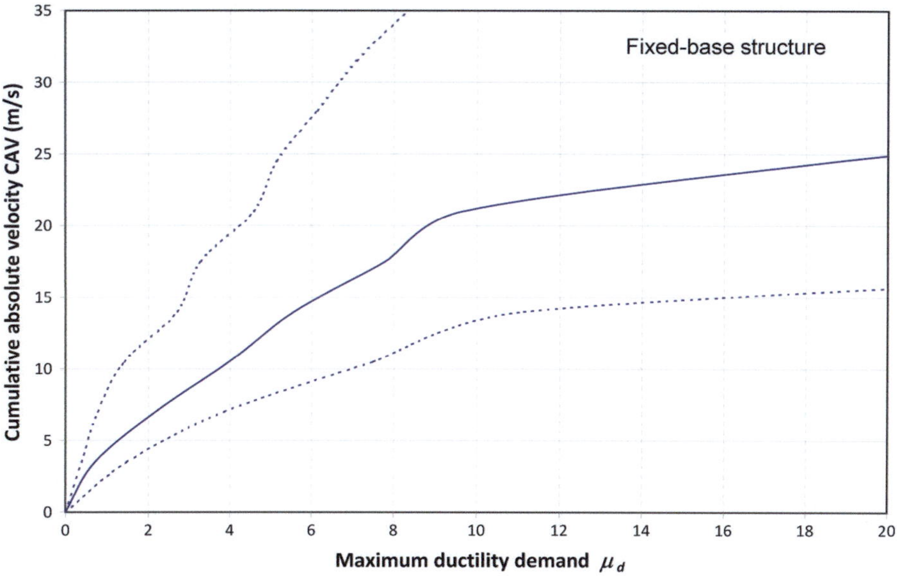

Fig. 4.8 IDA curves for ductility demand versus CAV for the fixed-base structure. Thick curve: median, dotted curves: 16 and 84% fractiles

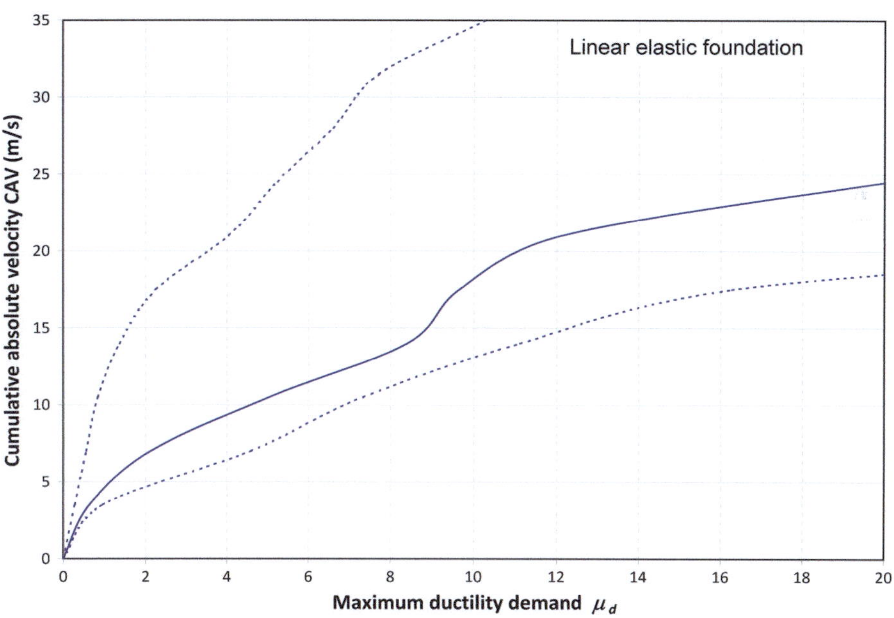

Fig. 4.9 IDA curves for ductility demand versus CAV for the linear elastic foundation. Thick curve: median, dotted curves: 16 and 84% fractiles

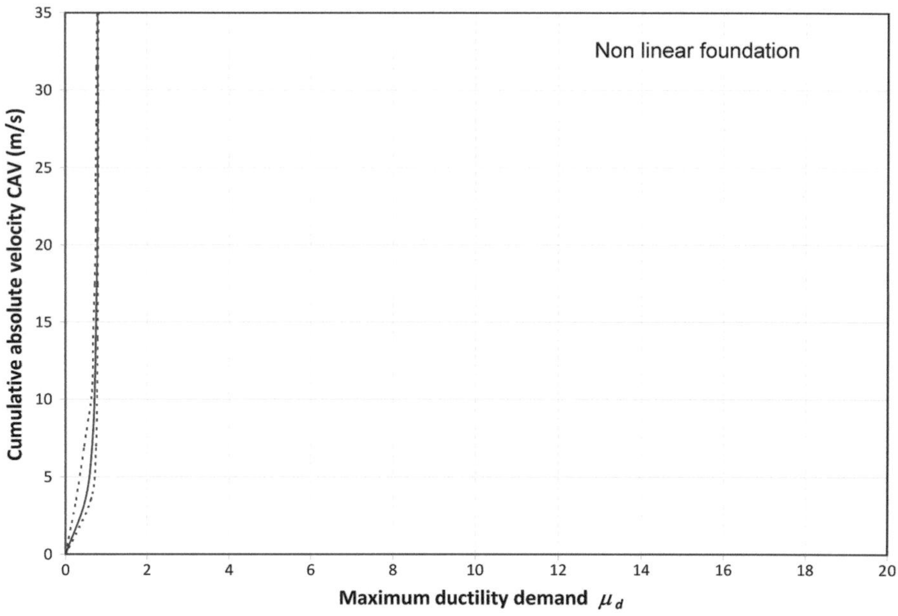

Fig. 4.10 IDA curves for ductility demand versus CAV for non-linear foundation. Thick curve: median, dotted curves: 16 and 84% fractiles

remains limited, of the order of 1.0, with no evidence so far of instability in the bridge pier. The explanation for such a different behavior lies in the yielding of the foundation that protects the structure, as pointed out in Anastasopoulos et al. (2009); the structure is prevented from yielding but permanent settlement and rotation are developed at the foundation. This is evidenced in Fig. 4.11 showing the residual foundation displacement; obviously for the fixed-base structure and the linear SSI system no such values exist. The median maximum displacement remains limited but the variability increases drastically as the CAV increases; at the maximum CAV value the 84% fractile is 2.5 times the median. Therefore, foundation settlement may become highly unpredictable and can easily go from an acceptable quantity to an unacceptable one depending on the probability of exceedance the designer is ready to accept. This factor requires in depth consideration before accepting significant foundation yielding.

Residual displacements are not the only issue in the seismic response of the structure. For instance, total displacement at the deck level may also be important for the design of connections. Figure 4.12 presents for the three examined systems the median maximum horizontal displacement at the deck level. Once again, the same salient features are evidenced: beyond a $S_A(T_{SSI})$ of 0.50–0.70 g, both the fixed-base structure and the linear SSI system fail. In contrast, the non linear SSI system does not show lack of stability (only beyond $S_A(T_{SSI}) \approx 1.1$ g), not represented in the figure); the displacement increases steadily to large values. However, the displacement is always larger than for the two others systems and the differences become

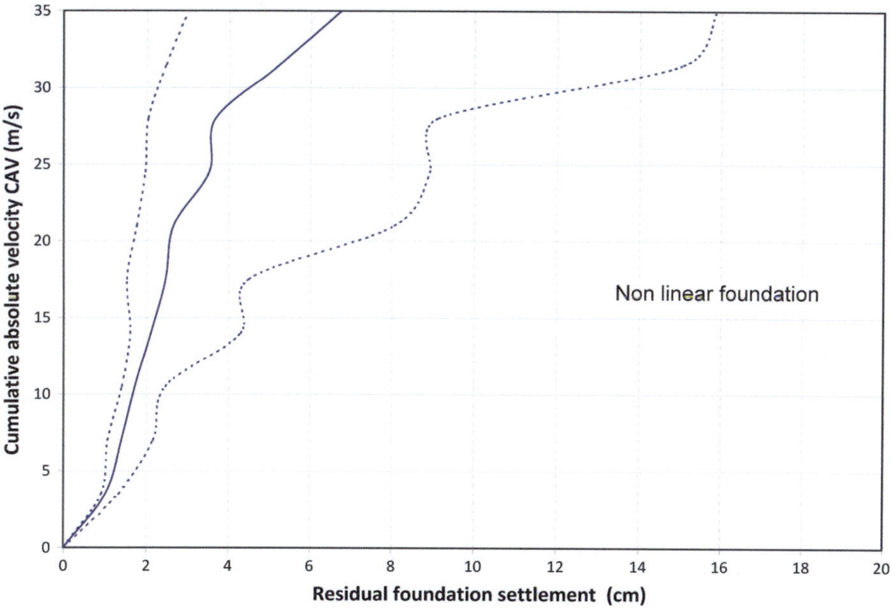

Fig. 4.11 IDA curves for residual foundation settlement versus CAV for non-linear foundation. *Thick curve*: median, *dotted curves*: 16 and 84% fractiles

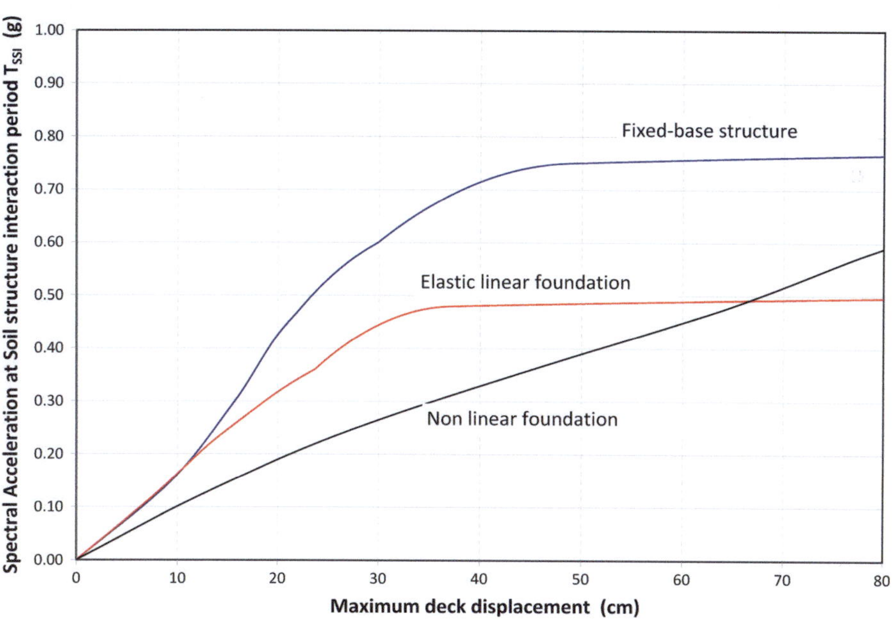

Fig. 4.12 IDA curves for maximum deck displacement versus spectral acceleration at the SSI period of vibration for three cases of base conditions

significant close to the IM level corresponding to failure of the linear or fixed-base systems. Stability of the structure, defined here with respect to the maximum horizontal deck displacement, is ensured provided large displacements, mainly caused by foundation rotation, are acceptable. How large can the displacement be is beyond the scope of this paper.

Other results not shown herein exhibit the same trend: yielding of the foundation "protects" the structure but the price to pay is an increase of the maximum or permanent displacements and rotations of the foundation; more importantly, the variability in the computed response becomes large for the non linear SSI system when the IM is approaching the value for which the fixed-base structure or the linear SSI system shows instabilities.

4.9 Conclusions

The development of a dynamic macroelement renders possible the use of extensive time history analyses to analyze the effect of foundation compliance and non linearity on the structural response of a non linear structure. The approach followed in this paper is based on the concept of incremental dynamic analyses which allows the derivation of statistical properties of the response. A simple bridge pier modeled either as a fixed-base structure, or founded on a foundation, for which linear or non linear soil structure interaction is considered, has served to illustrate the most salient features of the response. On a whole, consideration of non linear soil structure interaction appears beneficial to drastically reduce the ductility demand in the structure; however, this positive effect is counterbalanced by larger displacements and rotations at the foundation which may become unacceptable. Furthermore, it has been noticed that the variability in the response becomes large as more demand is placed on the foundation. Therefore, care must be exercised before accepting to transfer the ductility demand from the structure to the foundation. This implies a careful definition of acceptable criteria for the foundation displacement and rotation, and a thorough investigation of the variability of the response. As demonstrated in the paper the variability is conveniently handled with incremental dynamic analyses, which can be further incorporated in a performance based design approach. Nevertheless, this concept of allowing non linearities to develop in the foundation shows some promise as already pointed out in Gazetas (2006). A final interesting finding of this study is that, as already shown in Moghaddasi Kuchaksarai et al. (2010), consideration of linear soil structure interaction may not be always as beneficial as considered in practice.

References

Anastasopoulos I, Gazetas G, Loli M, Apostolou M, Gerolymos N (2009) Soil failure can be used for seismic protection of structures. Bull Earthquake Eng. DOI: 10.1007/s10518-009-9145-2

Chatzigogos CT (2007) Comportement sismique des fondations superficielles: vers la prise en compte d'un critère de performance dans la conception. PhD thesis, Ecole Polytechnique

Chatzigogos CT, Figini R, Pecker A, Salencon J (2009a) A macroelement formulation for shallow foundations on cohesive and frictional soils. Int J Numer Anal Meth Geomech (accepted for publication)

Chatzigogos CT, Pecker A, Salençon J (2007) Seismic bearing capacity of a circular footing on a heterogeneous cohesive soil. Soils Found 47(4):783–797

Chatzigogos CT, Pecker A, Salencon J (2009b) Macroelement modeling of shallow foundations. Soil Dyn Earthquake Eng 29(6):765–781

Dafalias YF, Hermann LR (1982) Bounding surface formulation of soil plasticity. In: Pande GN, Zienkiewicz OC (eds) Soil mechanics – transient and cyclic loading. Wiley, New York, NY

EC8 (2000) Design provisions for earthquake resistance of structures, part 5: foundations, retaining structures and geotechnical aspects, EN, 1998–2005. European Committee for Standardization, Brussels

Figini R (2010) Nonlinear dynamic soil-structure interaction: application to seismic analysis and design of structures on shallow foundations. PhD thesis, Politecnico di Milano

Gajan S, Kutter BL (2008) Capacity, settlement, and energy dissipation of shallow footings subjected to rocking. J Geotech Geoenviron Eng ASCE 134(8):1129–1141

Gazetas G (1991) Foundation vibrations. In: Fang HY (ed) Foundation engineering handbook, 2nd edn. Van Reinhold Rostrand, New York, NY

Gazetas G (2006) Seismic design of foundations and soil–structure interaction. In: Proceedings of the 1st European conference on earthquake engineering and seismology, Geneva

Gazetas G, Apostolou M, Anastasopoulos I (2003) Seismic uplifting of foundations on soft soil, with examples from Adapazari (Izmit 1999, Earthquake). BGA international conference on foundation, innovation, observations, design & practice, University of Dundee, Scotland, pp 37–50

Kausel E (2009) Early history of soil–structure interaction. Soil Dyn Earthquake Eng doi: 10.1016/j.soildyn.2009.11.001

Kausel E, Roësset JM (1974) Soil structure interaction problems for nuclear containment structures. In: Proceedings of the ASCE power division conference, Boulder, CO

Martin GR, Lam IP (2000) Earthquake resistant design of foundations: retrofit of existing foundations. In: Proceedings of the GeoEngineering 2000 conference, Melbourne

Moghaddasi Kuchaksarai M, Cubrinovki M, Chase J, Pampanin S, Carr A (2010) Probabilistic evaluation of soil-foundation-structure interaction effects on seismic structural response. Earthquake Eng Str D (accepted for publication)

Paolucci R (1997) Simplified evaluation of earthquake-induced permanent displacement of shallow foundations. J Earthquake Eng 1(3):563–579

Pecker A (1998) Capacity design principles for shallow foundations in seismic areas. In: Proceedings of the 11th European Conference Earthquake Engineering, AA Balkema

Pecker A (2003) Aseismic foundation design process, lessons learned from two major projects: the Vasco de Gama and the Rion Antirion bridges. ACI International Conference Seismic Bridge Design and Retrofit, University of California, San Diego, CA

Prevost JH (1985) A simple plasticity theory for frictional cohesionless soils. Soil Dyn Earthquake Eng 4(1): 9–17

Prevost JH (2008) DYNAFLOW v02 release 08.B. Department of Civil and Environmental Engineering, Princeton, NJ

Priestley MJN, Calvi GM, Kowalski MJ (2007) Displacement-based seismic design of structures. IUSS Press Pavia, Pavia, PV

Vamvatsikos C, Cornell CA (2002) Incremental dynamic analysis. Earthquake Eng Str D 31: 491–514

Wolf JP, Deeks AJ (2004) Vibration analysis: a strength-of-materials approach. Elsevier, Oxford

Chapter 5
From Non-invasive Site Characterization to Site Amplification: Recent Advances in the Use of Ambient Vibration Measurements

P.-Y. Bard, H. Cadet, B. Endrun, M. Hobiger, F. Renalier, N. Theodulidis, M. Ohrnberger, D. Fäh, F. Sabetta, P. Teves-Costa, A.-M. Duval, C. Cornou, B. Guillier, M. Wathelet, A. Savvaidis, A. Köhler, J. Burjanek, V. Poggi, G. Gassner-Stamm, H.B. Havenith, S. Hailemikael, J. Almeida, I. Rodrigues, I. Veludo, C. Lacave, S. Thomassin, and M. Kristekova

Abstract A series of investigations has been carried out over the last decade in Europe aimed at deriving quantitative information on site amplification from non-invasive techniques, based principally on surface wave interpretations of ambient noise measurements. The present paper focuses on their key outcomes regarding three main topics. First, methodological, hardware and software developments focusing on the acquisition and the processing of both single point and array microtremor measurements, led to an efficient tool with in situ control and processing, giving rise to robust and reproducible results. A special attention has been devoted to the derivation and use of the Rayleigh wave ellipticity. Second, the reliability of these new tools has been assessed through a thorough comparison with borehole measurements for a representative – though limited – set of sites located in Southern Europe, spanning from stiff to soft, and shallow to thick. Finally, correlations between the site parameters available from such non-invasive techniques, and the actual site amplification factors as measured with standard techniques, are derived from a comprehensive analysis of the Japanese KIKNET data. This allows to propose alternative, simple site characterization providing an improved variance reduction compared with the "classical" V_{S30} classification. While these results could pave the road for the next generation of building codes, they can also be used now for regulatory site classification and microzonation studies, in view of improved mapping and estimation of site amplification factors, and for the characterization of existing strong motion sites.

P.-Y. Bard (✉)
LGIT, Maison des Geosciences, Joseph Fourier University, 38041 Grenoble Cedex 9, France
e-mail: bard@obs.ujf-grenoble.fr

M. Garevski, A. Ansal (eds.), *Earthquake Engineering in Europe*, Geotechnical, Geological, and Earthquake Engineering 17, DOI 10.1007/978-90-481-9544-2_5,

5.1 Introduction

Shear wave velocity is the most important material property controlling amplification phenomena during earthquakes, and the need for reliable, affordable site survey techniques has been often emphasized in engineering seismology. Amongst a wide variety of direct applications, one may mention the drastic lack of quantitative information on subsurface structure for most of seismic stations in the EURO-MED area, microzonation studies at the city scale (i.e., from a few to 100 km^2), and the identification of site classes as required by building codes. Such survey techniques should combine cost efficiency and physical soundness in order to provide reliable, quantitative estimates of the relevant site parameters over wide areas or numerous sites.

In that aim, the use of ambient noise recordings is indeed very appealing: its non-invasive character makes it well suited for dense urban environments, the required equipment (sensitive seismometers and data acquisition systems) is available at affordable cost, and the processing techniques have been the topic of many developments in recent years. However, the wide variety of processing techniques (from very simple to highly sophisticated), and the existence of different interpretation viewpoints (for instance on the use of H/V information) results in legitimate questions and doubts in both geotechnical and end user communities.

Given this background situation, a series of investigations has been launched over the last decade in Europe in order to explore the actual capabilities of noise-based techniques in view of deriving quantitative information on site amplification. This has been achieved mainly within the framework of two European projects: SESAME (Site Effects aSsessment from AMbient noisE, a FP5 project # EVG1-CT-2000-00026, 2001–2004, see Bard et al., 2004) and NERIES – JRA4 (NEtwork of Research Infrastructures for European Seismology, a FP6 I3 project # RII3-CT-2006-026130, 2006–2010), with complementary funding from various national projects and agencies in France, Germany, Greece, Italy, Portugal, Switzerland and Turkey. It included methodological, hardware and software developments, which led to an efficient tool combining in situ control and preliminary processing, with robust and reproducible results. It also included a comprehensive data analysis in order to derive statistically meaningful correlations between site amplification characteristics and the site parameters that can reliably be derived from such non-invasive, noise-based techniques. The following sections briefly summarize the main outcomes of this work, addressing successively the software and hardware developments, a careful comparison with results of borehole soundings for a representative series of sites, and the derivation of correlations between site amplification factors and site parameters.

5.2 Array Measurements and Processing of Ambient Vibrations

The base idea schematically illustrated in Fig. 5.1 is to deploy temporary, small aperture (typically from a few meters up to kilometric scale), 3-component, high sensitivity seismological arrays, to record the ambient vibrations, to extract the

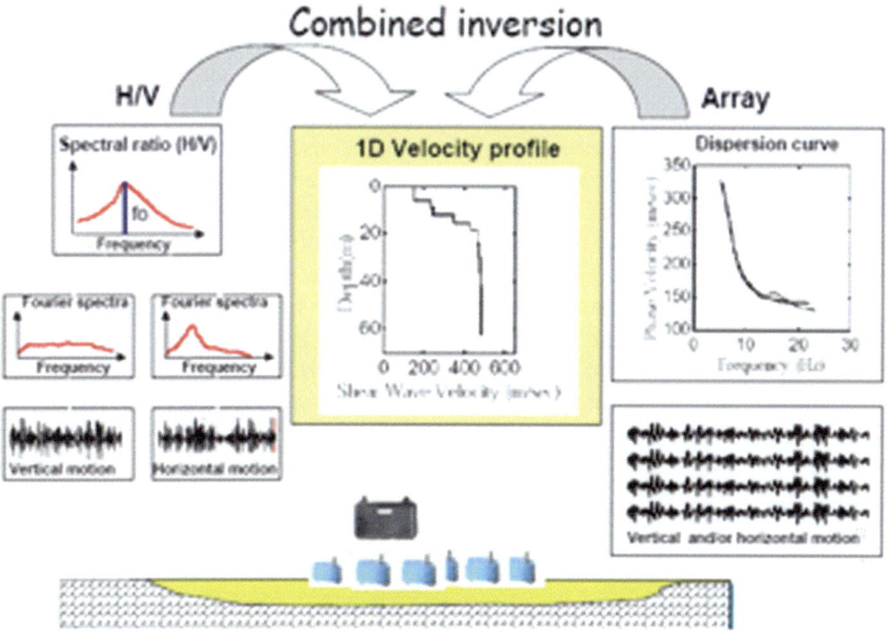

Fig. 5.1 Principle of H/V and array processing

dispersion characteristics of Rayleigh – and possibly Love – waves, from which to derive either detailed velocity profiles or average velocity values. These ambient vibration measurements may also be complemented with active measurements (MASW[1] type, cf. Park et al., 1999) allowing a better resolution of very shallow layers, while the array processing is also usefully enlightened by the classical H/V analysis. Recent developments addressed improvements in both hardware and software tools to help the field and processing work, and methodological developments as well to investigate new, complementary processing techniques.

5.2.1 Hardware

The target is to perform wireless, synchronous recording of microtremors on 10 m to 1–2 km wide arrays within urban environments, with real time array processing for in situ control. About a decade ago, there did not exist any equipment meeting these requirements. A series of hardware developments and tests were thus carried out, first by the Potsdam University group (F. Scherbaum, M. Ohrnberger, D. Vollmer), and later in Grenoble (M. Wathelet), ITSAK (A. Savvaidis, N. Theodulidis, H. Cadet), and SED-ETHZ (D. Fäh, J. Revilla, S. Marano, V. Poggi). In the early

[1] Multi-channel analysis of surface waves.

phase of the project (2006–2008), the developed hardware (thanks to a national, complementary funding at U. Potsdam), was fully dedicated to array measurements of ambient vibrations. All measurements at all European sites (about 25 in total, see the next section) were performed with this instrumentation. Later however, acknowledging the fact that the total cost of this specific tool could look prohibitive, new reflections were initiated to design an alternative "add-on" system that could be implemented on existing mobile seismological stations at a much lower cost, and provide the same field efficiency and user-friendliness without altering reliability and robustness. Prototypes have been developed at LGIT and UP; their cost is about 1,000 Euros/station. It includes precise, real time GPS positioning, wireless automatic meshing and data transmission to a central unit, and it is flexible enough to fit different acquisition systems. However, the technology is evolving extremely rapidly, and manufacturers of seismic stations are now proposing material – or announce it for very soon – that meets some – not always all – of the above requirements; the proposed costs remain nevertheless significantly higher than plain seismic stations.

5.2.2 Software

The target was to develop and document reliable software tools to derive shear wave velocities from non-invasive, surface measurements (microtremor array recordings as well as active MASW recordings). This requires (a) to extract the dispersion characteristics (DC) of Rayleigh and Love waves, and (b) to derive either detailed velocity profiles or average velocity values from DC curves or SPAC (Spatial Autocorrelation) processing.

Retrieving reliable information from complex ambient vibrations is indeed a non-trivial issue, which is not satisfactorily addressed by most of the black-box software packages already available on the market. Their main limitations are basically two-fold:

- the use of one single, specific array processing technique to derive the dispersion curves DC (ex.: FK only, or SPAC only) does not allow cross-checking which is always useful in ambiguous cases;
- the inversion part (deriving velocity profiles from DC) generally neglects, or at best only poorly addresses, the non-uniqueness of solutions. This is very often witnessed by the apparent high-resolution of the resulting profiles, with rather thin layers often including one or several velocity inversions at depth: this is an easy, but highly non unique, way to reach an excellent fit with measured DC, which however proves most often to be fictitious.

As a consequence, a specific, multiplatform software tool, named "geopsy", was developed, which has now reached a satisfactory maturity level, and is freely available on line (http://www.geopsy.org). Its development first emerged as a side product of the SESAME project especially between LGIT and University of Potsdam. The initial objective of this joint effort has been to centralize in one unique framework all state-of-the-art techniques for processing ambient vibrations and to provide the tools for their necessary integration. Very rapidly however, though built around ambient vibrations, its design was extended to cover most of the non-invasive methods used in site characterization: for instance, refraction and active surface wave experiments. With the NERIES project, geopsy has evolved a lot, including a number of new modules developed with a graphical user interface, and also accepting real-time feeding with data streams for *in situ* checks. The array processing modules include standard and high resolution frequency-wavenumber analysis ("FK"/"HRFK": Capon, 1969; Lacoss et al., 1969), spatial autocorrelation analysis ("SPAC": Aki, 1957; Bettig et al., 2001; Köhler et al., 2007; Ohrnberger et al., 2004, 2005), active body and surface wave experiments (reflection, refraction, MASW, see Renalier, 2010). All techniques may be applied to 3-component recordings, therefore addressing Love waces as well as Rayleigh waves. The inversion module is based on the neighborhood algorithm proposed by Sambridge (1999), with various adaptations and improvements as detailed in Wathelet et al. (2004, 2005, 2008) and Wathelet (2008), Di Giulio et al. (2006); a special attention has been devoted to the non-uniqueness of solutions and to the sensitivity to the initial model parameterization, which led to various recommendations : combining inversions using different types of information, using a-priori knowledge whenever available, visualizing the uncertainties on velocity profiles, balancing the model complexity with the gain in misfit reduction through the Akaike information criteria (Akaike, 1974; Savvaidis, 2009). A more detailed description of the above-mentioned developments and improvements can be found in the referenced papers, with a global synthesis in the deliverable D9 of the NERIES-JRA4 project (Fäh et al., 2010). An open source model has been definitely adopted for the distribution of these codes, which lets all doors open for further developments and improvements. Open source and free accessibility offer a quick distribution to a wide community world-wide which in turn accelerates the debug and stabilization processes (variety of environments and user opinions).

The NERIES project thus allowed to transform the geopsy package from a small software distributed within a limited group of highly specialized research and industrial individuals, into a reference software distributed all around the globe in a wide range of scientific and engineering communities. Substantial efforts were made to include more processing techniques and to integrate them in a comprehensive package.

In parallel, considering the complexity of this versatile software, extended training 1-week long seminars are organized around the world to teach ambient vibration fundamentals and explain how to use geopsy in this context. The corresponding course material is presently being used as a basis for an in-depth documentation to be distributed with the software, and is completed by an on-line wiki – type documentation.

5.2.3 Derivation and Inversion of Rayleigh Wave Ellipticity

The derivation of dispersion curves with these array techniques needs however a large number of seismic sensors and is somewhat time-consuming (e.g., ½–1 day of field work per site). It is therefore tempting to search for simpler alternatives. The ellipticity of Rayleigh waves, i.e. the ratio between the horizontal and the vertical movement, strongly depends on the local soil structure (e.g. Fäh et al., 2001). As a result, it can be inverted to retrieve the underground structure, i.e., the shear wave velocity profile and sediment thickness. Special attention was thus devoted to attempts to extract Rayleigh wave ellipticity from single point or array, 3-component measurements, and to its direct inversion in terms of velocity profile.

Two methods were proposed and tested during the NERIES project for retrieving ellipticity from single-station measurements. The first method was initiated during the SESAME project and is based on time-frequency analysis with continuous wavelet transform. It reduces the SH-wave influence by identifying P-SV-wavelets along the signal and computing the spectral ratio from these wavelets only. The second method is the so-called "RAYDEC" technique (Hobiger et al., 2009a; Fig. 5.2), which is adapted from the random decrement technique commonly used to characterize dynamic parameters of buildings, and is indeed tightly connected with the autocorrelation analysis (see Asmussen, 1997 for a comprehensive review). It is basically looking for the optimal cross-correlation between the vertical motion and one direction of horizontal component, with due consideration for the fact that for Rayleigh waves, vertical and horizontal components exhibit a 90° phase shift.

Tests on synthetic noise with both single station methods proved very encouraging, with resulting ellipticity estimates much closer to the theoretical ones than the raw H/V curves. Reliable results were obtained for the right flank of the H/V curve, between the first peak at the fundamental frequency of resonance and the first trough at higher frequency. The procedures eliminate efficiently most of the Love and body wave contributions.

As ellipticity alone cannot fully constrain the velocity profile, it has to be coupled with some scaling measurements, for instance MASW or small aperture SPAC, which allow to estimate the very shallow velocity. A few test applications on real data sets also proved very encouraging when compared with "classical" array analysis (Hobiger et al., 2010).

For recordings of ambient vibrations on large arrays, there are two ways to retrieve ellipticity information. Such techniques were developed within the NERIES project. The first strategy curves considers a reduction factor to be applied on the raw H/V ratio, so as to eliminate the contribution of Love waves on the H component (see Bonnefoy-Claudet et al., 2008). This ratio is related to the Rayleigh/Love wave ratio that can be derived from three-component SPAC analysis (Köhler et al., 2007) as a function of frequency. A combination with the H/V curve computed by the classical method (simple spectral ratios) should then produce a good estimate of the Rayleigh wave ellipticity. The second strategy proposed by Poggi and Fäh (2010) is using high-resolution frequency-wavenumber array analysis. The technique is applied to the three components of motion and is based on the assumption that amplitude

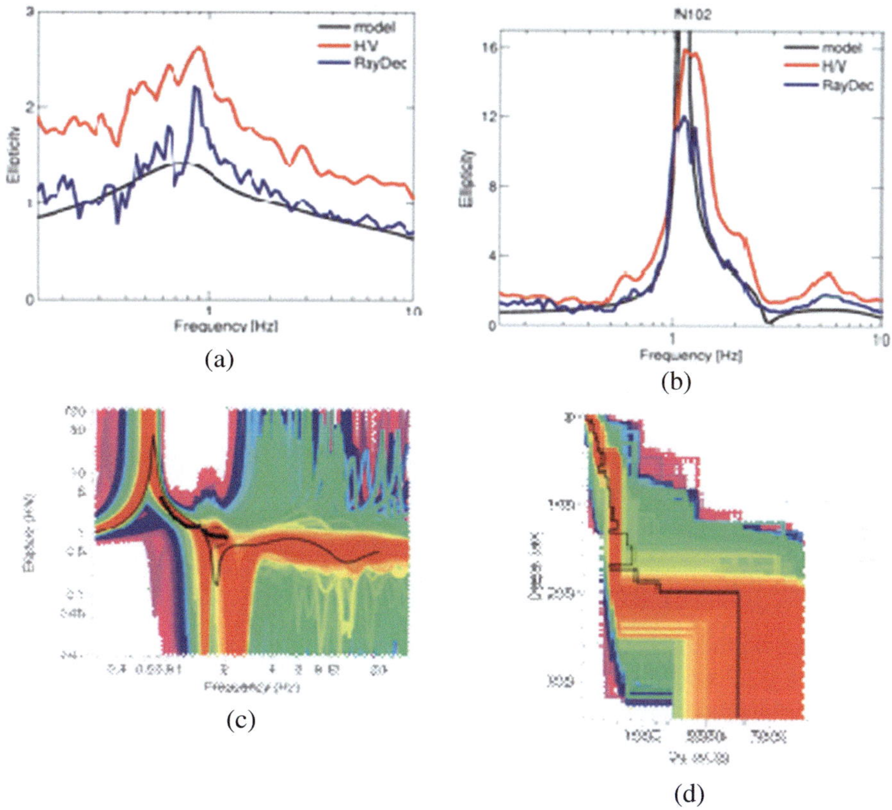

Fig. 5.2 Example use of ellipticity for velocity profile inversion. The *top* two plots (Hobiger et al., 2009b) display, for two cases with a moderate (*left*, **a**) and large (*right*, **b**) impedance contrast the comparison between the actual Rayleigh wave ellipticity (*black curve*), the H/V curve derived from a standard processing (*red*) and the estimated ellipticity with the RAYDEC method (*blue*). The *bottom* plots (Hobiger et al., 2010) display an example inversion of ellipticity using additional information to constrain the shallow velocity: the continuous broad band *black curve* represents the theoretical ellipticity (**c**, *left*) and the actual velocity profile (**d**, *right*), the limited band *black curve* with *vertical bars* represent the estimated ellipticity and its uncertainty used for the inversion, and the *colored curves* display the ellipticity (*left*) for many inverted velocity profiles (*right*); the *red color* corresponds to the lower misfit, while other *colors* correspond to increasing misfit values, from *yellow* to *magenta* through *green* and *blue*

maxima in the f-k cross-spectrum must represent the true power amplitude of the corresponding signal. In the case of Rayleigh waves, the ratio between maxima obtained from the horizontal (radial-polarized) and vertical components of motion will thus also represent the frequency-dependent ellipticity function. Consequently, if the Rayleigh dispersion curves of the different modes can be identified on the f-k plane, then the corresponding modal ellipticity patterns can also be separated and extracted. This second method also offers the possibility of estimating the

Rayleigh/Love ratio. Testing all these single-station and array methods in real cases is part of on-going research, and not yet implemented in the geopsy software.

5.3 Testing of Ambient Vibration Array Techniques

The present state of practice in geotechnical engineering considers borehole techniques (i.e., Cross-hole – "CH"–, Down-hole – "DH" –, and sonic logging), as the "ground truth", i.e. the most widely accepted survey techniques. Even though the results of such borehole investigations do include some non-negligible uncertainties related to both the measurement and the processing/interpretation steps, any new technique needs to be validated through a comparison with the well-established practice. Therefore, the careful testing and comparison with standard borehole techniques was considered a key issue for an objective assessment of the reliability of these non-invasive techniques and tools.

5.3.1 Technical and scientific considerations

A first step was achieved in 2006 with the organization of a blind test (ESG2006, Cornou et al., 2009) about the retrieval of velocity profiles from array recordings. The main learnings have been the very good consistency of all derived dispersion curves (they agreed within ±10% in most cases), contrasting with the much larger variability of the inverted velocity profiles, in relation with (a) the difficulties of proper mode identification and (b) the very heterogeneous quality of inversion algorithms.

The second step was carried out within the NERIES project. A set of about 20 representative sites was selected in Italy, Greece and Turkey, spanning from stiff to soft, and thick to shallow, for which prior borehole velocity measurements – either CH or DH, or sometimes both – were available (Fig. 5.3a). Ambient vibration (AMV) array measurements were performed together with active seismics (refraction and MASW) at each site. The specific scientific targets were:

Fig. 5.3 (continued) Comparison between non-invasive techniques and borehole measurements. *Top* **a** location of measured sites. *Bottom left* **b** ratio between the borehole velocity profiles and admissible inverted profiles, for all sites; *Bottom right* **c** comparison of V_{S30} values from borehole (abscissa) and the range of values derived from non-invasive techniques (ordinate). *Top* **d** Range of dispersion curves obtained with different non-invasive techniques (FK, SPAC, MASW) as mapped in the (velocity/wavelength) plane. The *red* (resp., *green*) *horizontal lines* correspond to maximum (resp., minimum) wavelengths; *Bottom left* **e** comparison between the V_{S30} ranges derived from admissible inverted profiles and the range of phase velocities $V_{\lfloor 30}$ corresponding to a wavelength of 30 m (*upper dotted black line* on Fig. 5.3d); *Bottom right* **f** similar comparison between $V_{\lfloor 30}$ and V_{S30} derived from borehole measurements

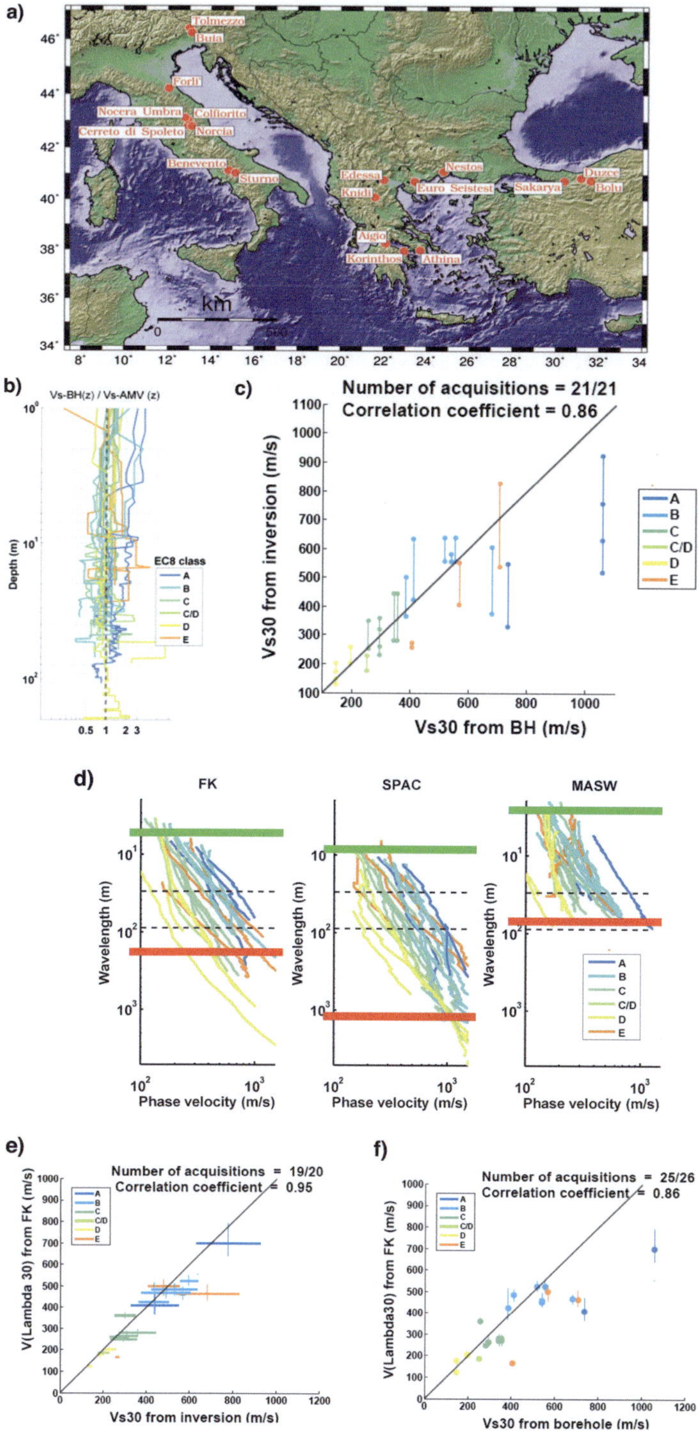

Fig. 5.3 (continued)

- to assess the validity of the AMV technique by comparing it to borehole measurements and to Multichannel Analysis of Surface Waves (MASW);
- to evaluate the typical wavelength ranges derived by the different methods;
- to evaluate the necessity and sensitivity of the inversion for the evaluation of V_{Sz} (time-averaged velocity down to depth z).

In order to stay affordable and feasible, active seismic experiments involved 24–4.5 Hz geophones (both horizontal and vertical) for recording the signals generated with hammer and plate (the use of explosives or Vibroseis was deliberately avoided because of the practical and logistic difficulties in urban environments), with 24–115 m long lines. They were analyzed with the MASW technique to compute the dispersion curves. Passive seismic was acquired with 8 stations linked with wireless connections and monitored with near real-time processing software allowing the on site adaptation of the acquisition. Dispersion curves were computed both with the frequency wave-number (FK) and with the Spatial AutoCorrelation (SPAC) techniques for Rayleigh and Love waves. They were inverted with the neighborhood algorithm as implemented in the geopsy software. Measured dispersion curves and admissible inverted Vs profiles, together with V_{S30} values, were finally compared to results of borehole tests available at all Italian and Greek sites, and to previous MASW results at Turkish sites.

The answers to the targeted questions are summarized below:

- The comparison proved good for all sites with V_{S30} lower than 500 m/s; at stiffer sites, velocity values estimated with surface wave techniques (both passive and active) are smaller than those derived from boreholes measurements (Fig. 5.3b, c). This trend is consistent with the previous comparison results reported by Moss (2008). Incidentally, one outcome of the non-invasive techniques with the geopsy processing is to provide an estimate of the measurement uncertainties, indicated by the error bars in Fig. 5.3c.
- Minimum and maximum wavelengths are in average around 10 and 1,000 m for the array measurements, and are around 6 and 45 m for MASW (Fig. 5.3d). With the same array geometry, SPAC processing generally allows to reach larger depth than FK processing, which in turn allows a better resolution of shallow wavelengths. The corresponding penetration depths, corresponding to about one-third to one half of the maximum wavelength, are typically in the range 10–30 m for MASW, while they exceed 100 m for most of AMV cases, and often 200 m. This is to be compared to the borehole depths, typically of a few tens of meters, with a cost significantly increasing with depth. The analysis of the high frequency part of DC, corresponding to shallow velocities, showed that the AMV Rayleigh wave results were good at high frequencies, especially from FK techniques. Even though Love waves estimated from AMV and MASW covered complementary frequency ranges, including the MASW Love wave dispersion curve did not improve much the inversion results because of the good performance of FK processing at high frequency.
- Considering the fact that inversion step is the most tricky one, it is useful to look for ways to skip it, at least for a site classification purpose. The starting point is

again to map the dispersion curves in the (Rayleigh wave velocity/wavelength) plane displayed in Fig. 5.3d. One may directly compare the measured Rayleigh wave velocity corresponding to a wavelength of 30 m ($V\lambda30$), and the VS30 values derived either from inverted velocity profiles (Fig. 5.3e), or from borehole measurements (Fig. 5.3f). The correlation between $V\lambda30$ and VS30 derived from inversions proved rather good; indeed, when considering larger data sets, the best correlation with inverted VS30 is observed when considering $V\lambda40$ or $V\lambda45$ (Cornou, personal communication; Zor et al., 2010): this indicates that the inversion step, which is the most subjective, is not needed to derive VS30 values. The correlation between $V\lambda30$ and borehole estimates of VS30 thus exhibits the same characteristics as discussed above and displayed on Fig. 5.3c, i.e., a good agreement for soft and intermediate sites (VS30 < 500 m/s), and an underestimation trend for stiff sites (VS30 > 600 m/s). However, given the limited size of the site sample considered here, these trends should be considered only as indicative and should be checked with further studies.
- Finally, another valuable outcome of this series of measurements concerns the robustness of the results. As detailed in Endrun et al. (2009), array microtremor measurements performed on the same sites at different periods (day, night, different years and seasons) by different teams with different instruments, did yield the same dispersion characteristics. In addition, even though a wide variety of individual velocity profiles are compatible with these dispersion curves, the estimates of average parameters such as V_{S30} also exhibit a very satisfactory robustness whatever the implicit or explicit assumptions considered in the inversion step.

The difference for stiff sites is somewhat intriguing, and can have several origins:

- the first one is the frequency range of the measurements: borehole techniques provide S-wave velocities for high-frequency/short wavelength waves (typically 500 Hz to 1 kHz for cross-hole techniques, and 100–300 Hz for down-hole techniques), while non-invasive techniques operate in the engineering seismology frequency range, i.e. typically 0.5–20 Hz. The “effective propagation medium” may therefore greatly differ from one technique to the other, since low frequency techniques sample larger wavelengths and may be affected by intermediate-size heterogeneities (fractures, joints, . . .) which are not affecting short distance travel times. Examples of such differences are mentioned in Havenith et al. (2002). In such a case, wave velocities identified from non-invasive techniques should be more representative of the actual dynamic behavior during earthquakes, because of the more appropriate frequency range of the measurements.
- the second one – which is linked to the first one – is the volume sampled by each technique: borehole techniques represent essentially point measurements, while non-invasive, surface measurements represent average velocities over tens to hundreds of meters. While the spatial and depth resolution is without any doubt much finer for borehole techniques than for surface-wave techniques, the averaging effects of the latter provide a very complementary image of the subsoil, at a much more affordable cost than the multiplication of borehole measurements.

5.3.2 Cost Considerations

Another important topic for comparison between invasive (borehole) and non-invasive surface wave techniques is the cost.

The needed equipment for ambient vibration measurements now amounts to about 60–80 k Euros for a complete array system consisting of 8–10 sensors, an acquisition system, wireless connexions and a precise positioning system. At least half of this cost corresponds to intermediate to broad band, sensitive sensors. There is presently no fully suited material available from the manufacturers; would this type of measurements become a routine engineering practice, one may anticipate a significant cost decrease. This is slightly to significantly more expensive than a MASW 24–48 sensor equipment, but it allows to reach much larger depths (see Fig. 5.3d). This amount should be compared with the equipment cost required by borehole techniques, consisting in the drilling device (generally installed on a truck), and the borehole tool (including the processing software).

The most important component is the marginal cost of measurements, which is mainly consisting in work days. Borehole techniques typically require 1 day of work for 2 persons simply for drilling down to 30 m, which thus results in 4–6 work-days for the cross-hole technique (depending on whether 2 or 3 close boreholes are used), and 2 for the down-hole one, followed by another 3 work days for the measurements and routine processing. Non-invasive techniques, especially ambient vibration array techniques, require slightly more time for the measurement and processing (about 4 work days), but do not need any preparatory work. As a consequence, even though the initial equipment cost is still more expensive than borehole equipment, the measurement cost is significantly lower, especially when compared with cross-hole techniques.

5.4 Usefulness for Routine Applications: Derivation of Noise-Compatible Site Amplification Prediction Equations (SAPE)

Over the last decade, the site classifications used in seismic regulations have been increasingly based on the use of the V_{S30} parameter, following the works of Borcherdt (1994) and colleagues in the early nineties. However, many seismologists and engineers (e.g., Mucciarelli and Gallipoli, 2006; Castellaro et al., 2008) have expressed some reluctancy since this single parameter does not capture the physics of 1D site amplification, even in the simple 1D case: the amplification characteristics should indeed be related both to the impedance contrast between the shallow soil and the underlying bedrock (and also to the damping characteristics), and to the thickness of the surface layers. As a consequence, the single parameter V_{S30} can only be considered as a proxy to such more physical parameters, and its correlation to the actual amplification characteristics should therefore be at least adjusted regionally to correspond to the local geology. This adaptation work is nevertheless only rarely performed, mainly because of the lack of reliable data (absence of strong motion recordings, or missing geotechnical information on recording sites).

The simplicity of this site classification, its satisfactory performance on the original available data, together with the relative low cost of the background site survey (SPT down to 100 ft/30 m which could be performed within 1 day), made it very popular and led to its spreading in many earthquake regulations throughout the world, since no alternative could be proposed combining cost effectiveness, simplicity, and physical relevance. This challenge is addressed here as a continuation of the previous developments on noise-based site surveys, by investigating the correlations between some alternative, twin-parameter site categorization that may be derived from non-invasive techniques, and the site amplification factors on high quality data.

The work briefly summarized here is described in more detail in Cadet (2007) and Cadet et al. (2008, 2010a, b, c): it involved an extensive analysis of a subset of the Japanese KIKNET data consisting of about 4,000 3-component recordings from a total of 375 sites. Only events with a moment magnitude (Mw) higher than 4.0 and a depth less than 25 km were considered. The range of hypocentral distances for the selected records is 0.5–343 km and the range of magnitudes (Mw) is 4–7.3. The range of recorded peak ground acceleration PGAs is 0.4–927 cm/s^2. The records were band-passed filtered between 0.25 and 25 Hz (Pousse, 2005).

The investigated site parameterization is based on the time-averaged shear wave velocity over the top z meters, V_{Sz}, and the site fundamental frequency f_0. V_{Sz} parameters were derived for the KIKNET sites from the measured velocity profile (down-hole technique) for four different depths (z = 5, 10, 20 and 30 m), while the fundamental frequency was obtained from surface to down-hole spectral ratios, and checked for consistency both with the theoretical 1D transfer function based on the down-hole velocity profile, and the surface H/V ratios. As displayed on Fig. 5.4, these two parameters are shown to be complementary and to provide independent information on the overall impedance contrast or shallow soil softness

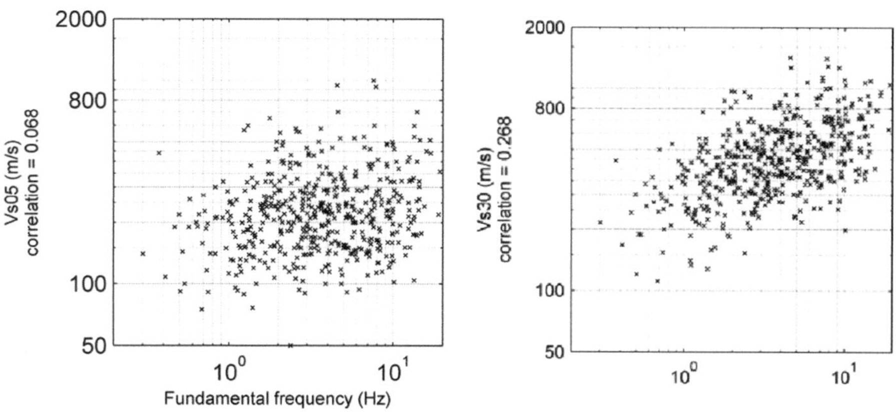

Fig. 5.4 Distribution of the considered KIKNET site set in the (f_0, V_{S05}) and (f_0, V_{S30}) planes, (*left* and *right*, respectively)

(V_{Sz}), and the overall thickness of the surface layers responsible for the amplification (f_0). Most importantly, both parameters may be derived in a robust and inexpensive way from single point ambient noise measurements (H/V processing, Haghshenas et al., 2008), and array microtremor processing or even SASW/MASW techniques for very shallow V_{Sz}, i.e., V_{S05}, V_{S10} and sometimes V_{S20}.

The site amplification factors were derived empirically from the average surface/downhole ratios between response spectra (BHRSR): considering the wide scatter in the S-wave velocities and depths of down-hole sites (300–3,300 m/s, 8–900 m), a correcting procedure was established with two main goals:

- to normalize the raw BHRSR ($BHRSR_{raw}$ in Fig. 5.5b) to a standard reference corresponding to the "generic rock profile" proposed by Boore and Joyner (1997) with $V_{S30} = 800$ m/s,
- and to remove high frequency amplification artefacts associated with the location of reference sites at depth.

More details can be found in Cadet et al. (2010b) on this impedance and depth correction procedure. As displayed in Fig. 5.5b, the so corrected $BHRSR_{cn}$ values exhibit a significantly reduced scatter compared to the original amplification factors $BHRSR_{raw}$.

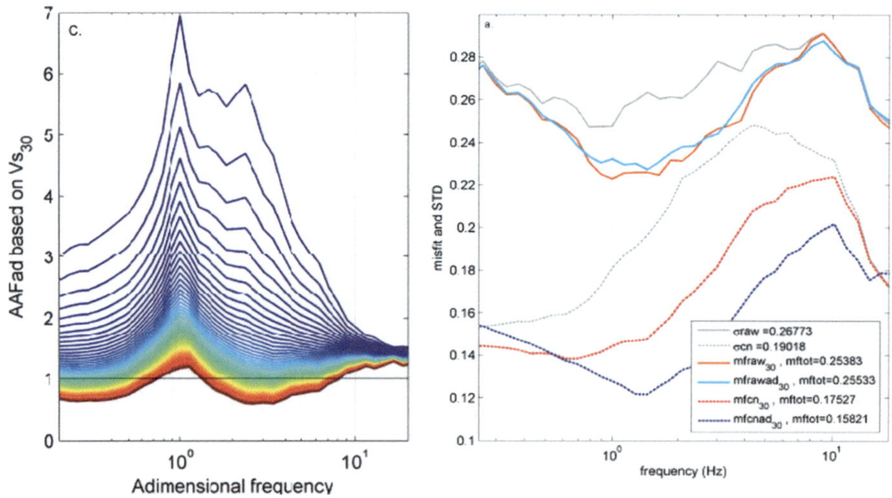

Fig. 5.5 Example results on new site amplification prediction equations ("SAPE", adapted from Cadet et al., 2010c) derived from KIKNET data. *Left*: dependence of the amplification function on the dimensionless frequency and V_{S30} (color code on *left side*, related to V_{S30} value). *Right*: Comparison in the real frequency space between standard deviation of the initial family $BHSR_{raw}$ (*gray*), standard deviation of the corrected $BHSR_{cn}$ family (*dot gray*, with correction for depth and impedance). The misfit obtained by correlating raw (*solid line*) and corrected (*dotted line*) BHSR to V_{S30} only and to the couple (V_{S30}, f_0) are shown in *blue* and *orange*, respectively)

The final step consisted in establishing correlations between these corrected amplification factors and the site parameters $V_{Sz,i}$ and $f_{0,i}$ for all sites i. This correlation has been performed in two steps (Cadet et al., 2010c):

1. the corrected amplification functions were first expressed as a function of dimensionless frequency $\nu = f/f_{0i}$. The underlying idea is that, when $f < f_{0i}$, the amplification should remain small, while it should be significantly larger around f_{0i}, and more scattered for $f > f_{0i}$. It results in new, "shifted" amplification functions $A_i(\nu)$, which exhibit in general a maximum around $\nu = 1$.
2. the second step was to correlate, for each discrete value of the imensionless frequency ν_k, the corresponding amplifications $A_i(\nu_k)$ with the site velocity $V_{Sz,i.}$ The rationale behind this correlation is simply that the lower V_{Sz}, the larger should be the amplification at the fundamental frequency. This is done by a least-square fitting of the following, NGA-like functional form

$$\log(A_i(\nu_k)) = a_k + b_k \log(V_{Sz,i})$$

or, in other terms,

$$A_i(\nu_k) = (V_{ref,k}/V_{Sz,i})^{\alpha_k}$$

Such a procedure has been performed for each of the four parameters V_{Sz}, with $z = 5$, 10, 20 and 30 m, and for both the original $BHRSR_{raw}$ values, and the depth-impedance corrected $BHRSR_{cn}$ ratios. A similar correlation has been looked for also with the fundamental frequency, having in mind that f_0 might be a proxy to the soil softness in a way similar to V_{S30}. Five different such "SAPE" (Site Amplification Prediction Equations) based on (V_{S5}, f_0), (V_{S10}, f_0), (V_{S20}, f_0), (V_{S30}, f_0) or f_0 alone, were obtained, an example of which is illustrated in Fig. 5.5 for the couple (V_{S30}, f_0).

The quality of such correlations is quantified through the resulting "misfit" between the actually measured amplification factors and the predicted ones. As displayed in Fig. 5.5b and Table 5.1, the main variance reduction is coming (a) from the depth-impedance correction and (b) from the transformation to dimensionless frequency. Once these steps are carried out, the best explanations of the amplitude variations are associated with the parameter couple (V_{S30}, f_0); however, very shallow velocities such as V_{S05} and V_{S10} also provide a non-negligible variance reduction. It is worth also to notice that, amongst the single parameter correlations, the best variance reduction is not obtained with the routinely used V_{S30} parameter, but with the f_0 parameter: the fundamental frequency thus appears once more as the key parameter, and should be preferred to an impedance index. Beyond their possible use and/or further testing in the derivation of ground motion prediction equations, these results could prove very valuable and easy to use for the next generation of building codes.

Table 5.1 Standard deviation and misfits resulting from the correlation between amplification functions and various site parameters

Parameters	Non corrected surface-downhole response spectra ratios: $BHRSR_{raw}$	Depth-impedance corrected surface-downhole response spectra ratios: $BHRSR_{cn}$
Initial standard deviation	0.268	0.202
V_{S30} only	0.255	0.174
V_{S20} only	0.257	0.177
V_{S10} only	0.260	0.184
V_{S05} only	0.264	0.190
f_0 only	0.254	0.159
(V_{S30}, f_0)	0.255	0.158
(V_{S20}, f_0)	0.255	0.159
(V_{S10}, f_0)	0.254	0.164
(V_{S5}, f_0)	0.255	0.168

The standard deviation and misfits are computed from the $\log_{10}$ values of observed and computed amplifications, and averaged over the whole frequency range from 0.25 to 20 Hz. For more details see Cadet et al. (2010c).

5.5 Conclusions

The outcomes of these series of investigations can be summarized briefly as follows:

- Non-invasive, surface wave methods do provide a reliable, lower cost alternative to existing and widely accepted borehole techniques. This is especially true for soft and intermediate stiffness sedimentary sites, which bear a particular importance since they favor higher amplifications.
- In particular, in view of simple site characterization, surface-wave techniques do provide reliable estimates of the time-averaged velocities V_{Sz}. In that aim, it has been shown that the inversion step may not be mandatory, and that EC8-type site classes could be derived with an acceptable accuracy directly from dispersion curves in the (velocity/wavelength) plane.
- However, when the target is the velocity profile $V_S(z)$ in view of forward computations of site amplification, surface wave-techniques can provide only smoothed estimates of $V_S(z)$, as they definitely cannot resolve thin layers. They nevertheless offer the non-negligible capacity to investigate large depth (up to several hundred meters) for very thick sedimentary deposits, and could then be viewed as a useful complement to (shallow) borehole measurements.
- When combined with the analysis of comprehensive, high quality strong motion data sets, these techniques pave the way for a simple two-parameter site classification, that performs significantly better than the classical one based on V_{S30} in terms of prediction of site amplification. The main improvements are that it relies on parameters which are easily available with simple, non-invasive, passive or active survey techniques, and that these parameters provide a more satisfactory link to the physics of site amplification, at east in the 1D case.

The initial goal of the NERIES research activity was the development of a reliable, low cost characterization of strong motion sites in Europe. It turns out however that these results can be extended to site characterization required by the majority of building codes in relation with the seismic design of constructions, and an improved estimation of the associated site amplification factors, with special emphasis on microzonation studies.

However, it must also be clearly emphasized that such *low-cost* tools should not be associated with *low-expertise* analysis. On the contrary, the acquisition, processing and interpretation of ambient vibration measurements should not yet be viewed as a routine elementary practice, and do require a rather high level of expertise. One of the key goals of the geopsy software tools, the associated on-line documentation and training courses, and all SESAME and NERIES reports, has definitely been to help the user in building his own expertise, and sharing his experience with a broad community. We hope the availability of open-source, fully understood software tools will progressively contribute to disseminate and generalize the use of non-invasive techniques, as a cost-effective complement and/or sometimes a substitute to the well-established borehole techniques.

Acknowledgments The developments reported here were made possible through several European grants (SESAME # EVG1-CT-2000-00026, NERIES (NEtwork of Research Infrastructures for European Seismology, # RII3-CT-2006-026130, and the ITSAK-GR, Transfer of Knowledge Marie-Curie action, # MTKD-CT-2005-029627), complemented with several national research grants in France (ANR QSHA), Germany, Greece and Switzerland. Seismograms and geotechnical information used for the derivation of "SAPE" were collected from the Japanese KiK-net network (http://www.kik.bosai.go.jp); thanks are due to the KiK-net network staff and to F. Bonilla and G. Pousse for providing the ready-to-use data.

References

Akaike H (1974) A new look at the statistical model identification. IEEE Trans Automat Contr 19:716–723

Aki K (1957) Space and time spectra of stationary stochastic waves, with special reference to microtremors. Bull Earth Res Inst Tokyo Univ 25:415–457

Asmussen JC (1997) Modal analysis based on the random decrement technique – application to civil engineering structures. PhD thesis, University of Aalborg, Denmark, 227p

Bard P-Y, SESAME Participants (2004) The SESAME project: an overview and main results. In: Proceedings of the 13th world conference in earthquake engineering, Vancouver, BC, Aug 2004, Paper No 2207

Bettig B, Bard P-Y, Scherbaum F, Riepl J, Cotton F, Cornou C, Hatzfeld D (2001) Analysis of dense array noise measurements using the modified spatial auto-correlation method (SPAC). Application to the Grenoble area. Boll Geof Teor Appl 42:281–304

Bonnefoy-Claudet S, Kohler A, Cornou C, Wathelet M, Bard PY (2008) Effects of love waves on microtremor H/V ratio. Bull Seism Soc Am 98(1):288–300

Boore DM, Joyner WB (1997) Site amplifications for generic rock sites. Bull Seism Soc Am 74(5):2035–2039

Borcherdt RD (1994) Estimates of site-dependent response spectra for design (methodology and justification). Earthquake Spectra 10:617–653

Cadet H (2007) Utilisation combinée des méthodes basées sur le bruit de fond dans le cadre du microzonage sismique. Ph.D. thesis, Joseph Fourier University, 31 Oct 2007 (301p, in French)

Cadet H, Bard P-Y, Duval A-M (2008) A new proposal for site classification based on ambient vibration measurements and the kiknet strong motion data set. In: Proceedings of the 14th world conference on earthquake engineering, Beijing (China), Oct 2008, 8p, Paper No 03-01-0036

Cadet H, Bard P-Y, Rodriguez-Marek A (2010a) Defining a standard rock. Propositions based on the KiK-net data. Bull Seism Soc Am 100(1):172–195, Feb 2010. doi: 10.1785/0120090078

Cadet H, Bard P-Y, Rodriguez-Marek A (2010b) Site effect assessment using KiK-net data – Part 1 – Normalizing site over down-hole reference spectral ratios: a proposal for correction procedures for depth and impedance effects. Bull Earthquake Eng (submitted)

Cadet H, Bard P-Y, Duval AM, Bertrand E (2010c) Site effect assessment using KiK-net data – Part 2 – site amplification prediction equation SAPE based on f_0 and Vsz. Bull Earthquake Eng (submitted)

Capon J (1969) High-resolution frequency – wavenumber spectrum analysis. Proc IEEE 57(8):1408–1418

Castellaro S, Mulargia F, Rossi PL (2008) Vs30: proxy for seismic amplification? Seism Res Lett 79(4):540–543. doi: 10.1785/gssrl.79.4.540

Cornou C, Ohrnberger M, Boore D, Kudo K, Bard P-Y (2009) Derivation of structural models from ambient vibration array recordings: results from an international blind test. ESG2006 2:1127–1219

Di Giulio G, Cornou C, Ohrnberger M, Wathelet M, Rovelli A (2006) Deriving wavefield characteristics and shear-velocity profiles from two-dimensional small-aperture arrays analysis of ambient vibrations in a small-size alluvial basin, Colfiorito, Italy. Bull Seism Soc Am 96(5):1915–1933

Endrun B, Ohrnberger M, Savvaidis A (2009) On the repeatability and consistency of three-component ambient vibration array measurements. Bull Earthquake Eng 8(3):535–570. doi: 10.1007/s10518-009-9159-9

Fäh D, Kind F, Giardini D (2001) A theoretical investigation of average H/V ratios. Geophys J Int 145:535–549.

Fäh D, Poggi V, Marano S, Michel C, Burjanek J, Bard P-Y, Cornou C, Wathelet M, Renalier F, Hobiger M, Cadet H, Ohrnberger M, Endrun B, Savvaidis A, Theodulidis N, Kristekova M, Hailemikael S, Sabetta F et al (2010) Guidelines for the implementation of ambient vibration array techniques: measurement, processing and interpretation. Neries deliverable JRA4-D9, http://www.neries-eu.org

Haghshenas E, Bard P-Y, Theodulidis N, SESAME WP04 Team (2008) Empirical evaluation of microtremor H/V spectral ratio. Bull Earthquake Eng 6:75–108. doi: 10.1007/s10518-007-9058-x

Havenith HB, Jongmans D, Faccioli E, Abdrakhmatov K, Bard P-Y (2002)Site effect analysis around the seismically induced Ananevo rockslide, Kyrgyzstan. Bull Seism Soc Am 92(8):3190–3209

Hobiger M, Bard P-Y, Cornou C, Le Bihan N (2009a) Single station determination of Rayleigh wave ellipticity by using the random decrement technique (RayDec). Geophys Res Lett 36:L14303. doi: 10.1029/2009GL038863

Hobiger M, le Bihan N, Cornou C, Bard P-Y (2009b) Rayleigh wave ellipticity estimation from ambient seismic noise using single and multiple vector-sensor techniques, accepted for EUSIPCO 2009 (Seventeenth European Signal Processing Conference, Glasgow, 24–28 Aug 2009)

Hobiger M, Cornou C, Bard P-Y, Le Bihan N, Renalier F, Endrun B (2010) Inversion of Rayleigh wave ellipticity measurements in preparation for BSSA

Köhler A, Ohrnberger M, Scherbaum F, Wathelet M, Cornou C (2007) Assessing the reliability of the modified three-component spatial autocorrelation technique. Geophys J Int 168(2):779–796

Lacoss RT, Kelly EJ, Toksöz MN (1969) Estimation of seismic noise structure using arrays. Geophysics 34:21–38

Moss RES (2008) Quantifying measurement uncertainty of thirty-meter shear-wave velocity. Bull Seism Soc Am 98(3):1399–1411, June 2008. doi: 10.1785/0120070101

Mucciarelli M, Gallipoli MR (2006) Comparison between Vs30 and other estimates of site amplification in Italy. In: Proceedings of the 1st European conference on earthquake engineering and seismology, Geneva, Switzerland, 3–8 Sept, Paper No 270

Ohrnberger M (2005) Report on the FK/SPAC capabilities and limitations. SESAME Deliverable D19.06, 43 pp, http://sesame-fp5.obs.ujf-grenoble.fr/Delivrables/Del-D19-Wp06.pdf

Ohrnberger M, Schissele E, Cornou C, Bonnefoy-Claudet S, Wathelet M, Savvaidis A, Scherbaum F, Jongmans D (2004) Frequency wavenumber and spatial autocorrelation methods for dispersion curve determination from ambient vibration recordings. In: Proceedings of the 13th world conference on earthquake engineering, Vancouver, BC, Paper No 0946

Park CB, Miller RD, Xia J (1999) Multi-channel analysis of surface waves (MASW). Geophysics 64:800–808

Poggi V, Fäh D (2010) Estimating Rayleigh wave particle motion from three-component array analysis of ambient vibrations. Geophys J Int 180(1):251–267

Pousse G (2005) Analyse des données accélérométriques de K-net et KIK-net: implications sur la prédiction du mouvement sismique – accélérogrammes et spectres de réponse – et la prise en compte des effets de site non-linéaires. Ph.D. Thesis, University Joseph Fourier (in French)

Renalier F (2010) Caractérisation sismique de sites hétérogènes à partir de méthodes actives et passives: variations latérales et temporelles. Ph.D. Thesis, Joseph Fourier University, Grenoble, 224p

Sambridge M (1999) Geophysical inversion with a neighbourhood algorithm: I. Searching a parameter space. Geophys J Int 138:479–494. doi: 10.1046/j.1365-246X.1999.00876.x

Savvaidis A, Ohrnberger M, Wathelet M, Cornou C, Bard P-Y, Theodoulidis N (2009) Variability analysis of shallow shear wave velocity profiles obtained from dispersion curve inversion considering multiple model parameterization. Abstr Seism Res Lett 80(2):354, SSA Meeting, Monterey, Apr 2009

Wathelet M (2008) An improved neighborhood algorithm: parameter conditions and dynamic scaling. Geophys Res Lett 35:L09301. doi: 10.1029/2008GL033256

Wathelet M, Jongmans D, Ohrnberger M (2004) Surface wave inversion using a direct search algorithm and its application to ambient vibration measurements. Near Surf Geophys 2: 211–221

Wathelet M, Jongmans D, Ohrnberger M (2005) Direct inversion of spatial autocorrelation curves with the neighborhood algorithm. Bull Seism Soc Am 95:1787–1800

Wathelet M, Jongmans D, Ohrnberger M, Bonnefoy-Claudet S (2008) Array performance for ambient vibrations on a shallow structure and consequences over vs inversion. J Seism 12:1–19. doi: 10.1007/s10950-007-9067-x

Zor E, Özalaybey S, Karaaslan A, Tapırdamaz MC, Özalaybey SÇ, Tarancioglu Al, Erkan B (2010) Shear-wave velocity structure of the Izmit Bay area (Turkey) Estimated from active-passive array surface wave and single-station microtremor methods. Geophys J Int under revision (submitted)

Chapter 6
Effects of Non-plastic Fines on Liquefaction Resistance of Sandy Soils

Misko Cubrinovski, Sean Rees, and Elisabeth Bowman

Abstract The effects of non-plastic fines on the liquefaction resistance of sandy soils are examined using results from laboratory studies and re-interpretation of well-known SPT-based criteria. Given that fines significantly affect both the density of sand-fines mixes and penetration resistance of sandy soils, one of the key problems in the evaluation of the influence of fines on sand behaviour is establishing a common basis for comparison of clean sands and sands with fines. Effects of fines on the liquefaction resistance of fines-containing sands observed in laboratory tests are first presented using three different density measures as a basis for comparison. In the second part of the paper, conventional SPT-based criteria for liquefaction resistance are re-interpreted and presented in a form allowing direct evaluation of the influence of fines on the liquefaction resistance. The paper shows that the effects of fines on liquefaction resistance observed in the laboratory and those derived from field-based correlations are consistent.

6.1 Introduction

Over the past two decades, the focus in the liquefaction studies and procedures for evaluation of liquefaction has gradually shifted from the behaviour of uniform clean sands to the liquefaction characteristics of fines-containing soils (e.g. Boulanger and Idriss, 2006, 2007; Bray and Sancio, 2006) and well-graded gravelly soils (e.g., Kokusho, 2007), and issues around complex phenomena associated with void redistribution (e.g. Kokusho, 2003; Idriss and Boulanger, 2008) and stratified structure of soils (e.g. Amini and Sama, 1999; Yoshimine and Koike, 2005). Much of these efforts have resulted from evidence from strong earthquakes such as the 1995 Kobe earthquake (Japanese Geotechnical Society, 1998), 1999 Kocaeli earthquake (Bardet et al., 2000) and 1999 Chi-Chi earthquake (Stewart et al., 2001), in which

---

M. Cubrinovski (✉)
Department of Civil and Natural Resources Engineering, University of Canterbury, Christchurch 8140, New Zealand
e-mail: misko.cubrinovski@canterbury.ac.nz

M. Garevski, A. Ansal (eds.), *Earthquake Engineering in Europe*, Geotechnical, Geological, and Earthquake Engineering 17, DOI 10.1007/978-90-481-9544-2_6,

well-graded fines-containing sands and gravelly soils extensively liquefied causing significant damage to engineering structures.

The liquefaction studies on fines-containing sands in relation to the composition of soils have focused largely on two factors: the amount of fines (or fines content) and the nature of fines with regard to their plasticity. Systematic laboratory studies based on considerations of skeleton structure (matrix) of granular mixes have identified three general groups of soils: (i) clean sands and fines-containing sands with relatively small amount of fines (typically $F_C \leq 30\%$) which are characterized by a sand-matrix (or load transfer mechanism predominantly through sand particles) (ii) fine-grained soils with $F_C \geq 50\%$ which are controlled by a fines-matrix, and (iii) fines-containing sands in the transition zone between 30 and 50% fines content which have a sand-fines-matrix. Note that the abovementioned percentages of fines are used for general guidance only, and that the threshold fines content between different matrix-structures may deviate from the above values depending on the grain-size composition and particle characteristics of the soil. Given that plasticity of fines is known to affect liquefaction resistance of fines-containing sands, one may distinguish six general groups of soils: three groups defined with respect to the fines content as above, for non-plastic fines, and three respective groups of soils for plastic fines. In this context, Boulanger and Idriss (2006, 2007) recently proposed liquefaction susceptibility criteria for fine-grained soils for which the fines content (particles with a diameter of less than 0.075 mm) is greater than 50%.

This paper focuses on sandy soils containing 0–30% non-plastic fines. These soils are characterized by a sand-matrix, and hence, the clean sand behaviour ($F_C \leq 5\%$) makes a rational reference for comparison and evaluation of the effects of fines. Results from laboratory studies on such soils are first examined in order to investigate the effects of fines on the liquefaction resistance when using density measures (such as the void ratio, relative density and equivalent void ratios) as a basis for comparison. The results are used to systematically examine the effects of fines on liquefaction resistance and to illustrate the important influence of the adopted basis for comparison (density measure) in this assessment. In the second part of the paper, conventional SPT-based criteria for liquefaction evaluation are re-interpreted and presented in a form that allows direct evaluation of the effects of fines on liquefaction resistance. Note that in their original form, the SPT-criteria include combined effects of fines on the liquefaction resistance and penetration resistance, and hence, they do not indicate whether the liquefaction resistance increases or decreases with increased fines content (Youd and Idriss, 1998). The proposed interpretation of the SPT criteria addresses the effects of non-plastic fines specifically, and provides consistent effects of fines on liquefaction resistance with those observed in laboratory studies.

6.2 Observations from Laboratory Studies

Over the past two decades, systematic laboratory studies have been conducted investigating the effects of fines on the undrained behaviour and liquefaction resistance of sandy soils (e.g. Lade and Yamamuro, 1997; Cubrinovski and Ishihara, 2000; Polito and Martin, 2001; Thevanayagam and Martin, 2002; Cubrinovski and Rees, 2008). The approach taken in these studies has been to directly compare the behaviour of sand mixed with a specific amount of fines (say $F_C = 10$, 20 or 30% by weight) to that of the clean sand ($F_C \leq 5\%$). Since these studies were conducted on reconstituted soil samples, they do not provide means for quantification of the liquefaction resistance of field deposits and hence cannot be directly applied to the geotechnical practice. However, the results from these laboratory studies do provide a good basis for understanding the essential influence of fines on the microstructure of sand-fines mixtures and their respective stress-strain behaviour.

6.2.1 Influence of Fines on Packing (Soil Skeleton Structure)

In several comprehensive laboratory studies, sand-fines mixtures have been produced by mixing a specific amount of fines with host sand in order to produce and examine deformational behaviour of sand-fines mixtures with different fines content. An example of limiting void ratios obtained for such sand-fines mixtures is shown in Fig. 6.1a where relationships between the maximum void ratio and fines content ($e_{max} - F_C$) and minimum void ratio and fines content ($e_{min} - F_C$) are shown for mixtures of Cambria sand and Nevada fines (Lade et al., 1998). There is a clear link between the fines content and limiting void ratios (e_{max} and e_{min}) which also implies a significant effect of fines on the relative density of the mixtures. Hence, when adding fines to a clean sand, the density state of the mixture will be significantly affected both in terms of void ratio and relative density.

Note that the mixtures considered in Fig. 6.1 are gap-graded since the smallest particles of Cambria sand were nearly 11 times greater than the largest particles

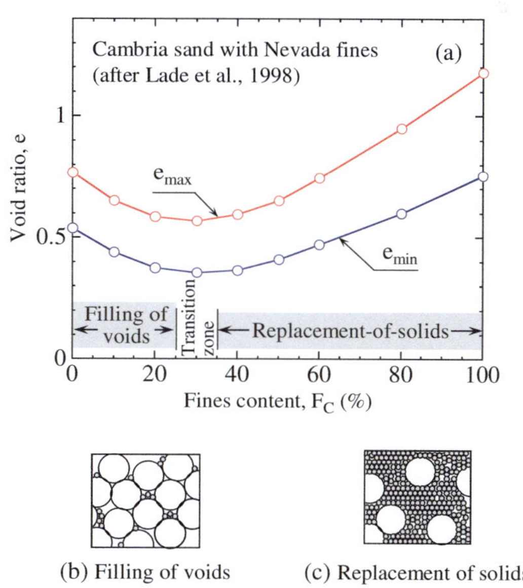

Fig. 6.1 Variation of limiting void ratios with fines content for gap-graded sand-fines mixes (Cubrinovski and Ishihara, 2002; after Lade et al., 1998)

of Nevada fines. This grading feature of the sand-fines mixtures is reflected in the variation of limiting void ratios with the fines contents in Fig. 6.1a. Namely, when adding a small amount of fines to the sand (say 10% of fines), the fines (because of their very small size relative to the sand particles) fill-in the voids between the sand particles and essentially do not influence the sand skeleton structure (the fines particles replace the voids but do not participate in the inter-particle load transfer). For this reason, a decrease in the limiting void ratios (increase in density) is seen with increasing fines content from 0% to about 30%.

Conceptually, when the fines content is relatively small (F_C < 20% in Fig. 6.1), the microstructure of the granular mix is defined (and the deformational behaviour is controlled) by the sand matrix, as illustrated schematically for an idealized binary packing of spherical particles in Fig. 6.1b. On the other hand, at high fines content (F_C > 40% in Fig. 6.1), the microstructure is effectively controlled by the fines matrix, in which case the coarse grains (sand particles) are separated by finer grains (fines particles), as depicted in Fig. 6.1c. As indicated in Fig. 6.1a, there is a transition in the microstructure from a sand-controlled-matrix to a fines-controlled-matrix as the fines content increases from 20 to 40% approximately. There are a number of variations in the possible arrangements and hence role of the finer grains even for idealized binary mixtures (e.g. Thevanayagam et al., 2002). By and large however, Fig. 6.1 conceptually illustrates the link between the fines content, microstructure of granular mixes, and consequent effects on the void ratio and relative density of fines-containing sands.

Unlike the gap-graded sand-fines mixtures, natural sands have a more or less gradual change in the grain-size distribution from coarser sand particles to finer particles (fines). The absence of gaps in the gradation for the majority of natural sands practically eliminates (or reduces the effects of) the filling-of-voids phase discussed above since it is physically impossible to fit the fine particles in the available voids between the coarser sand particles. Thus, the relationship between the minimum void ratio and fines content ($e_{min} - F_C$) does not show a drop in the void ratio with increased fines content from 0 to 20%, as illustrated in Fig. 6.2 where respective data for over 300 natural sandy and silty soils are presented (Cubrinovski and Ishihara, 2002).

6.2.2 Basis for Comparison

When comparing the deformational behaviour or liquefaction resistance of fines-containing sand and clean sand, one encounters the problem of establishing a proper basis for comparison. It is well known that stress-strain behaviour of sands is significantly affected by the density of the sand. Hence, ideally, the behaviour of the clean sand and sand with fines should be compared at an identical density state. As described in the previous section, however, the addition of fines to a clean sand significantly affects both void ratio ($e = V_v/V_s$; V_v = volume of voids; V_s = volume of solids) and relative density ($D_r = [(e_{max} - e)/(e_{max} - e_{min})]^* 100$, in percent) of the soil. Thus, it is not obvious what would constitute an *identical density state* for

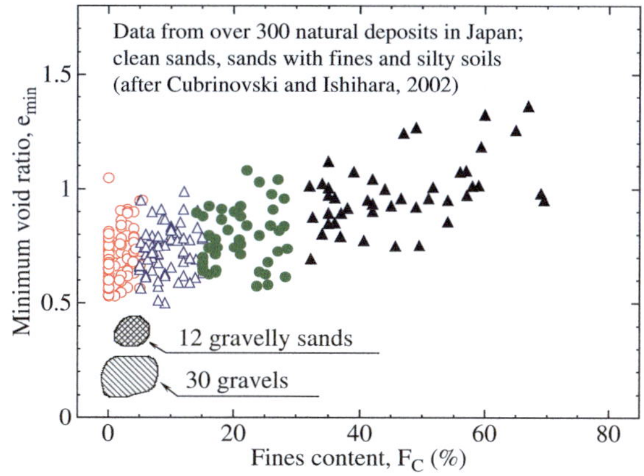

Fig. 6.2 Relationship between e_{min} and fines content for natural sandy soils (Cubrinovski and Ishihara, 2002)

a clean sand and sand with fines. The lack of a clear and sound basis for comparison has resulted in different density measures being used in the evaluation of effects of fines on sand behaviour, often leading to different trends in behaviour where fines may either increase or decrease the liquefaction resistance as compared to that of the clean sand.

Over the past decade, some alternative density measures have been scrutinized such as the *equivalent intergranular void ratio, e^** (Thevanayagam, 2000)

$$e^* = \frac{e + (1-b)f_C}{1-(1-b)f_C} \tag{1}$$

where e is the void ratio, f_C is the fines content (expressed as a ratio, $f_C = F_C/100$) and b ($0 \leq b \leq 1$) is a parameter determining the proportion of fines considered as solids in the calculation of e^*. Thus, $b = 1$ indicates that all fines are considered to be solids, whereas $b = 0$ indicates that all fines are considered as voids in the calculation of e^*. In the latter case, the definition of e^* reduces to the intergranular void ratio definition of Mitchell (1976), $e^* = e_g = (e + f_C)/(1 - f_C)$. Note that, on the other hand, $b = 1$ yields $e^* = e$. In essence, one may interpret b as a measure for the participation of fines as *active solids* (or particles contributing to the load transfer) in the microstructure of sand-fines mixes, relative to the coarser sand particles. $b = 1$ indicates that all fines are treated equally with the coarser sand particles, as solids, whereas $b = 0.6$ for example, implies that 60% of the fines are treated as solids while the remaining 40% are considered *inactive* and hence are treated as voids.

In what follows, results from a comprehensive laboratory study investigating the effects of fines on the undrained behaviour and liquefaction resistance of sandy soils are presented using different density measures as a basis for comparison. Our focus here is on sand-fines mixtures with $F_C \leq 30\%$ of non-plastic fines. These limitations were adopted based on the reasoning that for sandy soils with up to 30%

non-plastic fines the microstructure is sand-dominated and hence the clean sand behaviour makes a rational reference for comparison. These constraints allow a single (and relatively simple) framework to be used in the evaluation of effects of fines with reference to the clean sand behaviour. A fines content over the threshold value ($F_C > 30\%$) and/or plasticity of fines would introduce additional complexities and need for an alternative evaluation framework which is beyond the scope of this study.

6.2.3 Effects of Fines on Liquefaction Resistance

Four soils were tested in the laboratory all derived from the same source material (host soil), a sandy soil sampled from a single layer at the Fitzgerald Bridge Avenue site in Christchurch, New Zealand. The host soil as well as all mixes produced from it are referred to as Fitzgerald Bridge Mixes (FBM) and are differentiated amongst them only by the fines content. The host soil was obtained by mixing "undisturbed" samples obtained from the site, resulting in a mixture of sand with 10% non-plastic fines (FBM-10). This material was dry-sieved to separate the clean-sand fraction that happened to have 1% fines thus creating the clean sand (FBM-1), and then two additional sand-fines mixes with 20% fines (FBM-20) and 30% fines (FBM-30) respectively were produced by reintroducing the fines to the clean sand. The fines content and limiting void ratios of the tested soils are summarized in Table 6.1 while their grain size curves are shown in Fig. 6.3.

Cyclic undrained triaxial tests were performed on reconstituted specimens of FBM soils to investigate the effects of fines on liquefaction resistance (Rees, 2010). All specimens were prepared using the moist-tamping technique in which the soil at a moisture content of 9% was placed in 6 equal layers and subjected to an appropriate amount of tamping required for achieving the target soil density. In this way specimens were prepared over a wide range of densities with void ratios of $e = 0.592 - 0.888$ and relative densities of $D_r = 7 - 80\%$. The specimens were fully saturated (Skempton's B-value greater than 0.95 was measured in all tests) and isotropically consolidated to an effective stress of 100 kPa.

For each soil (FBM-1, FBM-10, FBM-20 and FBM-30) four or five specimens were prepared at a nearly identical target relative density, as described above, and then each specimen was subjected to cyclic axial stresses at different cyclic stress ratios, $CSR = \sigma_d/2\sigma'_c$ where σ_d is the single amplitude cyclic axial stress and σ'_c is the effective confining stress. Using the results from these tests, liquefaction resistance curves were generated defining a relationship between CSR and

Table 6.1 Limiting void ratios of tested sand-fine mixtures

Soil	Fines content (%)	e_{min}	e_{max}
FBM-1	1	0.628	0.907
FBM-10	10	0.597	0.945
FBM-20	20	0.511	0.895
FBM-30	30	0.527	0.860

Fig. 6.3 Grain-size distribution curves of tested sand-fines mixes

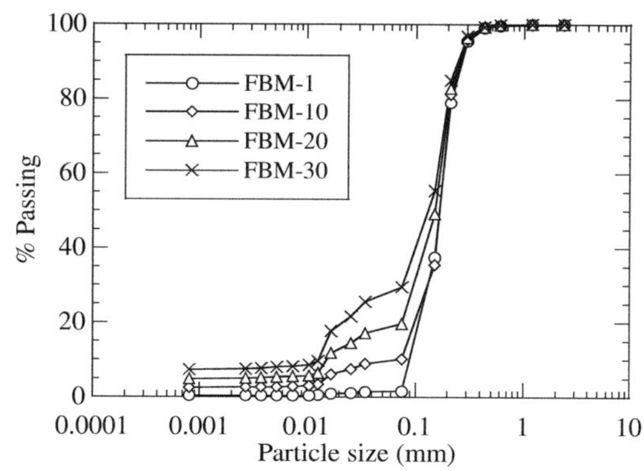

the number of loading cycles N_C required to cause 5% double amplitude (DA) strain. Figure 6.4 shows the liquefaction resistance curves obtained for FBM-1 and FBM-10 for a range of different relative densities. In what follows, the effects of fines on the liquefaction resistance measured in the above tests are examined using three different density measures as a basis for comparison: void ratio, relative density and equivalent intergranular void ratio.

6.2.3.1 Void Ratio as a Reference State

Figure 6.5 shows liquefaction resistance curves (CSR-N_C relationships) for the tested soils with different fines content, for specimens at nearly identical void ratio of $e \approx 0.70$. An alternative presentation of the measured liquefaction resistance is shown in Fig. 6.6 where the cyclic stress ratio causing 5% DA strain in 15 cycles (CSR_{15}) is plotted against the void ratio. Both plots clearly illustrate that, at a given void ratio, the liquefaction resistance decreases with increase in the fines content.

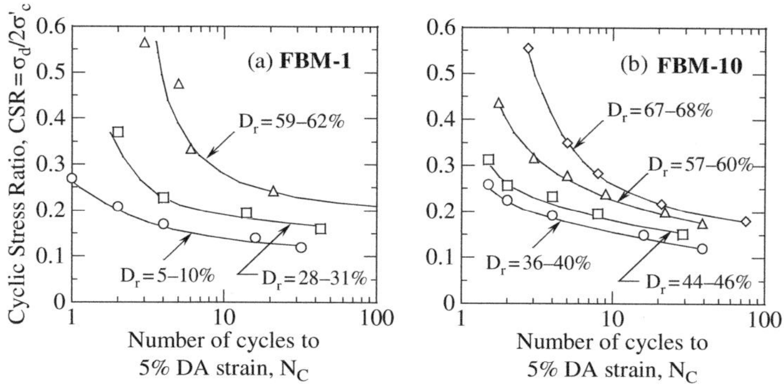

Fig. 6.4 Liquefaction resistance curves for different relative densities: **a** Clean sand (FBM-1). **b** Sand with 10% fines (FBM-10)

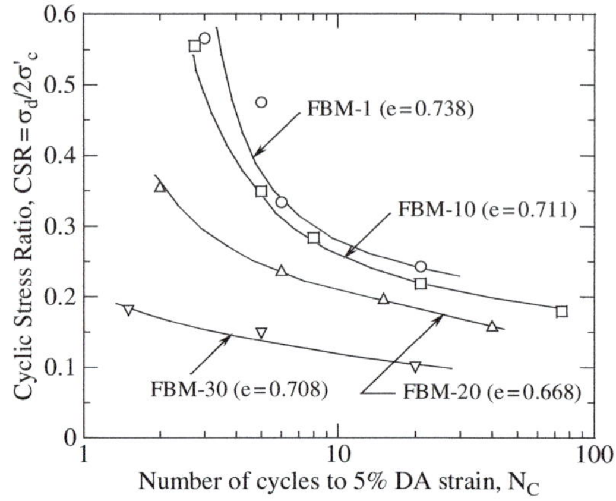

Fig. 6.5 Liquefaction resistance curves of FBM soils for specimens with similar void ratios (target void ratio of $e \approx 0.70$)

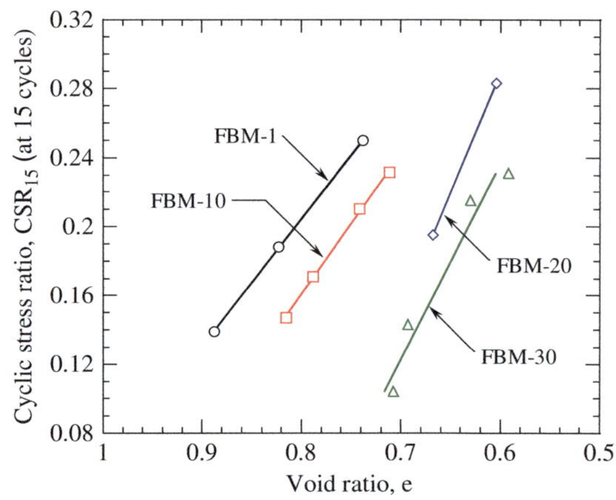

Fig. 6.6 Cyclic stress ratios causing liquefaction (5% DA strain) in 15 cycles for FBM soils with different fines content, as a function of void ratio

This trend for a decrease in the liquefaction resistance with increased fines content has been consistently reported in the literature when clean sand and sand with non-plastic fines of $F_C \leq 30\%$ have been compared at the same void ratio (e.g. Polito and Martin, 2001; Thevanayagam and Martin, 2002; Ueng et al., 2004). This is depicted in Fig. 6.7 where data from six studies are plotted in terms of a normalized liquefaction resistance (CSR_{15}/CSR_{15-CS}) against the fines content; here, CSR_{15-CS} is the cyclic stress ratio at $N_C = 15$ for the clean (host) sand. Note that for any given sand (symbol), the void ratios of specimens with different fines content are nearly identical, whereas the void ratios between different sands (symbols) could be quite different.

6.2.3.2 Relative Density as a Reference State

The liquefaction resistance data obtained from the tests on FBM soils are re-plotted in Fig. 6.8 in terms of cyclic stress ratios at 5 cycles (CSR_5) or 15 cycles (CSR_{15})

Fig. 6.7 Effects of fines on liquefaction resistance when using void ratio as a basis for comparison (results for six different sands are shown; all sand-fines mixes of a given sand are at nearly identical void ratios)

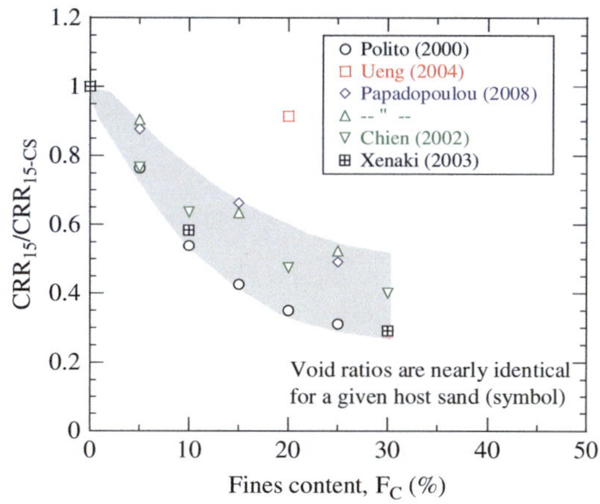

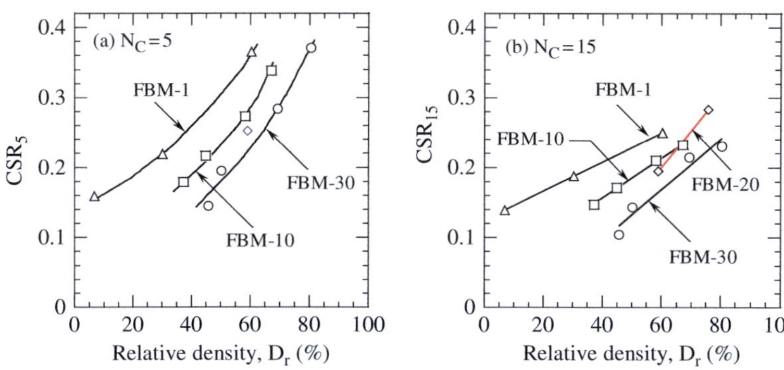

Fig. 6.8 Cyclic stress ratios for FBM-soils required to cause 5% DA strain in 5 cycles (**a**) and 15 cycles (**b**) as a function of relative density

versus the relative density of the soil. These plots allow comparison of the liquefaction resistance of soils that have different fines content but same relative density. Clearly, for a given relative density, the liquefaction resistance of FBM soils decreases with increased fines content. An identical trend showing a decrease in the liquefaction resistance with the fines content, based on D_r, has been reported in most of the other studies (e.g. Chien et al., 2002; Kokusho, 2007). However, there are also studies in which the effects of fines on liquefaction resistance are either not conclusive or do not show a definite trend when D_r is used as a basis for comparison (e.g. Polito and Martin, 2002; Carraro et al., 2003). A larger scatter in the data and less consistent effects of fines on liquefaction resistance based on D_r could be attributed to the difficulties in determining the limiting void ratios (e_{max} and e_{min}) for fines-containing sands. Typically, the testing procedures for evaluation of e_{max} and e_{min} are specified for clean sands and their applicability to fines-containing sands is questionable (not proven), though some standard procedures have been found to produce consistent e_{max} and e_{min} values for sands with fines of up to 30% (e.g. JGS procedures, Cubrinovski and Ishihara, 2002). In summary, the liquefaction resistance decreases with increased fines content even when using the relative density as a basis for comparison, though, exceptions to this trend are possible.

6.2.3.3 Equivalent Intergranular Void Ratio as a Reference State

It was illustrated in the previous sections that when using the void ratio as a basis for comparison, there is a clear tendency for a decrease in the liquefaction resistance with increasing fines content. This tendency is similar, though less consistent and showing more scatter, when the effects of fines are evaluated based on D_r. In both cases, however, the quantification of effects of fines on liquefaction resistance is difficult and could not be generalized across sands with different material properties (e.g. grain-sizes and angularity of particles). In this context, the key contribution of the equivalent intergranular void ratio is that it provides an effective parameter for characterization of the effects of fines on sand behaviour while allowing for effects of the abovementioned sand characteristics.

In addition to the void ratio and fines content (as evident from Eq. 1), a value for the parameter b is required for the calculation of e^*. As described in the previous sections, b allows considering either all fines particles as voids ($b = 0$), all fines particles as solids ($b = 1$) or any proportion in between ($0 < b < 1$). Using the liquefaction resistance data for the four FBM soils, a best-fit value was back-calculated for the parameter b in order to establish a unique correlation between the liquefaction resistance and equivalent intergranular void ratio irrespective of fines content. Such a correlation for the FBM soils is shown in Fig. 6.9 in terms of CSR_{15} versus e^* for the adopted best-fit value of $b = 0.65$. This correlation allows the estimate of the liquefaction resistance (CSR_{15}) of FBM soils for any fines content between 0 and 30%. Note that $e^* = e$ for a clean sand with $F_C = 0\%$, and hence e_{max} and e_{min} in Fig. 6.9 provide reference to the *equivalent relative density* defined using e^*, i.e. $D_r^* = [(e_{max} - e^*)/(e_{max} - e_{min})] \times 100$.

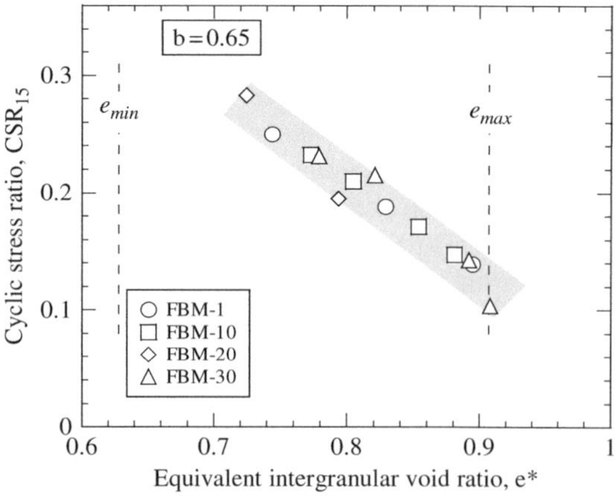

Fig. 6.9 Cyclic stress ratios for FBM-soils required to cause 5% DA strain in 15 cycles as a function of equivalent intergranular void ratio

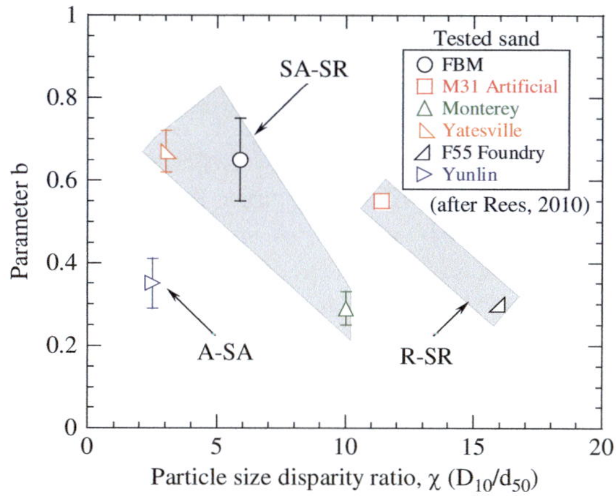

Fig. 6.10 Effects of relative size of sand and fines particles (disparity ratio χ), and angularity of sand particles on the parameter *b* (Rees, 2010; summary of results from six studies using moist-tamped specimens); A = angular, SA = subangular, SR = subround; R = round particles

Other researchers have also used e^* as a density measure capturing the effects of fines on liquefaction resistance in the above manner, suggesting either a constant value for b (e.g. $b = 0.35$ for Ottawa sand-silt mixtures, Thevanayagam and Martin, 2002) or defining b as a function of fines content (Rahman et al., 2008). These studies have indicated that b depends on the disparity ratio, $\chi = D_{10}/d_{50}$ of the sand-fines mixture, where D_{10} and d_{50} are the diameters of sand and fines particles at 10 and 50% passing respectively. Rees (2010) has recently shown that the relationship between b and χ is also affected by the angularity of the sand particles, as illustrated in Fig. 6.10 where back-calculated values for the parameter b are plotted against χ using results from several independent studies (Rees, 2010). The data show a clear tendency for reduction in the value of b with increasing disparity ratio and angularity of sand particles which is consistent with the overall effects of fines and their role in the microstructure of sand-fines mixes, since a large value of χ implies a grading gap between the sand and fines particles, and larger angularity of particles is associated with higher void ratios or more voids in the sand matrix (skeleton).

Through the parameter b, the equivalent intergranular void ratio (e^*) provides means for quantifying the *activity* of fines in the sand-fines microstructure in relation to fundamental soil characteristics such as grading and shape of sand particles (angularity). This in turn allows evaluation of the effects of fines on the behaviour of different soils (sand-fines mixes) and comparison of these effects on a relative basis. Further studies with regard to quantification of the above effects on the parameter b are required including relationships for different fabrics.

6.3 Interpretation of Field-Based Criteria for Liquefaction Resistance of Sands with Fines

In the conventional liquefaction evaluation procedure, the liquefaction resistance of sandy soils is estimated using empirical criteria based on field (in-situ) tests (Youd and Idriss, 1998; Idriss and Boulanger, 2008). SPT, CPT and shear wave velocity

(V_s) criteria for liquefaction resistance have been established, each having particular advantages and disadvantages in the evaluation of liquefaction. One important advantage of the SPT-based criteria is that the SPT blow count is the most sensitive to changes in the relative density of sands. For example, a change in the relative density of clean sand from 30 to 80% would be expected to increase the SPT blow count by a factor of about 7, the CPT resistance by a factor of about 3.3 and the shear wave velocity by a factor of only 1.4 (e.g., Idriss and Boulanger, 2008). Given that the liquefaction resistance of saturated sands is strongly affected by D_r (as demonstrated in Fig. 6.4 for the FBM soils), the SPT-based criteria are the most appropriate for investigating the effects of fines especially when a conversion between the field parameter and relative density is attempted, as described in the following sections.

6.3.1 SPT-Criteria for Liquefaction Resistance

Youd and Idriss (1998) presented the well-known SPT-criteria for liquefaction resistance of clean sands and sands with fines (NCEER Workshops 1996 and 1998; modified from the original work of Seed et al., 1985) in an empirical chart correlating the cyclic resistance ratio ($CRR_{7.5}$) with the normalized SPT blow count $(N_1)_{60}$. Here, $CRR_{7.5}$ is the cyclic resistance ratio for a magnitude $M = 7.5$ earthquake while $(N_1)_{60}$ is the SPT blow count corresponding to an effective vertical stress of $\sigma'_v = 100$ kPa and energy ratio of 60% (delivered energy as a percentage of the theoretical free-fall energy, e.g. Seed et al. 1985, Skempton, 1986). Note that $CRR_{7.5}$ in essence is identical to the previously introduced cyclic stress ratio causing liquefaction in 15 uniform cycles, CSR_{15}.

In addition to the empirical chart depicting $CRR_{7.5} - (N_1)_{60}$ curves for $F_C \le 5\%$ (clean sand), $F_C = 15\%$ and $F_C = 35\%$, Youd and Idriss (1998) provided a set of expressions approximating the $CRR_{7.5} - (N_1)_{60}$ curves for sands with fines from 0 to 35% as follows:

$$CRR_{7.5} = \frac{1}{34 - (N_1)_{60cs}} + \frac{(N_1)_{60cs}}{135} + \frac{50}{\left[10 \cdot (N_1)_{60cs} + 45\right]^2} - \frac{1}{200} \qquad (2)$$

where

$$(N_1)_{60cs} = (N_1)_{60}, \quad \mathrm{F_c} \le 5\% \qquad (3a)$$

and

$$(N_1)_{60cs} = \exp\left(1.76 - \frac{190}{F_c^2}\right) + \left(0.99 + \frac{F_c^{1.5}}{1000}\right) \cdot (N_1)_{60}, \quad 5\% < \mathrm{Fc} < 35\% \qquad (3b)$$

Using Eqs. (2) and (3), $CRR_{7.5} - (N_1)_{60}$ curves for $F_C = 0\%$ (clean sand), $F_C = 15\%$ and $F_C = 35\%$ were generated, as shown in Fig. 6.11. It is evident

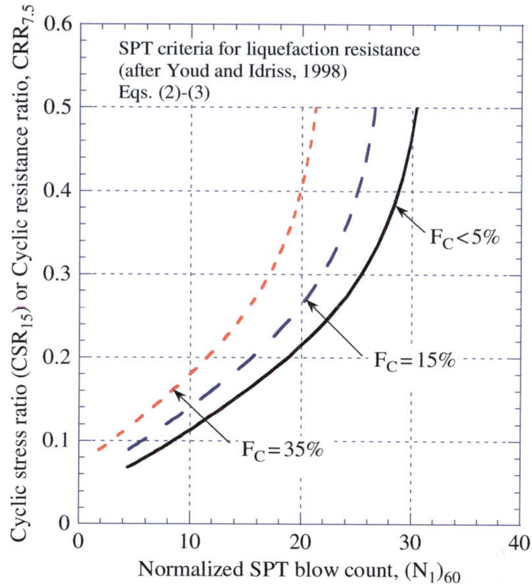

Fig. 6.11 SPT-criteria for liquefaction resistance of clean sands and sands with fines (after Youd and Idriss, 1998)

from this plot that for a given SPT blow count $(N_1)_{60}$, the liquefaction resistance increases with increased fines content. Youd and Idriss (1998) pointed out that it was not clear whether this increase of *CRR* is caused by an increase of liquefaction resistance or a decrease of penetration resistance due to the effects of fines. In essence, this problem in the interpretation of the effects of fines on the liquefaction resistance is equivalent to that discussed in the previous sections where the effects of fines could either decrease (when using e or D_r) or increase the liquefaction resistance (when using e_g, for example) depending on the parameter used as a basis for comparison. In what follows it is shown that the increase of *CRR* in the SPT-based criteria is due to a decrease in the penetration resistance of fines-containing sands and that in essence the effects of fines on liquefaction resistance observed in the laboratory and those derived from field-based correlations are consistent.

This study is limited to sandy soils with up to 30% non-plastic fines, as elaborated in the previous sections. Thus, the effects of fines depicted in the SPT-based criteria (Fig. 6.11; Eqs. 2 and 3) need to be considered within these constraints, particularly if a rigorous comparison between the SPT criteria and laboratory studies is to be attempted. In this context, it is important to recall that for any given fines content (say $F_C = 15\%$), the curve shown in Fig. 6.11 represents a boundary line that separates case histories in which liquefaction was observed (*CRR*-N_1 combinations above or to the left of the curve) from case histories in which liquefaction was not observed (*CRR*-N_1 combinations below or to the right of the curve). This empirical boundary encompasses all cases for which liquefaction was observed and hence it essentially represents a lower bound value for the liquefaction resistance of a particular soil group (sand with specific fines content). The case histories considered in the development of the SPT-based criteria (Seed et al., 1985; Tokimatsu and Yoshimi,

1983) include sands with non-plastic and sands with plastic fines. However, given that the plasticity of fines increases the liquefaction resistance of fines-containing sands (e.g. Ishihara and Koseki, 1989), it could be argued that the above SPT-based criteria, as lower-bound values of the liquefaction resistance, in effect are defined by and are representative of the liquefaction resistance of sands with non-plastic fines. Hence, these criteria, when limited to a fines content in the range between 0 and 30%, do conform to the limitations adopted in this study with respect to the amount (0–30%) and nature (non-plastic) of fines. In order to further scrutinize the liquefaction resistance criteria based on the SPT blow-count and allow their rigorous comparison with results form laboratory studies, these criteria are expressed in terms of the relative density as described below.

6.3.2 Correlation Between Relative Density and SPT Blow Count

Cubrinovski and Ishihara (1999) used data of high-quality undisturbed samples of clean sands, silty sands and gravely soils recovered (by ground freezing) from natural deposits to establish an empirical correlation between the SPT blow count and relative density of granular soils. The proposed correlation is shown in Fig. 6.12 in terms of $(N_1)_{78}/D_r^2$ ratio plotted against the void ratio range ($e_{max} - e_{min}$) and can be expressed as:

$$(N_1)_{78} = D_r^2 \frac{9}{(e_{\max} - e_{\min})^{1.7}} \tag{4}$$

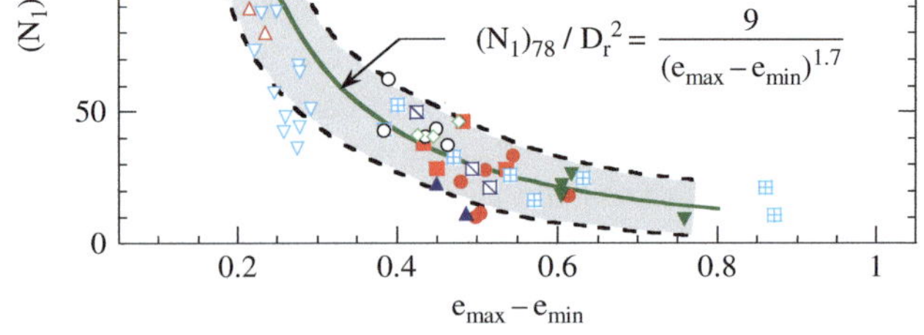

Fig. 6.12 Empirical relationship between SPT blow count and relative density of granular soils (Cubrinovski and Ishihara, 1999)

The correlation is expressed in terms of $(N_1)_{60}$ as:

$$(N_1)_{60} = D_r^2 \frac{11.7}{(e_{\max} - e_{\min})^{1.7}} \tag{5}$$

where $(N_1)_{60} = (78/60) \cdot (N_1)_{78}$.

In addition to the high-quality data used, the key quality of the $N_1 - D_r$ correlation is that through the use of the void ratio range ($e_{max} - e_{min}$) it allows for the combined effects of grain-size composition of soils, fines content and angularity of particles on the SPT resistance. Within the same study, Cubrinovski and Ishihara (2002) examined the characteristics of limiting void ratios (e_{min} and e_{max}) and void ratio range ($e_{max} - e_{min}$) of over 300 natural sandy and silty soils in relation to the mean grain size of soils (D_{50}), fines content (F_C) and angularity of sand particles. The established relationship between the void ratio range and fines content of sandy and silty soils is shown in Fig. 6.13 where a clear trend for an increase in ($e_{max} - e_{min}$) with increased fines content is seen.

The relationship between ($e_{max} - e_{min}$) and F_C for sands with fines between 0 and 30% is approximated as:

$$(e_{\max} - e_{\min}) = 0.43 + 0.0087 \cdot F_C, \quad 0\% \leq \mathrm{F_C} \leq 30\% \tag{6}$$

where F_C is the fines content in percent. Substituting Eq. (6) into Eq. (5), the relationship between $(N_1)_{60}$ and D_r given in Eq. (5) can be expressed in terms of the fines content as:

$$(N_1)_{60} = D_r^2 \frac{11.7}{(0.43 + 0.0087 \cdot F_c)^{1.7}}, \quad 0\% \leq \mathrm{F_c} \leq 30\% \tag{7}$$

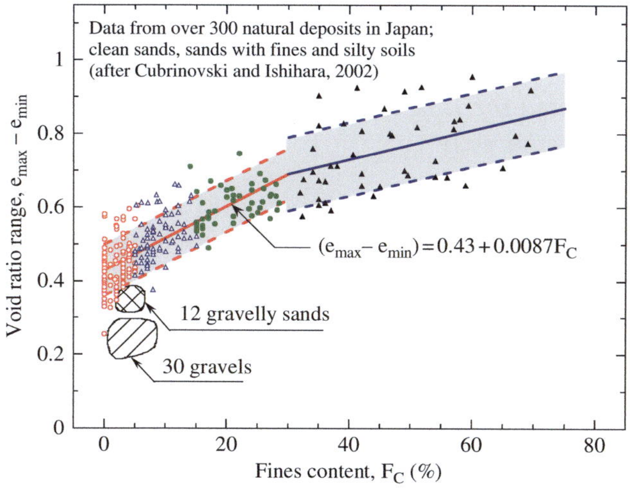

Fig. 6.13 Correlation between void ratio range ($e_{max} - e_{min}$) and fines content for sandy and silty soils (Cubrinovski and Ishihara, 2002)

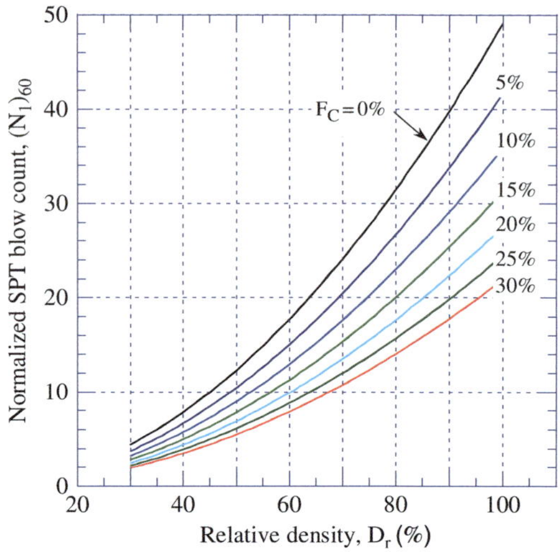

Fig. 6.14 Relationship between SPT blow count and relative density of sandy soils illustrating a decrease in the penetration resistance with increased fines content (generated using Eq. 7)

Note that in Fig. 6.13 there is a clear effect of the grain-size of soils and fines content on the void ratio range. Gravelly soils have the lowest $(e_{max} - e_{min})$ values in the range between 0.2 and 0.35; $(e_{max} - e_{min})$ of clean sands is around 0.36–0.50, and the void ratio range increases with the fines content and reaches values of about 0.62–0.76 for sands with 30% fines. With these $(e_{max} - e_{min})$ characteristics in mind, it is apparent that the relationship between N_1 and D_r shown in Fig. 6.12 in essence indicates a significant influence of grain size and fines content on the penetration resistance. The latter effect is depicted in Fig. 6.14 where $(N_1)_{60} - D_r$ relationships, generated using Eq. (7), for sands with different fines content show a pronounced decrease of penetration resistance with increased fines content. In other words, at a given relative density, penetration resistance decreases with increased fines content.

6.3.3 SPT-Criteria for Liquefaction Resistance Expressed in Terms of Relative Density

The above manipulations enable us to express the SPT-based criteria for liquefaction resistance of Youd and Idriss (1998) in terms of the relative density D_r. Substituting Eq. (7) into Eq. (3), and then Eq. (3) into Eq. (2), an expression is obtained for the liquefaction resistance representing the SPT-based criteria of Youd and Idriss (1998) in terms of the relative density and fines content, i.e. $CRR_{7.5} = f(D_r, F_c)$. The expression is not explicitly given because of its lengthiness. Figure 6.15 shows three $CRR_{7.5}$–D_r relationships generated using this expression, for clean sand ($F_C = 0\%$) and sand with 15 and 30% fines content respectively.

The plot shows a pronounced decrease in liquefaction resistance with increased fines content, at a given relative density, which is consistent with the effects of fines observed in laboratory studies. Note that this decrease is very pronounced for high

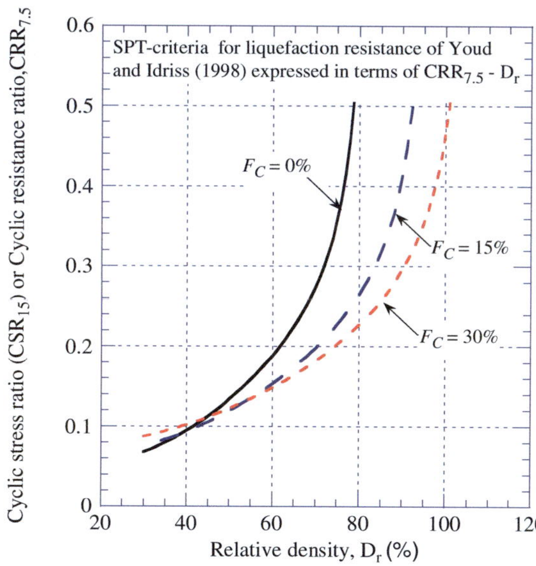

Fig. 6.15 SPT-based criteria for liquefaction resistance of Youd and Idriss (1998) expressed in terms of relative density ($CRR_{7.5} - D_r$ relationships)

relative densities while it is relatively insignificant for relative densities below 50%. In fact, the trend may even reverse for low relative densities below 45% which simply reflects the nature of the $(N_1)_{60} - D_r$ relationship where the ratio $(N_1)_{60}/D_r^2$ is constant for a given soil. These results clearly suggest that the increase of CRR with the fines content seen in the empirical relationships ($CRR - (N_1)_{60}$ plot) in Fig. 6.11 is caused by a reduction in the penetration resistance with increased fines content.

A set of empirical relationships containing significant scatter have been used to express the SPT criteria for liquefaction resistance in terms of the relative density, thus allowing direct comparison of the effects of fines on liquefaction resistance between laboratory-based and field-based studies (parameters). Uncertainties associated with the empirical relationships and application of the above methodology to recently revised $CRR - (N_1)_{60}$ correlations proposed by Idriss and Boulanger (2004, 2008) are discussed elsewhere (Cubrinovski et al., 2010).

6.4 Summary and Conclusions

The effects of fines on liquefaction resistance of sandy soils have been investigated using results from laboratory studies and a re-interpretation of well-established empirical criteria based on penetration resistance (SPT blow count). The key findings can be summarized as follows:

1. The study was limited to sands containing 0–30% non-plastic fines based on the reasoning that such soils are characterized by a sand-matrix, and hence, the evaluation of effects of fines could be made with reference to the clean sand behaviour.

2. Fines significantly affect the density of sand-fines mixes both in terms of void ratio and relative density. Thus, it is not obvious what would constitute an identical density state for a clean sand and sand with fines in the laboratory. Much of the inconsistencies in the interpretation of the effects of fines on sand behaviour result from difficulties in establishing a common basis (reference state or parameter) for comparison of clean sands and fines-containing sands.
3. When using the void ratio as a basis for comparison, laboratory studies consistently show that liquefaction resistance decreases with increased fines content. A similar trend, though less consistent and with more scatter, is observed when evaluating the effects of fines on liquefaction resistance based on the relative density.
4. The equivalent intergranular void ratio provides a means for normalization of the effects of fines and establishing a unique correlation (for a given fabric) between liquefaction resistance and modified void ratio e^* irrespective of fines content. Back-calculation of the parameter b suggests that this parameter is affected by both relative size of sand and fines particles (D_{10}/d_{50}) and angularity of sand particles.
5. Conventional SPT-based criteria for liquefaction resistance of sandy soils (Youd and Idriss, 1998) were re-interpreted and expressed in terms of the relative density. This interpretation shows that the liquefaction resistance of sandy soils decreases with increased fines content and that the increase of $CRR_{7.5}$ with fines content seen in the empirical relationships of Youd and Idriss (1998) (Fig. 6.11) is caused by a decrease in penetration resistance due to increased fines content. At a given relative density, both empirical SPT-based criteria and results from laboratory studies are consistent and show a decrease of liquefaction resistance with increased (non-plastic) fines content.

Acknowledgments The authors would like to acknowledge the long-term support of earthquake geotechnics research programmes at the University of Canterbury provided by the Earthquake Commission (EQC), New Zealand.

References

Amini F, Sama KM (1999) Behaviour of stratified sand-silt-gravel composites under seismic liquefaction conditions. Soil Dyn Earthquake Eng 18:445–455

Bardet JP et al (2000) Ground failure and geotechnical effects: soil liquefaction, landslides and subsidences. Earthquake Spectra 16:141–162

Boulanger RW, Idriss IM (2006) Liquefaction susceptibility criteria for silts and clays. J Geotech Geoenviron Eng 132(11):1413–1426

Boulanger RW, Idriss IM (2007) Evaluation of cyclic softening in silts and clays. J Geotech Geoenviron Eng 133(6):641–652

Bray JD, Sancio RB (2006) Assessment of liquefaction susceptibility of fine-grained soils. J Geotech Geoenviron Eng 132(9):1165–1177

Carraro JAH, Bandini P, Salgado R (2003) Liquefaction resistance of clean and nonplastic silty sands based on cone penetration resistance. J Geotech Geoenviron Eng 129(11):965–976

Chien L-K, Oh Y-N, Chang C-H (2002) Effects of fines content on liquefaction strength and dynamic settlement of reclaimed soil. Canad Geotech J 39:254–265

Cubrinovski M, Ishihara K (1999) Empirical correlation between SPT N-value and relative density for sandy soils. Soils Found 39(5):61–71
Cubrinovski M, Ishihara K (2000) Flow potential of sandy soils with different grain compositions. Soils Found 40(4):103–119
Cubrinovski M, Ishihara K (2002) Maximum and minimum void ratio characteristics of sands. Soils Found 42(6):65–78
Cubrinovski M, Rees SD (2008) Effects of fines on undrained behaviour of sands. ASCE Geotech Spec Publ 181:1–11
Cubrinovski M, Rees, SD, Rahman MD, Bowman ET (2010) (manuscript in preparation for Soil Dynamics and Earthquake Engineering)
Idriss IM, Boulanger RW (2004) Semi-empirical procedures for evaluating liquefaction potential during earthquakes. In: Proceedings of the 11th international conference on soil dynamics and earthquake engineering, vol 1, pp 32–55
Idriss IM, Boulanger RW (2008) Soil liquefaction during earthquakes. Earthquake Engineering Research Institutes, MNO-12.
Ishihara K, Koseki J (1989) Cyclic strength of fines-containing sands. In: Proceedings of the discussion session on influence of local conditions on seismic response, 12th ICSMFE Rio de Janeiro, pp 101–106
Japanese Geotechnical Society (1998) Special issue on geotechnical aspects of the January 17, 1995 Hyogoken-Nambu Earthquake, No. 2. Soils and Foundations, September 1998.
Kokusho T (2003) Current state of research on flow failure considering void redistribution in liquefied deposits. Soil Dyn Earthquake Eng 23:585–603
Kokusho T (2007) Liquefaction strength of poorly-graded and well-graded granular soils investigated by lab tests. In: Pitilakis KD (ed) Earthquake geotechnical engineering. Springer, Dordrecht, pp 159–184
Lade PV, Yamamuro JA (1997) Effects of nonplastic fines on static liquefaction of sands. Canadian Geotech J 34:918–928
Lade PV, Liggio CD, Yamamuro JA (1998) Effects of non-plastic fines on minimum and maximum void ratio of sand. Geotech Test J 21(4):336–347
Mitchell JK (1976) Fundamentals of soil behaviour. Wiley, New York, NY
Papadopoulou A, Tika T (2008) The effect of fines on critical state and liquefaction resistance characteristics of non-plastic silty sands. Soils Found 48(5):713–725
Polito CP, Martin JR (2001) Effects of nonplastic fines on the liquefaction resistance of sands. J Geotech Geoenviron Eng 127(5):408–415
Rahman MM, Lo SR, Gnanendran CT (2008) On the equivalent granular void ratio and steady state behaviour of loose sand with fines. Canad Geotech J 45(10):1439–1455
Rees SD (2010) Effects of fines on the undrained behaviour of Christchurch sandy soils. PhD Thesis University of Canterbury, New Zealand
Seed HB, Tokimatsu K, Harder LF Jr, Chung R (1985) Influence of SPT procedures in soil liquefaction resistance. J Geotech Eng ASCE 111(12):1425–1445
Skempton AW (1986) Standard penetration test procedures and the effects in sands of overburden pressure, relative density, particle size, ageing and overconsolidation. Geotech 36(3):425–447.
Stewart JP et al. (2001) Chi-Chi earthquake reconnaissance report: soil liquefaction. Earthquake Spectra 17:37–60
Thevanayagam S (2000) Liquefaction potential and undrained fragility of silty soils. In: Proceedings of the 12th world conference on earthquake engineering. Paper 2383, pp 1–11
Thevanayagam S, Martin GR (2002) Liquefaction in silty soils – screening and remediation issues. Soil Dyn Earthquake Eng 22:1035–1042
Thevanayagam S, Shenthan T, Mohan S, Liang J (2002) Undrained fragility of clean sands, silty sands and sandy silts. J Geotech Geoenviron Eng 128(10):849–859
Tokimatsu K, Yoshimi Y (1983) Empirical correlation of soil liquefaction based on SPT N-value and fines content. Soils Found 23(4):56–74
Ueng T-S, Sun C-W, Chen C-W (2004) Definition of fines and liquefaction resistance of Mulou River soil. Soil Dyn Earthquake Eng 24:745–750

Xenaki VC, Athanasopoulos GA (2003) Liquefaction resistance of sand-silt mixtures: an experimental investigation of the effects of fines. Soil Dyn Earthquake Eng 23:183–194

Yoshimine M, Koike R (2005) Liquefaction of clean sand with stratified structure due to segregation of particle size. Soils Found 45(4):89–98

Youd TL, Idriss IM (1998) Liquefaction resistance of soils: summary report from the 1996 NCEER and 1998 NCEER/NSF Workshops on evaluation of liquefaction resistance of soils. J Geotech Geoenviron Eng 127(4):297–313

Part III
Seismic Performance of Buildings

Chapter 7
Performance Based Seismic Design of Tall Buildings

Farzad Naeim

Abstract An overview of current performance based methodologies utilized for design of tall buildings is presented. The reasons why common prescriptive code provisions are incapable of addressing the needs of tall building design engineers are explained. The performance objectives commonly associated with tall building design are identified and the evolution of current component-based performance objectives to a more rigorous and fully probabilistic approach to performance based design is discussed. Modeling and acceptance criteria associated with various performance based design guidelines are explained and special issues such as selection and scaling of ground motion records, soil-foundation-structure interaction issues, and seismic instrumentation and peer review needs are elaborated on.

7.1 Introduction

This paper provides an overview of standard-of-practice for performance based seismic design of tall buildings. The paper has a United States based tilt simply because that is where the author is based and most of his practice and research is focused on. However, the concepts and methods reviewed in this paper are equally applicable to any seismic region in the world if proper modifications are made to incorporate regional or local seismicity, engineering practices, and construction quality issues.

We begin by attempting to answer a number of frequently asked questions with respect to tall buildings. Then we proceed with an overview of various performance criteria, modeling procedures, acceptance criteria, ground motion issues, peer review requirements and finish with a brief discussion of seismic instrumentation needs.

F. Naeim (✉)
Earthquake Engineering Research Institute, Oakland, CA, USA; John A. Martin & Associates, Inc., Los Angeles, CA, USA
e-mail: FARZAD@johnmartin.com

M. Garevski, A. Ansal (eds.), *Earthquake Engineering in Europe*, Geotechnical, Geological, and Earthquake Engineering 17, DOI 10.1007/978-90-481-9544-2_7,

7.2 What Is a Tall Building?

In 1964 when Justice Potter Stewart of United States Supreme Court faced the dilemma of defining what obscene material was, he uttered the now famous statement that "I know it when I see it (Stewart, 1964)." We face the same dilemma in defining what constitutes a "tall buildings" today. Although such buildings are relatively easy to identify, there is no universally accepted definition for a tall building.

One way to classify a tall building is to use the overall height, or the height of the highest occupied floor measured from the ground surface. The taller the building, according to these definitions, the larger gravity load the vertical members of the structural system have to carry, the more shortening the columns may experience, and for a given cross-section, the member will have less reserve capacity to carry the additional forces and moments imposed by lateral load (i.e., wind or seismic). There can be, however, no clear cut-off point to designate a building as "tall" using this definition. Using the number of floors as an indicator of "tallness" suffers from the same limitations. Council on Tall Buildings and Urban Habitat (CTBUH) maintains an up to date database of "tall buildings" according to the height criteria and number of floors (Council on Tall Buildings and Urban Habitat, 2010). Many building codes or standards limit application of certain structural systems to certain height. For example, ASCE 7-05 (American Society of Civil Engineers, 2006) prohibits the use of reinforced concrete shear wall only systems (and several other bracing systems) in regions of high seismicity for buildings taller than 160 ft (about 50 m).

Another way to classify a building as tall is to measure its largest aspect ratio (the ratio of its overall height to its smaller plan dimension at the base. This approach has the engineering advantage of providing a crude insight as to the importance of design to resist overturning moments caused by lateral forces when the aspect ratio becomes large. However, there is no established aspect ratio barrier for defining tall buildings.

Dynamic characteristics of a building such as prevalence of higher modes in seismic response and/or fundamental period longer than a certain value (say 1.0 s) are more useful to engineers; however, they have not resulted in a generally accepted definition of a tall building either.

7.3 Are Tall Buildings Particularly Vulnerable to Earthquake Ground Motions?

The debate over whether tall and flexible building structures are particularly vulnerable to earthquake ground motions has persisted for a long time (Heaton et al., 1995; Naeim and Garves, 2005) What works in favor of tall (flexible) structures is that earthquakes generally release significantly less energy at the longer periods (i.e., 3 s and longer) associated with the fundamental periods of taller, more flexible structures, than they do in the short period range (i.e., 0.2–1.0 s) associated with the fundamental period of stiffer, shorter buildings (Naeim and Garves, 2005). What works against them is that earthquakes with larger magnitude release more energy in the long period range than smaller earthquakes and if these earthquakes occur in close proximity of tall buildings then significant directivity and near-source and basin effects could amplify the impact of the earthquake on taller, more flexible, structures. Both sides of the argument probably agree that, generally speaking, if the earthquake problem is viewed from a probabilistic perspective, tall buildings will perform better than similarly designed short (stiff) buildings while if the earthquake problem is looked at from a worst-case scenario, deterministic, perspective for large earthquakes occurring nearby, tall buildings may not be in such a good position and some of them could suffer partial or full collapse.

7.4 Should Tall Buildings Be Treated Like Other Buildings?

A tall building represents a significant investment of human and material resources and may be occupied by hundreds, if not thousands, of occupants. Building codes' reaction to this fact, at least in the United States, has been twofold. First, application of certain structural systems has been limited to certain heights. For example, ASCE 7-05 (American Society of Civil Engineers, 2006) does not permit the use of certain lateral load resisting systems for buildings taller than code prescribed heights in regions of high seismicity. Second, buildings with high occupancy are required to be designed for larger lateral forces via the use of an importance factor (I) which is taken as unity for ordinary buildings. For example, ASCE 7-05 requires that buildings "where more than 300 people congregate in one area" must be designed using $I = 1.25$. This provision, however, is commonly interpreted in a way that indicates an $I = 1.0$ for most tall buildings because it is often successfully argued that if the building does not have an auditorium or an assembly hall with a capacity of 300 occupants or larger then not more than 300 people will congregate in one area. Therefore, you could have a tall building occupied by thousands of people yet designed with $I = 1.0$.

Many, if not most, code imposed height limits on lateral systems are difficult, if not impossible, to rationally justify. Numerous studies and evaluations (Los Angeles Tall Buildings Structural Design Counsil, 2007, 2008, 2009) have shown that it is possible and economical to design tall buildings as safe or safer than code designed buildings while ignoring code imposed height limits. Furthermore, even a strict imposition of an arbitrary 25% or higher premium on elastic design forces does not do much to address the issues of damage and potential collapse which are inherently inelastic and nonlinear phenomena.

Prescriptive codes by in large contain a collection of empirical rules and experimental results that have evolved over many years of practice. While these rules and procedures, when followed, have resulted in buildings that have been generally safe and have exhibited a margin of safety larger than that indicated by design analysis calculations, there is no way to quantify the margin of safety provided by following code procedures. By following prescriptive rules that are not tied to a particular performance level we are closing are eyes to detailed building and region specific demand and capacity concerns. Codes provide a "one size fits all" approach to seismic design. Tall buildings as small class of specialized structures will perform better during earthquakes if special attention is afforded to their individual seismic behavior and engineers are provided with ample opportunities to explore new frontiers, utilize state of the art technologies and latest research results in order to improve the performance, feasibility, and constructability of their designs. Prescriptive building codes are simply incapable of offering such attributes.

7.5 Why Performance Based Design Is a Necessity for Tall Buildings?

For over half a century the implicit objective of prescriptive building codes, at least in the United States, has been to produce buildings which resist minor earthquakes with little or no structural damage; moderate earthquakes with repairable structural damage; and major earthquakes with severe structural damage but no loss of life or limb. While statement of this objective can be found in either preamble or commentary of most prescriptive codes, there is hardly any provisions in the body of the codes that is in one way or another tied to the stated performance objectives. Engineers have been brought up to believe that if they follow the code prescriptive rules, the objectives will be automatically achieved. After many major earthquakes, this collective engineering psyche has suffered a blow and resulted in changes in the code prescriptive rules. For example, in the aftermath of the 1971 San Fernando earthquake massive changes were made to code provisions for design of reinforced concrete members and the 1994 Northridge earthquake caused a rethinking of code's approach to design or steel moment-resisting beam-column connections. Perhaps the unexpected poor performance of welded steel beam-column connections during the 1994 Northridge earthquakes provided the last blow to the prevalent attitude of separating performance objectives from design provisions. The rules for design of new steel connections where clearly performance specific (i.e., such connections should be capable of resisting a minimum rotation of 0.04 rad without substantial reduction in their load carrying capacity). Many such connections after being conceptualized have been subjected to intense analytical and experimental evaluations before being implemented on actual projects.

There are other reasons why performance-based design of tall buildings has gathered momentum. The overwhelming majority of construction in United States and worldwide consists of low-rise buildings. According to Portland Cement Association (Portland Cement Association, 2000), buildings with one to three floors represent 93% of floor area of construction in United States while buildings with 14 floors or more represent only 1% of floor area of construction. With so much of the construction effort concentrated on low-rise construction it is not surprising that the code writers have these buildings in mind when crafting code provisions. As a result many of the provisions included in a typical building code either do not have relevance to tall building design, or even worse, do not make much sense for design of tall buildings. For example, until just a few years ago, the Los Angeles Building Code had a very peculiar drift design provision (International Conference of Building Officials, 2002) requiring story drift not to exceed $0.020/T^{1/3}$ where T is the fundamental vibration period of the building. This provision, which was later retracted, probably did not have a serious effect on design of low-rise buildings but was a huge straightjacket for design of tall buildings with long vibration periods.

Another shortcoming of current prescriptive codes is they do not distinguish between the racking component of interstory drift which can lead to significant damage and/or collapse and the rigid body displacement associated with the "rotation" of a tall building as a whole at upper levels caused by axial deformation of columns and walls which generally and induces no damage (see Fig. 7.1).

Tall buildings present significant monetary investments and command higher engineering fees. Their owners and developers are kin to maximize the benefit of their investment by making the structural system of their building as cost effective as possible. Larger engineering fees and usually more sophisticated design engineers open the window for critical evaluation of prescriptive code provisions and use of advanced systems and technologies which are either not permitted or rewarded by a typical prescriptive building code.

Last but not least, damage or collapse of a tall building has far more adverse consequences to life and well-being of communities that they are located at compare damage or collapse of a small low-rise building. Therefore, owners, developers, potential occupants, and building officials are increasingly more receptive to the idea of requiring a better performance from tall buildings than that expected from ordinary construction.

It was the culmination of all of the above factors, reliance on validated experimental data, and recent advances in our analytical and computational capabilities that made the performance based seismic design of tall buildings a reality in the first decade of the twenty-first century.

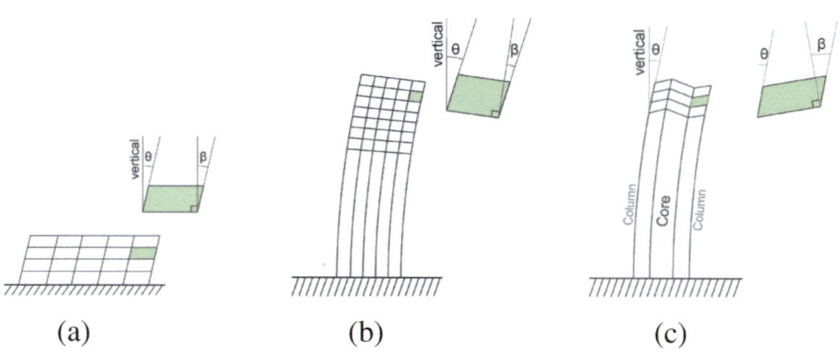

Fig. 7.1 Differences in overall story drift and racking story drift in a tall building (Council on Tall Buildings and Urban Habitat, 2008). **a** Low rise building: racking deformation angle equals storey drift ratio, **b** "tube" high-rise building: racking deformation angle is smaller than the storey drift ratio, and **c** wall-frame high-rise building: racking deformation angle can exceed storey drift ratio (Reproduced with permission from the Council on Tall Buildings and Urban Habitat)

7.6 What Is Involved in Performance Based Design of Tall Buildings?

Several procedures need to be established to achieve a meaningful performance based seismic design of a tall building, namely:

1. A set of reasonable performance objectives must be defined
2. A set of rational design procedures must be contemplated
3. A set of sound performance evaluation procedures must be in place, and
4. A set of earthquake ground motion records, consistent with the hazard levels considered, must be selected and processed so that they can be applied towards building performance evaluations

We will discuss each if the above requirements in the subsections that follow.

7.6.1 Establishment of Performance Objectives

In 1999, Structural Engineers Association of California (SEAOC) developed perhaps the first conceptual frame work for establishment of performance objectives for design of new buildings (Structural Engineers Association of California (SEAOC), 1999). According to this document, the earthquake levels identified in Table 7.1 were considered proper for performance based design and evaluation.

Efforts spearheaded by the Federal Emergency Management Agency (FEMA) and carried out by the Applied Technology Council (ATC) culminated in publication of seminal prestandard for performance based seismic rehabilitation of existing structures (Federal Emergency Management Agency (FEMA), 2000). This document, which is commonly referred to as FEMA-356 was later modified and republished by the American Society of Civil Engineers (ASCE) as a performance based standard commonly referred to as the ASCE 41-06 (American Society of Civil Engineers, 2006). FEMA-356 and ASCE 41-06 further refined performance objectives in terms of acceptable performance of structural systems and components as well as nonstructural systems, attachments, and contents. These documents utilized three basic performance levels termed Immediate Occupancy (IO), Life Safety (LS)

Table 7.1 Earthquake levels and associated performance objectives suggested by the 1999 SEAOC document

Event	Recurrence interval	Probability of exceedance	Performance objective
Frequent	43 years	50% in 30 years	Fully operational
Occasional	72 years	50% in 50 years	Operational
Rare	475 years	10% in 50 years	Life safe
Very rare	975 years	10% in 100 years	Near collapse

and Collapse Prevention (CP) and ranges of performance bridging the identified performance levels. For each performance level, the anticipated behavior of structural and nonstructural components and contents were identified. As far as seismic hazard was concerned, these documents placed their emphasis on two probabilistic earthquake levels: (1) a 10% in 50 years (475 year mean recurrence interval) event usually associated with the LS performance objective, and (2) a 2% in 50 years (2,475 year mean recurrence interval) event usually associated with the CP performance objective.

FEMA-356 and ASCE 41-06 recognized four distinct analytical procedures: (1) the linear elastic static analysis procedure (LSP) (2) linear dynamic procedure (LDP) commonly carried out in terms of response spectrum analysis (3) the nonlinear static procedure (NSP) commonly referred to as the push-over analysis, and (4) the dynamic nonlinear response analysis (NDP). Due to limitations of applicability of LSP and NSP to tall buildings, current performance based guidelines for design of tall buildings, which we will discuss later, generally permit application of LDP for evaluation of performance where building is anticipated to remain essentially elastic such as the IO performance level and NDP for evaluation of performance for nonlinear stages of building behavior (LS and CP).

7.6.1.1 The Current Approach

The current approach to performance based design in the United States relies on component-based evaluation as delineated in the FEMA-356 and ASCE 41-06 documents. In the component-based approach, each component of the building (beam, column, wall segment, etc.) is assigned a normalized force/moment – deformation/rotation relation such as the one shown in Fig. 7.2 where segment AB indicates elastic behavior, point C identifies the onset of loss of capacity, segment DE identifies the residual capacity of the component, and point E identifies the ultimate inelastic deformation/rotation capacity of the component. Components are classified as primary (P) or secondary (S) and assigned with different deformation limits corresponding to various performance objectives. The vertical axis in this figure represents the ratio of actual force or moment to the yield force or moment. Primary

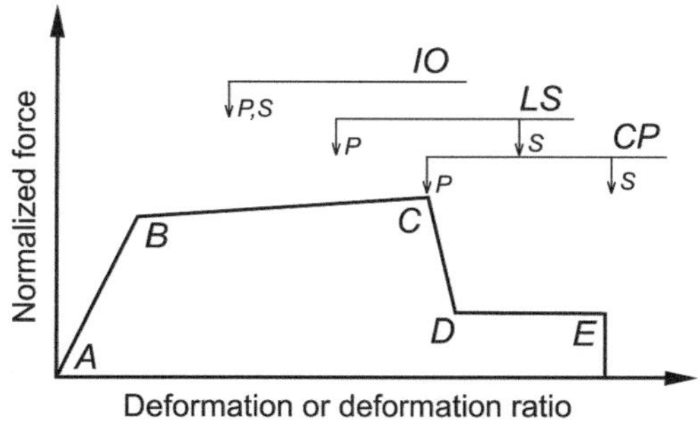

Fig. 7.2 Generalized component force-deformation relations for depicting modeling and acceptance criteria in FEMA-356 and ASCE 41-06 documents (Federal Emergency Management Agency (FEMA), 2000)

components are those whose failure results in loss of vertical load carrying capacity and endanger the safety of occupants (i.e., columns). Secondary elements are those whose failure can be sustained by the system via redistribution of forces to adjacent components without endangering the occupants (i.e., link beams in a coupled wall system).

Although ASCE 41-06 is officially intended for seismic rehabilitation of existing structures, its component-based performance limits for NDP are routinely referenced by guidelines for performance based design of tall buildings. Engineers, who believe that ASCE 41-06 tabulated limits are not applicable or too conservative for their intended component, perform laboratory testing and obtain confirmation of behavior for their component subject to approval by peer reviewers and approval agencies.

The Los Angeles Tall Buildings Structural Design Council (LATBSDC) was the first professional group in the United States to publish a performance-based alternative seismic analysis and design criteria specifically intended for tall buildings (Los Angeles Tall Buildings Structural Design Council (LATBSDC), 2005, 2009) and to obtain approval by the city's building officials in 2010 for its use in lieu of using prescriptive code provisions for buildings of all types and heights. The 2008 edition of LATBSDC criteria (Los Angeles Tall Buildings Structural Design Council (LATBSDC), 2009) sets two performance objectives: (1) serviceable behavior when subjected to frequent earthquakes defined as events having a 50% probability of being exceeded in 30 years (43 year return period); and (2) a very low probability of collapse under extremely rare earthquakes defined as events having a 2% probability of being exceeded in 50 years (2,475 year return period) with a deterministic cap. This earthquake is the Maximum Considered Earthquake (MCE) as defined by ASCE 7-05.

The intent of LATBSDC's serviceability performance objective is to make sure that the building structural and nonstructural components retain their general functionality during and after frequent events. Repairs, if necessary, are expected to be minor and could be performed without substantially affecting the normal use and functionality of the building. Under frequent earthquakes the building structure and nonstructural components associated with the building are expected to remain essentially elastic. Essentially elastic response may be assumed for elements when force demands generally do not exceed provided strength. When demands exceed provided strength, this exceedance shall not be so large as to affect the residual strength or stability of the structure.

The intent of LATBSDC's collapse prevention objective is to validate that collapse does not occur when the building is subjected to MCE ground motions. Demands are checked against both structural members of the lateral force resisting system and other structural members. Claddings and their connections to the structure must accommodate MCE displacements without failure.

Performance based tall building design guidelines published by other entities such as CTBUH (Council on Tall Buildings and Urban Habitat, 2008) and PEER (Pacific Earthquake Engineering Research Center (PEER), 2010) have followed performance objectives similar to those expressed in the 2008 LATBSDC document.

While component-based approaches are convenient and widely used, they are incapable of assessing the performance of the building as a whole. There is a real difference between a building with hundreds of columns where one column exceeds the CP limit and where all columns in a particular level exceed the CP limit and create a potential for collapse. Component-based methods are incapable of making that distinction and leave such a crucial distinction to the so-called "engineering judgment."

7.6.1.2 The Rigorous Approach

The Pacific Earthquake Engineering Research Center (PEER) performance based design framework has made it possible to evaluate the seismic performance of buildings and their attachments and consents in a system-wide rigorous and probabilistic approach (Cornell and Krawinkler, 2000) first suggested this approach by stating that "the basis for assessing adequacy of the structure or its design will be a vector of certain key *Decision Variables, DV*, such as the annual earthquake loss and/or the exceedance of one or more limit states (e.g., collapse). These can only be predicted probabilistically. Therefore the specific objectives of engineering assessment analyses are in effect quantities such as $\lambda_{\$(x)}$, the mean annual frequency (MAF) of the loss exceeding x dollars, or such as λ_{Icoll}, the MAF of collapse."

Later, this definition was expanded (Krawinkler and Miranda, 2004; Moehle and Deierlein, 2004) to include the vector of engineering demand parameters, *EDP*, resulting in the now famous triple integral of Eq. (1) with the process shown conceptually in Fig. 7.3

$$\lambda(DV) = \int\int\int G\langle DV|DM\rangle dG\langle DM|EDP\rangle dG\langle EDP|IM\rangle d\lambda(IM) \quad (1)$$

A comprehensive implementation of this approach is currently being undertaken by ATC in its ATC-58 project (Applied Technology Council, 2009) with funding from FEMA and a companion software tool (Performance Assessment Calculation

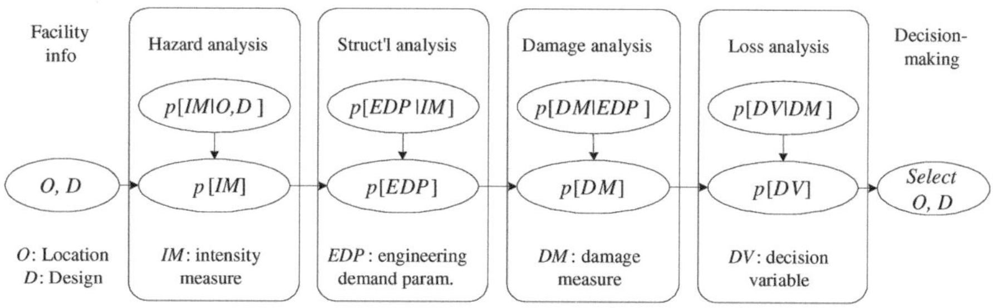

Fig. 7.3 PEER's probabilistic frame work for performance based design (after Cornell and Porter, Courtesy of Moehle and Deierlein)

Tool – PACT 2.0) is currently being developed and tested (Naeim, and Hagie, 2010; Hagie and Naeim, 2010). Using this tool and the methodology developed by ATC it is possible to assess probabilities of various outcomes for a given scenario, intensity of ground motion, or on an annualized time-based basis. Using this tool, once the fragility specifications for the building core, shell, improvements, and contents (including fragility curves and repair cost, repair time, and casualty consequences) are defined, seismic hazard curve, and earthquake demand parameters in terms of a series of linear and nonlinear structural analyses are introduced, the program carries out the necessary computations and provides a set of probabilistic outcomes.

Unfortunately, as currently envisioned by the ATC-58 project, a time-based evaluation requires consideration of 8 distinct hazard level and 11 sets of dynamic nonlinear response analyses per hazard level. That totals to 88 sets of nonlinear analyses! It is hard to imagine a design firm having the necessary time, budget, and resources to perform these many nonlinear analyses for performance based design. Particularly given the fact that with today's software and computer speeds each of these analyses could take somewhere between a 2 and 10 of days of computer time for a typical tall building.

7.6.2 Design Procedures

Unfortunately, none of the documents guiding the practice of performance-based design of tall buildings contain specific instructions on how to design a tall building so that it will satisfy the delineated performance objectives. For example, the 2008 LATBSDC criteria requires that the building be designed based on capacity design principles described in a project-specific seismic design criteria clearly describing how the structural system will achieve a well defined inelastic behavior where nonlinear actions and members are clearly defined and all other members are stronger than the elements designed to experience nonlinear behavior. Nonlinear action should be limited to the clearly defined members and regions. Yielding due to compression and bending at the base of columns (top of foundation or basement podiums) for steel structures is usually permitted. Other commonly designated zones of nonlinear behavior are listed in Table 7.2.

The document, however, does not explain how the engineer is supposed to achieve this capacity design. Since design by application of nonlinear response history analyses is neither feasible or advisable in a design office environment, most engineers select either a set of lateral forces or a push-over profile, based on their experience, to perform a preliminary capacity design of the building and then subject the building to the evaluation criteria contained in these documents and refine their design as necessary to achieve the specified performance objectives. Fortunately, methodologies for direct design of buildings which will likely satisfy performance criteria based on nonlinear dynamic response evaluations are emerging (Priestley et al., 2006; Goel and Chao, 2008).

Table 7.2 Zones and actions commonly designated for nonlinear behavior

Structural system	Zones and actions
Special moment resisting frames (steel, concrete, or composite)	• Flexural yielding of beam ends (except for transfer girders) • Shear in beam-column panel zones
Special concentric braced frames	• Braces (yielding in tension and buckling in compression)
Eccentric braced frames	• Shear link portion of the beams (shear yielding preferred but combined shear and flexural yielding permitted)
Unbonded braced frames	• Unbonded brace cores (yielding in tension and compression)
Special steel-plate shear walls	• Shear yielding of web plates • Flexural yielding of beam ends
R/C shear walls	• P-M-M yielding at the base of the walls (top of foundation or basement podiums) or other clearly defined locations with plastic hinge region permitted to extend to a reasonable height above the lowest plane of nonlinear action as necessary • Flexural yielding and/or shear yielding of link beams
Foundations	• Controlled rocking • Controlled settlement

7.6.3 Evaluation Procedures

In this section we provide an overview of evaluation procedures recommended by current performance based seismic design guidelines for tall buildings. We will identify the situations where the guidelines differ on their recommendations.

7.6.3.1 Analysis Methods

A three-dimensional mathematical model of the physical structure is used that represents the spatial distribution of the mass and stiffness of the structure to an extent that is adequate for the calculation of the significant features of the building's dynamic response. Structural models are required to incorporate realistic estimates of stiffness and damping considering the anticipated levels of excitation and damage. Generally, expected material properties (see Table 7.3) are used throughout except when calculating the capacity of brittle elements where specified strength values are used. For serviceability analyses, realistic values of stiffness should be used such as those listed in Table 7.4. Given the current state of modeling capabilities and available software systems, there is no reason to estimate the actual three-dimensional behavior of tall buildings by relying on approximate two-dimensional models.

For evaluation of performance under service level earthquakes either linear elastic response spectrum analyses or nonlinear analyses may be used. For evaluation of performance under MCE event, nonlinear dynamic response history analysis is required. In both types of analyses, P-Δ effects should be explicitly included as inclusion of P-Δ effects is crucial for establishing the onset of collapse because

Table 7.3 Suggested expected material strengths

Material	Expected strength	Strength
Structural steel		
	Hot-rolled structural shapes and bars	$1.5F_y$
	ASTM A36/A36M	$1.3F_y$
	ASTM A572/A572M Grade 42 (290)	$1.1F_y$
	ASTM A992/A992M	$1.1F_y$
	All other grades	
	Hollow structural sections	$1.3F_y$
	ASTM A500, A501, A618 and A847	
	Steel pipe	$1.4F_y$
	ASTM A53/A53M	$1.1F_y$
	Plates	$1.1F_y$
	All other products	
Reinforcing steel		1.17 times specified F_y
Concrete		1.3 times specified f'_c

Table 7.4 Suggested effective component stiffness values

Component	Flexural rigidity	Shear rigidity	Axial rigidity
Structural steel beams, columns and braces	$E_S I$	$G_S A$	$E_S A$
Composite concrete metal deck floors	$0.5E_c I_g$	$G_c A_g$	$E_c A_g$
R/C beams – nonsprestressed	$0.5E_c I_g$	$G_c A_g$	$E_c A_g$
R/C beams – prestressed	$E_C I_g$	$G_c A_g$	$E_c A_g$
R/C columns	$0.5E_c I_g$	$G_c A_g$	$E_c A_g$
R/C walls	$0.75E_c I_g$	$G_c A_g$	$E_c A_g$
R/C slabs and flat plates	$0.5E_c I_g$	$G_c A_g$	$E_c A_g$

Notes: E_c shall be computed using expected material strength; G_c shall be computed as $E_c/(2(1+\nu))$, where ν is taken as 0.20.

collapse of tall buildings is ultimately P-Δ related. The push-over curves shown in Fig. 7.4 are illustrative of this fact.

In elastic response spectrum analyses at least 90% of the participating mass of the structure is to be included in the calculation of response for each principal horizontal direction and modal responses are combined using the Complete Quadratic Combination (CQC) method.

There are various ways to model the behavior of nonlinear elements (Fig. 7.5). Most commonly, concentrated plasticity models are used for beams and columns and distributed plasticity or fiber models are used for modeling walls and floor diaphragms.

Inherent torsional properties of the structural system should always be considered. The 2008 LATBSDC Guideline (Los Angeles Tall Buildings Structural Design

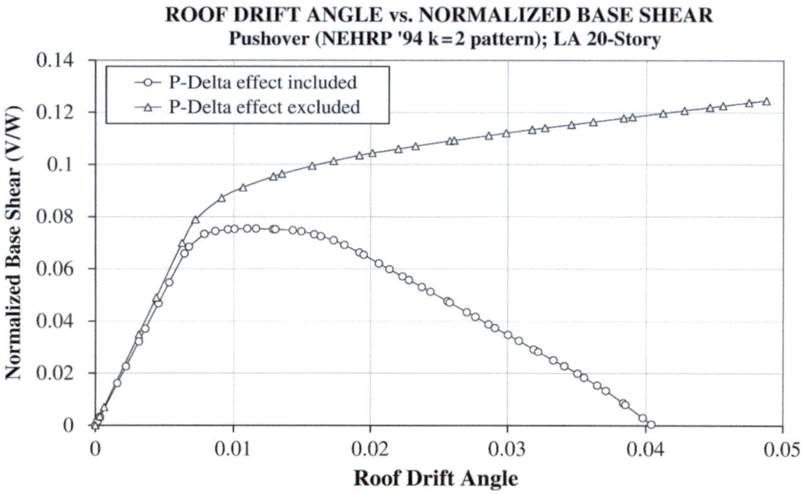

Fig. 7.4 Base shear – roof displacement pushover curves for a 20-story steel moment frame building. Note that reduction in load carrying capacity does not occur if P-Δ effects are not considered (Courtesy of Prof. Helmut Krawinkler)

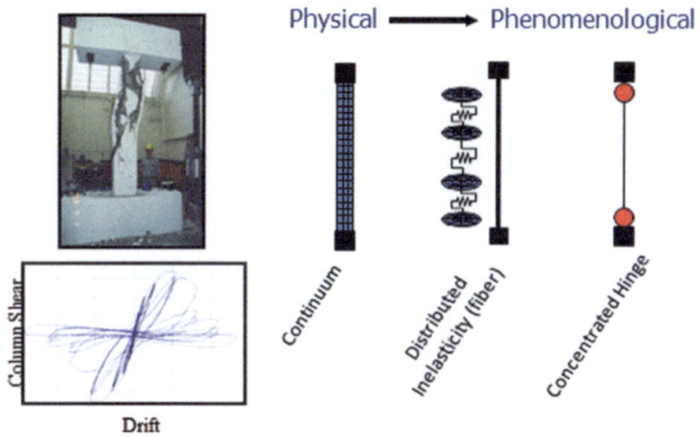

Fig. 7.5 Nonlinear component model types (Courtesy of Prof. Greg Deierlein)

Council (LATBSDC), 2009) requires assessment of accidental eccentricities during the service level building evaluation and if the impact of such eccentricities is proven to be significant, then it requires addressing them in one way or another during the MCE level evaluation. The PEER Guideline (Pacific Earthquake Engineering Research Center (PEER), 2010) which is not published in the final form at the time of writing of this paper does not require consideration of accidental eccentricities. Consideration of accidental eccentricities, particularly during nonlinear dynamic analyses, substantially complicates the evaluation process and little is gained by this increased effort. Therefore, the author tends to agree with the PEER Guideline on this issue which is consistent with the previous version of the LATBSDC Guideline (Los Angeles Tall Buildings Structural Design Council (LATBSDC), 2005).

In addition to the designated elements and components of the lateral force resisting system, all other elements and components that in combination significantly contribute to or affect the total or local stiffness of the building should be included in the mathematical model. Axial deformation of gravity columns in a core-wall system is one example of effects that should be considered in the structural model of the building (Applied Technology Council, 2008; Wallace, 2010).

7.6.3.2 Modeling Criteria

The 2008 LATBSDC Guideline limits itself to a general statement with respect to modeling techniques and consideration of strength degradation in nonlinear analysis. According to this document, all structural elements for which demands for any of the response-history analyses are within a range for which significant strength degradation could occur, should be identified and the corresponding effects appropriately considered in the analysis.

The PEER Guideline is a bit clearer in its requirements but goes significantly further and identifies four possible ways for modeling nonlinear components in its commentaries.

According to this document, deformation capacities may be taken equal to the corresponding CP values for primary elements published in ASCE 41 (with Supplement 1) for nonlinear response procedures, or may be based on analytical models validated by experimental evidence. When applicable, the ASCE 41 component force versus deformation curves may be used as modified backbone curves, with the exception that the drop in resistance following the point of peak strength should not be as rapid as indicated in the ASCE 41 curves. In the commentary section, this document states the commonly accepted fact that the rapid post-peak drop in resistance indicated in the ASCE-41 curves is not realistic (unless fracture occurs) and is likely to cause numerical instabilities in the analysis process. The four possible ways for modeling nonlinear components identified by this document are (see Fig. 7.6):

1. Explicitly incorporate cyclic deterioration in the analytical model.
2. Use cyclic envelope curve as a modified backbone curve if cyclic deterioration is not considered explicitly.
3. Use factors for modification of backbone curve if cyclic deterioration is not considered explicitly.
4. Limit deformation capacities so that no deterioration occurs in the analytical model.

For steel moment resisting frame systems, the contribution of panel zone (beam-column joint) deformations is to be included. If linear models are used for service level evaluations, then in lieu of explicit modeling of beam-column panel zone behavior, center-to-center beam dimensions may be used for such evaluations.

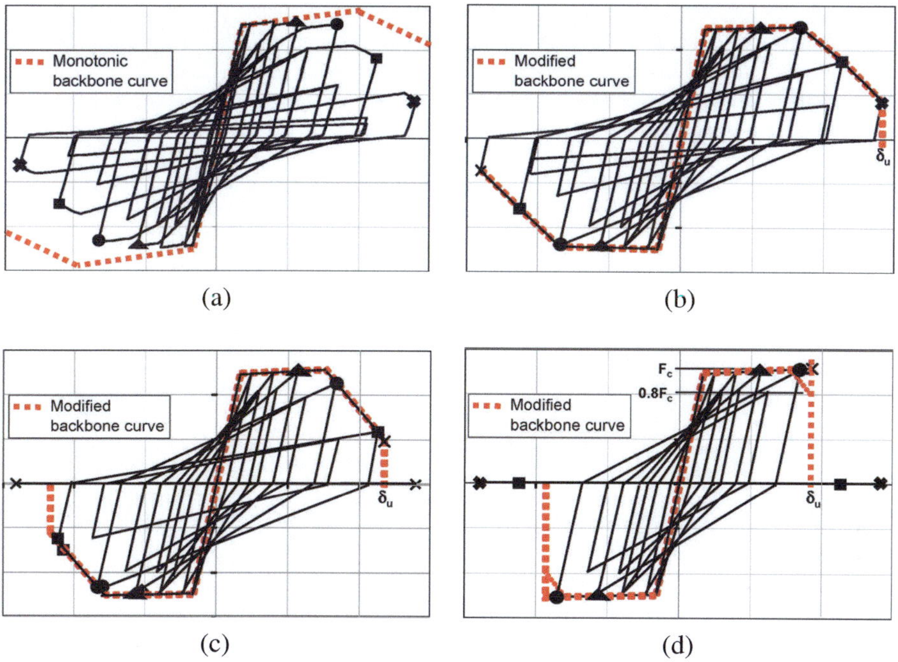

(a) (b) (c) (d)

Fig. 7.6 Illustration of the four options for analytical component modeling explained in PEER Guidelines commentary. (a) Option 1 – with cyclic deterioration, **b** option 2 – modified backbone curve = envelope curve, **c** option 3 – modified backbone curve = factored monotonic backbone curve, and **d** option 4 – no strength deterioration (Courtesy of Prof. Helmut Krawinkler)

Large compressive forces on columns reduce their ductility. To address this issue Supplement to 2008 LATBSDC limits the MCE compressive force demand on reinforced concrete columns to $0.4f'_cA_g$, where f'_c is the compressive strength of the concrete and A_g is the gross cross sectional area of the column. PEER Guidelines limit MCE compressive force demand to the balanced load which may be taken as $0.3f'_cA_g$.

The 2008 LATBSDC is silent on the issue of modeling soil-foundation-structure interaction (SFSI) and issues involved in modeling of subterranean floors common in tall building construction. Naeim et al. (2008) have identified the extreme difficulties involved in accurate modeling SFSI and subterranean floors for tall buildings with software tools used in a typical design office (Figs. 7.7 and 7.8). For models excited with base displacements as shown in Fig. 7.7, such software often report an erroneous and huge acceleration spike at the first time step of response-history analysis (see trace to the left on Fig. 7.8). If the spike at the first time step is removed, the trace shown on the right of Fig. 7.8 is obtained which still contains smaller spikes which in author's opinion are not real and are caused by solution instability. Realizing these difficulties and studying the results of numerous approximate procedures, Naeim et al. suggested that as long as explicit SFSI modeling in a design

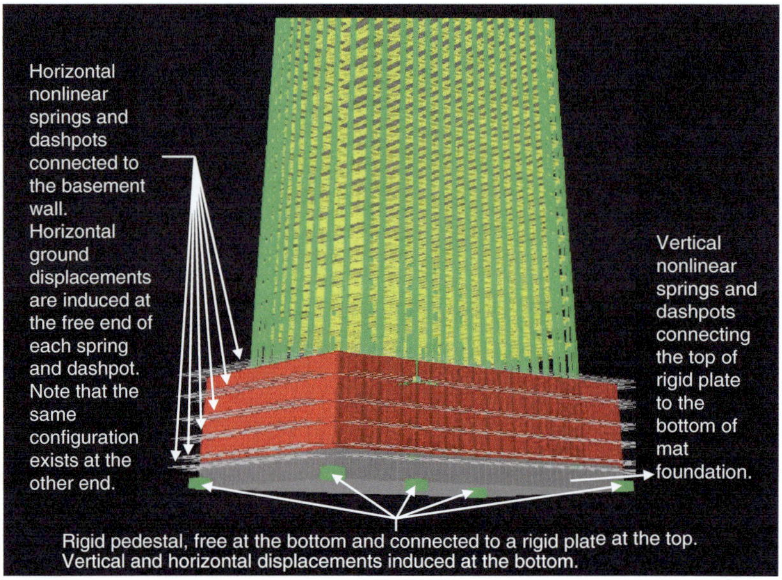

Fig. 7.7 A soil-foundation-structure interaction modeling technique for tall buildings with subterranean floors

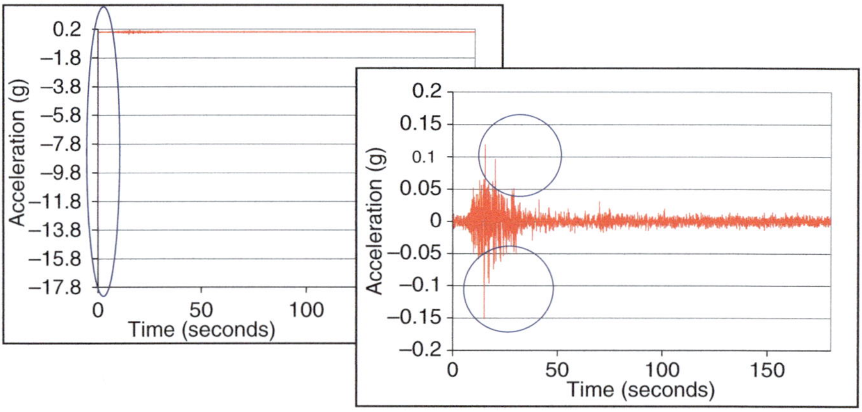

Fig. 7.8 Issues related with computed accelerations obtained from software usually used in a design office environment

office environment is unattainable, effects of SFSI may conservatively enveloped by using two simplified models where in one the soil media around the subterranean portions of the structure is ignored and in the other the building is considered fixed at the ground level. Based on Naeim et al. findings, PEER Guidelines recommends the use of the first approximation suggested by Naeim et al. for service level modeling and the use of a "bathtub" model (in its commentary) for MCE evaluations (Fig. 7.9).

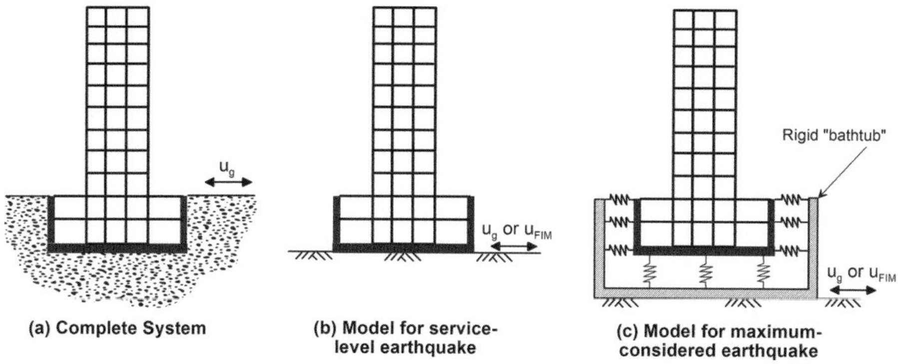

Fig. 7.9 Schematic illustration of tall building with subterranean levels and simple models for analysis in which soil-foundation interaction effects are neglected (**Part b**) and included in an approximate manner (**Part c**). **Part (c)** only shows springs but parallel dashpots are generally also used (Courtesy of Prof. Jonathan Stewart)

In this author's opinion, the suggested "bathtub" model is not practically implementable in a design office environment given the limitations of existing software tools.

Damping is a particularly thorny issue and perhaps the most comprehensive coverage of damping as far as tall building design is concerned can be found in the ATC-72 draft (Applied Technology Council, 2008). While 5% critical damping is universally used, right or wrong, in linear analysis of structures for design according to prescriptive codes, Guidelines for performance based design of tall buildings currently contain little guidance, if any, on proper level of damping to consider and how to model it. The 2008 LATBSDC document is silent on the issue of damping and thereby leaving it to the discretion of design engineers and oversight of the project's peer review panel and building officials. The PEER Guidelines specifies 2.5% damping to be used for serviceability evaluations. For nonlinear analyses, most of energy dissipation occurring in the structural components is directly modeled through definition of hysteretic force/moment – displacement/rotation formulations. Some viscous damping is usually included in nonlinear analyses to account for energy dissipation occurring in the nonstructural components and parts of the structural system which is not included in the nonlinear model. Furthermore, a small amount of viscous damping usually alleviates numerical stability issues that are once in a while encountered during such analyses. With these considerations in mind, the ATC-72 draft recommends using a equivalent viscous damping value (D) equal to α/N, where D is the maximum percent critical damping, N is the number of stories in a tall building (>30), and α is a coefficient with a recommended range of $\alpha = 70$–150 for nonlinear analyses with interstory drift amplitudes of 0.005–0.03. System specific recommended values for α are as follows:

- Dual systems (RC core wall plus RC or steel frame): $\alpha = 130$
- RC moment frame systems: $\alpha = 100$
- RC core wall systems: $\alpha = 80$

- Steel moment frame systems: $\alpha = 80$
- Steel braced frame systems: $\alpha = 70$

Finally, for use with the mass and stiffness proportioned damping (Rayleigh damping), ATC-72 draft recommends using $\alpha_M = 9\zeta/T_1$ and $\alpha_K = \zeta T_1/15$ where T_1 is the fundamental period of vibration of the building and ζ is percentage of critical damping.

7.6.3.3 Acceptability Criteria

Acceptability criteria for both serviceability and MCE usually contain an absolute ceiling on the permitted drift. For serviceability earthquakes this is intended to minimize damage. For MCE event this is to limit the P-Δ effects. Both 2008 LATBSDC and PEER Guidelines limit overall drift ratio to one half of 1% (0.005) for serviceability and 3% (0.030) for MCE level earthquakes. In addition the maximum interstory drift at any story is limited to 41/2% (0.045) for MCE level events.

Important component level serviceability criteria in 2008 LATBSDC Peer Guidelines may be briefly summarized as follows:

- Force demands do not exceed the capacities for brittle actions (i.e., shear, axial force, etc.).
- Inelastic deformation demand ratios do not exceed $\Delta_e + 0.15\Delta_p$ for ductile actions per 2008 LATBSDC or the relevant ASCE-41 IO values per Peer Guidelines (see Fig. 7.10). If elastic response spectrum analysis is performed, this limited nonlinear behavior for ductile actions is accommodated by permitting a maximum demand to capacity ratio to 1.2 by 2008 LATBSDC and 1.5 by PEER Guidelines. Notice that 2008 LATBSDC assumes 5% damped spectrum compared to 2.5% used by PEER Guidelines. Therefore the difference between 1.2 and 1.5 in the two documents is not as significant as it may appear.

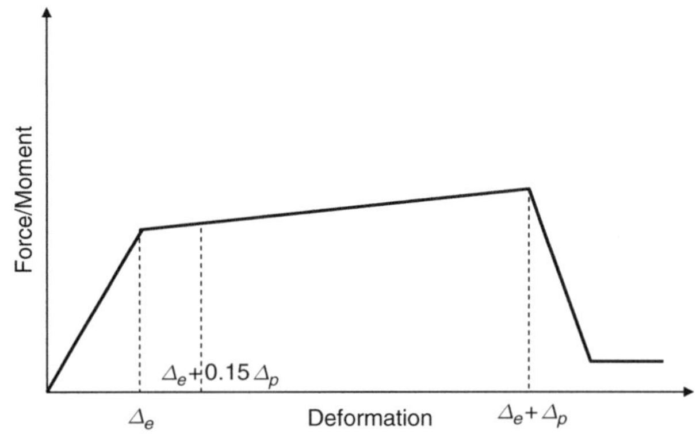

Fig. 7.10 Schematic illustration of permitted inelastic behavior for ductile elements under serviceability earthquake according to 2008 LATBSDC

PEER Guideline is a more comprehensive document and provides much more details with respect to MCE acceptability criteria. The essence of the matter, however, is pretty simple: demand should be less than capacity provided. For ductile action this means deformation demands should be less than deformation capacities (2008 LATBSDC assumes that capacity is exhausted when strength drops below 80% of maximum strength). For all other actions it means that force demands are less than component's nominal strength. Diaphragm chords and collectors are designed to deliver the force generated at the MCE with little, if any, inelasticity.

Current best practice for performance based design of tall buildings requires nonlinear dynamic analyses for MCE evaluations using a minimum of seven pairs of earthquake ground motions. Maximum average earthquake demand parameters obtained from such analyses is used to verify the capacity for ductile action. Nominal (unreduced) capacities are unusually used. According to 2008 LATBSDC nonductile actions are evaluated based on capacity design principles. PEER Guidelines contain more stringent requirements where the nonductile (force-controlled) actions are divided into two categories. The first category of actions referred to as *Critical Actions* are those in which failure mode pose severe consequences to structural stability under gravity and/or lateral loads. The second category, referred to as *Noncritical Actions* consists of all other force-controlled actions. For Critical Actions, if the computed demand for an action is not limited by a well defined yielding mechanism, 1.5 times the maximum average demand is to be used. If, however, the computed demand for an action is limited by a well defined yield mechanism, then the mean plus 1.3 times the standard deviation obtained from the individual response history analyses but not less than 1.2 times the maximum average values. It is generally accepted that use of 7 pairs of time histories provides reasonable estimates of mean values of response parameters but does not provide an adequate representation of dispersion in the data. The 1.3 and amplification 1.5 factors recommended by PEER Guidelines are intended to address this issue.

7.6.4 Ground Motion Record Selection and Scaling

While 2008 LATBSDC simply incorporated by reference the provisions of Section 16.1.3 and Chapter 21 of ASCE 7-05 for construction of site-specific uniform hazard spectra and selection and scaling of earthquake records, PEER Guidelines devotes an entire detailed chapter to this subject. 2008 LATBSDC permits either scaling (time-domain manipulation) or spectral matching (frequency-domain) manipulation of earthquake records as long as the requirements of Section 16.1.3 of ASCE 7-05 are satisfied. PEER Guidelines only speaks of spectral matching although it is this author's opinion that the Guidelines writers did not intend the term matching to be interpreted in the strict sense as described above. According to this document, records may be *matched* either to the uniform hazard spectrum or conditional mean spectrum (CMS). If the CMS approach is used, then a suite of CMS, each matched to one of the key periods described in the document because the use of CMS for only the fundamental period is not recommended for tall buildings. Key building periods are defined to consist at least, the first three translational periods of structural response in each of the structure's two principal orthogonal response directions. This would mean that if the CMS approach is used, the engineer should perform at least 21 sets on nonlinear dynamic response analyses instead of 7. This author does not believe that the use of CMS approach as recommended by the PEER Guidelines will be adopted anytime soon in practice of tall building design as this imposes a substantial computational burden on the design engineers. In addition, the author agrees with Moehle that much is not gained by performing such an onerous exercise (Moehle, 2010).

There is a shortage of earthquake records from large magnitude events at close distances to the source of energy release. To overcome this deficiency, sometimes simulated (synthetic) ground motion records are used to augment the recorded set of ground motions used for design. The users of synthetic ground motions should be aware that at the time of writing of this paper, at least some of the procedures used to develop simulated ground motions result in records that exhibit faster attenuation at short periods and exhibit less dispersion compared to the recorded ground motions (Stewart, 2009).

7.6.5 Peer Review Requirements

Stringent peer review conducted during the entire process of structural design and not limited to a review of the end product is an essential quality control and assurance necessity for performance based seismic design of tall buildings. The generally accepted peer review requirements are as follows:

- Each project needs a Seismic Peer Review Panel (SPRP). The SPRP is to provide an independent, objective, technical review of those aspects of the structural design of the building that relate to seismic performance and to advise the Building Official whether the design generally conforms to the intent of the design criteria established for the project.
- The SPRP participation is not intended to replace quality assurance measures ordinarily exercised by the engineer of record (EOR) in the structural design of a building. Responsibility for the structural design remains solely with the EOR, and the burden to demonstrate conformance of the structural design to the intent of this document and other requirements set forth by the Building Official resides with the EOR. SPRP is not a plan checking entity and the responsibility for conducting structural plan checking resides with the Building Official.
- The SPRP should include a minimum of three members with recognized expertise in relevant fields, such as structural engineering, earthquake engineering research, performance-based earthquake engineering, nonlinear response history analysis, tall building design, earthquake ground motion, geotechnical engineering, geological engineering, and other such areas of knowledge and experience relevant to the issues the project poses.

- The SPRP members shall be selected by the Building Official based on their qualifications applicable to the Seismic Peer Review of the project. The Building Official may request the opinion of the Project Sponsor and EOR on proposed SPRP members, with the Building Official making the final decision on the SPRP membership.
- SPRP members shall bear no conflict of interest with respect to the project and shall not be part of the design team for the project.
- The SPRP provides their professional opinion to and acts under the instructions of the Building Official.

7.6.6 Instrumentation and Structural Health Monitoring

Performance based design of tall buildings is in its early stages of application and development. The funding necessary to experimentally validate performance of various components and systems utilized in tall buildings probably will not be available for a long time. Analytical simulations, as detailed and elaborate as they may be, cannot replace the need for experimental results and observed performances. It is imperative that we maximize every opportunity at our disposal to learn as much we can and as quickly as possible about performance of tall buildings designed according to these procedures during major earthquakes so that we can improve our design practices and produce more efficient and safe buildings.

Seismic instrumentation can provide valuable insight into performance of structures and help us assess the validity, or lack thereof, of assumptions used and methods applied. It is precisely for this reason that 2008 LATBSDC mandates extensive seismic instrumentation of tall buildings designed according to its provisions. According to 2008 LATBSDC, a tall building must be instrumented with the minimum number of sensors shown in Table 7.5. Each sensor records a single response quantity of interest (e.g., unidirectional floor acceleration, interstory displacement, etc.).

Instrumentation is inexpensive and nonintrusive if planned during the design process and implemented during the construction of a tall building. Given the recent advances in sensor technology, now it is possible to install sensors not only to measure accelerations but to measure and record relative or overall displacements (building tilt), and various stresses and strains throughout the structure. Modern information technology has made real-time or near real-time measurements and remote transmission of sensor data and engineering interpretation of them to remote

Table 7.5 Minimum tall building instrumentation levels

Number of stories above ground	Minimum number of sensors
10–20	15
20–30	21
30–50	24
>50	30

locations via the Internet not only possible, but feasible. Integration of seismic mentoring with broad building health monitoring which includes monitoring buildings during more frequent events and malfunctions (such as wind storms, fires, floor vibrations, flooding, elevator functions, HVAC problems), may finally produce enough tangible benefits for tall building owners and developers to cause them to willingly and enthusiastically embrace the modes cost of building instrumentation and health monitoring.

7.7 Conclusion

An overview of current state of practice for application of performance based design methodologies to tall building design was presented. The contents of leading guidelines currently used for performance based design of tall buildings were introduced and compared. Performance based design has already established itself as the methodology of choice for design of tall buildings located in seismic regions as it has shown its capability to produce safer more cost-effective tall building structures compared to conventional design techniques. Performance based design of tall buildings is in its early stages of development. A series of elaborate analytical and experimental research is needed in order to fulfill its ultimate potential. Extensive seismic instrumentation and structural health monitoring have the potential of significantly accelerating our path forward and transforming performance based design of tall buildings from its current state to a fully matured seismic design methodology.

Acknowledgments The author is indebted to his colleagues at LATBSDC and PEER. Special gratitude is due to Professors Jack Moehle, Helmut Krawinkler, Greg Deierlein, Jonathan Stewart, Farzin Zareian, Mr. Tony Ghodsi and Mr. Ronald Hamburger for graciously providing the author with access to their research results and permission to reproduce their artwork as needed.

References

American Society of Civil Engineers (2006) ASCE/SEI standard 41-06, seismic rehabilitation of existing buildings, Reston, VA

American Society of Civil Engineers (2006) Minimum design loads for buildings and other structures, ASCE 7-05 January

Applied Technology Council (2009) Guidelines for seismic performance assessment of buildings. ATC-58 50% Draft, Redwood City, CA

Applied Technology Council (Sept 2008) Interim guidelines on modeling and acceptance criteria for seismic design and analysis of tall buildings. PEER Tall Buildings Initiative, ATC-72-1, 95% Draft

Cornell CA, Krawinkler H (2000) Progress and challenges in seismic performance assessment. PEER Center News 3:2 Spring 2000

Council on Tall Buildings and Urban Habitat (2008) Recommendations for the seismic design of high-rise buildings – a consensus document – CTBUH seismic working group, Chicago, IL

Council on Tall Buildings and Urban Habitat (2010) CTBUH tall building database. http://www.ctbuh.org/HighRiseInfo/TallestDatabase/tabid/123/language/en-US/Default.aspx

Federal Emergency Management Agency (FEMA) (November 2000) Prestandard and commentary for the seismic rehabilitation of buildings. FEMA 356, Washington, DC

Goel SC, Chao S-H (2008) Performance-based plastic design: earthquake-resistant steel structures. ICC Press, Boca Raton, FL

Hagie S, Naeim F (2010) PACT 2.0 technical manual, a report submitted to Applied Technology Council. John A. Martin & Associates, Inc.

Heaton TH, Hall JF, Wald DJ, Halling MW (1995) Response of high-rise and base-isolated buildings to a hypothetical Mw 7·0 blind thrust earthquake. Science 267:206–211

International Conference of Building Officials. (2002) City of Los Angeles building code. Sec. 1630.10.2, Whittier, CA

Krawinkler H, Miranda E (2004) Performance based earthquake engineering. In: Bozorgnia Y, Bertero VV (eds) Earthquake engineering: From engineering seismology to performance-based engineering. CRC Press, Boca Raton, FL

Los Angeles Tall Buildings Structural Design Council (2007) In: Proceedings of the 2007 annual meeting, Los Angeles, CA, May 2007

Los Angeles Tall Buildings Structural Design Council (2008) In: Proceedings of the 2008 annual meeting, Los Angeles, CA, May 2008

Los Angeles Tall Buildings Structural Design Council (LATBSDC) (May 2005) An alternative procedure for seismic analysis and design of tall buildings located in the Los Angeles region, 2005 edn.

Los Angeles Tall Buildings Structural Design Council (LATBSDC) (May 2009) An alternative procedure for seismic analysis and design of tall buildings located in the Los Angeles region with supplement #1, 2008 edn.

Los Angeles Tall Buildings Structural Design Council (2009) In: Proceedings of the 2009 annual meeting, Los Angeles, CA, May 2009

Moehle J (2010) Ground motion selection and scaling for tall building design. In: Proceedings of the University of Tokyo symposium on long-period ground motion and urban disaster mitigation, 17–18 Mar 2010

Moehle J, Deierlein G (2004) A framework methodology for performance-based earthquake engineering. In: Proceedings of the 13th world conference on earthquake engineering, Vancouver, BC, paper no. 679

Naeim F, Garves R (2005) The case for seismic superiority of well-engineered tall buildings. Struct Des Tall Spec Build 14(5):401–416, Wiley InterScience, London

Naeim F, Hagie S (2010) PACT 2.0 user manual, a report submitted to applied technology council. John A. Martin & Associates, Inc. Los Angeles, CA

Naeim F, Tileylioglu S, Alimoradi A, Stewart JP (2008) Impact of foundation modeling on the accuracy of response history analysis of a tall building. In: Proceedings of the SMIP2008 seminar on utilization of strong motion data, California strong motion instrumentation program, Sacramento, CA, pp 19–55

Pacific Earthquake Engineering Research Center (PEER) (February 2010) Seismic design guidelines for tall buildings final draft. An Educational CD-Rom Disc published by PCA, Illinois.

Portland Cement Association (2000) Concrete structural floor systems and more. CD013

Potter S (1964) Concurring opinion in Jacobellis v. Ohio, 378 U.S. 184

Priestley MJN, Calvi GM, Kowalsky MJ (2006) Displacement-based seismic design of structures. IUSS Press, Italy

Stewart JS (2009) Tall buildings initiative: comparison of recorded and simulated ground motions for tall buildings. In: Proceedings of the SMIP09 seminar on utilization of strong-motion data, San Francisco, CA, 19 Nov 2009

Structural Engineers Association of California (SEAOC) (1999) Recommended lateral force requirements and commentary, 7th edn. SEOAC, Whittier, CA Appendices G and I

Wallace J (2010) Performance-based design of tall reinforced concrete core wall buildings, Theme Lecture. In: Proceedings of the 14th European conference on earthquake engineering, contained in this book.

Chapter 8
Evaluation of Analysis Procedures for Seismic Assessment and Retrofit Design

M. Nuray Aydınoğlu and Göktürk Önem

Abstract Analysis procedures developed in the last two decades for performance-based seismic assessment and retrofit design of building structures are critically evaluated. Nonlinear analysis procedures within the framework of deformation-based seismic assessment process are classified with respect nonlinear modeling and acceptance criteria. The critical transition from linear engineering to nonlinear, performance-based engineering practice is addressed. Specifically, the need for enhancing engineers' knowledge on nonlinear behavior and analysis methods in university education and professional training is highlighted. Rigorous as well as practice-oriented nonlinear analysis procedures based on pushover analysis are treated where special emphasis is given to the latter. All significant pushover analysis procedures developed in the last two decades are summarized and systematically assessed on the basis of a common terminology and notation. Each procedure is evaluated in terms of its practical use as a *capacity estimation tool* versus *capacity-and-demand estimation tool.*

8.1 Introduction

With rapidly growing urbanization in earthquake prone areas in various parts of the world and the consequent increase in urban seismic risk, seismic performance assessment of existing buildings continues to be one of the key issues of earthquake engineering. This contribution is devoted to the evaluation of developments took place in the last two decades in seismic assessment and retrofit design of existing buildings in terms of progress achieved in analysis philosophy and implementation procedures.

M.N. Aydınoğlu (✉)
Kandilli Observatory and Earthquake Research Institute, Boğaziçi University, 34684 Istanbul, Turkey
e-mail: aydinogn@boun.edu.tr

M. Garevski, A. Ansal (eds.), *Earthquake Engineering in Europe*, Geotechnical, Geological, and Earthquake Engineering 17, DOI 10.1007/978-90-481-9544-2_8,

In spite of the rationalization of the strength-based design of new buildings in 1980s with the publication of ATC-03 report (ATC, 1978), it was realized in time that this procedure was not suitable for the seismic assessment of existing, old structures, which remained as a critical problem to be resolved until 1990s. Eventually it was concluded that structural behavior and damageability of structures during strong earthquakes were essentially controlled by the inelastic deformation capacities of ductile structural elements. Accordingly, earthquake engineering inclined towards a new approach where seismic evaluation and design of structures are based on nonlinear deformation demands, not on linear stresses induced by reduced seismic forces that are crudely correlated with an assumed overall ductility capacity of a given type of a structure, the starting point of the strength-based design.

The footsteps of performance-based seismic engineering were heard in 1995 with the publication of Vision 2000 document (SEAOC, 1995). This paved the way for two major documents, ATC-40 (ATC, 1996) and FEMA 273-274 (FEMA, 1997), which pioneered the implementation of practice-oriented nonlinear analysis procedures for seismic evaluation and rehabilitation of buildings within the framework of performance-based seismic engineering. In the last decade, such practice was started to be codified in the USA (FEMA, 2000; ASCE, 2007), in Europe (CEN, 2004), in Japan (BCJ, 2009) and in Turkey (MPWS, 2007).

8.2 Analysis Procedures for Seismic Assessment and Design

Analysis procedures for seismic assessment can be broadly broken down into two main categories, namely, linear analysis procedure for strength-based design and assessment and, nonlinear analysis procedures for deformation-based assessment.

8.2.1 Linear Analysis Procedure for Strength-Based Assessment and Design: Traditional Procedure for "Linear Engineers"

Strength-based design is the traditional code procedure, which is still being used throughout the world for the seismic design of new structures. It is the extension of the historical *seismic coefficient method*, which is rationalized and re-defined in 1978 with the publication of ATC-03 document (ATC, 1978). In this procedure, elastic equivalent seismic loads are reduced by certain load reduction factors (response modification factors) and applied to the building in each mode for a linear elastic analysis. The concept of load reduction is based on a single-valued global ductility capacity assumed for the entire structure. This is a judgmental assumption based on several factors, including structural material behavior, redundancy, inherent over strength as well as past experience obtained from case studies with nonlinear analyses, laboratory tests and post-earthquake observations. A typical reduction factor (R) is estimated through the so-called $R_y - \mu - T$ relationships and augmented by an over-strength factor.

Essentially, estimation of reduced seismic load is nothing but a more elegant way of directly assuming a seismic coefficient for a given type of a building. The most significant progress in strength-based design was realized with the introduction of *capacity design principles* in 1970s (see Paulay and Priestley, 1992). Although such principles were included in ATC-03 document (1978), their adoption by the seismic codes came relatively late, starting in 1988 with the Uniform Building Code (ICBO, 1988).

With the design of sections according to section forces obtained from the linear analysis under reduced seismic loads combined with an appropriate implementation of capacity design principles have led to a relatively simple and successful prescriptive design approach for new buildings, which is still being used worldwide by almost all seismic design codes.

Regarding the use of strength-based approach for the seismic assessment of existing buildings, however, there are several obstacles. The first major problem is the estimation of a target ductility factor for the building to be assessed. Even the terminology is at odds, as nothing can be targeted for an existing building. Essentially the strength-based approach was developed as a *design approach*, not an *assessment approach*. Estimation of an existing ductility capacity applicable to the entirety of an existing building is impossible. On the other hand, linear behavior assumption under reduced seismic loads gives the engineer no indication about the real, inelastic response of the structural system and the possible damage distribution.

In spite of serious drawbacks of strength-based approach in seismic assessment, we have to admit that until recently it was the only approach that practicing engineers could possibly apply to existing buildings. We have to confess that we, civil and structural engineers all over the world, are all "linear engineers" by mentality as a result of our education. Even though we have learned through ultimate strength design that materials behave nonlinearly at section basis, still we are accustomed to analyze our structural systems based on linear system behavior for any action under "prescribed loads" defined by the codes. Actually there is nothing wrong with it, because we are sure that our systems would remain more or less in the linear range under almost all actions. But alas, practicing engineers learned rather lately that the seismic action was an exception.

In view of the fact that majority of design engineers in practice still remain as "linear engineers", a group of code writers attempted to develope an *assessment version* of the strength-based design approach. In this scheme, demand to capacity ratios (DCR's) are calculated by dividing the elastic section forces (elastic demands) to the corresponding section capacities of the existing sections. Thus in an equivalent sense, strength reduction factors of the individual sections are obtained. This is followed by applying the well-known *equal displacement rule* by assuming that those factors are equal to the corresponding section ductility demands, and finally such demands are compared, at each section, to the prescribed section ductility capacities.

It is seen that the system-based approach applied in the traditional strength-based design is now being imitated in a kind of section-based approach. The question is whether such an imitation is theoretically viable. First of all, the parameters and relationships including the *displacement ductility ratio*, μ, are all valid only for the equivalent (modal) SDOF system. It is not clear how section ductility ratio can be defined. Moreover equal displacement rule is verifiable only through comparisons of linear and nonlinear peak displacement responses of SDOF systems and such a relationship can hardly be extended to the section responses of linear and nonlinear multi-degree-of-freedom (MDOF) systems. Finally this section-based approach eventually relies on judgmentally prescribed section ductility capacities as acceptance criteria, as for the global target ductility capacity prescribed in the system-based approach.

It is thus clear that section-based approach of strength-based design cannot be considered as a viable and reliable seismic assessment procedure. Yet, this procedure is now contained in a number of codes, including the USA (ASCE, 2007) and Turkey (MPWS, 2007). Considering the fact that the engineering community worldwide is in a transition stage from linear to nonlinear engineering practice, such linear procedures should be considered to be temporary applications until practicing engineers become fully familiar with the new, modern procedures.

8.2.2 Nonlinear Analysis Procedures for Deformation-Based Seismic Assessment: A New Era in Earthquake Engineering

The last decade of the previous millennium witnessed the development of the concept of performance-based seismic assessment, where seismic demand and consequent damage are estimated on a quantifiable basis under given levels of seismic action and such damage is then checked to satisfy the acceptable damage limits set for the specified performance objective(s). Seismic action levels and performance objectives can be found in the relevant literature (ASCE, 2007).

Since damage to be estimated at the component level is generally associated with the nonlinear behavior under strong ground motion, the concept of performance-based seismic assessment is directly related to nonlinear analysis procedures and deformation-based seismic assessment concept.

There is no doubt that development of deformation-based seismic assessment procedure represents a new era in earthquake engineering. For the first time, practicing engineers have become able to *calculate* the plastic deformation quantities under a given seismic action as seismic demand quantities, which are the direct attributes of the seismic damage. They have realized that such demand quantities should be within the limits of deformation capacities of the ductile structural elements. They better understood the significance of brittle demand quantities and brittle failure modes. They learned how they could identify the weaknesses of a given structural system and thus they began feeling themselves better equipped to construct well-behaved ductile structural systems, either in new construction or in retrofit design of existing structures. We should admit that before the advent of deformation-based assessment, any of the above-mentioned topics was hardly on the agenda of the structural engineer who had no idea of a seismic design with an alternative approach other than strength-based approach.

Nonlinear analysis procedures for deformation-based seismic assessment can be grouped into two main categories, namely, Nonlinear Response History Analysis (NLRHA) and Practice Oriented Nonlinear Analysis (PONLA) procedures.

NLRHA is the rigorous analysis procedure based on the numerical integration of nonlinear equations of motion of MDOF structural system in the domain. Currently NLRHA appears to remain less popular in engineering community compared to PONLA due to a number of understandable reasons, such as the difficulties in selection and scaling of ground motion input data, excessive computer hardware and run time requirements, lack of availability of reliable software in sufficient numbers to choose from, difficult post-processing requirements of excessive amount of output data, etc.

Practice Oriented Nonlinear Analysis (PONLA) procedures, however, have become extremely popular during the last decade due to practical appeal of the *pushover* concept by structural engineers and the direct use of elastic response spectrum tool in a nonlinear assessment practice. In fact, it is much easier to explain the essentials of deformation-based assessment to the engineers with a pushover concept. Deformation-based approach can be presented as an opposite to the strength-based approach, where the capacity curve is obtained from the nonlinear pushover analysis followed by an appropriate coordinate transformation. Thus yield strength of the equivalent SDOF system and the yield reduction factor is directly obtained from a nonlinear analysis, not as a result of target ductility assumption as in the strength-based approach. Once the yield reduction factor is calculated, peak inelastic displacement response of the nonlinear equivalent SDOF system, i.e., inelastic spectral displacement can be readily obtained from the elastic spectral displacement by utilizing an appropriate $\mu - R_y - T$ relationship. Following the calculation of the peak inelastic displacement response of the nonlinear equivalent SDOF system, peak inelastic response quantities of MDOF system, i.e., plastic hinge rotations or plastic strains are readily obtained from the corresponding pushover analysis output. Internal forces associated with the brittle failure modes are also calculated.

It is clear that deformation-based seismic assessment procedure is particularly suitable for the seismic evaluation of existing buildings, bridges and other structures, as the engineer is directly able to calculate the inelastic seismic demand quantities corresponding to the seismic damage to occur in the structural system. The assessment is finalized by checking whether seismic demand quantities remain within the limits of acceptance criteria, i.e., limiting values of plastic hinge rotations or plastic strains specified for various performance objectives.

Practice Oriented Nonlinear Analysis (PONLA) procedures are based on single-mode or multi-mode pushover analysis, which will be covered in the subsequent parts of this contribution. When rigorous Nonlinear Response History Analysis (NLRHA) procedure is used, inelastic seismic demand quantities, e.g., plastic hinge rotations, are directly obtained from the analysis output.

8.2.2.1 Reshaping Engineers' Minds for Nonlinear Seismic Behavior: From University Education to Professional Training

The above-given arguments and recent developments in earthquake engineering dictate that we can no longer escape from the nonlinear seismic performance assessment of existing structures. However as pointed out above, we, civil and structural engineers all over the world, are all "linear engineers", who generally feel themselves at odds with the analytical aspects of the nonlinear analysis methods in spite of the worldwide availability of pushover-based simple nonlinear analysis software. Most engineers often think of even the simple single-mode pushover analysis as a complex and difficult-to-understand procedure. In other words, they are not aware how the procedure is actually handled in the computer programs and hence the pushover analysis is still treated as a *black box.*

It is clear that we need a conceptual transformation towards nonlinear response analysis in both university education and professional training. A rational university curriculum needs to be developed. In the short run, a straightforward professional training model may be based on a simple plastic hinge concept for nonlinear modeling and piecewise linear hinge-by-hinge *incremental analysis* of the system under equivalent seismic loads without nonlinear iteration, which will be demonstrated subsequently in this contribution.

8.2.2.2 "Linear Engineers" Strikes Back: Fallacy of Equivalent Linear Response with a Fictitious Damping

While the need for a conceptual transformation is stressed for a realistic understanding of the nonlinear behavior in seismic response, there are counter attacks from the proponents of "linear engineering", who prefer to hide the nonlinear nature of the seismic response and present it as if it were a linear response based on a secant stiffness combined with a fictitious damping to imitate the nonlinear response. In fact many people with minds resting on this fallacy interpret the effect of an increasing nonlinearity as nothing but an increase in viscous damping of the structural system.

The underlying theory is based on Jacobsen's well known *equivalent damping* concept (see standard textbooks, e.g. Chopra, 2001). To some people this is perfectly legitimate. But many see it as the distortion of the concept of real nonlinear behavior in seismic response. This artificial treatment of nonlinearity may be viewed as the imprisonment of the engineer's mind behind the bars of linearity. It is ironic to note that a recently introduced new seismic design methodology by Priestley et al. (2007) is totally based on equivalent linear response concept where design response spectra are defined for very long periods and very high artificial damping factors.

8.2.2.3 Nonlinear Modeling and Acceptance Criteria in Deformation-Based Seismic Assessment

The first critical stage of any nonlinear analysis is the modeling of nonlinear properties. In principle, the same nonlinear model can be used in both Nonlinear Response History Analysis (NLRHA) and Practice Oriented Nonlinear Analysis (PONLA).

Concentrated Plasticity Approach

Concentrated (lumped) plasticity and *distributed plasticity* are the two main approaches used for nonlinear modeling. The former is represented by the simplest and most popular model based on *plastic hinges*, which are zero-length elements through which the nonlinear behavior is assumed to be concentrated or lumped at predetermined sections. A typical plastic hinge is ideally located at the centre of a plastified zone called *plastic hinge length*, which are generally defined at the each end of a clear length of a beam or column. A one-component plastic hinge model with or without strain hardening can be appropriately used to characterize a bi-linear moment-curvature relationship (Filippou and Fenves, 2004). The so-called *normality criterion* of the classical plasticity theory can be used to account for the interaction between plastic axial and bending deformation components (Jirasek and Bazant, 2001).

Plastic hinge concept is ideally suited to the piecewise linear representation of concentrated nonlinear response. Linear behavior is assumed in between the pre-determined plastic hinge sections as well as temporally in between the formation of two consecutive plastic hinges. As part of a piecewise linearization process, the yield surfaces of plastic hinge sections may be appropriately linearized, i.e., they may be represented by finite number of yield lines and yield planes in two- and three-dimensional hinge models, i.e., in the so-called PM hinges and PMM hinges, respectively.

In Practice Oriented Nonlinear Analysis (PONLA) procedures based on pushover analysis, modeling of backbone curves of typical moment-rotation relationships of plastic sections are sufficient for nonlinear modeling. Typical bi-linear back-bone curves with and without strength-degradation have been specified in the applicable codes (e.g., ASCE, 2007). In the Nonlinear Response History Analysis (NLRHA), cyclic hysteretic behavior of plastic hinge is additionally required to be defined. Standard bi-linear model with parallel loading and unloading branches, peak-oriented model with or without pinching and Takeda type models are the most well-known hinge hysteretic models. The so-called Ibarra-Krawinkler model is recently developed as the most advanced general model to simulate the hysteretic behavior of reinforced concrete and steel hinges (Ibarra and Krawinkler, 2005).

Acceptance criteria for plastic hinge response is generally defined in terms *plastic rotation capacities*, which are specified in the relevant codes, e.g., ASCE 41 (2007) and Eurocode 8-, Part 3 (CEN, 2005). On the other hand, concrete compressive strain and rebar steel strain capacities have been specified in the recent Turkish Code (MPWS, 2007) as acceptance criteria, which are to be compared with strain demands obtained from plastic hinge rotation demands.

Distributed Plasticity Approach

The *fiber model* is the most popular distributed plasticity model being used for the nonlinear modeling, where cross section of the structural element is subdivided into concrete fibers and steel fibers (Spacone et al., 1996; Filippou and Fenves, 2004;

CSI, 2006). Since the response is obtained in terms of uniaxial deformation of the fibers, the load versus deformation response of a fiber model is defined in terms of uniaxial stress-strain relations specified for concrete and reinforcement. Various material models are available in the literature, e.g., Orakcal and Wallace (2004) for concrete, Menegotto and Pinto (1973) for steel.

Although fiber model is more advanced compared to plastic hinge model, its use in beams and columns is not warranted from practical viewpoint. In such elements, abundance of test data regarding stiffness modeling and rotation capacities leads instead to a wider use of the plastic hinge modeling. Fiber model, however, is more appropriate in flexural walls of rectangular and U or L shapes in plan. Acceptance criteria may be specified in terms of concrete and steel strain capacities or compatible plastic hinge rotations.

In addition to plastic hinge and fiber models, as briefly described above, more rigorous nonlinear finite element models are also available. However for the time being, such models are not very suitable for practical applications.

8.3 Rigorous Nonlinear Analysis Procedure: Nonlinear Response-History Analysis

Nonlinear Response-History Analysis (NLRHA) procedure is the most advanced and precise procedure to obtain inelastic demand quantities. The procedure is based on the direct, step-by-step integration of coupled equations of motion of the MDOF structural system.

It has to be admitted that NLRHA is still far from a routinely used procedure in the practical seismic assessment and design process. As indicated above, the obstacles of a wider use of NLRHA in engineering practice include the difficulties in selection and scaling of ground motion input data, excessive computer hardware and run time requirements, lack of availability of reliable software in sufficient numbers to choose from, problems associated with the construction of damping matrix, difficult post-processing requirements of excessive amount of output data, etc. However, a very rapid progress is currently taking place to reduce, if not completely eliminate, such obstacles. It appears that in a few years time, the problems related to excessive run-time requirements will be avoided with the significant developments to occur in computer hardware industry. Although currently only a couple of reliable software is available, it is expected that the growing demand would accelerate the competition in this field.

Selecting and scaling the input motion still appear to be a problematic area where more research is needed. In the current practice, at least three or seven ground motion records are needed to be run. In the former case the maximum results and in the latter case mean values obtained from seven analyses are considered as the governing seismic demand quantities. There is no doubt that such number of selected ground motions is not sufficient to obtain statistically meaningful results. Maximum or mean values of response quantities may be very unreliable, depending on the ground motion selection and scaling process. It is expected that numbers of selected ground motions will be significantly increased in the near future with the developments in computer hardware for a meaningful statistical evaluation.

8.4 Practice-Oriented Nonlinear Analysis Procedures Based on Pushover Analysis

In the last two decades, a significant progress has been achieved in the seismic assessment and design of structures with the development of Practice-Oriented Nonlinear Analysis (PONLA) procedures based on the so-called pushover analysis.

All pushover analysis procedures can be considered as approximate extensions of the modal response spectrum method to the nonlinear response analysis with varying degrees of sophistication. For example, Nonlinear Static Procedure – NSP (ATC, 1996; ASCE, 2007) may be looked upon as a single-mode inelastic response spectrum analysis procedure where the peak response can be obtained through a nonlinear analysis of a modal single-degree-of-freedom (SDOF) system. In practical applications, modal peak response can be appropriately estimated through inelastic displacement spectrum (ASCE, 2007; CEN, 2005).

In the following sections, pushover analysis methods will be evaluated in detail. It is believed that in spite of its shortcomings, pushover analysis is a practice-oriented procedure in the right direction to help familiarize the practicing engineers with rational estimation of seismic damage.

8.4.1 Historical Evolution of Pushover Analysis: From "Capacity Analysis" to "Capacity-and-Demand Analysis"

From a historical perspective, pushover analysis has always been understood as a nonlinear capacity estimation tool and generally called as capacity analysis. The nonlinear structure is monotonically pushed by a set of forces with an invariant distribution until a predefined displacement limit at a given location (say, lateral displacement limit at the roof level of a building) is attained. Such predefined displacement limit is generally termed target displacement. The structure may be further pushed up to the collapse state in order to estimate its ultimate deformation and load carrying capacities. It is for this reason that pushover analysis has been also called as collapse analysis.

However, in view of performance-based seismic assessment and design requirements, the above definition is not sufficient. According to the improved concept introduced by Freeman et al. (1975) and Fajfar and Fischinger (1988), which was subsequently adopted in ATC 40 (1996), FEMA 273 (1997), FEMA 356 (2000) and Eurocode 8 (CEN, 2004, 2005), pushover analysis with its above-given historical definition represents only the first stage of a two-stage nonlinear static procedure, where it simply provides the nonlinear capacity curve of an equivalent single-degree-of-freedom (SDOF) system. The peak response, i.e., seismic demand is then estimated through nonlinear analysis of this equivalent SDOF system under a given earthquake or through an inelastic displacement spectrum. In this sense the term *pushover analysis* now includes as well the estimation of the so-called *target displacement*. Eventually, controlling seismic demand parameters, such as plastic hinge rotations, are obtained and compared with the specified limits (acceptance criteria) to verify the performance of the structure according to a given performance objective under a given earthquake. Thus according to this broader definition, pushover analysis is not only a capacity estimation tool, but at the same time it is a *demand estimation tool*.

It is interesting to note that majority of the new pushover analysis procedures developed in the last decade, particularly those dealing with multi-mode response, belong to *capacity estimation* category, which will be briefly evaluated later in this contribution.

8.4.2 Piecewise Linear Relationships for Modal Equivalent Seismic Loads and Displacements

Piecewise linear representation of pushover analysis, which provides a non-iterative solution technique with an adaptive load or displacement pattern, has been introduced by Aydınoğlu (2003, 2005, 2007). At each pushover step in between the formation of two consecutive plastic hinges, structural system can be considered to be piecewise linear. Accordingly, relationships between the coordinates of modal capacity diagrams (i.e., modal displacement and modal pseudo-acceleration of modal SDOF systems) versus the corresponding response quantities of the MDOF system can be expressed as in the following.

Piecewise linear relationship between n'th modal displacement increment, $\Delta d_n^{(i)}$, and the corresponding displacement increment of MDOF system, $\boldsymbol{\Delta u}_n^{(i)}$, at (i)'th pushover step is

$$\boldsymbol{\Delta u}_n^{(i)} = \boldsymbol{\Phi}_n^{(i)}\,\Gamma_{xn}^{(i)}\,\Delta d_n^{(i)} \tag{1}$$

where $\boldsymbol{\Phi}_n^{(i)}$ represents the instantaneous mode shape vector and $\Gamma_{xn}^{(i)}$ denotes the participation factor for the n'th mode at the (i)'th step for an earthquake in x direction, which is expressed as

$$\Gamma_{xn}^{(i)} = \frac{L_{xn}^{(i)}}{M_n^{*(i)}} = \frac{\boldsymbol{\Phi}_n^{(i)T}\mathbf{M}\,\boldsymbol{\imath}_\mathbf{x}}{\boldsymbol{\Phi}_n^{(i)T}\mathbf{M}\,\boldsymbol{\Phi}_n^{(i)}} \tag{2}$$

in which $\mathbf{M}$ is the lumped mass matrix and $\boldsymbol{\imath}_\mathbf{x}$ represents the ground motion influence vector for an x direction earthquake. Instantaneous mode shape vector $\boldsymbol{\Phi}_n^{(i)}$ is obtained from the solution of the eigenvalue problem:

$$(\mathbf{K}^{(i)} - \mathbf{K}_G^{(i)})\,\boldsymbol{\Phi}_n^{(i)} = (\omega_n^{(i)})^2\mathbf{M}\,\boldsymbol{\Phi}_n^{(i)} \tag{3}$$

in which the elements of the second-order stiffness matrix, i.e., $\mathbf{K}^{(i)}$ and $\mathbf{K}_G^{(i)}$ represent the first-order stiffness matrix and geometric stiffness matrix, respectively. $\omega_n^{(i)}$ is the instantaneous natural circular frequency. The equivalent modal seismic load increment, $\Delta\mathbf{f}_n^{(i)}$, representing the n'th mode capacity increment of the structure is given by

$$\Delta\mathbf{f}_n^{(i)} = (\mathbf{K}^{(i)} - \mathbf{K}_G^{(i)})\,\Delta\mathbf{u}_n^{(i)} \tag{4}$$

from which piecewise linear relationship between n'th modal pseudo-acceleration increment and the corresponding equivalent seismic load increment of MDOF system at (i)'th pushover step can be expressed in the familiar form as

$$\Delta\mathbf{f}_n^{(i)} = \mathbf{M}\boldsymbol{\Phi}_n^{(i)}\,\Gamma_{xn}^{(i)}\,\Delta a_n^{(i)} \tag{5}$$

where $\Delta a_n^{(i)}$ represents modal pseudo-acceleration increment, defined as

$$\Delta a_n^{(i)} = (\omega_n^{(i)})^2\,\Delta d_n^{(i)} \tag{6}$$

In monotonic pushover response, cumulative values of modal displacement and modal pseudo-acceleration, i.e., the coordinates of modal capacity diagrams (see Figs. 8.1 and 8.2) can be written for the end of the (i)'th step as

$$\begin{aligned} d_n^{(i)} &= d_n^{(i-1)} + \Delta d_n^{(i)} \\ a_n^{(i)} &= a_n^{(i-1)} + \Delta a_n^{(i)} \end{aligned} \tag{7}$$

Implementation of pushover analysis can be based on either a monotonic increase of displacements given by Eq. (1) or equivalent seismic loads given by Eq. (5). These correspond to displacement-controlled and force-controlled pushovers, respectively.

8.4.3 Single-Mode Pushover Analysis: Piecewise Linear Implementation with Adaptive and Invariant Load Patterns

Single-mode piecewise linear pushover procedure is applicable to low-to-medium rise regular buildings whose response is effectively controlled by the first (predominant) mode. Slight torsional irregularities may be allowed provided that a 3-D structural model is employed.

Single-mode pushover analysis can be performed as an *incremental analysis*, for which no iterative solution is required. This is the analysis scheme that can be easily grasped by "linear" practicing engineers.

8.4.3.1 Adaptive Load or Displacement Patterns

In the case of adaptive patterns, first-mode counterpart of equivalent seismic load increment given in Eq. (5) can be written for the (i)'th pushover step as

$$\mathbf{\Delta f}_1^{(i)} = \bar{\mathbf{m}}_1^{(i)} \Delta a_1^{(i)} \qquad ; \qquad \bar{\mathbf{m}}_1^{(i)} = \mathbf{M}\mathbf{\Phi}_1^{(i)} \Gamma_{x1}^{(i)} \tag{8}$$

where $\bar{\mathbf{m}}_1^{(i)}$ represents the vector of *participating modal masses* effective in the first mode. Superscript (i) on the participating modal mass and mode shape vectors as well as on the modal participation factor indicates that instantaneous first-mode shape corresponding to the current configuration of the structural system is considered following the formation of the last plastic hinge at the end of the previous pushover step. In adaptive case, a fully compatible modal expression can be written from Eq. (1) for the increment of displacement vector as well:

$$\mathbf{\Delta u}_1^{(i)} = \bar{\mathbf{u}}_1^{(i)} \Delta d_1^{(i)} \qquad ; \qquad \bar{\mathbf{u}}_1^{(i)} = \mathbf{\Phi}_1^{(i)} \Gamma_{x1}^{(i)} \tag{9}$$

Since both $\mathbf{\Delta u}_1^{(i)}$ and $\mathbf{\Delta f}_1^{(i)}$ are based on the same instantaneous modal quantities, there is a one-to-one correspondence between them. Thus, adaptive implementation of the single-mode pushover analysis can be based on either a monotonic increase of displacements or equivalent seismic loads, leading to displacement-controlled or load-controlled analyses.

8.4.3.2 Invariant Load Pattern

In the case of invariant load pattern, Eq. (8) is modified as

$$\mathbf{\Delta f}_1^{(i)} = \bar{\mathbf{m}}_1^{(1)} \Delta a_1^{(i)} \qquad ; \qquad \bar{\mathbf{m}}_1^{(1)} = \mathbf{M}\mathbf{\Phi}_1^{(1)} \Gamma_{x1}^{(1)} \tag{10}$$

where the vector of first-mode participating modal masses, $\bar{\mathbf{m}}_1^{(1)}$,is defined at the first *linear* pushover step (i = 1) and retained *invariant* during the entire course of pushover history. Note that inverted triangular or even height-wise constant amplitude mode shapes have been used in practice (FEMA, 2000) in place of $\mathbf{\Phi}_1^{(1)}$. Currently ASCE 41 (2007) solely requires the use of $\mathbf{\Phi}_1^{(1)}$.

8.4.3.3 Load-Controlled Piecewise Linear Pushover-History Analysis

In load-controlled piecewise linear pushover history analysis, equivalent seismic load vector of the MDOF system, which could have either adaptive or invariant pattern, is increased monotonically in increments of $\mathbf{\Delta f}_1^{(i)}$ where modal pseudo-acceleration increment, $\Delta a_1^{(i)}$, is calculated as the single unknown quantity at each (i)'th pushover step leading to the formation of a new hinge.

In the case of using adaptive load pattern, once $\Delta a_1^{(i)}$ is calculated for a given step, the corresponding modal displacement increment, $\Delta d_1^{(i)}$ can be obtained from

Eq. (6) for n = 1. Cumulative modal displacement and modal pseudo-acceleration at each step can then be calculated from Eq. (7). Thus, modal capacity diagram can be directly plotted without plotting and then converting the pushover curve.

In the case of an invariant pattern, modal equivalent loads and resulting displacement increments are not compatible with respect to modal parameters. In this case, modal displacement increment is calculated through Eq. (9), by specializing it for the roof displacement increment with the corresponding first-mode shape amplitude of the first pushover step.

8.4.3.4 Displacement-Controlled Piecewise Linear Pushover-History Analysis

In displacement-controlled piecewise linear pushover history analysis, adaptive displacement vector of the MDOF system is increased monotonically in increments of $\Delta\mathbf{u}_1^{(i)}$ where modal displacement increment, $\Delta d_1^{(i)}$, is calculated as the single unknown quantity at each (i)'th pushover step leading to the formation of a new hinge. Once $\Delta d_1^{(i)}$ is calculated for a given step, the corresponding modal pseudo-acceleration increment, $\Delta a_1^{(i)}$ can be obtained from Eq. (6) for n = 1. Cumulative modal displacement and modal pseudo-acceleration at each step can then be calculated from Eq. (7). As in load-controlled adaptive analysis, modal capacity diagram can be directly plotted without plotting the pushover curve.

8.4.3.5 Estimation of Modal Displacement Demand: Inelastic Spectral Displacement

When the last pushover step is reached, the modal displacement at the end of this step, $d_1^{(p)}$ (indicated by superscript p for *peak*), is effectively equal to first-mode inelastic spectral displacement, $S_{di,1}$, which may be calculated for a given ground motion record through nonlinear analysis of the modal SDOF system. The analysis is performed by considering the hysteresis loops defined according to modal capacity diagram taken as the backbone curve.However for practical purposes, inelastic first-mode spectral displacement, $S_{di,1}$, can be appropriately defined through a simple procedure based on *equal displacement rule*:

$$d_1^{(p)} = S_{di,1} = C_{R,1}\, S_{de,1} \tag{11}$$

in which $S_{de,1}$ represents the elastic spectral displacement of the corresponding linear SDOF system with the same period (stiffness) of the initial period of the bilinear inelastic system. $C_{R,1}$ refers to *spectral displacement amplification factor*, which is specified in seismic codes through empirical formulae (FEMA, 2000; MPWS, 2007; ASCE, 2007).

Note that in practice cracked section stiffnesses are used in reinforced concrete systems throughout the pushover analysis and therefore the fundamental period of the system calculated at the first *linear* pushover step (i = 1) is taken as the initial period of the bilinear inelastic system. This is contrary to the traditional approach

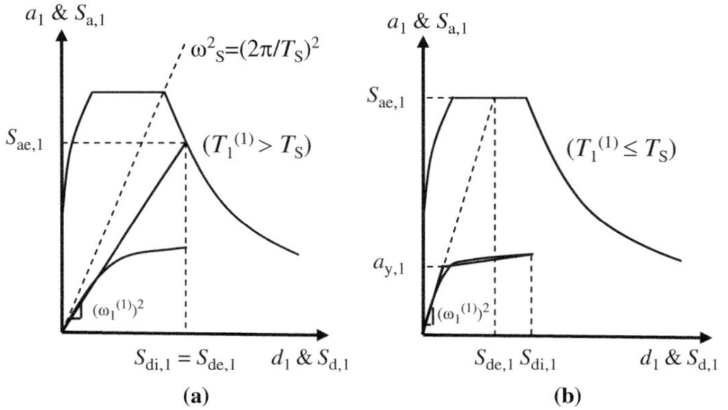

Fig. 8.1 Estimating modal displacement demand

where fundamental period is further lengthened excessively due to bi-linearization of modal capacity diagram. In Fig. 8.1, modal capacity diagram and the elastic response spectrum are combined in a *displacement – pseudo-acceleration* format, where T_S refers to characteristic spectrum period at the intersection of constant velocity and constant acceleration regions.

8.4.4 Multi-Mode Pushover Analysis

Single-mode pushover analysis can be reliably applied to only two-dimensional response of low-rise building structures regular in plan or simple regular bridges, where the seismic response is essentially governed by the fundamental mode. There is no doubt that application of single-mode pushover to high-rise buildings or any building irregular in plan as well as to irregular bridges involving three-dimensional response would lead to incorrect, unreliable results. Therefore, a number of improved pushover analysis procedures have been offered in the last decade in an attempt to take higher mode effects into account (Gupta and Kunnath, 2000; Elnashai, 2002; Antoniou et al., 2002; Chopra and Goel, 2002; Kalkan and Kunnath, 2004; Antoniou and Pinho, 2004a, b). In this context, Incremental Response Spectrum Analysis (IRSA) procedure has been introduced as a direct extension of the traditional linear Response Spectrum Analysis (RSA) procedure (Aydınoğlu, 2003, 2004).

With reference to the above-mentioned differentiation in identifying pushover analysis as a capacity estimation tool only or a capacity-and-demand estimation tool, it is observed that among the various multi-mode methods appeared in the literature during the last decade, only two procedures, i.e., Modal Pushover Analysis (MPA) introduced by Chopra and Goel (2002) and Incremental Response Spectrum Analysis (IRSA) by Aydınoğlu (2003, 2004) are able to estimate the seismic demand under a given earthquake ground motion. Others have actually dealt with structural capacity estimation only, although this important limitation has been generally overlooked. Those procedures have generally utilized the elastic response spectrum of a specified earthquake, not for the demand estimation, but only for scaling the relative contribution of vibration modes to obtain seismic load vectors (Gupta and Kunnath, 2000; Elnashai, 2002; Kalkan and Kunnath, 2004; Antoniou and Pinho, 2004a) or to obtain displacement vectors (Antoniou and Pinho, 2004b) through modal combination. Generally, building is pushed to a selected target displacement that is actually predefined by a nonlinear response history analysis (Gupta and Kunnath, 2000; Kalkan and Kunnath, 2004). Alternatively a pushover analysis is performed for a target building drift and the earthquake ground motion is scaled in nonlinear response history analysis to match that drift (Antoniou and Pinho, 2004a, b). Therefore results are always presented in a relative manner, generally in the form of story displacement profiles or story drift profiles where pushover and nonlinear response history analysis results are superimposed for a matching target displacement or a building drift. Thus, such pushover procedures are able to estimate only the relative distribution of displacement and deformation demand quantities, not their magnitudes and hence their role in the deformation-based seismic evaluation/design scheme is questionable.

Various multi-mode pushover analysis procedures developed in the last decade are systematically evaluated in the following sub-sections with consistent terminology and notation.

8.4.4.1 Modal Scaling

Referring to piecewise linear relationships for modal equivalent seismic loads and displacements given in Section 8.4.2, it is clear that in order to define modal MDOF response, modal displacement increments $\Delta d_{\mathrm{n}}^{(\mathrm{i})}$ or modal pseudo-acceleration increments $\Delta a_{\mathrm{n}}^{(\mathrm{i})}$ have to be determined in all modes at each pushover step, depending on whether displacement- or force-controlled pushover is applied. Since just a single plastic hinge forms and therefore only one yield condition is applicable at the end of each piecewise linear step, a reasonable assumption needs to be made for the relative values of modal displacement or modal pseudo-acceleration increments, so that the number of unknowns is reduced to one. This is called *modal scaling*, which is the most critical assumption to be made in all multi-mode pushover procedures. In this respect the only exception is the Modal Pushover Analysis – MPA (Chopra and Goel, 2002) where modal coupling is completely disregarded in the formation of plastic hinges and therefore modal scaling is omitted.

Modal Scaling Based on Instantaneous or Initial Elastic Spectral Quantities

As it is mentioned above, modal scaling is probably the most critical and at the same time one of the most controversial issues of the multi-mode pushover analysis. In a number of studies, such as Gupta and Kunnath (2000), Elnashai (2002), Antoniou et al. (2002), Antoniou and Pinho (2004a), force-controlled pushover is implemented based on Eq. (5) where modal scaling is performed on *instantaneous* modal pseudo-accelerations. Using consistent terminology and notation, such a modal scaling can be expressed as

$$\Delta a_{\mathrm{n}}^{(\mathrm{i})} = S_{\mathrm{aen}}^{(\mathrm{i})} \, \Delta \overset{\frown}{F}^{(\mathrm{i})} \tag{12}$$

where $S_{\mathrm{aen}}^{(\mathrm{i})}$ represents the instantaneous n'th mode *elastic* spectral pseudo-acceleration at the (i)'th pushover step and $\Delta\overset{\frown}{F}^{(\mathrm{i})}$ refers to an incremental scale factor, which is independent of the mode number. Thus Eq. (12) means that modal pseudo-acceleration increments are scaled in proportion to the respective elastic spectral accelerations. Note that the above defined modal scaling is essentially identical to scaling of modal displacement increments in proportion to respective instantaneous elastic spectral displacements in a displacement-controlled pushover implementation based on Eq. (1), which may be expressed as

$$\Delta d_{\mathrm{n}}^{(\mathrm{i})} = S_{\mathrm{den}}^{(\mathrm{i})} \, \Delta \overset{\frown}{F}^{(\mathrm{i})} \tag{13}$$

where $S_{\mathrm{den}}^{(\mathrm{i})}$ represents the instantaneous n'th mode *elastic* spectral displacement corresponding to above-given $S_{\mathrm{aen}}^{(\mathrm{i})}$, i.e., $S_{\mathrm{aen}}^{(\mathrm{i})} = (\omega_{\mathrm{n}}^{(\mathrm{i})})^2 \, S_{\mathrm{den}}^{(\mathrm{i})}$. Such a scaling has been used in a displacement-controlled pushover procedure (Antoniou and Pinho, 2004b).

It is doubtful whether this type of modal scaling should be implemented for a nonlinear response. In fact instantaneous elastic spectral parameters have no relation at all with the instantaneous nonlinear modal response increments. When the structure softens due to accumulated plastic deformation, the instantaneous *elastic* spectral displacement of the first mode would increase disproportionately with respect to those of the higher modes, leading to an exaggeration of the effect of the first-mode in the hinge formation process prior to reaching the peak response.

Note that a modal scaling scheme similar to the one expressed by Eq. (12) have been prescribed in FEMA 356 document (2000) where modal pseudo-acceleration increments were scaled in proportion to the *initial elastic* spectral accelerations:

$$\Delta a_{\mathrm{n}}^{(\mathrm{i})} = S_{\mathrm{aen}}^{(1)} \, \Delta \hat{F}^{(\mathrm{i})} \tag{14}$$

where $S_{\mathrm{aen}}^{(1)}$ represents the instantaneous n'th mode *elastic* spectral pseudo-acceleration at the *first linear* pushover step, which is retained invariant during pushover analysis along with the invariant mode shapes and participation factors defined for the first pushover step. Such a multi-mode invariant scheme is even more controversial than the adaptive scheme explained above with the instantaneous modal parameters. This highly approximate scheme was removed from the practice with ASCE 41 (2007), however it is still important as it represents one of the early schemes in the development of multi-mode pushover procedures, which will be summarized in Sections 8.4.4.2, 8.4.4.3, 8.4.4.4, 8.4.4.5, 8.4.4.6, and 8.4.4.7.

Modal Scaling Based on Instantaneous Inelastic Spectral Displacements and Application of "Equal Displacement Rule"

Displacement-controlled pushover based on Eq. (1) is the preferred approach in Incremental Response Spectrum Analysis – IRSA (Aydınoğlu, 2003, 2004), in which modal pushovers are implemented simultaneously by imposing instantaneous displacement increments of MDOF system at each pushover step.

In principle, modal displacements are scaled in IRSA with respect to *inelastic spectral displacements*, $S_{\mathrm{din}}^{(\mathrm{i})}$, associated with the *instantaneous* configuration of the structure (Aydınoğlu, 2003). This is the main difference from the other studies referred to above where modal scaling is based on *instantaneous elastic* spectral pseudo-accelerations or displacements. IRSA's adoption of *inelastic spectral displacements* for modal scaling is based on the notion that those spectral displacements are nothing but the peak values of modal displacements to be reached.

In practice, modal scaling based on *inelastic spectral displacements* can be easily achieved by taking advantage of the *equal displacement rule*. Assuming that seismic input is defined via *smoothed elastic response spectrum*, according to this simple and well-known rule, which is already utilized above for the estimation of modal displacement demand in single-mode pushover, *peak displacement* of an inelastic SDOF system and that of the corresponding elastic system are assumed practically equal to each other provided that the effective initial period is longer than the *characteristic period* of the elastic response spectrum. The characteristic period is approximately defined as the transition period from the constant acceleration segment to the constant velocity segment of the spectrum. For periods shorter than the characteristic period, elastic spectral displacement is amplified using a displacement modification factor, i.e., C_1 coefficient given in FEMA 356 (2000). However such a situation is seldom encountered in mid- to high-rise buildings and long bridges involving multi-mode response. In such structures, effective initial periods of the first few modes are likely to be longer than the characteristic period and therefore those modes automatically qualify for the equal displacement rule. On the other hand, effective post-yield slopes of the modal capacity diagrams get steeper and steeper in higher modes with gradually diminishing inelastic behavior (Fig. 8.2). Thus it can be comfortably assumed that inelastic spectral displacement response in higher modes would not be different from the corresponding spectral elastic response.

Hence, smoothed elastic response spectrum may be used in its entirety for scaling modal displacements without any modification. As in single-mode analysis, in reinforced concrete buildings elastic periods calculated at the first pushover step may be considered in lieu of the initial periods obtained from bi-linearization of modal capacity diagrams (see Fig. 8.1(b)).

In line with the equal displacement rule, scaling procedure applicable to n'th mode increment of modal displacement at the (i)'th pushover step is expressed as

$$\Delta d_{\mathrm{n}}^{(\mathrm{i})} = S_{\mathrm{den}}^{(1)}\, \Delta \tilde{F}^{(\mathrm{i})} \tag{15}$$

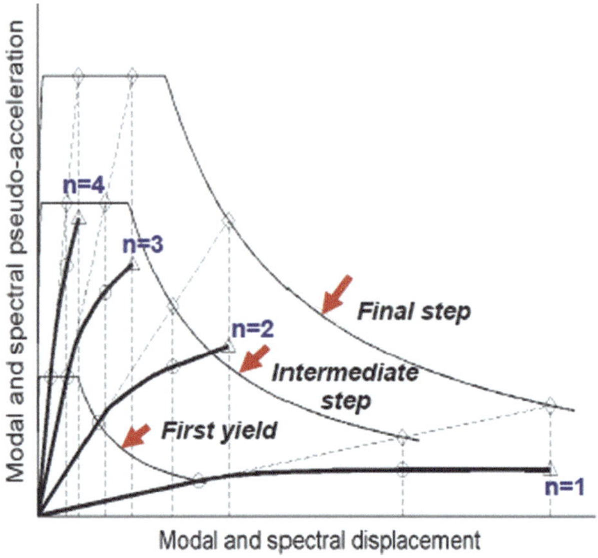

Fig. 8.2 Scaling of modal displacements through monotonic scaling of response spectrum

where $\Delta\tilde{F}^{(i)}$ is an *incremental scale factor*, which is applicable to all modes at the (i)'th pushover step. $S_{\text{den}}^{(1)}$ represents the *initial elastic spectral displacement* defined at the first step (Fig. 8.2), which is taken equal to the *inelastic spectral displacement* associated with the *instantaneous* configuration of the structure at any pushover step. Cumulative modal displacement and the corresponding *cumulative scale factor* at the end of the same pushover step can then be written as

$$d_{\text{n}}^{(i)} = S_{\text{den}}^{(1)}\,\tilde{F}^{(i)} \quad ; \quad \tilde{F}^{(i)} = \tilde{F}^{(i-1)} + \Delta\tilde{F}^{(i)} \leq 1 \tag{16}$$

Note that modal scaling expressions given above correspond to a monotonic increase of the elastic response spectrum progressively at each step with a cumulative scale factor starting from zero until unity. Physically speaking, the structure is being pushed such that at every pushover step modal displacements of all modes are increased by increasing elastic spectral displacements defined at the first step (i = 1) in the same proportion according to equal displacement rule until they simultaneously reach the target spectral displacements on the response spectrum. Shown in Fig. 8.2 are the scaled spectra corresponding to the first yield, to an intermediate pushover step ($\tilde{F}^{(i)} < 1$) and to the final step ($\tilde{F}^{(i)} = 1$), which are plotted in ADRS (Acceleration-Displacement Response Spectrum) format and superimposed onto modal capacity diagrams. It is worth warning that equal displacement rule may not be valid at near-fault situations with forward directivity effect.

8.4.4.2 Single-Run Pushover Analysis with Invariant Combined Single-Load Pattern

Pushover analysis performed with a single-load pattern that accounts for elastic higher mode effects is one of the procedures recommended in FEMA 356 (2000). In this procedure, the equivalent seismic loads defining the load pattern to be applied to a structure is calculated from the story shears determined by linear response spectrum analysis (RSA) through modal combination. Resultant load pattern is assumed invariant throughout the pushover analysis.

In this case modal pseudo-acceleration increments are scaled in proportion to the *initial elastic* spectral accelerations as given in Eq. (14). Substituting into Eq. (5) and considering initial elastic modal properties,

$$\Delta \mathbf{f}_{\mathrm{n}}^{(\mathrm{i})} = \hat{\mathbf{f}}_{\mathrm{n}}^{(1)} \Delta \hat{F}^{(\mathrm{i})} \quad ; \quad \hat{\mathbf{f}}_{\mathrm{n}}^{(1)} = \mathbf{M} \boldsymbol{\Phi}_{\mathrm{n}}^{(1)} \Gamma_{\mathrm{xn}}^{(1)} S_{\mathrm{aen}}^{(1)} \tag{17}$$

Combined story shear vector obtained from RSA and a typical j'th story shear calculated by SRSS rule can then be expressed as

$$\Delta \mathbf{q}^{(\mathrm{i})} = \hat{\mathbf{q}}^{(1)} \Delta \hat{F}^{(\mathrm{i})} \quad ; \quad \hat{q}_{\mathrm{j}}^{(1)} = \sqrt{\sum_{\mathrm{n}} (\hat{q}_{\mathrm{jn}}^{(1)})^2} \tag{18}$$

and finally combined invariant load pattern defined is defined as

$$\Delta \mathbf{p}^{(\mathrm{i})} = \hat{\mathbf{p}}^{(1)} \Delta \hat{F}^{(\mathrm{i})} \quad ; \quad \hat{p}_{\mathrm{j}}^{(1)} = \hat{q}_{\mathrm{j}}^{(1)} - \hat{q}_{\mathrm{j-1}}^{(1)} \tag{19}$$

Pushover analysis is run under the above-defined combined single-load invariant pattern of the equivalent seismic loads, and the resultant single pushover curve is plotted as usual. Note that definition of combined invariant load pattern based on elastic seismic loads yields to a highly controversial result. While on one hand the resultant capacity diagram of the equivalent SDOF system is deemed to represent all modes considered, on the other hand it is treated as if it was a capacity diagram of a single mode (first – dominant mode) in the estimation of inelastic peak displacement of the equivalent SDOF system through C_1 displacement modification coefficient of FEMA 356 (2000). It is clear that this procedure can estimate neither SDOF nor MDOF system inelastic response accurately. It can be identified only as a highly approximate *capacity estimation tool.* For this reason the procedure specified in FEMA 356 (2000) was later removed from ASCE 41 (2007) document.

8.4.4.3 Single-Run Pushover Analysis with Adaptive Combined Single-Load Patterns

Similar to above-given invariant single-run pushover procedure, an alternative multi-mode pushover analyses based on adaptive single-load patterns have been proposed by Elnashai (2002), Antoniou et al. (2002) and Antoniou and Pinho (2004a). In these force-controlled pushover procedures, equivalent modal seismic load increments, consistent with the instantaneous mode shapes, are calculated at each pushover step based on instantaneous stiffness state of the structure, and modal pseudo-acceleration increments are scaled in proportion to the respective instantaneous elastic spectral accelerations as expressed by Eq. (12):

$$\mathbf{\Delta f}_{\mathrm{n}}^{(\mathrm{i})} = \widehat{\mathbf{f}}_{\mathrm{n}}^{(\mathrm{i})} \, \Delta \widehat{F}^{(\mathrm{i})} \quad ; \quad \widehat{\mathbf{f}}_{\mathrm{n}}^{(\mathrm{i})} = \mathbf{M}\mathbf{\Phi}_{\mathrm{n}}^{(\mathrm{i})} \Gamma_{\mathrm{xn}}^{(\mathrm{i})} S_{\mathrm{aen}}^{(\mathrm{i})} \tag{20}$$

Then equivalent seismic loads are combined (for example with SRSS rule) at each step to obtain single load patterns. Combined load vector and a typical j'th story load can then be expressed as

$$\mathbf{\Delta p}^{(\mathrm{i})} = \widehat{\mathbf{p}}^{(\mathrm{i})} \, \Delta \widehat{F}^{(\mathrm{i})} \quad ; \quad \widehat{p}_{\mathrm{j}}^{(\mathrm{i})} = \sqrt{\sum_{\mathrm{n}} (\widehat{f}_{\mathrm{jn}}^{(\mathrm{i})})^2} \tag{21}$$

As similar to above-described invariant single-run pushover procedure, pushover analysis is run under the combined single-load adaptive pattern of the equivalent seismic loads until a prescribed target roof displacement. The analysis is terminated by plotting the resultant single pushover curve.

The main drawback of this procedure is that it is short of providing the essential output required for seismic performance assessment, i.e., it cannot estimate the inelastic seismic demand quantities under a given earthquake ground motion. Instead, as indicated in Section 8.4.4, target displacements are predefined through inelastic time history analyses of the MDOF systems. Alternatively pushover analysis is performed for a target building drift and the earthquake ground motion is scaled in nonlinear response history analysis to match that drift (Antoniou and Pinho, 2004a, b). Thus this procedure can only be treated as an improved *capacity estimation tool*, rather than a demand estimation tool under a given earthquake ground motion, as required in seismic performance assessment.

On the other hand, when P-delta effects are included in pushover analysis, pushover curve gradually descends after a certain step. Previously formed plastic hinges and P-delta effects lead to a negative-definite second-order stiffness matrix starting with that step, where eigenvalue analysis results in a negative eigenvalue and eventually an imaginary natural vibration frequency for the first mode or first few modes. As there is no physical meaning of an imaginary natural frequency, natural vibration periods and corresponding instantaneous spectral accelerations can not be estimated. Thus, multi-mode pushover analyses utilizing modal scaling procedure based on instantaneous spectral quantities have to be terminated without reaching the target displacement. This is an important limitation of the modal scaling procedure based on the instantaneous elastic spectral quantities when P-delta effects are considered in the analysis.

8.4.4.4 Single-Run Pushover Analysis with Adaptive Combined Single-Displacement Patterns

Antoniou and Pinho (2004b) presented a displacement-controlled adaptive pushover procedure (DAP) to avoid the drawbacks of force-controlled adaptive pushover procedure (FAP) described in the previous sub-section. In this procedure, displacement increments of the MDOF system, consistent with the instantaneous mode shapes, are calculated at each pushover step based on instantaneous stiffness state of the structure, and modal displacement increments are scaled in proportion to the respective instantaneous elastic spectral displacements as expressed by Eq. (13):

$$\mathbf{\Delta u}_{\mathbf{n}}^{(i)} = \overset{\frown}{\mathbf{u}}_{n}^{(i)} \, \Delta \overset{\frown}{F}^{(i)} \quad ; \quad \overset{\frown}{\mathbf{u}}_{n}^{(i)} = \mathbf{\Phi}_{n}^{(i)} \Gamma_{xn}^{(i)} S_{den}^{(i)} \tag{22}$$

This is followed by the calculation of story drifts as,

$$\mathbf{\Delta}\delta_{n}^{(i)} = \overset{\frown}{\boldsymbol{\delta}}_{n}^{(i)} \, \Delta \overset{\frown}{F}^{(i)} \quad ; \quad \overset{\frown}{\delta}_{jn}^{(i)} = \overset{\frown}{u}_{jn}^{(i)} - \overset{\frown}{u}_{(j-1)n}^{(i)} \tag{23}$$

which are then combined (for example with SRSS rule) at each step for a story drift pattern:

$$\mathbf{\Delta}\boldsymbol{\delta}^{(i)} = \overset{\frown}{\boldsymbol{\delta}}^{(i)} \, \Delta \overset{\frown}{F}^{(i)} \quad ; \quad \overset{\frown}{\delta}_{j}^{(i)} = \sqrt{\sum_{n} (\overset{\frown}{\delta}_{jn}^{(i)})^{2}} \tag{24}$$

Finally a single-displacement pattern is obtained by successively summing up the story drift pattern:

$$\mathbf{\Delta x}^{(i)} = \overset{\frown}{\mathbf{x}}^{(i)} \, \Delta \overset{\frown}{F}^{(i)} \quad ; \quad \overset{\frown}{\mathbf{x}}_{j}^{(i)} = \overset{\frown}{\mathbf{x}}_{(j-1)}^{(i)} + \overset{\frown}{\delta}_{j}^{(i)} \tag{25}$$

In this case, pushover analysis is run by imposing the combined single-displacement adaptive pattern to the building until a prescribed target roof displacement. The analysis is terminated by plotting the resultant single pushover curve.

Although displacement-controlled DAP appears to be more meaningful than its force-controlled counterpart (FAP), it suffers from exactly the same problems indicated in the previous sub-section, which will not be repeated here.

8.4.4.5 Simultaneous Multi-Mode Pushover Analyses with Adaptive Multi-Mode Load Patterns: Adaptive Spectra-Based Pushover Procedure

In a force-controlled adaptive pushover procedure developed by Gupta and Kunnath (2000), multi-modal load patterns are defined exactly in the same way the other force-controlled procedures. It means equivalent modal seismic load increments, consistent with the instantaneous mode shapes, are calculated at each pushover step based on instantaneous stiffness state of the structure, and modal pseudo-acceleration increments are scaled in proportion to the respective instantaneous elastic spectral accelerations as expressed by Eq. (12):

$$\Delta \mathbf{f}_n^{(i)} = \widehat{\mathbf{f}}_n^{(i)} \, \Delta \widehat{F}^{(i)} \quad ; \quad \widehat{\mathbf{f}}_n^{(i)} = \mathbf{M}\boldsymbol{\Phi}_n^{(i)} \Gamma_{xn}^{(i)} S_{aen}^{(i)} \tag{26}$$

which is nothing but the modal load expression given by Eq. (20). The main difference of this procedure is that the above-defined modal loads are not combined as opposed to other force-controlled procedures. Instead they are applied incrementally in each mode individually to the structural system and the increments of modal response quantities of interest including the coordinates of the pushover curve are calculated at each step, followed by modal combination by SRSS.

This approach is more meaningful compared to other procedures described above, as the conventional response spectrum analysis is actually being applied at each step. However, use of instantaneous spectral accelerations in modal scaling, as given in Eq. (12) and applied to Eq. (26), impairs the procedure for the consistent estimation of the inelastic seismic demands. For this reason, Gupta and Kunnath (2000) had to compare the story drifts estimated by this procedure at an equal roof displacement obtained from a nonlinear response-history analysis. As in the others, this procedure suffers as well from the improper representation of P-Delta effects.

As with the others given above, this procedure also can be treated as an improved *capacity estimation tool* only, rather than a demand estimation tool under a given earthquake ground motion.

8.4.4.6 Simultaneous Multi-Mode Pushover Analyses with Adaptive Multi-Mode Displacement Patterns: Incremental Response Spectrum Analysis (IRSA)

In a displacement-controlled adaptive procedure developed by Aydınoğlu (2003, 2004, 2007), piecewise linear relationship between n'th modal displacement increment, $\Delta d_n^{(i)}$, and the corresponding displacement increment of MDOF system, $\Delta \mathbf{u}_\mathbf{n}^{(\mathbf{i})}$, at (i)'th pushover step is expressed as given by Eq. (1). At the same time modal displacement increment, $\Delta d_n^{(i)}$ is scaled at each step with an instantaneous inelastic spectral displacements. As explained in sub-section "Modal Scaling Based on Instantaneous Inelastic Spectral Displacements and Application of "Equal Displacement Rule"" above, this is achieved by utilizing the well-known *equal displacement rule*, which simplifies the modal scaling as given by Eq. (15). Substituting into Eq. (1) leads to the following expression for the displacement vector increment in the n'th mode at the (i)'th pushover step:

$$\Delta \mathbf{u}_n^{(i)} = \tilde{\mathbf{u}}_n^{(i)} \Delta \tilde{F}^{(i)} \quad ; \quad \tilde{\mathbf{u}}_n^{(i)} = \boldsymbol{\Phi}_n^{(i)} \Gamma_{xn}^{(i)} S_{den}^{(1)} \tag{27}$$

Piecewise-Linear Pushover-History Analysis and Estimation of Peak Response (Seismic Demand)

IRSA is performed at each pushover step (i), by monotonically imposing MDOF system displacement increments $\Delta \mathbf{u}_n^{(i)}$ defined in Eq. (27) simultaneously in all modes considered. In this process, the increment of a generic response quantity of interest, such as the increment of an internal force, a displacement component, a story drift or the plastic rotation of a previously developed plastic hinge etc, is calculated in each mode as

$$\Delta r_n^{(i)} = \tilde{r}_n^{(i)} \Delta \tilde{F}^{(i)} \tag{28}$$

where $\tilde{r}_n^{(i)}$ represents the generic response quantity to be obtained in each mode for $\Delta \tilde{F}^{(i)} = 1$, i.e., by imposing the displacement vector $\tilde{\mathbf{u}}_n^{(i)}$ given in Eq. (27). This quantity is then combined by an appropriate modal combination rule, such as Complete Quadratic Combination (CQC) rule to obtain the relevant response increment for $\Delta \tilde{F}^{(i)} = 1$:

$$\Delta r^{(i)} = \sqrt{\sum_{m=1}^{N_m} \sum_{n=1}^{N_m} (\tilde{r}_m^{(i)} \rho_{mn}^{(i)} \tilde{r}_n^{(i)})} \tag{29}$$

where $\rho_{mn}^{(i)}$ is the cross-correlation coefficient of the CQC rule. Thus, generic response quantity at the end of the (i)'th pushover step can be estimated as

$$r^{(i)} = r^{(i-1)} + \Delta r^{(i)} = r^{(i-1)} + \tilde{r}^{(i)} \Delta \tilde{F}^{(i)} \tag{30}$$

in which $r^{(i)}$ and $r^{(i-1)}$ are the generic response quantities to develop at the end of current and previous pushover steps, respectively. In the first pushover step (i = 1), response quantities due to gravity loading are considered as $r^{(0)}$.

Incremental scale factor $\Delta \tilde{F}^{(i)}$ is the only, single unknown quantity at each pushover step, which is obtained without any iteration from piecewise linearized yield condition of the currently weakest plastic hinge. Once it is calculated, all response quantities of interest are obtained from the generic expression of Eq. (30).

Essentially IRSA is the extension of the single-mode pushover history analysis described earlier. Indeed, instead of running a static analysis under first-mode displacements or adaptive equivalent seismic loads, a multi-mode response spectrum analysis is performed at each step where seismic input data is specified in the form of initial spectral displacement in each mode, $S_{den}^{(1)}$, which is calculated in the first pushover step and remains unchanged at all pushover steps.

Peak response, i.e., seismic demand quantities are obtained when cumulative modal displacements, $d_n^{(i)}$, which are calculated by Eqs. (15) and (7), simultaneously reach the respective initial elastic spectral displacements $S_{den}^{(1)}$, which are assumed equal to the corresponding inelastic displacements according to *equal displacement rule*. IRSA has been comprehensively tested by Önem (2008).

Treatment of P-Delta Effects in IRSA

P-delta effects are rigorously considered in IRSA through straightforward consideration of geometric stiffness matrix in each increment of the response spectrum analysis performed. Along the pushover-history process, accumulated plastic deformations result in negative-definite second-order stiffness matrices, which in turn yield negative eigenvalues and hence negative post yield slopes in the modal capacity diagrams of the lower modes. The corresponding mode shapes are representative of the post-buckling deformation state of the structure, which may significantly affect the distribution of internal forces and inelastic deformations of the structure.

Analysis of inelastic SDOF systems based on bilinear backbone curves with negative post-yield slopes indicate that such systems are susceptible to *dynamic instability* rather than having amplified displacements due to P-delta effects. Therefore the use of P-delta amplification coefficient (C_3) defined in FEMA 356 document (FEMA, 2000) is no longer recommended (FEMA, 2005; ASCE, 2006). The dynamic instability is known to depend on the yield strength, initial stiffness, negative post-yield stiffness and the hysteretic model of SDOF oscillator as well as on the characteristics of the earthquake ground motion. Accordingly, practical guidelines have been proposed for minimum strength limits in terms of other parameters to avoid instability (Miranda and Akkar, 2003; ASCE, 2006; FEMA, 2005, 2009). For the time being, equal displacement rule is used in IRSA even P-delta effects are present as long as an imminent danger of dynamic instability is not expected according to the above-mentioned practical guidelines.

8.4.4.7 Individual Multi-Mode Pushover Analysis with Invariant Multi-Mode Load Patterns: Modal Pushover Analysis (MPA)

Modal Pushover Analysis (MPA), which was developed by Chopra and Goel (2002) based on earlier studies by Paret et al. (1996) and Sasaki et al. (1998), ignores the joint contribution of the individual modes to the section forces in the formation of plastic hinges. Nonlinear response is estimated independently for each mode with a single-mode pushover analysis based on an invariant load pattern proportional to initial linear elastic mode shape of a given mode (see Eq. (17)).

Since joint contribution of modes to response quantities during pushover is ignored, modal scaling is not required at all in MPA. Peak modal response quantities are obtained for each mode from the corresponding equivalent SDOF system analysis independently and then combined (exactly as in the linear response spectrum analysis) with an appropriate modal combination rule. It is reported that MPA procedure is able to estimate story drifts with a reasonable accuracy (Chopra and Goel, 2002; Chintanapakdee and Chopra, 2003). However it fails to estimate the locations of plastic hinges as well as the plastic hinge rotations and section forces, the essential demand quantities for the performance assessment in ductile and brittle behavior modes. Such quantities, which are already calculated by MPA, are completely disregarded and recalculated approximately by indirect supplementary analyses (Goel and Chopra, 2004, 2005). Certain refinements have been made on MPA through energy based development of modal capacity diagrams (Hernandeez-Montes et al., 2004; Kalkan and Kunnath, 2006).

8.5 Concluding Remarks

Rapid urban development all around the world including earthquake prone areas in the last 50 years resulted in higher seismic risk for existing buildings and lifelines. This created a need for improved tools in assessment of existing structures for seismic performance and retrofit design, which in turn prompted the development of performance-based assessment and design concept.

Performance-based assessment and design concept, which is essentially based on deformation-based evaluation approach, has evolved as a reaction to the traditional strength-based design approach. Towards the end of the last century engineers began realizing that seismic evaluation and design of structures should be based on nonlinear deformation demands, not on linear stresses induced by reduced seismic forces that are crudely correlated with an assumed overall ductility capacity of a given type of a structure, the starting point of the strength-based de-sign.

Development of performance-based assessment and design concept supported by deformation-based assessment approach is a totally new era in earthquake engineering, where practicing engineers have become able to *quantify* the nonlinear deformations as *seismic demand* quantities under a given seismic action, which are the direct attributes of the seismic damage.

The significant developments took place in the last two decades in practice-oriented nonlinear analysis procedures based on the so-called *pushover analysis*. It appears that some more time is needed before the use of rigorous nonlinear response history analysis, which is the ultimate tool of performance-based seismic assessment, is accepted by the engineering profession. On the contrary, engineers liked the idea behind and physical appeal of the pushover analysis. Yet they have a serious barrier in their front: By education and professional inheritance they are all "linear engineers". A new transformation is needed in university education and professional training to incorporate the nonlinear behavior of materials and systems into the curricula. Another barrier is the attempts by some linear-minded method developers and code writers to replace the nonlinear response with a fictitious "equivalent linear" response concept.

It can be argued that single-mode pushover based on predominant mode response may be considered to reach a maturity and engineers (although currently in small numbers) enjoy using this new and exciting tool for the performance assessment of existing structures. However, regarding the progress in pushover analysis, the challenge is in the development of rational and practical multi-mode pushover procedures. Although a great deal of effort has been spent during the last decade by many researchers to develop new and reliable procedures, the outcome is not satisfactory. The majority of the newly developed procedures have confined themselves merely with plotting the *single* multi-mode pushover curve, as if it were the ultimate goal of such an analysis. This tendency limits the role of pushover analysis of being no more than a "capacity estimation tool". Yet some researchers could not stop themselves to try the illogical: To convert the *single* multi-mode pushover curve to an equivalent SDOF system (basically of the dominant mode) in an attempt to estimate the seismic demand.

The number of multi-mode pushover procedures that are able to estimate the seismic demand under a given earthquake is very limited and their accuracy has not been fully tested. More research and practical application studies are still needed towards achieving the ultimate goal: Performing seismic assessment of existing structures as well as the design of new structures with improved performance-based analysis procedures.

References

Antoniou S, Rovithakis A, Pinho R (2002) Development and verification of a fully adaptive pushover procedure. Paper No. 822. In: Proceedings of the 12th European conference on earthquake engineering, London

Antoniou S, Pinho R (2004a) Advantages and limitations of adaptive and non-adaptive force-based pushover procedures. J Earthquake Eng 8:497–552

Antoniou S, Pinho R (2004b) Development and verification of a displacement-based adaptive pushover procedure. J Earthquake Eng 8:643–661

ASCE – American Society of Civil Engineers (2007) Seismic rehabilitation of existing buildings (ASCE 41-6), 1st edn, Washington, DC

ATC – Applied Technology Council (1978) Tentative provisions for the development of seismic regulations for buildings (ATC – 03), Redwood City, CA

ATC – Applied Technology Council (1996) Seismic evaluation and retrofit of concrete buildings (ATC – 40), Redwood City, CA

Aydınoğlu MN (2003) An incremental response spectrum analysis based on inelastic spectral displacements for multi-mode seismic performance evaluation. Bull Earthquake Eng 1:3–36

Aydınoğlu MN (2004) An improved pushover procedure for engineering practice: Incremental Response Spectrum Analysis (IRSA). International workshop on "performance-based seismic design: concepts and implementation", Bled, Slovenia, PEER Report 2004/05 pp 345–356

Aydınoğlu MN (2005) A code approach for deformation-based seismic performance assessment of reinforced concrete buildings. International workshop on "seismic performance assessment and rehabilitation of existing buildings", Joint Research Centre (JRC), ELSA Laboratory, Ispra, Italy

Aydınoğlu MN (2007) A response spectrum-based nonlinear assessment tool for practice: incremental response spectrum analysis (IRSA), Special issue: response spectra (Guest Editor: Trifunac MD). ISET J Earthquake Technol 44(1):481

BCJ (2009) Building center of Japan: the building standard law of Japan, Tokyo

CEN (2004) European committee for standardization: Eurocode 8: design of structures for earthquake resistance, Part 1: general rules, seismic actions and rules for buildings. European Standard EN 1998-1, Brussels

CEN (2005) European committee for standardization: Eurocode 8: design of structures for earthquake resistance, Part 3: assessment and retrofitting of buildings. European Standard EN 1998-3, Brussels

Chintanapakdee C, Chopra AK (2003) Evaluation of the modal pushover analysis procedure using vertically regular and irregular generic frames. Report No. EERC-2003/03, Earthquake Engineering Research Center, University of California, Berkeley, CA

Chopra AK (2001) Dynamics of structures, 2nd edn. Prentice Hall, Englewood Cliffs, NJ
Chopra AK, Goel RK (2002) A modal pushover analysis for estimating seismic demands for buildings. Earthquake Eng Struct Dyn 31:561–582
CSI (2006) Computers and Structures Inc.: perform – components and elements for perform-3D and perform-collapse, Version 4, Berkeley, CA
Elnashai AS (2002) Do we really need inelastic dynamic analysis? J Earthquake Eng 6:123–130
Fajfar P, Fischinger M (1988) N-2 A method for nonlinear seismic analysis of regular structures. In: Proceedings of the 9th world conference on earthquake engineering, Tokyo
FEMA – Federal Emergency Management Agency (1997) NEHRP guidelines for the seismic rehabilitation of buildings (FEMA 273). Washington, DC
FEMA – Federal Emergency Management Agency (1997) NEHRP Commentary on the guidelines for the seismic rehabilitation of buildings (FEMA 274). Washington, DC
FEMA – Federal Emergency Management Agency (2000) Prestandard and commentary for the seismic rehabilitation of buildings (FEMA 356). Washington, DC
FEMA – Federal Emergency Management Agency (2005) Improvement of nonlinear static seismic analysis procedures (FEMA 440). Washington, DC
FEMA – Federal Emergency Management Agency (2009) Effects of strength and stiffness degradation on seismic response (FEMA 440A). Washington, DC
Filippou FC, Fenves GL (2004) Methods of analysis for earthquake-resistant design. In: Bozorgnia Y, Bertero VV (eds) Earthquake engineering – from engineering seismology to performance-based engineering. CRC Press, Boca Raton, FL
Freeman SA, Nicoletti JP, Tyrell JV (1975) Evaluations of existing buildings for seismic risk – a case study of puget sound naval shipyard, Bremerton, WA. In: Proceedings of the 1st U.S. national conference on earthquake engineering, EERI, Berkeley, pp 113–122
Goel RK, Chopra AK (2004) Evaluation of modal and FEMA pushover analyses: SAC buildings. Earthquake Spectra 20:225–254
Goel RK, Chopra AK (2005) Extension of modal pushover analysis to compute member forces. Earthquake Spectra 21:125–139
Gupta B, Kunnath SK (2000) Adaptive spectra-based pushover procedure for seismic evaluation of structures. Earthquake Spectra 16:367–391
Hernandeez-Montes E, Kwon OS, Aschheim MA (2004) An energy based formulation for first and multiple-mode nonlinear (pushover) analyses. J Earthquake Eng 8:69–88
Ibarra LF, Krawinkler H (2005) Global collapse of frame structures under seismic excitations. PEER Report 2005/06, Berkeley, CA
ICBO – International Conference of Building Officials (1988) Uniform building code, Whittier, CA
Kalkan E, Kunnath SK (2004) Method of modal combinations for pushover analysis of buildings. Paper No. 2713. In: Proceedings of the 13th world conference on earthquake engineering, Vancouver, BC
Kalkan E, Kunnath SK (2006) Adaptive modal combination procedure for nonlinear static analysis of building structures. J Struct Eng ASCE 132:1721–1731
Jirasek M, Bazant ZP (2001) Inelastic analysis of structures. Wiley, Ne York, NY
Menegotto M, Pinto PE (1973) Method of analysis for cyclically loaded R.C. plane frames including change in geometry and non-elements behavior of elements under combined normal force and bending. In: Proceedings of the IABSE symposium, vol 13, Lisbon, Portugal, pp 15–22
Miranda E, Akkar SD (2003) Dynamic instability of simple structural systems. J Struct Eng 129:1722–1726
MPWS – Ministry of Public Works and Settlement (Turkish Government) (2007) Specification for buildings to be built in earthquake zones (in Turkish), Ankara
Orakcal K, Wallace JW (2004) Modeling of slender reinforced concrete walls. In: Proceedings of the 13th world conference on earthquake engineering, Vancouver, BC
Önem G (2008) Evaluation of practice-oriented nonlinear analysis methods for seismic performance assessment. Ph.D. Thesis, Boğaziçi University Kandilli Observatory and Earthquake Research Institute, Istanbul

Paret TF, Sasaki KK, Eilbeck DH, Freeman SA (1996) Approximate inelastic procedures to identify failure mechanisms from higher mode effects. Paper No. 966. In: Proceedings of the 11th world conference on earthquake engineering, Acapulco, Mexico

Paulay T, Priestley MJN (1992) Seismic design of reinforced concrete and masonry buildings. Wiley, New York, NY

Priestley MJN, Valvi GM, Kowalsky MJ (2007) Displacement-based seismic design of structures. IUSS Press, Pavia

Sasaki KK, Freeman SA, Paret TF (1998) Multimode pushover procedure (MMP) – a method to identify the effects of higher modes in a pushover analysis. In: Proceedings of the 6th U.S. national conference on earthquake engineering, Seattle, WA

SEAOC – Structural Engineers Association of California (1995) Vision 2000: performance based seismic engineering of buildings. San Francisco, CA

Spacone E, Filippou FC, Taucer FF (1996) Fibre bean-column model for nonlinear analysis of R/C frames. Earthquake Eng Struct Dyn 25:711–725

Chapter 9
Reflections on the Rehabilitation and the Retrofit of Historical Constructions

Carlos Sousa Oliveira and Aníbal Costa

Abstract The objective of this paper is to address techniques for the seismic rehabilitation and strengthening of historical constructions, especially those that are part of old urban agglomerations, focusing attention on the existing "pombaline" buildings in the heart of Lisbon and on the traditional masonry construction inserted in urban areas of the Azores Islands. This paper further details a previous review by Oliveira (2003a, Bull Earthquake Eng 1(1):37–82) made on the seismic vulnerability of historical constructions. After an initial reference to the major damage caused to old urban agglomerations by recent earthquakes, with special emphasis on the 1980 and the 1998 Azores earthquakes, on the 2009 L'Aquila earthquake and on the 2010 Haiti earthquake, the main reasons for what has happened are discussed, and solutions to mitigate damage in future events are proposed, by focusing the attention on two construction types. The first one is the in Lisbon "pombaline" construction, consisting of a timber cage inside old masonry walls, which was developed in the reconstruction of the downtown area after the large destruction caused by the 1755 earthquake. The second case is the traditional construction in stone masonry of the Azores Islands. For each case, a brief description of the materials and of the existing construction systems is made, followed by a discussion of the various possible rehabilitation and retrofit techniques, presenting the pros and cons of each one and finalizing with a synthesis of the most effective ones.

9.1 Introduction

The rehabilitation of old structures is clearly a public issue which deserves as much attention as possible by the scientific and technical communities. Whether by deliberate intent or by the obligation of preserving a certain memory and constructive

---

C.S. Oliveira (✉)
Department of Civil Engineering and Architecture/ICIST Lisbon,
Instituto Superior Técnico, Lisbon, Portugal
e-mail: csoliv@civil.ist.utl.pt

M. Garevski, A. Ansal (eds.), *Earthquake Engineering in Europe*, Geotechnical, Geological, and Earthquake Engineering 17, DOI 10.1007/978-90-481-9544-2_9,

heritage, by the need or the intention to profit from existing built up areas where new construction is not permitted, or even due to the current architectural-cultural trend "make use of the old", it is clear that the various players in the fields of architecture and construction are now particularly sensitive to the option of rehabilitating existing buildings, strengthening them if necessary. In the case of old buildings located in places where seismic activity is important, the preservation of these buildings requires a compulsory seismic strengthening.

Unfortunately, in many of these situations, building materials and construction techniques used in retrofitting are not consistent with the traditional techniques used in the original buildings. Therefore, there is the urgent need to discuss the issues of cultural identity and preservation of the built heritage together with the current knowledge in the rehabilitation fields, addressing the unsuitability and the compatibility problems of the retrofitting solutions, Oliveira (2003a).

As an example of these issues, reference is made to the use of structural solutions involving the destruction of the core of the buildings leaving only the façades which, although against all rules of rehabilitation, is a common practice in many urban areas, Fig. 9.1 Another example corresponds to rehabilitation solutions that use a wire mesh associated to the plaster covering the existing masonry which could correct the existing lack of seismic resistance of the buildings. However, this solution introduces a change in the building system which may be considered intrusive, Fig. 9.2.

In this context, we discuss various possible techniques for rehabilitation of the building stock, without trying to establish the most appropriate ones. The paper

Fig. 9.1 Standing façade after the demolition of an interior core

Fig. 9.2 Wire mesh in plaster covering

presents some of the techniques to strengthen two different building systems: the "pombaline" cage and the traditional Azorean masonry. A variety of situations is focussed, including actual cases of structural intervention (which, in general, try to be as minimalist as possible, respecting the existing conditions) and cases of seismic strengthening of buildings after the occurrence of earthquakes.

9.2 Damage Caused by Recent Earthquakes in Old Buildings and Monumental Structures

The 1980 and 1998 earthquakes of the Azores strongly affected the housing stock in terms of buildings and of monumental structures. As far as buildings are concerned, some parishes were completely destroyed such as the parishes of Ribeirinha and Espalhafatos in the Faial Island, Fig. 9.3a, b.

In terms of monumental structures, the destruction of churches due to the 1980 earthquake was quite intense and most of the entire stock was rebuilt using a large amount of concrete elements (Fig. 9.4).

Similarly, the 2009 L'Aquila earthquake levelled down the centennial villages of Onna and Castelnuovo, Fig. 9.5a, b. With few exceptions, the existing buildings of these situations were made out of stone masonry walls, although with different characteristics, and timber floors and roofs. Many damages and partial collapses were also observed in several monuments.

Fig. 9.3 Açores 1998 earthquake: **a** Ribeirinha Parish; **b** Espalhafatos Parish (aereal view)

Fig. 9.4 Açores 1980 earthquake: damage to the monumental structures and reconstruction with reinforced concrete

Fig. 9.5 L'Aquila earthquake, April 2009: **a** Onna village (photo: http://latimes.image2.trb.com/lanews/media/photo/2009-04/italy-earthquake_46010577.jpg); **b** Castelnuovo village (photo: http://digidownload.libero.it/vocedellaventino/immagini/image502.png)

The earthquake of 1980 in Azores killed 60 people, as a direct result of the collapse of buildings (Teves-Costa et al., 2007). During the 1998 earthquake, the destruction of buildings in two towns resulted in 8 deaths also directly related to their collapse (Zonno et al., 2009). During the L'Aquila earthquake, the 40 casualties in the town of Onna and the 5 in the town of Castelnuovo were a direct result of the collapse of buildings. As can be seen, the number of deaths in each case can be regarded as limited when compared to the level of destruction that occurred.

During the Haiti earthquake of January 12, 2010, the death toll amounts to about 200,000 (a number which is not fully supported as estimates of March 12, 2010 put the casualties between 200,000 and 300,000). It should be noted that the construction system used in the more modern buildings consisted essentially of cement masonry blocks forming the bearing walls and supporting concrete slabs for the floors and roofs (Fig. 9.6). Almost no confining elements and no connections between the slabs and the walls were present, Fig. 9.7.

According to Eberhard et al. (2010), the walls of 90% of buildings are constructed of one of the four following materials: (1) cement/block; (2) earthen materials; (3) clisse ("clissage" can be translated as "intertwined sticks, twigs, and branches"); and (4) bricks/stone. In rural regions, walls are most commonly made of earthen materials while in urban regions cement/block walls are the most common. Metal roofs predominate in both rural and urban regions. Based on information from the Institute Haitien de Statistique et d'Informatique (2003), the materials used for the key building component are:

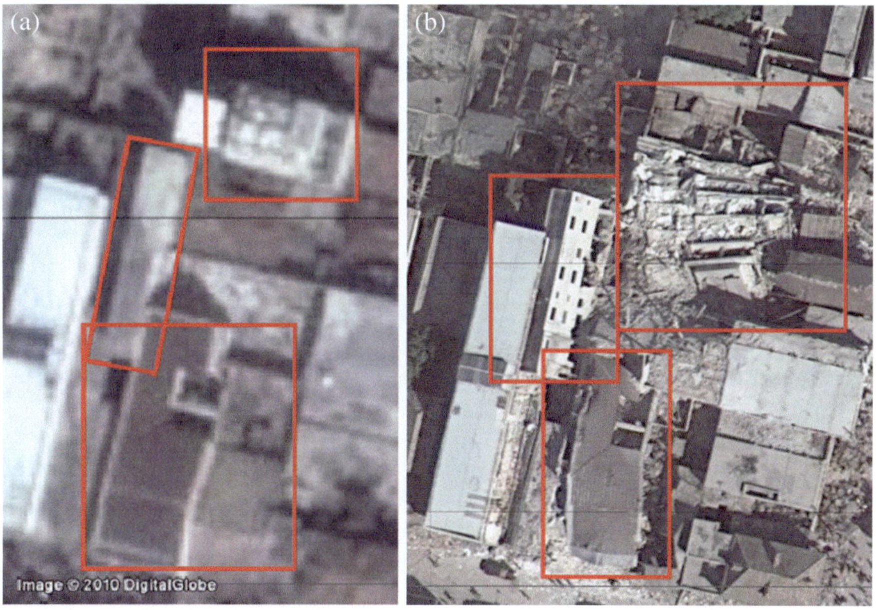

Fig. 9.6 Haiti earthquake 2010: **a** satellite image prior to event; **b** a satellite image after the event (Courtesy EERI, 2010, World Bank)

Fig. 9.7 Lack of reinforcement in columns (Photos by Peter Coats http://www.eqclearinghouse.org/20100112-haiti/general-information/photos-by-peter-coats)

- Roofs: Most one-storey houses have roofs made of sheet metal, but most multistorey houses and apartments have roofs made of concrete.
- Walls: Walls made of concrete/block/stone are common in ordinary one-storey houses, and predominate even more in ordinary multi-storey houses and apartments.
- Floors: Ordinary one-storey houses usually have floors made of concrete or compacted earth. Multistory houses and apartments usually have floors made of concrete or mosaic/planks.

These constructive solutions are known for their inadequacy for seismic regions and resulted in a very large death toll, Fig. 9.8. However, these solutions were recommended by the Haitian Government, supported by the World Bank, as being good

Fig. 9.8 Haiti earthquake. Overview of damage caused to buildings (photos: http://www.refinery29.com/pipeline/img/earthquake-haiti.jpg and http://3.bp.blogspot.com/_U54NM9QE5VY/S07Fcu30qDI/AAAAAAAAJY4/p06ii_gUrbI/s640/haiti+earthquake.jpg)

Fig. 9.9 Construction solutions: **a** confined walls; **b** reinforced concrete structure with block masonry infills

solutions to withstand the tornadoes which ravage the island each year. However, their seismic resistance was not correctly addressed.

Similar to what happened in Haiti, after the 1998 earthquake in the Azores, the new buildings were constructed with cement block walls confined with reinforced concrete columns at the corners and interior connections and with reinforced concrete slabs and roofs, Fig. 9.9a, b.

There is a definite trend to use these design solutions and they are considered to be "good solutions" because they are made of reinforced concrete which is also thought to be the best solution to resist earthquakes.

Without going into an extensive discussion regarding the various construction systems currently existing, it should be noted that, according to current code regulations, it is expected that the occurrence of a large magnitude earthquake, e.g. such as the 2010 Chile earthquake, may cause extensive structural damage to the buildings, but without collapsing in order to preserve human lives. Since the ultimate goal of seismic safety is to preserve human life, a structural solution involving the use of concrete slabs will endanger people's lives unless properly designed. In most cases, people are killed by the construction and not by the earthquake!

A building system involving lightweight floors and roofs is expected to have, in principle, a better performance since the mass reduction at the storey levels reduces the likelihood of casualties in the event of building collapse. The advantages of such type of structural systems are especially important in poor countries where much of the construction is carried out by the people themselves (non-engineered construction) without much technical and scientific expertise and, therefore, where the adoption of efficient and economic building systems that are less vulnerable to earthquakes is recommended.

9.3 The Cases in Portugal: Lisbon and Azores

To illustrate these ideas, two different construction systems, used in two separate individual sites subjected to intense earthquakes in the past, are analyzed in the following.

9.3.1 The "Pombaline" Construction

After the Lisbon earthquake of November 1, 1755, which devastated the city and caused a few tens of thousands of dead (Oliveira, 2008a), the Marquis of Pombal, Minister of the King José I, defined emergency measures for the reconstruction of the Lisbon downtown, imposing a specific urban design and establishing rules for the reconstruction of the city, including recommendations for the construction techniques to be used. The first rules imposed (i) the establishment of an urban grid between the Rossio and the Terreiro do Paço (the Front Square), open to the Tagus river; (ii) the widening of streets and (iii) the limitation of the height of the buildings which should have symmetrical and homogeneous façades along the streets. From the earthquake resistance viewpoint, the preservation of human life was the adopted principle, which was a new concept for its time in terms of structural safety. This principle would only be adopted much later as a fundamental principle in the philosophy of the structural rules of modern design codes (FEMA, 1998; EC8-1, 2005; RSAEEP, 1983).

The pombaline construction consists of a cage which is a wooden frame made up of horizontal, vertical and diagonal elements (forming the Saint Andrew Cross), strongly linked together, forming a robust and stable three-dimensional structural system, Fig. 9.10a, b. This system is anchored to the traditional masonry façade walls and infilled with ruble masonry of small dimensions or plywood to make the interior walls. All wood elements are covered by plaster, a noncombustible material, to reduce fire hazard. In the event of an earthquake, the exterior walls would fall to the outside of the building, leaving the structure intact and, consequently, preserving the lives of the building occupants. The pombaline structure introduced several other innovations, namely in terms of building use and in terms of safety, against earthquakes and fire. The reader is referred to França (1978), Mascarenhas (2005), and Cóias (2007) for additional details.

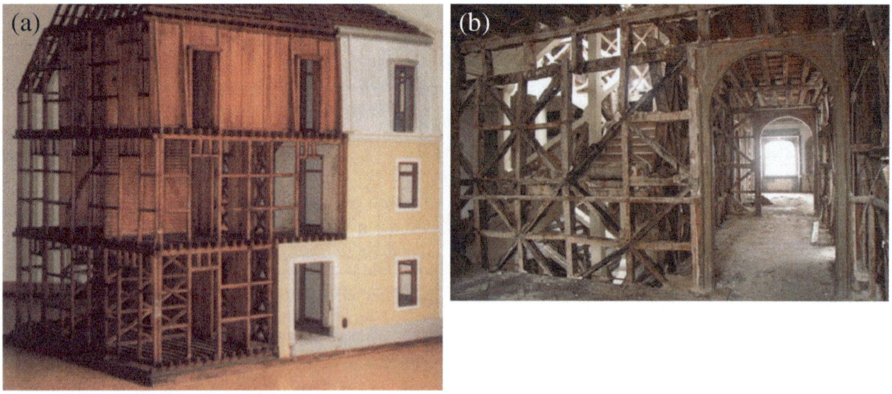

Fig. 9.10 "Pombaline" construction: **a** general view of a small scale model; **b** detail of an interior wall during rehabilitation works

9.3.2 The Traditional Construction in the Azores

The fact that the Azores are frequently struck by earthquakes leads to the existence of a wide variety of traditional masonry building typologies resulting from the several building reconstructions, strengthening and alterations. There is a general agreement among the affected population and authorities (Lamas, 2003) that the old stone houses have a reduced structural safety against the occurrence of earthquakes while reinforced concrete structures have a good performance. Such belief leads to the existence of many buildings having a mixture of different types of construction elements and materials that were not commonly used in traditional buildings in the Azores. It is interesting to note that a close observation of the building stock reveals that the building interventions throughout history work as a record log clearly reflecting the disasters that have affected the islands of the Azores.

9.3.2.1 Characterization of Buildings

Types of Constructions

Before the seismic event of July 9, 1998, the housing stock of the Faial and Pico Islands was essentially composed of traditional architecture houses. These traditional houses have a simple construction style and are made out of stone masonry and timber with 2–3 storeys, typically having apartments on the top storeys and shops in the ground floor. In some houses there are outsides ovens and water tanks.

A typological characterization of the building stock prior to 1998 can be made based on the type of (i) constructive system, (ii) roof structure, (iii) floor structure, (iv) inner wall and (v) outer wall. This characterization is important in order to be able to explain the type of damage suffered by each construction typology and to understand their structural behaviour. Based on the 5 items previously referred, the following constructive typologies can be defined: "current construction" (CC), "mixed construction" (MC), "traditional construction" (TC) and "altered traditional construction" (ATC), (Fig. 9.11). Table 9.1 (Zonno et al., 2009) shows a more detailed description of this classification referring to some of the items.

In this classification, a building is considered to be a traditional construction (TC) when the exterior walls (main façade, inside face and gable) are made of stone masonry. The main façade and the inside face have large openings for windows and doors with lintels and the foundation walls that support the structure are usually made by rubble stones. The thickness of the masonry walls is usually constant, with about 66 cm ("côvado" – an ancient measure of length). The horizontal support is made by wood planks supported by wood beams, usually of the "Acácia" and "Cryptomeria" types. The roof structure is made of wood and has usually two slopes. The inner walls are usually of wood or similar to frontal walls. The traditional rural constructions are essentially one-storey or two-storey high. The one-storey structures are generally more modest and are located in flat areas. Two-storey buildings usually take advantage of the slope of the terrain and have a basement (or, more commonly, a half basement) which is used for storage. The traditional urban construction

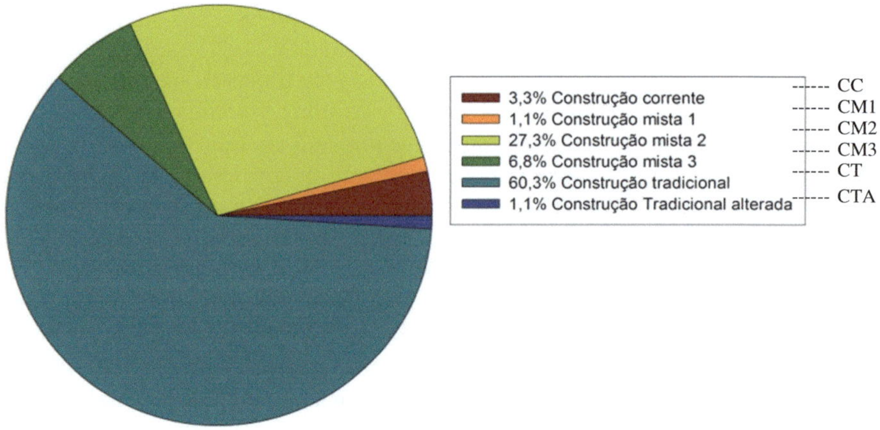

Fig. 9.11 Distribution of the structural systems in the Faial Island (sample of 2305 buildings, on July 1998)

Table 9.1 Common structural systems in the Faial and the Pico Islands (typologies)

Constructive system	Vertical structure	Floor	Roof
CC	Reinforce concrete	Reinforced concrete slab	Reinforced concrete or wood
MC1	Stone masonry stone	Reinforced concrete slab	Wood
MC2	Reinforce concrete and stone masonry	Wood and Reinforced concrete slab	Wood
MC3	Reinforce concrete	Reinforced concrete slab	Reinforced concrete or wood
TC	Stone masonry	Wood	Wood
ATC	Stone masonry	Half reinforced concrete slab (kitchen or WC)	Wood

is organized in city blocks or arranged in line along the street, usually with two to three storeys, rarely exceeding four. Regardless of the size, this type of construction, although having a more complex internal organization, has a secondary body perpendicular to the inside face where the kitchen is located. The altered traditional construction (ATC) is a construction type where the wooden floor (planks and beams) was replaced by a reinforced concrete slab (plate) which is supported by the stone masonry. This replacement is essentially only made in a small area of the house, typically in service areas such as the kitchen and the bathroom. As in traditional construction (TC), the inner walls are made by wood elements, such as the structure of the roof. This type of building resembles a traditional building without any extension carried out. The current constructions (CC) are characterized by having a resistant reinforced concrete structure with masonry walls made of cement blocks (confined or not) with reinforced concrete slabs. The structural system of the roof in this type of construction may also involve a reinforced concrete slab. In the mixed construction type 1 (MC1) the timber floor (planks and wooden beams) was replaced by a reinforced concrete slab which is supported by the stone masonry walls and include the possible existence of columns in the middle of the slab. In these cases, the inner walls are made of blocks of cement or wood and the wooden roof is maintained. The mixed construction 2 (MC2) refers to constructions where an intervention related to an extension occurred. In this new extension the resistant elements are columns, beams and a reinforced concrete slab. The structural elements of the original construction (roof, floor and walls) are kept. The mixed construction 3 (MC3) exhibits a larger intervention, usually characterized by a complete change of the floor, and of the outer and innerwalls. The floor is a massive reinforced concrete slab supported by columns and beams. The inner and exterior walls are made of cement blocks. With respect to the original construction, only the structural elements of the roof are maintained.

As can be seen in Fig. 9.11, a query to the database with the records of a building survey showed that most of the existing buildings on the Faial and Pico Islands were traditional construction when the July 9, 1998 earthquake occurred.

9.4 Strengthening Techniques

When defining a specific strengthening intervention, one usually seeks to stabilize the structure with respect to vertical and horizontal loads (e.g. earthquakes), to foundation settlements and to physical degradation. One may define the stabilization of a given structure or of its components in order to restore the original strength characteristics only or one may want to improve their characteristics in terms of strength (compressive, bending and/or shear) and deformability (e.g. the available ductility which is closely linked to confinement issues).

It is important to note that, in any structure, the strengthening can be seen at the global level or at a more local level, depending on the type of damage found and also on the main causes for such damage. However, in older structures, usually made of masonry, and given the weak characteristics of the connections between the members, it is always recommended to strengthen the structure globally in order to ensure that it can function as a whole. This way, the forces can be properly distributed by defining a suitable force path towards structural elements that are more resistant. However, this attitude does not relieve the need for a few local strengthening that may be necessary in elements or areas of insufficient resistance.

The underlying reason to define any strengthening operation is the need to establish or to restore appropriate safety conditions, which are usually met by complying with the appropriate safety factors. However, this is perhaps one of the more complex and disturbing issues associated to the strengthening of old buildings since it requires an adequate characterization of the strength of the existing of materials which, in most cases, is very difficult to obtain with a satisfactory degree. Furthermore, an adequate knowledge of the existing construction and of its behaviour is crucial, in particular with respect to the construction techniques used and to the changes it has undergone over time.

9.4.1 The Pombaline Construction

The rehabilitation of the Pombaline construction, in which we seek to maintain the structural cage system, requires, in the first place, a full comprehension of its structural behaviour under earthquake loads and, secondly, the definition of appropriate energy dissipation devices in order to improve the seismic resistance of this type of construction, without endangering human lives.

This combination of interests is not easy to accomplish, since, in many situations encountered in practical cases, the earthquake resistance of the existing masonry walls is small, i.e. in the occurrence of an earthquake, the masonry is assumed to fall to the outside leaving the structural cage intact. On the other hand, the structural changes made throughout the time in many of these buildings (changes that include increasing the number of storeys, opening spaces by destroying important walls, introducing elevators, etc.), the aging of the materials, the degradation of the connections and the deterioration of existing wood elements inside masonry walls, cast many doubts about the efficiency of the structural performance in case of an earthquake of larger magnitude.

The strengthening techniques for these structures point out to the need of all structural elements working together as a whole, thus providing a more efficient connection between the outside walls and the inner cage so that the entire system resists the earthquake. The original cage already provided some links to the masonry wall in the form of wood devices called "hands" (connectors) and other metallic devices. However, the proposed strengthening incorporates an effective link between the timber floors and the outer walls along all their length and, at the same time, the use of a wire, fiber glass or plastic mesh to strengthen the walls, Fig. 9.12a. This technique, where a wire mesh covers the masonry walls, has been used in many situations, prior to the occurrence of earthquakes or as a strengthening technique, and laboratory experiments confirm that, if well applied, a considerable increase in the strength and especially in the ductility of the walls can be expected, Fig. 9.12b (Figueiredo et al., 2010). The utility of this increase is, however, questionable since it undermines the principle of preserving human life by preventing the walls from falling and letting the core construction (the cage) intact to protect people. With a strengthening technique of this type the overall resistance of the building increases but, on the other hand, since the structure now behaves like a whole, its originally safer collapse mode is now altered with potentially harmful consequences.

What to do in this situation, having in mind that this construction technique is widespread and that, in many situations, the number of floors has increased, the streets where they were built are narrow and, therefore, the overturning of the walls to the outside may jeopardize the safety of people? Is it possible to say that if an earthquake with a magnitude of 8.8 (such as the recent earthquake in Chile and of 1755 Lisbon earthquake) occurs, the existing buildings in Lisbon's old town will

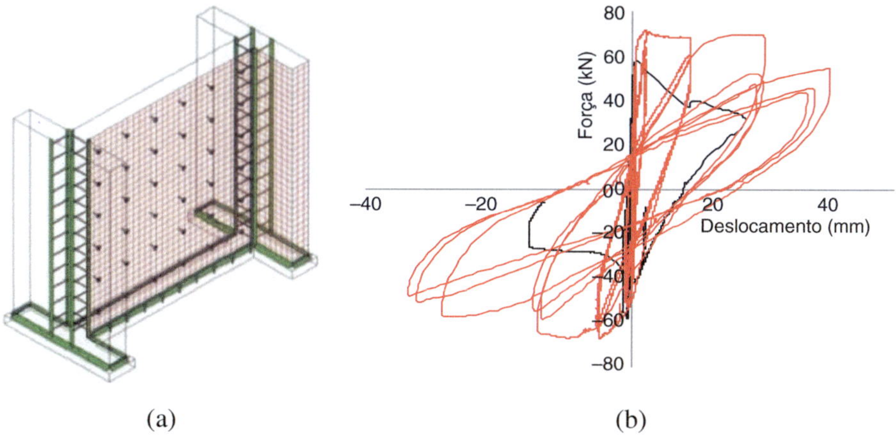

Fig. 9.12 **a** Schematics of wire mesh reinforcement; **b** Evolution of the applied force versus the displacement measured at the top of the wall: *black line*-without mesh; *red line*-with mesh

resist? Should the rehabilitation of these buildings involve the demolition of their interior, leaving the façades in an attempt to preserve some patrimonial memory, and the construction of the whole structure in reinforced concrete, in many situations using flat slabs to which the applicability of some concepts of seismic resistance is questionable, namely the weak-beam/strong-column type mechanism? Or should new city areas be built, such as in Shanghai, with modern buildings respecting up-to-date seismic codes?

What to do?

9.4.2 Rehabilitation in the Azores

In the case of the Azores several techniques have been used to strengthen the historical constructions, both the buildings as well as the monumental structures, which involve the previously referred concepts of (i) the need for a structural system working as whole and (ii) the need to avoid the disintegration of poor quality masonry walls. Such techniques involve the application of peripheral lintels at the top of the walls, the consolidation of masonry walls with wire mesh plaster applied on both sides of the wall, interconnected by linking elements, the application of wire mesh plaster only in one side of the wall, the application of fibreglass plaster applied over stone masonry walls, the construction of tie-rods, the incorporation of reinforced concrete columns, the injection of low pressure sandy cement inside the stone walls to fill the voids of rubble masonry, the vertical levelling and strengthening of house corners, the nailing of reinforced concrete members between adjacent walls, the rebuilding of confined block masonry walls, the construction of infill interior and exterior walls with masonry concrete blocks, the strengthening of the foundations, the use beams at the foundation level to anchor the wire mesh, the strengthening of the timber structures, the construction of links between the roof structure and the lintel, the bracing of the floors and of the roofs, etc.

As a summary, some of the solutions used for the rehabilitation in the Azores were defined in order to enhance the behaviour of the external walls as a whole and to consolidate them, in order to strengthen the foundations to increase further the consolidation and/or in order to consolidate other wood structures (roofs and floors).

To illustrate some of the referred solutions, three schemes of the adopted solutions are presented next. Figure 9.13 shows the reinforced concrete lintel located at the top of a peripheral masonry wall running throughout the whole width and well anchored to the wooden truss supporting the roof. In some cases, namely when, for practical reasons it was difficult to place the entire section at the top of the wall, an alternative solution was defined which consisted of a simpler reinforced concrete lintel which was 50 cm wide. Figure 9.14 presents another solution for the lintel together with the wire mesh plaster placed only in one face of the wall. This mesh is anchored to the wall by small connectors placed every 50 cm or so. The application

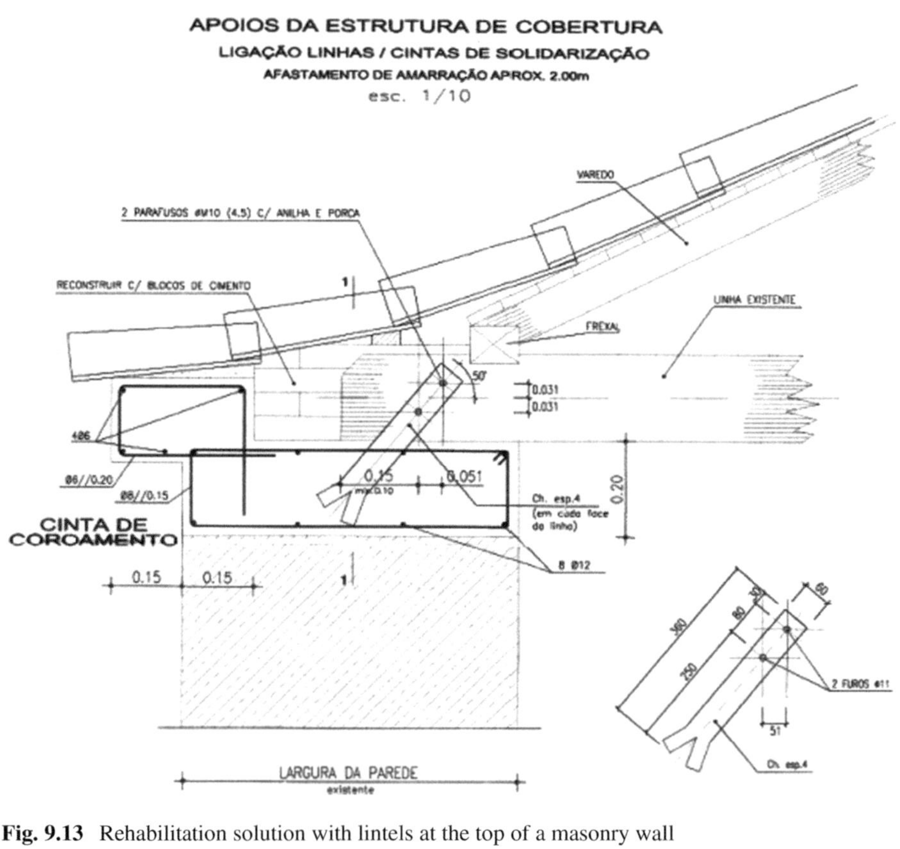

Fig. 9.13 Rehabilitation solution with lintels at the top of a masonry wall

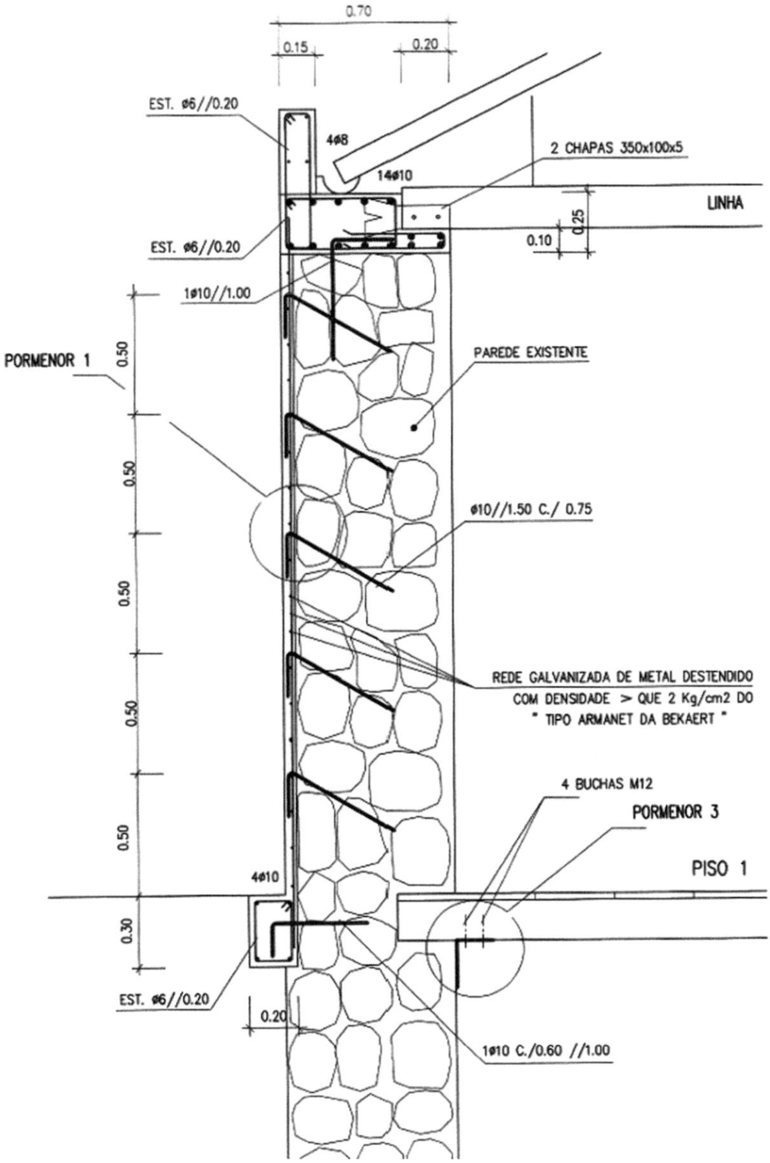

Fig. 9.14 Rehabilitation solution with a wire mesh plaster applied in one face

of the plaster was preceded by a detailed washing and cleaning of all the disintegrated material, and then by the filling of the joints between the stones. Since the weather conditions are very aggressive in the islands, and since salty sand is many times used in construction, all the wires used were made of stainless steel, in order to reduce corrosion to a minimum. Figure 9.14 also shows the beam around the

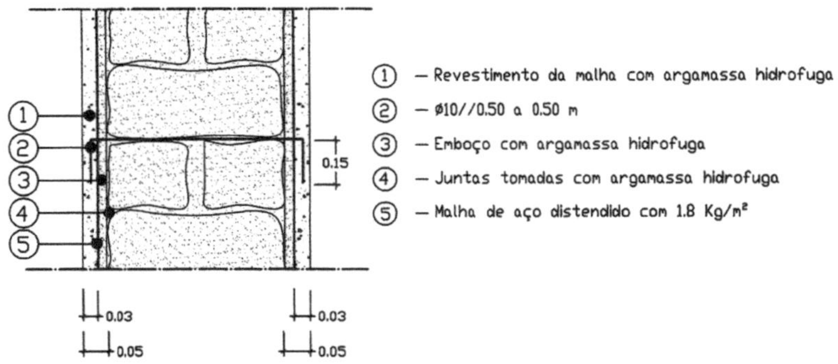

Fig. 9.15 Rehabilitation solution with a wire mesh plaster applied in both faces

foundation and the connection between the wooden floor and the wall. Figure 9.15 shows the wire mesh plaster applied to the two faces of the wall, the detail of the execution in the field and the connector linking the two faces.

Attempts to raise awareness have been made the population and contractors in order to avoid the use of reinforced concrete slabs and to keep the existing wooden floors. Although this is not an easy task, as reinforced concrete floors are seen as a symbol of modern times and of a higher social status, the idea was accepted in some parishes during the first years of reconstruction. However, in other situations, often imposed by the owners, reinforced concrete were used in slabs to make floors and roofs, creating situations of slabs supported (i) on walls of stone masonry only, (ii) on masonry walls strengthened with wire mesh, and (iii) on masonry walls with reinforced concrete columns inserted into them. Based on the consequences of the 1980 and the 1998 earthquakes, how to rehabilitate the building stock as well as the monumental heritage in a place such as the Azores? Should everything be demolished in order to use a modern reinforced concrete structural solution with walls of cement blocks? Should the existing architectural heritage be preserved, maintaining traditional building systems unchanged? Should the construction be strengthened in a more comprehensive way by connecting the walls in order for them to work together with the wooden floors, by improving the connections between the elements and by enhancing the seismic resistance of the walls?

What to do?

9.4.3 What to Do

Any strengthening intervention should be done under the framework of code regulations, namely the Eurocodes which are now beginning to be applied in different countries of the European Union. Unfortunately, codes are essentially aimed at new construction with little attention being devoted to rehabilitation and strengthening. Eurocode 8 (EC8-3, 2004) dedicates only a small part (Part 3) to this issue. Moreover, the design of old buildings requires knowledge about the present situation of the structure, an issue which involves detailed inspections and structural diagnoses that most designers are not able to do in due time. As can be seen, there is much to do in the field of rehabilitation and strengthening, not only in finding methods of inspection and diagnosis that are reliable and economic, as in the development of adequate research that would enable us to obtain a quick and safe adequate level of knowledge about the structure. Such objectives would imply the definition of material classes for old constructions having the mechanical characteristics and the parameters needed to carry out a safety analysis, as for the case of new constructions or parts of rehabilitation projects that use new materials.

To achieve this goal it is necessary to persevere in the use of experimental techniques, often without any standard or normative framework, in order to overcome the lack of knowledge about the materials and structural elements under analysis. In this context, a summary of the state-of-the-art on experimental testing of masonry structures is presented in the following.

Several works have been made concerning the seismic behaviour of masonry elements and structures, aiming to characterize their behaviour under horizontal loads. Both numerical simulations and experimental laboratory tests have been used. However, laboratory tests involve a series of difficulties that are very common when analyzing stone masonry constructions, namely those involving the correct reproduction of the materials of the original constructions and of the conditions that the original specimen is subjected to in-situ (e.g. the boundary conditions; the acting loads, etc.). Moreover, the majority of the dynamic tests on stone masonry structures performed on shaking tables are mainly carried out in specimens in a reduced scale, a fact which may strongly influence several issues of their seismic resistance (e.g. the aggregate interlock). These issues are also present in experiments performed on regular masonry structures, namely in tests made on masonry panels addressing their in-plane and out-of-plane behaviour (Anthoine et al., 1995; Magenes et al., 1995; Tomazevic et al., 1996; Willis et al., 2004; Abrams et al., 2007), for which the conditions that these specimens are subjected in-situ may not be correctly reproduced. The seismic behaviour of masonry walls (bearing and infill) is mainly characterized by the excitation of the dead load and not by the effect of concentrated loads as in frame structures with masses concentrated at the floor/roof levels.

With respect to the in-plane behaviour of masonry panels, the majority of the experiments are performed in laboratory conditions on specimens built to reproduce new/existing constructions (e.g. Anthoine et al., 1995). However, some in-situ tests have also been carried out on masonry panels in the form of compression, diagonal compression and in-plane shear tests (Corradi et al., 2002, 2003). Despite the importance of the performed experiments, the in-plane hysteretic behaviour, a crucial and important parameter concerning earthquake engineering, has not been identified yet and is very difficult to measure and study.

With respect to the out-of-plane behaviour, different test methodologies have been developed and are currently used, namely shaking table tests (e.g. Griffith et al., 2004; Hamed and Rabinovitch, 2008) distributed cyclic loads (Griffith et al., 2007; Mosallam, 2007) or concentrated loads in terms of point of load application (e.g. Maheri et al., 2008) or line loads (Willis et al., 2004; Papanicolaou et al.,

2008), all performed in laboratory conditions. Since real conditions have not been reproduced yet, the need for out-of-plane field tests must be emphasized. An example of such tests can be found in (Tumialan et al., 2003) where field tests were performed on brick masonry walls. However, the test setup that was used was not able to apply increasing cyclic reversal loads controlled through hydraulic actuators. Furthermore, attention is brought to the fact that the in-situ tests available in the literature only refer to monotonic loading tests. Therefore, some other testing methods enabling researchers to study the hysteretic characteristics through a feasible and simple manner are needed.

In this context, the authors have carried out an experimental campaign in the Azores involving tests in three houses of traditional masonry. Only one test is presented in the following.

The building, Fig. 9.16a, is a one-storey house made of stone masonry with irregular double leaf walls, with a surface mortar and a total thickness of 0.80 m. The

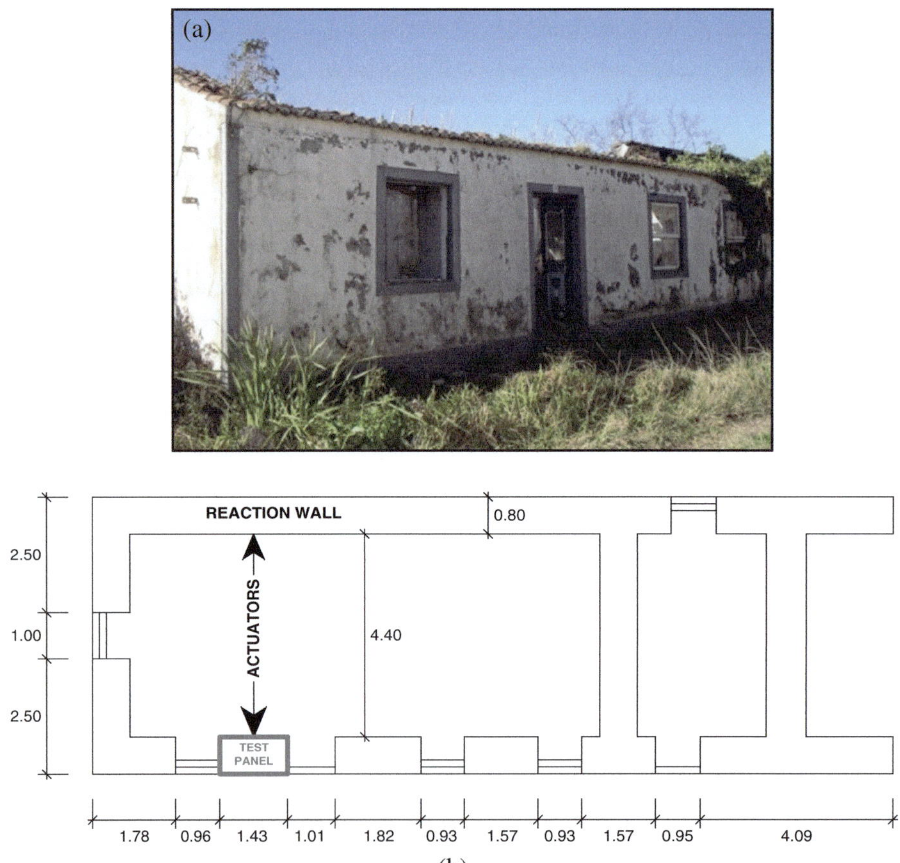

Fig. 9.16 Building 1: **a** main façade; **b** structural configuration of the bearing walls including the location of the tested element

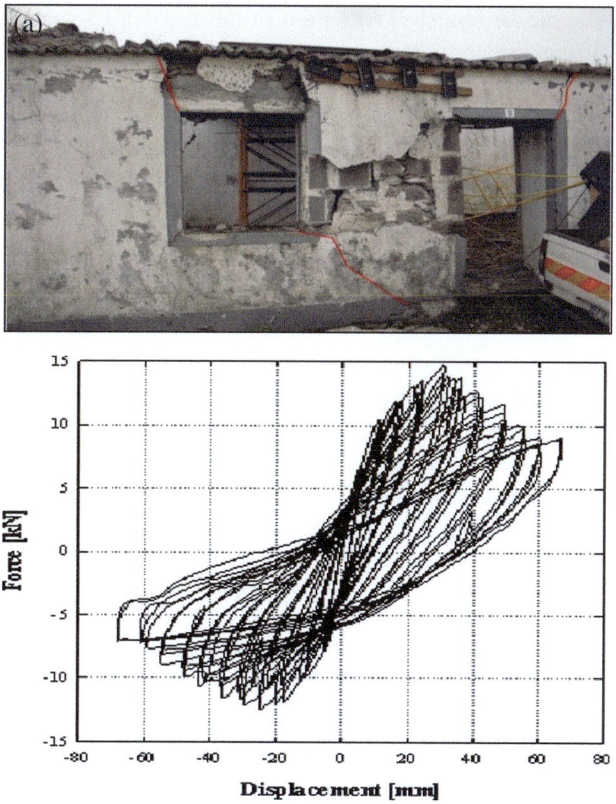

(b)

Fig. 9.17 Building 1 out-of-plane test: **a** hysteresis loop; **b** final cracking pattern and collapse mechanism

storey was 2.6 m high and was topped by a traditional wooden roof structure supported by the walls. Figure 9.16b shows the main façade of the house where it is possible to see the structural configuration of the bearing walls, including the location of the tested element.

Figure 9.17a presents the force vs. displacement curve of the out-of-plane tested panel subjected to increasing out-of-plane displacements up to a value of 6 mm applied at a height of 2.4 m. As can be seen, the out-of-plane behaviour can be divided in two different stages: a first stage until the achievement of maximum strength, which consists of a rigid body rocking motion exhibiting nonlinear elastic behaviour; a second stage which consists in a non ductile mechanism with significant post peak strength degradation. Figure 9.17b shows the final cracking pattern and collapse mechanism of the tested wall.

In the characterization of the mechanical behaviour of this type of structure, the difficulty in having adequate analytical models of such behaviour is another issue of considerable complexity. Although in the last 10–15 years we have been witnessing a strong research activity in this area, leading to the development of several behaviour models (Gambarrota et al., 1997a, b; Lourenço, 2004; Mistlet, 2004; Oliveira, 2003b), analytical modelling is still a difficult task and there is little consensus on the use of such models in general studies of traditional masonry strengthening. However, given the degree of development already achieved and the enthusiasm that currently exists around these issues, it is expected that, within a few years, the research efforts already undertaken will be of a considerable practical value to the engineering design of such structures. The recent focus on analysis strategies using macroscopic level modelling (Penna, 2002; Pegon et al., 2001) or using special homogenization techniques seems to be on the right path in the search of that goal.

To calibrate the knowledge of the material characteristics, involving the degradation of the buildings, with the results of the numerical models, it is essential to use modal identification techniques that allow for the fitting of the numerical models and can help to better understand the structural behaviour of the buildings when subjected to earthquake loading. In this field, there are several problems to solve. The first one is that identification techniques are essentially developed for ambient noise for which the dynamic behaviour of the structure model is still linear. A promising complement is the monitoring of structures which is able to record moderate to strong amplitude vibrations and check for the occurrence of nonlinear dynamical behaviour.

Also, since most of the existing masonry buildings have a relatively high first natural frequency (3–7 Hz), there have been some difficulties in the application of such procedures. It is recalled that ambient vibration tests are best suited for structures having natural frequencies that are closer to the frequency of the ambient excitation (which are usually low). Additional research is required in this field, namely in the development of identification techniques using more accurate numerical algorithms to deal with these higher frequencies.

Moreover, the fact that structures of stone masonry have a remarkable ability to adapt themselves to the external loads and have excellent durability characteristics, a fact corroborated by the existence of structures of this type with hundreds (and even thousands) of years old, should, by itself, justify the emphasis on research towards a better understanding of their behaviour and towards the definition of adequate strengthening techniques yielding appropriate levels of seismic safety while maintaining the characteristics of the existing structure.

For the above framework to be properly met, and because it is already commonly accepted (and required to), a good practice is to try to maintain the existing constructive memory through the adoption of reversible strengthening techniques, which requires, from the designer point of view, an adequate overview of the problem as a whole, backed up by a sound scientific and technical knowledge, and properly weighted by good engineering judgement. In fact, in many cases, it is ultimately concluded, after a careful and an adequately supported analysis, that the best intervention for these types of structures is the most minimalist and the least intrusive.

9.5 Final Notes

Rehabilitation and retrofit of historical or old constructions are very important subjects directly related to the high seismic vulnerability of these structures. To keep the patrimonial value they sustain, it is essential that the scientific and technical communities work together in order to develop optimal solutions. In most areas of minor engineering knowledge, the non-intrusive solutions presented above are recommended. However, other more sophisticated solutions using new technologies such as dissipators, "passive control", etc., should be viewed as alternative techniques. The level of seismic action to be used in conjunction with rehabilitation and strengthening depends not only on the population safety but also on the importance of the patrimonial value. The total amount of effort involved in any operation should be balanced through a multi-criteria evaluation where the different issues are weighted. Simulators using vulnerability concepts (Oliveira, 2008b) are good tools to obtain global diagnosis of the seismic behaviour of an historical complex, probably the most important parameter of that evaluation. "Low cost" interventions with "chirurgic" interventions are opposite to "heavy" interventions of doubtful results. And research needs to clarify these issues.

The proposed paper has addressed important questions regarding what should be done in the context of rehabilitating and strengthening older constructions. The issues emphasized were discussed based on two Portuguese construction systems which are found in Lisbon and in the Azores.

The proposed discussion clearly reflects that research has still a long way to go to be able to define rehabilitation procedures which preserve our built heritage. In very strict terms, such procedures are expected to fit the structures with adequate seismic resistance so that, in the event of an earthquake, people's lives are safe-guarded, as well as some of the constructions, if possible.

In the aftermath of a disaster, it is common to see the political class committing itself to implement seismic risk mitigation measures, namely by investing in aspects related to the quality of the construction. However, the memory of Man is short and, in the medium/long run, a change in the political priorities is usually witnessed, leaving behind the will to define strategic policies to mitigate such risks. Such strategic policies imply the implementation of measures to mitigate seismic risk in the long run and, therefore, should be independent of the political changes occurring in the central and regional administration bodies. Risk mitigation should be obtained by investing in research and development policies leading to an efficient transmission of the available technical and scientific knowledge (usually found in research units and universities) to the general population. Such knowledge transmission could be done in the form of guides or manuals for the construction of structures for earthquake resistance which should be easy to use and contain simple rules with low-cost and certified interventions, adequately supported by research. This type of information is especially important for economically underdeveloped areas where, many times, people build their own homes. Alternatively, to control the dissemination of the uncontrolled growth of non-engineered housing in such areas, government bodies would have to be responsible for the construction activities by using materials and constructive systems properly tested and proven. The importance of these issues cannot be overemphasized since the lack of such quality control policies and practical seismic resisting construction measures has been highlighted by recent earthquakes, namely the January 2010 Haiti earthquake.

Acknowledgments We acknowledge Fundação para a Ciência e a Tecnologia (FCT), through the "Programa Pluri-Annual", for partial support of the research referred in this paper. Credits are due to the colleagues from Instituto Superior Técnico (IST), Universidade de Aveiro (UA) and Faculdade de Engenharia da Universidade do Porto (FEUP), who participated in many projects connected with the topic. The authors also acknowledge X. Romão from FEUP for the revision of the manuscript.

References

Abrams D, Smith T, Lynch J, Franklin S (2007) Effectiveness of rehabilitation on seismic behaviour of masonry piers. Struct Eng 133(1):32–44

Anthoine A, Magonette G, Magenes G (1995) Shear-compression testing and analysis of brick masonry walls. In: Proceedings of the 10th European conference on earthquake engineering, Rotterdam, Holland

Cóias V (2007) Reabilitação Estrutural de Edifícios Antigos–Alvenaria, Madeira–Técnicas Pouco Intrusivas, Argumentum, GECoRPA, Lisboa (in Portuguese)

Corradi M, Borri A, Vignoli A (2002) Strengthening techniques tested on masonry structures struck by the Umbria-Marche earthquake of 1997–1998. Constr Build Mater 16:229–239

Corradi M, Borri A, Vignoli A (2003) Experimental study on the determination of strength of masonry walls. Constr Build Mater 17:325–337

Eberhard MO, Baldridge S, Marshall J, Mooney W, Rix GJ (2010) The M_W 7.0 Haiti earthquake of Jan 12, 2010; USGS/EERI Advance Reconnaissance Team report: U.S. Geological Survey Open-File Report 2010-1048, 58p

EC8-1 (2005) Eurocode 8: design of structures for earthquake resistance, Part 1: general rules, seismic actions and rules for buildings. CEN, Brussels

EC8-3 (2004) Eurocode 8: design of structures for earthquake resistance, Part 3: assessment and retrofitting of buildings. CEN, Brussels

FEMA (1998) Handbook for seismic evaluation of buildings. FEMA 310 (www.fema.gov)

Figueiredo A, Varum H, Costa A, Santos M (2010) Reforço de paredes de adobe: caracterização experimental de uma solução de reforço sísmico. 6º ATP (Seminário de Arquitectura em Portugal; 9º SIACOT–Seminário Ibero-Americano de Construção e Arquitectura em Terra, 20 a 23 de Fevereiro, Coimbra, Portugal (in Portuguese)

França JA (1978) A Reconstrução de Lisboa e a Arquitectura Pombalina. Instituto de Cultura Portuguesa, Lisboa (in Portuguese)

Gambarrota LS (1997a) Damage models for the seismic response of brick masonry shear wall. Part I: the Mortar joint model and its applications. Earthquake Eng Struct Dyn 26(4): 423–439

Gambarrota L, Lagomarsino S (1997b) Damage models for the seismic response of brick masonry shear wall. Part II: the continuum models and its applications. Earthquake Eng Struct Dyn 26(4):441–462

Griffith MC, Lam N, Wilson J, Doherty K (2004) Experimental investigation of unreinforced brick masonry walls in flexure. J Struct Eng 130(3):423–432

Griffith MC, Vaculik J, Lam NTK, Wilson J, Lumantarna E (2007) Cyclic testing of unreinforced Masonry walls in two-way bending. Earthquake Eng Struct Dyn 36:801–822

Hamed E, Rabinovitch O (2008) Nonlinear dynamic behaviour of unreinforced masonry walls subjected to out-of-plane loads. J Struct Eng 134(11):1743–1753

Institute Haïtien de Statistique et d'Informatique (2003) Enquête sure les Conditions de Vie de Haïti (Investigation of the Living Conditions in Haiti): Ministère de L'Economie et des Finances, p 62 (in French)
Lamas J (2003) Manual de Restauro e Recuperação/Guia do Construtor – Zona Antiga da Cidade da Horta. Edition Câmara Municipal da Horta (in Portuguese)
Lourenço P (2004) Current experimental and numerical issues in masonry research. International worshop on masonry walls and earthquake. Universidade do Minho, Guimarães (in Portuguese)
Magenes G, Kingsley GR, Calvi GM (1995) Static testing of a full scale, twos masonry building: test procedure and measured experimental response. University of Pavia, Pavia
Maheri MR, Najafgholipour MA, Rajabi AR (2008) The influence of mortar head joints on the in-plane and out-of-plane seismic strength of brick masonry walls. In: Proceedings of the 14th world conference on earthquake engineering, Beijing, China
Mascarenhas J (2005) O Edifício de Rendimento da Baixa Pombalina de Lisboa, in "Sistemas de Construção", vol V. Livros Horizonte, Lisboa (in Portuguese)
Mistlet M, Butenweg C, Anthoine A (2004) Evaluation of the failure criterion for masonry by homogenisation. In: Topping BHV, Soares CAM (eds) Proceedings of the 7th international conference on computational structures technology, Civil-Comp Press, Sctoland
Mosallam AS (2007) Out-of-plane flexural behaviour of unreinforced red brick walls strengthened with FRP composites. Composites B, 38:559–574
Oliveira CS (2003a) Seismic vulnerability of historical constructions: a contribution. Bull Earthquake Eng 1(1):37–82
Oliveira D (2003b) Experimental and numerical analysis of blocky masonry structures under cyclic loading. Tese de doutoramento, Universidade do Minho
Oliveira CS (2008a) Review of the 1755 Lisbon earthquake based on recent analyses of historical observations. In: Fréchet J et al (eds) Book evocating Jean Voigt in historical seismology. Springer, Dordrecht, pp 261–300
Oliveira CS (2008b) Lisbon earthquake scenarios: a review on uncertainties, from earthquake source to vulnerability modelling. Soil Dyn Earthquake Eng 28:890–913 (Special Issue on Urban Earthquake Hazard and Damage Assessment)
Papanicolaou CG, Triantafillou TC, Papathanasiou M, Karlos K (2008) Textile reinforced mortar (TRM) versus FRP as strengthening material of URM walls: out-of-plane cyclic loading. Mater Struct 41:143–157
Pegon P, Pinto AV, Geradin M (2001) Numerical modelling of stone-block monumental structures. Comput Struct 79(22–25):2165–2181
Penna A (2002) Una procedura a macroelementi per l' analisi dinamica non lineare di edifice in muratura. Tesi di Dottorato. Politecnico di Milano, Milan (in Italian)
RSAEEP (1983) Regulamento de Segurança e Acções em Estruturas de Edifícios e Pontes. Dec. Lei nº 235/83 de 31 de Maio de 1983. Casa da Moeda. Lisboa (in Portuguese)
Teves-Costa P, Oliveira CS, Senos ML (2007) Effects of local site and building parameters on damage distribution in Angra do Heroísmo–Azores. Soil Dyn Earthquake Eng 27:986–999
Tomazevic M, Lutman M, Petkovic L (1996) Seismic behaviour of masonry walls: experimental simulation. Struct Eng 122(9):1040–1047
Tumialan JG, Galati N, Nanni A (2003) Field assessment of unreinforced masonry walls strengthened with fiber reinforced polymer laminates. J Struct Eng 129(8):1047–1056
Willis CR, Griffith MC, Lawrence SJ (2004) Horizontal bending of unreinforced clay brick masonry. Masonry Int 17(3):109–122
Zonno G, Oliveira CS, Ferreira MA, Musacchio G, Meroni F, Mota-de-Sá F, Neves F (2009) Assessing seismic damage through stochastic simulation of ground shaking: the case of the 1998 Faial Earthquake (Azores Islands). Surv Geophys DOI: 10.1007/s10712-009-9091-1

Chapter 10
Engineers Understanding of Earthquakes Demand and Structures Response

Gian Michele Calvi

Graece magnificentiae vera admiratio exstat templum Ephesiae Dianae CXX annis factum a tota Asia. In solo id palustri fecere, ne terrae motus sentiret aut hiatus timeret rursus ne in lubrico atque instabili fundamenta tantae molis locarentur, calcatis ea substravere carbonibus, dein velleribus lanae.[1]

Abstract In this paper the author discusses the engineers understanding of the strength and displacement demands imposed to structures by earthquake motion, and of the structures capacities to withstand these demands, in a historical perspective. Without any claim of completeness or accurate critical assessment of the significance of various seismic events or of the scientific development of knowledge, the essential relevance of the lessons learnt from some seismic events is critically examined in parallel with the development of structural dynamics. The story moves from a claimed use of some base isolation measure in the temple of Diana at Ephesus in the sixth century B.C., continues with the renaissance treaties, where in one case only some emphasis is placed on how to build a safe structure and passing through the first understanding of dynamic equilibrium arrives to the ages of enlightenment

---

G.M. Calvi (✉)
Department of Structural Mechanics, University of Pavia, 27100 Pavia, Italy
e-mail: gm.calvi@eucentre.it

[1] Pliny, *Naturalis Historia*, Liber XXXVI, **xxi,** 95 *[Something that should be really admired of the Greek magnificence is the temple of Diana at Ephesus, constructed in 120 years with the contribution of all Asia. It was built on a marshy soil, locating charcoal and wool furs under its foundation, to reduce its sensitivity to earthquakes and to avoid locating such a big mass on unstable soil].*

This paper was the subject of the inaugural lecture for the beginning of the academic year at the Università degli Studi di Pavia on 18 January 2010, (1185th year from the *Capitolare* of Lotario, 649th from the *Studium Generale* institution), "without any claim of completeness or accurate critical assessment of the significance of various seismic events or of the scientific and technical development of knowledge, but instead illustrating a history of events and ideas seen from the point of view of the author, influenced by his training, the places in which he has lived, the master teachers he had"

M. Garevski, A. Ansal (eds.), *Earthquake Engineering in Europe*, Geotechnical, Geological, and Earthquake Engineering 17, DOI 10.1007/978-90-481-9544-2_10,

and the Lisbon earthquake of 1755. The breakthrough towards modern seismic analysis is clearly identified with the two earthquakes of San Francisco (1906) and Messina (1908). In particular it is discussed how most of the fundamental principles used for a century had already been stated after the second one. Spectra, ductility and performance based design are then identified as further milestones derived from earthquake evidence, to conclude with a critical appraisal of some major misunderstanding of structural response, with the merits of displacement based approaches and to eventually close the circle opened with the temple of Diana with modern base isolation techniques.

Pliny the Elder doesn't explain why the temple of Diana at Ephesus (Fig. 10.1) should not have feared earthquakes thanks to the fact that it had been built on marshy ground, or what function the layers of coal and animal hides laid underneath the foundations had. So it comes as no surprise that for centuries builders and scientists alike forgot about this passage and resigned themselves to seeing earthquakes essentially as divine punishment, humbly accepting deaths and collapses, without even wondering whether it was possible to build structures in such a way as to limit the damage, without realising that, eventually, it is houses and bridges that collapse and cause deaths, not woods and meadows.

There are a number of examples of this scientifically peculiar but basically useless approach in Renaissance treatises. For example, and this goes for all of them, it is useful to quote the treatise that Stefano Breventano, the caretaker of the Accademia degli Affidati, wrote in Pavia in 1576 following the Ferrara earthquake of 1570. The text discusses seismogenesis (W*hat an earthquake is and what causes it*), drastically concluding that *the principal cause of an earthquake is God*, wave motion (*How many kinds of earthquake there are*), warning signs (*Signs, which*

Fig. 10.1 A reconstruction of the image of the temple of Diana at Ephesus (sixth century B.C.)

Fig. 10.2 The seismic resistant house proposed by Pirro Ligorio[3]

precede the earthquake), duration (*The duration of earthquakes*), dangerousness and amplification of the movement (*Places which are more or less susceptible to earthquakes*), effects (*Under which weather conditions are earthquakes most likely to occur*, and *The effects that cause earthquakes*), but dedicates only half a page to the topic of reducing vulnerability and risk (*Remedies or protection against earthquakes*), drawing on some of the instructions given by Pliny (him again).[2]

It is curious to observe that there was a preference for debating theories on the causes of earthquakes, which had already been developed by Greek philosophers, rather than learning how to build.

The only exception can be found in a work by Pirro Ligorio[3] (which was also written after the Ferrara earthquake, which I mentioned earlier), in which the author, an architect, designed an seismic resistant building (Fig. 10.2). This was merely a matter of implementing geometric rules of proportion and structural details about the connection between the walls and the floor, which, moreover, would have considerably reduced the number of victims and the amount of damage caused if they had been systematically implemented in the construction of historic buildings.

The analytical concepts that would form the basis of seismic engineering until the second half of the twentieth century were set out in domains that had nothing to do with earthquakes, within the course of the 10 years from 1678 to 1687.

---

[2] Breventano S (2007) Trattato del terremoto [treatise on earthquakes]. In: Albini P (ed) IUSS Press, Pavia, p 24: "In buildings he says the archivolts, the corners of the walls, the doors and the cellars are extremely secure, because they are resistant to reciprocal impact. Brick walls suffer less damage than those made of stone or marble. [. . .]Emperor Trajan [. . .] ordered that houses be built no higher than the measurement of seventy feet, so that if there was another earthquake, they would not be damaged as easily. Propping up buildings on one side and the other with beams is not completely pointless".

[3] Ligorio P (2005) Libro di diversi terremoti [Book on various earthquakes]. In: Guidoboni E (ed) De Luca Editori d'Arte, Rome; the building is described on pages 93–97 of the edition cited, f. 58–61 of the work.

Between 1675 and 1679 Robert Hooke announced to the Royal Society that he had discovered the fundamental law of elasticity, published it in the form of an anagram (CEIIINOSSSTTVV)[4] and revealed the meaning of this with the formula "vt tensio sic vis" (Fig. 10.3). The theory of proportionality between force and elongation constituted a fundamental concept for progress in many areas of technology, including the science of construction, but, as we shall see, it subsequently became a heavy burden that hindered the logical development of seismic engineering.

In 1678 Newton, partially drawing on Galileo,[5] set out the three fundamental principles of dynamics,[6] and I hope physics teachers will forgive me if in this context I translate them by saying that he essentially expressed the concepts of balance and of proportionality between force and acceleration, in accordance with a constant property of the object to which force is applied, i.e. the mass.

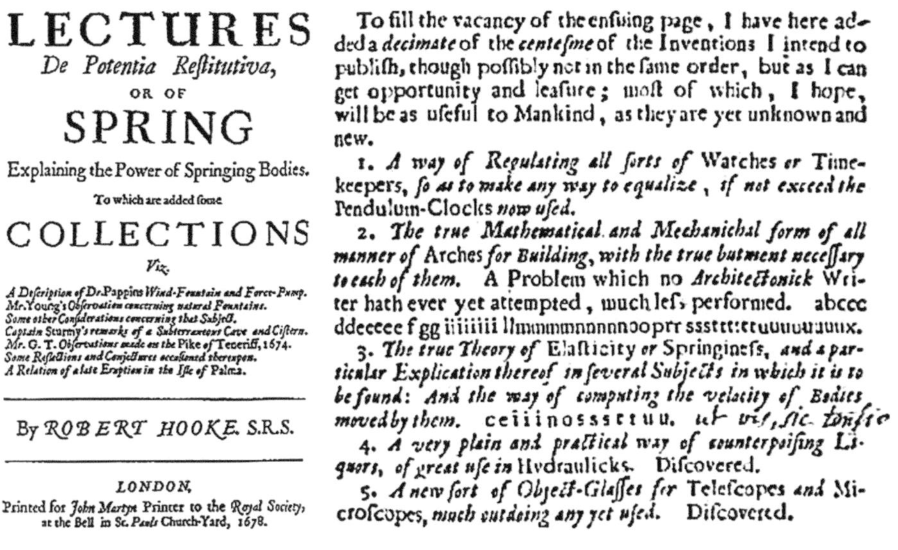
LECTURES
De Potentia Restitutiva,
OR OF
SPRING
Explaining the Power of Springing Bodies.
To which are added some
COLLECTIONS
Viz.
A Description of Dr. Pappins *Wind-Fountain and Force-Pump.*
Mr. Young's *Observation concerning natural Fountains.*
Some other Considerations concerning that Subject.
Captain Sturmy's *remarks of a Subterraneous Cave and Cistern.*
Mr. G. T. *Observations made on the* Pike *of* Teneriff, 1674.
Some Reflections and Conjectures occasioned thereupon.
A Relation of a late Eruption in the Isle of Palma.

By *ROBERT HOOKE.* S.R.S.

LONDON,
Printed for *John Martyn* Printer to the *Royal Society,* at the Bell in St. *Pauls* Church-Yard, 1678.

To fill the vacancy of the ensuing page, I have added a *decimate* of the *centesme* of the Inventions I intend to publish, though possibly not in the same order, but as I can get opportunity and leasure; most of which, I hope, will be as useful to Mankind, as they are yet unknown and new.

1. *A way of Regulating all sorts of* Watches *or* Time-keepers, *so as to make any way to equalize, if not exceed the* Pendulum-Clocks *now used.*

2. *The true Mathematical and Mechanichal form of all manner of* Arches *for Building, with the true butment necessary to each of them.* A Problem which no *Architectonick* Writer hath ever yet attempted, much less performed. abccc ddeeeee f gg iiiiiiii llmmmmnnnnnooprr sssttttttuuuuuuuux.

3. *The true Theory of* Elasticity *or* Springiness, *and a particular Explication thereof in several Subjects in which it is to be found: And the way of computing the velocity of Bodies moved by them.* ceiiinosssttuu. ut vis, sic tensio

4. *A very plain and practical way of counterpoising Liquors, of great use in* Hydraulicks. Discovered.

5. *A new sort of Object-Glasses for* Telescopes *and* Microscopes, *much outdoing any yet used.* Discovered.

Fig. 10.3 The fundamental law of elasticity as presented by Hooke, at point 3 (on the left the front page of the document). Note that at point 2 it is described the principle to design a perfectly compressed arch, whose geometry should correspond to the opposite of that derived from a flexible string hanging the loads. The meaning of the anagram "abcccddeeeeeeiiiiiiiillmmmmnnnnnprrssstttttttuuuuuuuuux" was revealed after his death, in 1705: "ut pendet continuum flexile, sic stabit contiguum rigidum inversum")

[4] Hooke R (1679) Lectiones Cutlerianæ, or A collection of lectures: physical, mechanical, geographical, & astronomical, Printed for John Martyn, London.

[5] Galilei G (1687) Dialogo sopra i due massimi sistemi del mondo [Dialogue concerning the two chief world systems], Florence.

[6] Newton I (1687) Philosophiae Naturalis Principia Mathematica. London.

Therefore, from the end of the 1600s, engineers, who had already existed for centuries,[7] though they did not yet have the formal training of university courses, had at their disposal all the tools required to design an antiseismic structure, in accordance with the criteria that would then be used for the majority of the twentieth century. They just didn't know it yet.

It is in this context that the "Age of Enlightenment" began, with the battle between a dynamic optimism that believed in the progression of knowledge, and the consequent possibility of improving every aspect of human life, and the theological optimism of Leibniz and Pope, who maintained that the created world was perfect and that it was impossible to improve any aspect of divine creation. The Lisbon earthquake of 1755 broadened and aggravated the debate. Voltaire, in what we now would call an "instant book",[8] wondered whether Pope would have dared to declare that *all that is is for the best* if he had lived in Lisbon. Rousseau noted that if people insisted on wanting to live in cities and build houses of six or seven storeys they should blame themselves, not God, for the consequences of earthquakes.[9]

It is again surprising to note that the only comment of any practical value came from a utopian philosopher. Earthquakes continued to be the subject of debate among thinkers, not builders.

Not surprisingly, there were now university courses for engineers, but these were generally taught in philosophy faculties (in Pavia from 1786[7]).

Two earthquakes once again came as a wake-up call, in San Francisco[10] in 1906 and in Messina[11] in 1908. In the first case it is interesting to analyse the logics and techniques applied in the reconstruction: the army build 5,610 small houses in a very short space of time, which were rented for 2 dollars a month, up to a maximum of 50 dollars to acquire ownership (Fig. 10.4).

The second was much more important from the point of view of the progress of science.

In the case of the Messina earthquake there was considerable debate regarding how to go about the reconstruction. Three weeks after the earthquake, in the *Monitore Tecnico*[12] it was stated: "An error, for example, that we believe to see appearing on the horizon as a great danger, is that which corresponds to the ideas set out by the hon. Mr Bertolini, Minister of Public Works, with regard to the reconstruction of the towns that have been destroyed. He has suggested ruling out

---

[7] Ingegneri a Pavia tra formazione e professione [Engineering in Pavia between training and profession] In: Cantoni V, Ferraresi A (eds) Cisalpino. Istituto Editoriale Universitario – Monduzzi Editore, Milan, 2007.

[8] Voltaire (1756) Poème sur le désastre de Lisbonne, ou examen de cet axiome: tout est bien.

[9] Rousseau JJ (1756) Letter to Voltaire about the Lisbon earthquake (also known as *Letter on Providence*).

[10] San Francisco, 5:12 a.m., 18 Apr 1906, $M_w = 7.8$ (estimated).

[11] Messina, 5:21 a.m., 28 Dec 1908, $M_w = 7.5$ (estimated).

[12] *Il Monitore Tecnico* (journal on engineering, architecture, mechanics, electronics, railways, agronomy, cadastre and industrial – official body of the association of former students of the Politecnico di Milano), 20 Jan 1909.

Fig. 10.4 One of the villages built after the San Francisco earthquake of 1906

temporary constructions and instead adopting permanent buildings. Just how wrong this idea is can be demonstrated by a complex series of reasons. Above all, it would be unwise to construct permanent buildings straightaway, before seriously and in an in-depth manner studying the construction methods that should be adopted in order to guarantee that new buildings will certainly be able to resist any future seismic movements. These construction methods need to be discussed at length by the experts, and the need cannot be met by the suggestions, no doubt mainly theoretical, that may come from the Commission appointed by the Minister [. . .]".

Clearly, everything had changed.

And while we will discuss how history has repeated itself in recent years a bit later, now I wish to point out that in less than 4 months a Royal Decree was published containing the new Technical Standards,[13] in which (article 24) it is explicitly stated that *when calculating the stability and resistance of buildings, the following must be taken into consideration: 1° the static actions due to the building's own weight and overloading, increased by a percentage that represents the effect of the vertical vibrations; 2° the dynamic actions due to the horizontal seismic movement, representing them with accelerations applied to the masses of the building in two directions [. . .].*

Subsequently, the various commissions created went well beyond the expected *predominantly theoretical suggestion*, publishing, among others, in the Giornale

[13] Royal Decree of 18 Apr 1909, no. 193, published in Official Gazette no. 95, on 22 Apr 1909.

del Genio Civile[14] a brief summary of modern seismic engineering, in which *the following concepts are cited as fundamental*:

1° *The theory that the effects of the dynamic actions on building elements can be compared to those produced by forces proportional to the masses both in the horizontal direction and the a vertical direction* (in other words, engineers had finally discovered Newton and the forces of inertia);
2° *The opportunity, for the calculation of these forces, to refer to the proportions of the buildings that have proven to satisfactorily bear seismic shocks with considerable destructive power* (in other words, since numerical data to be used to estimate the accelerations experienced by the structural masses are not available, let's rely on a kind of back analysis in order to decide on the design value);
3° *The suitability, imposed by considerations of an economic nature, of allowing, for horizontal seismic forces, a greater tolerance with respect to the limits normally adopted as safety levels, in light of the exceptional nature of the actions and of the advantages of avoiding excessive stiffness* (in other words, let's accept a certain level of damage for the design earthquake, since we cannot afford to plan for an event that may happen in several centuries' time);
4° *The preferred way to put this tolerance into action by undervaluing the horizontal seismic forces, reducing them to around 1/3 of their value* (in other words, a reasonable numerical estimate is given to what we now call the behaviour or force reduction factor);
5° *The confidence that the margin provided by these safely estimated loads offers [...] a sufficient guarantee of safety to people if not of absolute integrity to buildings* (in other words, the first performance based design logic was defined);

the text continues with numbers 6, 7, 8 and 9, but I'm going to stop there.

One of the commissions drew up, among others, a report (July 1909)[15] that can be considered to be the forerunner of the earthquake hazard maps and the geological and geotechnical criteria to be applied in order to reduce seismic risk. Professor Torquato Taramelli made a significant contribution to this work and, together with the younger Oddone and Baratta, formed the then great Pavia school of seismology.

It is also worth mentioning that the suggestions of a commission focused on seismological elements continually referred to the *Technical Standards*, offering the necessary support to the calculations, in accordance with a pragmatically effective logic which did not always form the basis of later studies, namely carried out in the second half of the last century.

---

[14] Instructions and examples of calculations for constructions that are stable against seismic actions, *Giornale del Genio Civile*, year LI, 1913 (the Commission that wrote this document comprised professors Ceradini, Canevazzi, Panetti, Reycend and Salemi Pace, and engineer Camerana).

[15] Report of the Royal Commission set up to identify the most suitable areas for the reconstruction of the urban areas affected by the earthquake that took place on 28 December 1908 or by other previous earthquakes. Printed by the R. Accademia dei Lincei, Rome, 1909.

But let's get back to the engineers.
It is now clear:

- that the action of the earthquake can be represented with a series of horizontal forces obtained by multiplying the masses (M) by the relative accelerations (a_M):

$$V_{\text{base}} = \Sigma M_i a_{Mi}$$

- that these forces must be balanced by the resistance of the structural elements, calculated using Hooke's law, or basically writing the equation of motion as an equilibrium balance equation in which the force of inertia (mass by acceleration) is countered by an elastic force (stiffness, K, by displacement, d):

$$M \times a = K \times d$$

The problem is defining the design ground acceleration (and seismologists will try to solve this problem) and the amplification factor required to go from the acceleration of the ground to that of the structural masses.

The mass acceleration values indicated in various technical circulars published together with the standards are first situated between 1/12 and 1/8 of gravity,[14] then increased to values between 1/8 and 1/6[16] (for vertical actions an increase in weight of 50% is suggested). These are low values, essentially based on sensations (*at the current state of knowledge [...] it appears that the stability of a building can be believed to be sufficiently guaranteed [...]*[14]).

Another earthquake occurred to teach us more.

This time it was the El Centro earthquake,[17] one of the first cases in which recorded experimental data were available, which were obviously analogical. It was a violent earthquake, with a magnitude around 7. Acceleration peaks of 0.319 g and maximum ground displacements of 212 mm were recorded. These values would remain significant as an indication of the demand of a strong earthquake for decades, in particular with regard to acceleration, which, as we have seen, was considered to be the fundamental design parameter.

Based on these instrumental data, Maurice Biot perfected the solution to the second problem, the passage from acceleration of the ground to acceleration of the structure, using the concept of a *response spectrum*,[18] which enables us to calculate an amplification coefficient using a single structural parameter, the fundamental vibration period of the structure.

---

[16] Lieutenant's Decree of 19 August 1917, Italian Official Gazette, 10 Sept 1917.

[17] El Centro, California, 8:37 p.m., 18 May 1940, $M_w = 6.9$. In reality the first accelerogram recording is from the Long Beach earthquake, in 1933, but which had a less significant impact on the development of knowledge.

[18] Biot MA (1934) Theory of vibration of buildings during earthquakes. Z Angew Matematik Mech 14(4):213–223. Critically discussed in: Trifunac MD (2006) Biot response spectrum. Soil Dyn Earthquake Eng 26:491–500.

A great deal of people would continue working on response spectra, from George Housner,[19] a doctoral student at Caltech in 1940, to Nathan Newmark and Bill Hall at Urbana,[20] moving from response spectra to design spectra, from linear responses to non-linear responses, from deterministic logics to probabilistic logics, from accelerations to velocities and displacements (Fig. 10.5).

Values typical of the maximum amplification were identified around 2.5/3.0, which, when applied to ground accelerations of 0.3/0.4 g, can mean horizontal forces greater than those due to gravity on a system with linear behaviour.

The wisdom of the 1908 Earthquake Commission seems evident when it suggested a reduction to a third of the estimated forces from mass and acceleration, which were then unknown. Nevertheless, conventional logics based essentially on the *sensations* of the experts continued to be applied the world over, and in Italy in particular, rather than proceeding in this direction, i.e. acknowledging the values that must be dealt with and accepting justified reductions, which obviously involve damage.

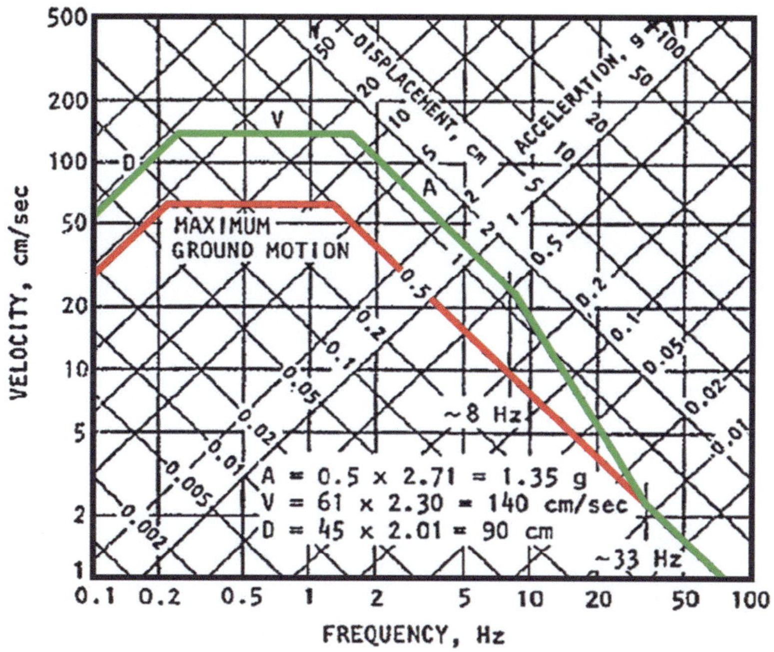

Fig. 10.5 A tripartite elastic response spectrum, as presented by Newmark and Hall (1982). Note that in this representation of the maximum response the displacement demand is constant for frequencies lower than 0.2 Hz, i.e. for period longer than 5 s (*red*: ground spectrum, *green*: mass spectrum)

[19] Housner G (9 Dec 1910, Saginaw, MI; 10 Nov 2008, Pasadena, CA) the teacher of all seismic engineers.

[20] Newmark NM, Hall WJ (1982) Earthquake Spectra and Design. Engineering Monographs, EERI, Oakland, CA.

Once again, the earthquake that started the new revolution in seismic engineering took place in California, not far from El Centro, in San Fernando.[21]

The most striking cases are a dam, two motorway junctions and two hospitals.

At the dam (*Pacoima Dam*) ground accelerations greater than 1 g were recorded.

The case of the Olive View UCLA Medical Center (Fig. 10.6), opened a month earlier, became the subject of studies in all seismic engineering centres, including Italy, which, however, didn't really wake up about this subject until 5 years later, with the Friuli earthquake.[22]

Fig. 10.6 Damage at the Olive View UCLA hospital, San Fernando earthquake (1971, photo USGS). Note the different response of columns with rectangular and spiral confinement, the shear collapse at the first floor and the relative displacement at the ground floor, estimated in 81 cm

[21] Sylmar, California, 6:01 a.m., 9 Feb 1971, $M_w = 6.6$.

[22] Gemona, 21:06, 6 May 1976, $M_w = 6.4$.

The essential element that characterises the research activity and its translation into design standards consists of recognising the inadequacy of the representation of the behaviour of structure through linear laws. In effect, the acceptance of a non-linear behaviour, and therefore of the manifestation of structural damage, was implicitly involved in both reducing the design resistance to a third of the forces of inertia and designing for conventional values, possibly ten times lower than the expected demand. The point was therefore rather to recognise that structures designed according to similar criteria can give rise to completely different results, can collapse or can resist the violence of an earthquake with no problems. Credit and blame were quickly attributed to the greater or smaller capacity of a structure to deform plastically after having reached its yield level.

The concept of displacement ductility (μ) was defined as the ratio between the displacement at collapse (Δ_u) and the yield displacement (Δ_y).

$$\mu = \Delta_u / \Delta_y$$

Ductility became the fundamental myth of every research study, every standard, every application.

Rather than simply acknowledging the possibility of comparing demand and capacity in terms of displacements, people turned to ductility as a corrective parameter of strength. Structures continued to be designed and checked with a comparison between forces, modified depending on the deformation capacity. The design logic typical of actions due to gravity, which is always present, continued to be applied to actions assumed to take place once in 500 years. After the myth of the linear response of structures, new myths were born, which were even harder to be demolished[23]:

The myth of ductility and the behaviour factor. In every part of the world studies on ductility were carried out and definitions and conventions were created. In reality the conventional value of displacement at the elastic limit and at collapse meant that the same term defined values that could differ from one another by as much as three or four times. On the basis of conventional ductility a force reduction factor was defined, which, on apparently only the most rational basis, drew on the reduction coefficient introduced after the Messina earthquake. The reduction factor was applied to the entire structure, without taking into account ductilities of different structural elements that can vary considerably. A typical and striking case is that of a bridge with piers of different heights, in which it is easy to demonstrate that it is impossible to attribute the same ductility to different piers (Fig. 10.7).

The myth of elastic stiffness. It was assumed that the fundamental vibration period (T) of the structural system, and therefore its stiffness (K), could be determined at the beginning of the design:

---

[23] Priestley MJN (2003) Myths and fallacies in earthquake engineering, revisited. The 9th Mallet Milne Lecture, IUSS Press, Pavia

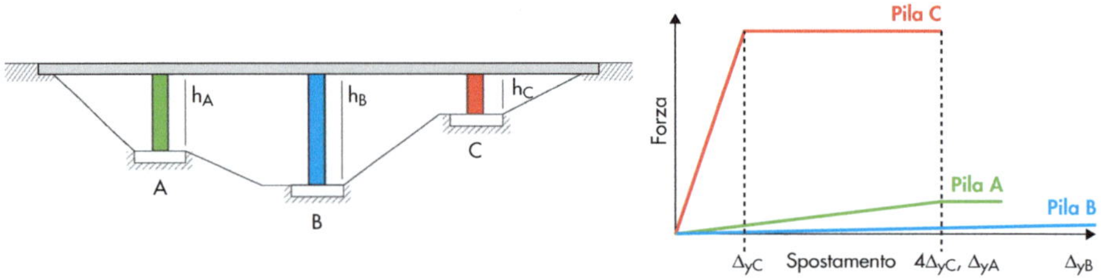

Fig. 10.7 Sketch of a bridge model and of expected force – displacement response of each pier

$$T = 2\pi\sqrt{M/K}$$

On the basis of the period the design spectrum was entered, the acceleration (S_a) at the structural mass (M) was assessed and the resistance (V_r) to be attributed to the structure was calculated, by multiplying mass by acceleration and dividing by the behaviour factor (q):

$$V_R = \frac{S_a M}{q}$$

Unfortunately, it is easy to prove that stiffness is not an independent variable, but depends on the strength, unknown at the beginning of the process. In addition, we must discuss which value of stiffness to use: the tangent initial one, that of a cracked structure, the secant to yield one, the secant to the design displacement?

The myth of refined analysis. The availability of complex calculation methods and more and more powerful calculators allowed geometrically refined representations of structures and the use of models with a large number of degrees of freedom. It became possible to carry out analyses that separately assess the various mode of vibration of the structure and then to combine them according to the mass participating in each mode. These approaches are extremely useful in many mechanical and structural applications, but can result in gross mistakes when we attempt to predict the non-linear response, which is characterised by a specific mechanism of damage and collapse, from the combination of a number of elastic mode of vibration, which have little or nothing to do with the post-elastic behaviour.

The myth of the conservation of displacements. Another fundamental theory on which essentially all codes are based consisted of the assumption that for structures with the same vibration period (in this context, with the same initial stiffness) a definite seismic action produces the same maximum displacement demand, regardless of the energy dissipation capacity of the system, and therefore of the form of its typical hysteresis loops. The antithesis of this myth, which is also false, can be found in the conviction of some researchers that the seismic problem can be tackled solely through energy balances. I could therefore call it *the myth of energy and its antithesis*.

Finally, *the myth of the engineer ("let's put one more re-bar").* I've left this one last, and I've given it a provocative and incomprehensible name, I know. But this is the most difficult taboo to overcome, the one that assumes that a greater strength, however distributed among the elements of a structure, in any case produces a greater safety against collapse. A false and dangerous conviction, discussed again later, pointing out the one true legacy from the seventies which nowadays can and should be used in design, i.e. the principle of hierarchy of strength or capacity design.

Once more, our teachers (earthquakes) made these and other problems clear.

Three events in quick succession (Loma Prieta[24], 1989, Northridge[25], 1994 and Kobe[26], 1995) again shook the engineers' certainties. In the first case a two-level viaduct collapsed in Oakland (Fig. 10.8), in the second considerable damage was caused to hundreds of kilometres of motorway viaducts, and the third shook the certainties of the country that considered itself to be the most advanced in the world in terms of seismic safety: Japan (Fig. 10.9).

After the Loma Prieta earthquake a report[27] was published for the Governor of California, George Deukmejian, coordinated by the legendary George Housner (50 years after El Centro he was no longer a doctoral student). The report was entitled *Competing Against Time*. Housner wrote:

> Future earthquakes in California are inevitable. Earthquakes larger than Loma Prieta with more intense ground shaking will occur in urban areas and have severe consequences – too large to continue "business as usual". [...] The Board of Inquiry has identified three essential challenges that must be addressed by the citizens of California, if they expect a future adequately safe from earthquakes:
>
> - *Ensure that earthquakes risks posed by new constructions are acceptable.*
> - *Identify and correct unacceptable seismic safety conditions in existing structures.*
> - *Develop and implement actions that foster the rapid, effective, and economic response to and recovery from damaging earthquakes.*
>
> [...] The State of California must not wait for the next great earthquake, and likely tens of billions of dollars damage and thousands of casualties, to accelerate hazard mitigation measures. [...] Earthquakes will occur – whether they are catastrophes or not depends on our actions.

Eventually, risk becomes the object of discussion, being understood that probabilistic logics have to be adopted, and that design and strengthening rules have to be coherent with the available resources, not with an ideal level of safety.

It seems to go back to the time of philosophers and thinkers, with streams of words to illustrate various theories of *performance based design* (Fig. 10.10), attempts to systematise refined logics, in which various design earthquakes were

---

24 Loma Prieta, 5:04 p.m., 17 Oct 1989, $M_w = 6.9$.

25 Northridge, 4:31 a.m., 17 Jan 2004, $M_w = 6.7$.

26 Kobe, 5:46 a.m., 17 Jan 2005, $M_w = 6.8$.

27 Competing against time, Report to Governor George Deukmejian from the Governor's Board of Inquiry on the 1989 Loma Prieta Earthquake, George W. Housner, Chairman, Department of General Service, North Highlands, CA, 1990.

Fig. 10.8 Collapse of the Cypress Viaduct of Intestate 880 at Oakland, Loma Prieta earthquake, 1989 (photos H.G. Wilshire, USGS)

defined, depending on the probability of occurrence (p) in a determined interval of time (T_L), or the average return period (T_R), and various performances to be required of structures, depending on their importance in the event of a catastrophe, and on the consequences in the event of damage and collapse:

$$p = 1 - e^{-T_L/T_R}, \quad \text{or} \quad T_R = \frac{-T_L}{\ln(1-p)}$$

For example, for an earthquake with a probability of 50% in 50 years, i.e. with an average period of return estimated in 72 years, it is required that the damage be basically insignificant, whereas for an event with a probability of 10% in the same interval of time ($T_R = 475$ years) it is required that an important bridge remain in full usable condition or that a hospital maintain full functionality, but even fairly considerable damages to a residential building may be accepted, and so on.

These are clear and logical concepts, but it is not immediate to translate them into the answers to the only two questions of interest for a builder: which resistance

Fig. 10.9 Collapse of the Hnshin Expressway, Kobe earthquake, 1995

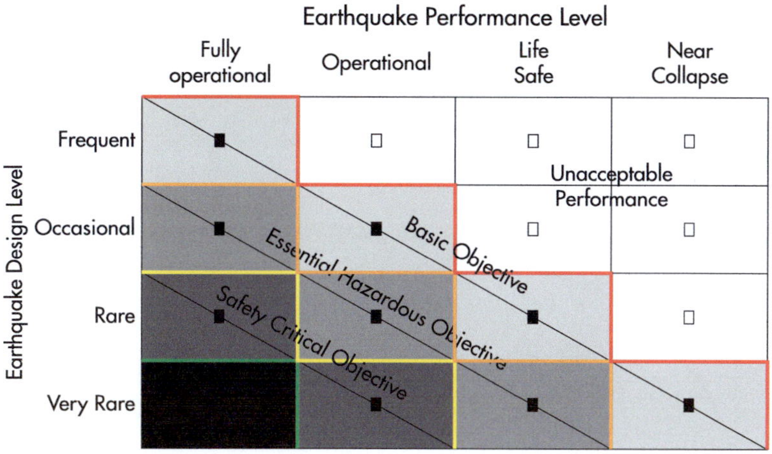

Fig. 10.10 Matrix earthquake – performance – exposure (*Vision 2000, Performance bases seismic engineering of buildings*, SEAOC, 1995)

should be assigned to a structure and how should this strength be distributed between the various structural elements.

There are no doubts about one fundamental aspect, however: the force to which a structure is subjected (and therefore its acceleration) is not a suitable variable to appropriately describe the expected damage, or the *performance* of the system. In fact, it is clear that very different levels of damage, corresponding for example to the usability of a building, the possibility of repairing it quickly, avoiding collapse, correspond to values of force, and of acceleration, which are not very different from one another (Fig. 10.11).

On the contrary, the various performances of interest are characterised by displacement values that differ greatly from one another, so that it is not hard to imagine forms of correspondence between expected performance and acceptable displacements, or, in a way that is much easier to apply, to define an adimensional displacement variable, to be used as a fundamental design parameter. A suitable parameter of this type is immediately identifiable in the relationship between horizontal displacement and height, is obviously an angle, and is normally called "drift".

For example, in the case of a building it can refer to the relative displacement between two storeys divided by the storey height, and we can talk about *interstorey drift*, whereas in the case of a bridge it can refer to the displacement of the deck divided by the height of the pier.

It is on the basis of this fundamental observation, and in the acknowledgement of the insurmountable limits of any design method that uses forces and accelerations as fundamental variables, that Nigel Priestley[28] developed what is to date the only method of displacement-based seismic design that can truly be implemented in practice, publishing a book[29] about which Graham Powell, Professor Emeritus at UC Berkeley, opened his review, published in *Earthquake Spectra*, with the words:

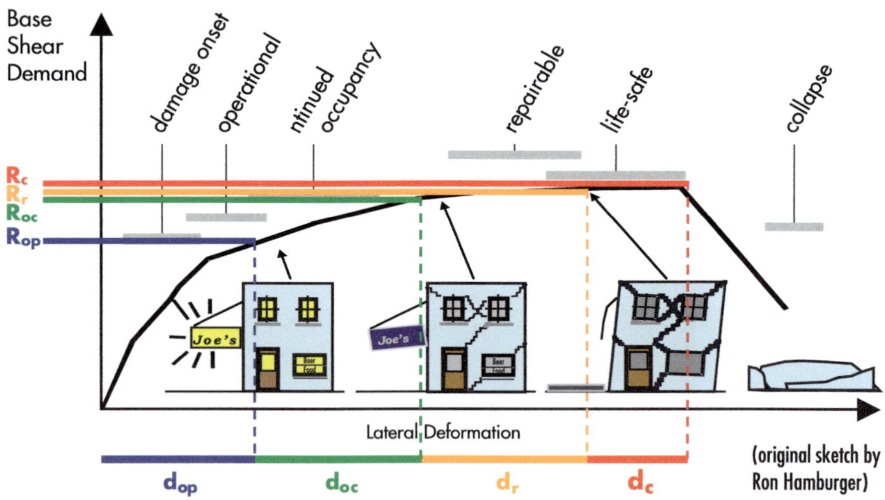

Fig. 10.11 Force and displacement levels characterizing different global performances

---

28 Michael John Nigel Priestley, 21 Jul 1943, Wellington, New Zealand, the true inventor of displacement-based design, friend and teacher.

29 Priestley MJN, Calvi GM, Kowalsky MJ (2007) Displacement based seismic design of structures. IUSS Press, Pavia.

It is rare for a book on structural engineering design to be revolutionary. I believe that this is such a book. If you are involved in any way with seismic resistant structural design, this should be on your bookshelf, and you should read at least the first three chapters.

The book is the culmination of 15 years of research, in which dozens of doctoral students were involved, though its fundamental logic is simple and direct (Fig. 10.12):

- design displacement (drift) values are defined for the various performances to be examined;
- displacement design spectra are defined for events corresponding to the various performances;
- the expected, or rather desired, structural behaviour is defined and a simplified model is created;

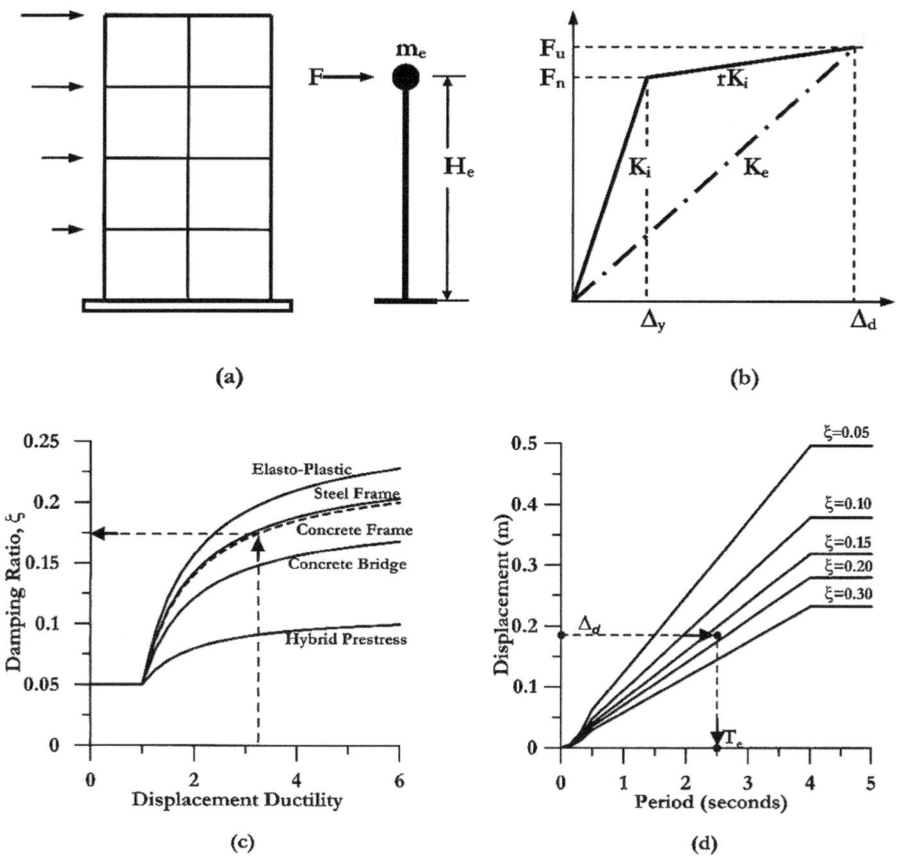

Fig. 10.12 Fundamentals of displacement based design of buildings. **(a)** SDOF simulation, **(b)** effective stiffness K_e, **(c)** equivalent damping vs. ductility, **(d)** design displacement spectra[29]

- for each performance a reliable value of equivalent damping is estimated, through which the spectrum that represents the displacement demand is reduced;
- the spectrum is entered with the design displacement and the corresponding stiffness is read;
- multiplying stiffness by displacement the value of strength to be attributed to the structure is found;
- the strength is distributed among the various resistant elements according to the assumed response.

Each step is further discussed below.

Design displacements. It is easy to approximately estimate the elastic limit of a structure using its geometry alone. For example, the secant yield rotation (θ_y) of a circular pier of a bridge (Fig. 10.13) can be estimated using the yield deformation of the steel (ε_y), its diameter (D) and its height (H), as:

$$\theta_y = 0.75\,\varepsilon_y H/D$$

Similarly, for a reinforced concrete frame with beams that are weaker than columns, the same variable can be estimated using only the span (l_b) and height (h_b) of the beams:

$$\theta_y = 0.5\,\varepsilon_y l_b/h_b$$

The design displacements may be similar to those calculated in this way to avoid significant structural damage, or considerably larger where the performance accepts to repair the structure after an event.

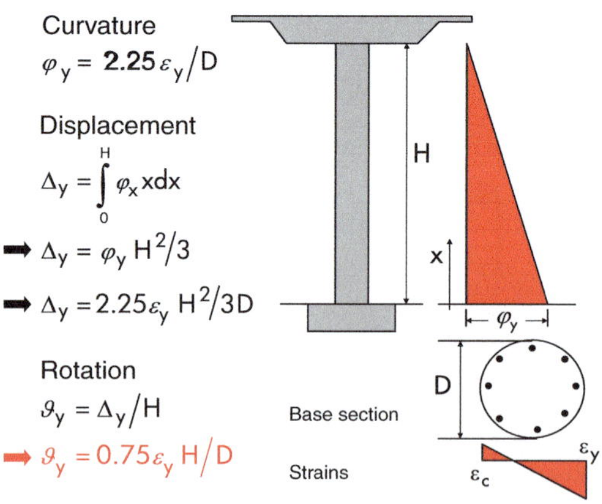

Fig. 10.13 Estimate of the yield curvature, rotation and displacement of a bridge pier

In any case interstorey displacement values able to limit non-structural damage must be taken into consideration, too. For example, stone walls could show significant damage, with drift values around 0.5%, whereas other non-structural elements, which are less sensitive to imposed displacements, could cope with values around 1% without particularly significant damage.

Displacement design spectra. The approximations that arose in the 1970s from the solution of Duhamel's integral enabled to express the ordinates of a displacement spectrum (S_d) using those of the corresponding acceleration spectrum (S_a) using only the vibration period of the structure (T):

$$S_d = \frac{T^2}{4\pi^2} S_a$$

In reality, more recent studies clearly indicate that the correlation between the two spectra is much weaker, and today it seems more reliable and efficient for practical purposes to express the displacement spectrum as a bilinear with a second constant branch, using a "corner" period value (T_c) and the corresponding displacement ($d_{\max}$), defined on the basis of the moment magnitude (M_w) and the distance from the epicentre (r, in kilometres):

$$T_c = 1.0 + 2.5\,(M_w - 5.7)$$

$$d_{\max} = \frac{10^{(M_w - 3.2)}}{r}$$

Equivalent viscous yield. In the equations of motion of a dynamic system, the term containing the damping coefficient, which, multiplied by the velocity, provided the third term of a balance equation, has always been a sort of free parameter, used to force numerical results to get closer to experimental evidence. In reality, the fact that a wider hysteresis loop tends to reduce the displacement demand is incontrovertible from experimental data. It is possible to express an equivalent viscous damping (ξ_e) using the hysteresis area of a loop (A_h) compared with the area of the triangle defined by maximum force (F_m) and displacement (d_m):

$$\xi_e = \frac{A_h}{2\pi F_m \Delta_m}$$

And on the basis of this equivalent damping value to define a displacement demand reduction factor (η_ξ):

$$\eta_\xi = \sqrt{\frac{0.07}{0.02 + \xi_e}}$$

It is easy to check that reasonable equivalent damping values seldom exceed 0.2/0.3 and that consequently the correction factor never exceeds 0.5. Furthermore, an error of 20% in the damping estimate would anyway produce an error in the displacement demand estimate of less than 10%, and in this business of engineers an error of 10% in the displacement estimate is a good approximation of an infinitesimal.

On the contrary, numerical analyses that overestimate the viscous damping contribution, by maintaining the proportionality to the initial stiffness even when this is considerably reduced, can lead to estimates that make the difference between a structure that survives and one that collapses.

Response and model. This approach has retained the old trick, or rather the brilliant idea, of Tom Paulay[30], which I referred to as capacity design, or hierarchy of resistances.

Tom explained how, in order to make ductile a chain a made of brittle links, it is sufficient to replace one of them with a ductile one, provided that its strength is slightly lower than that of the others, therefore yielding first, preventing the force from growing any further, and thus protecting all the other links.

Translated for seismic engineers: how to prevent brittle failure modes due to shear stresses by ensuring that the flexure failure modes are weaker, how to form plastic hinges in the beams by making them weaker than the connected columns, how to prevent collapses in the foundations by making them stronger than the vertical elements supported by them, and so on.

Distribution of strength. Acknowledging that stiffness and resistance are not independent variables implies the possibility of modifying the distribution of the horizontal forces amongst various elements simply by increasing the reinforcement of those which are geometrically less stiff. This results in the possibility of making more intelligent structures, reducing torsion problems, and bringing the centre of mass and the centre of resistance closer.

To those who are wisely wondering whether structures that are designed with an approach based on forces or on displacements are really different, I will only say here that the relationship between any parameter of intensity of motion and the strength to be assigned to a structure varies linearly when we refer to forces, whereas it varies quadratically when we refer to displacements (Fig. 10.14). I will also say that the reinforcement ratio in piers or walls with different geometry is constant when using displacements but variable with the square of the height or with the depth of the base when using forces. Only space prevents me from discussing how the displacement method is the only rational one when we wish to assess the safety of existing structures, in which there is no doubt that the perceived acceleration (S_a) is merely the ratio between resistance (V_R) and mass (M), regardless of the ground acceleration, thereby overturning the logic of elastic response:

$$S_a = V_R / M$$

[30] Paulay T (26 May 1923, Sofron, Hungary; 28 Jun 2009, Christchurch, New Zealand), count, cavalry officer, refugee in Germany and New Zealand, famous professor, author of successful books, fine gentleman, kind and affectionate teacher.

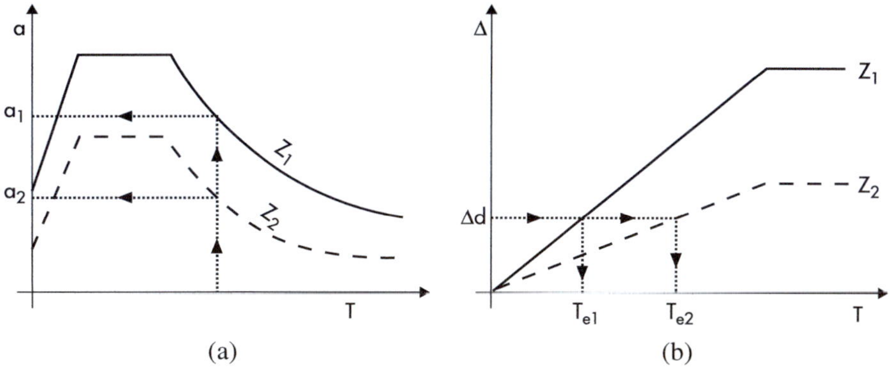

Fig. 10.14 (**a**) Acceleration spectra: mass acceleration and inertia force are proportional to an intensity parameter and (**b**) displacement spectra: period is proportional to an intensity parameter, therefore design strength varies with its square

We could stop here, but I can't conclude without mentioning the two most recent earthquakes, the one that took place in China in May 2008[31] and the L'Aquila earthquake[32] (Fig. 10.15).

I mention the first one only to establish a relationship of scale between the events. The L'Aquila earthquake seemed dramatic to us Italians, and it was dramatic. But

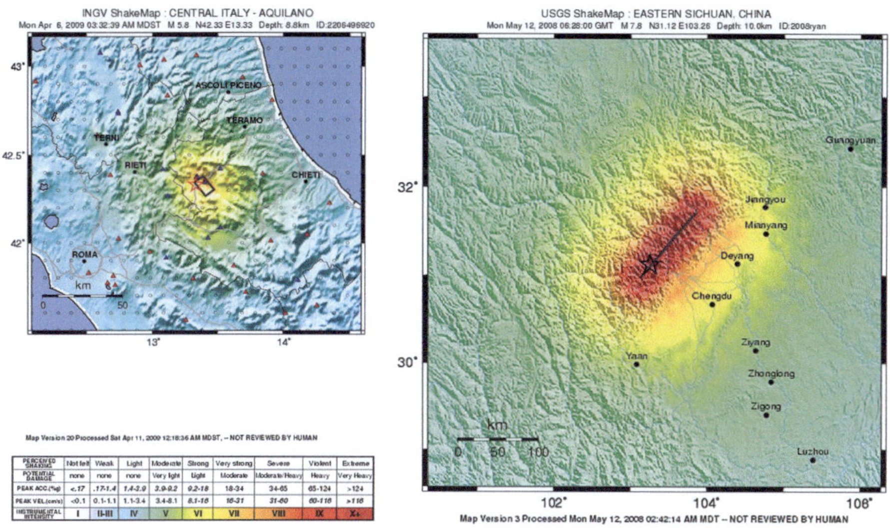

Fig. 10.15 Comparison between the earthquakes of L'Aquila ($M_w = 6.3$, 300 deaths and 70,000 homeless) and of Sichuan ($M_w = 7.9$, 70,000 deaths and 11 millions homeless)

[31] Wenchuan, 14:28, 12 May 2008, $M_w = 7.9$.

[32] L'Aquila, 3:32, 6 Apr 2009, $M_w = 6.2$.

how would we have coped with seventy thousand deaths instead of three hundred? How would we have dealt with eleven million homeless people instead of seventy thousand? Italy isn't China? Of course, if we overlook Messina, 1908.

Less than a year passed and a great deal of research took place between the two events. And it is significant that while the Sichuan earthquake was taking place, in Pavia an earthquake simulator was being used to test buildings showing damage and collapse mechanisms which would then be reproduced in the real-life laboratory of L'Aquila.

Which brings me on to L'Aquila.

The first earthquake for more than a century in Italy whose focus was exactly below an important city.

From a seismology point of view, the new data were striking and opened up new debates about the reliability of the recently adopted earthquake hazard map, which no doubt represents a state of the art at an international level.[33] Some people question whether it is acceptable for ground accelerations in excess of 0.6 g to be recorded in an area where the expected acceleration with a 10% probability in 50 years, or with a period of return of 475 years, is equal to around 0.25 g, thus demonstrating that they have no understanding of the concept of a uniform probability spectrum.

If once again we limit ourselves to lessons for engineers, the most interesting topic relates to reconstruction, and in particular to the technical choices that have enabled us to construct 4,500 permanent houses of high quality in around 8 months, with an average production of around 3 million euros per day (Fig. 10.16). This is neither the time nor the place to discuss the extraordinary organisational machine,[34] the non-profit consortium led by the Eucentre foundation, which permitted to operate without a general contractor, the choice related to town-planning, architecture, energy, sustainability, installations.

Now a few words about the fundamental structural choices. It was crucial to use various technologies and different materials, such as wood, steel and concrete, adopting in all cases a high level of prefabrication. It was fundamental to start designing, preparatory works and calls for bids before knowing the construction sites, the characteristics of the ground and the morphology of the areas.

The solution was identified in the construction of two plates, one working as a foundation, the other supporting the buildings, separated by a series of columns and by a seismic isolation system made up of sliding devices on a spherical surfaces. Friction pendulum devices are derived from a brilliant idea, developed in its current

---

33 Crowley H, Stucchi M, Meletti C, Calvi GM, Pacor F (2009) Revisiting Italian design code spectra following the L'Aquila earthquake. Progettazione Sismica, 03/English, IUSS Press, Pavia, pp 73–82.

34 Calvi GM, Spaziante V Reconstruction between temporary and definitive: The CASE project, ibidem, pp 221–250.

Fig. 10.16 Reconstruction after the earthquake in L'Aquila, the C.A.S.E. project

technology in Berkeley in the early 1980s,[35] which is based on the behaviour of a pendulum (like that of cuckoo clocks, Fig. 10.17).

It is interesting to point out that in 1909 (after the Messina earthquake) seismic isolation technologies were proposed (one of which was conceptually similar to the isolators we are talking about[36]), discussed and assessed, but ruled out for reliability reasons. Arturo Danusso[37] wrote: *we immediately understand that if we could practically put a house on springs, like an elegant horse-drawn carriage, an earthquake would come and go like a peaceful undulation for the happy inhabitants of that house*, but concluded: *I think that a certain practical sense of construction alone is sufficient by itself to dissuade from choosing mechanical devices to support stable houses.*

In this specific case the radius of oscillation is determined by the curvature of the sliding surface, which acts like the length of the suspension arm, and in a completely

[35] Zayas V, Low S (1990) A simple pendulum technique for achieving seismic isolation. Earthquake Spectra 6(2).

[36] A double-slide isolator on curved surfaces, patented by M. Viscardini in 1909, described in: Barucci C, La casa antisismica [The antiseismic house], Gangemi, 1990.

[37] Il monitore Tecnico, 10 Aug 1909.

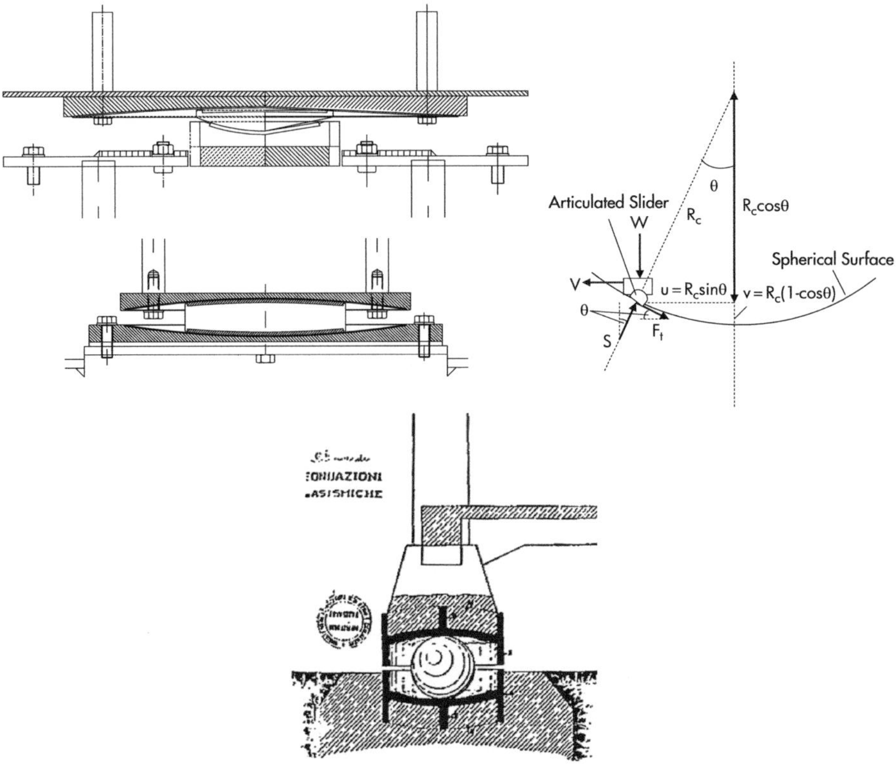

Fig. 10.17 Isolators sliding on a spherical surface used in L'Aquila and the Viscardini patent of 1909

analogous way the vibration period (T) is a function of the radius of curvature alone (r) and of the acceleration of gravity (g), and therefore the stiffness (K) of the system is exclusively a function of the weight of the building ($W = Mg$) and of the radius.

$$T = 2\pi\sqrt{\frac{r}{g}}, \quad K = W/r$$

If a seismic event occurs the upper plate may slide over the lower one, and the building constructed on this experiences accelerations, therefore forces, limited to values around a tenth of the acceleration of gravity, regardless of the violence of the earthquake. In the hypothetical event of a repeat of the L'Aquila earthquake, the acceleration experienced by the buildings would be reduced by around ten times.

So, a technology that is 30 years old, or a 100 years old, implemented with success in several projects, was for the first time employed systematically in one hundred and eighty five residential buildings, constructed in a few months on more than seven thousand isolators and tested reproducing a credible seismic motion on site on eleven buildings.

A much older idea, perhaps not decades, but millennia old, if we wish to interpret the *calcatis ea substravere carbonibus, dein velleribus lanae*, which Pliny attributed to the builders of the temple of Diana at Ephesus, as a seismic isolation measure.

And the circle is complete.

Now we wonder about the next earthquake, when and where it will happen, the energy it will release, the damage it will cause, the collapses, the victims. We don't know and we will never know, but we can try to use in the best way the scarce resources that humanity can afford to dedicate to prevention.

I will finish by reminding once again the words that concluded the preface of the report *Competing against time*[25]: *Earthquakes will occur – whether they are catastrophes or not depends on our actions.*

(On 12 January 2010 at 4.53 a 7.0 magnitude earthquake struck Haiti. On 27 February 2010 at 3.34 a 8.8 magnitude earthquake struck Chile. The lesson continues.)

Chapter 11
Current Trends in the Seismic Design and Assessment of Buildings

Andreas J. Kappos

Abstract Current trends in the seismic design and assessment of buildings are discussed, with emphasis on two procedures that merit some particular attention, displacement-based procedures and deformation-based procedures. A number of selected case-studies are summarised, involving reinforced concrete (R/C) buildings designed to the aforementioned procedures. Then, an overview of the currently available procedures for seismic assessment is presented and the different designs are assessed using state-of-the-art methods involving inelastic analysis of the static and/or dynamic type; alternative designs are compared in terms of economy and seismic performance, and some general conclusions are drawn regarding the feasibility of introducing the new procedures in seismic codes.

11.1 Introduction

A critical overview and discussion of the various seismic design procedures available for buildings is provided in the first part of this chapter, with a view to assessing whether currently adopted procedures are adequate and also whether new (or relatively new) proposals for improved design methods could be useful within the frame of the "new generation" of codes. After a quick reference to the seismic design procedures for buildings adopted by current leading codes for earthquake-resistant design, i.e. Eurocode 8 (CEN, 2004), and the American IBC (International Conference of Building Officials, 2009), the current trends for performance-based seismic design are then presented and discussed; emphasis is placed on two procedures that merit some particular attention, namely direct displacement-based design (Priestley et al., 2007) and deformation-based design (Kappos and Stefanidou, 2010).

A.J. Kappos (✉)
Laboratory of Concrete and Masonry Structures, Department of Civil Engineering, Aristotle University of Thessaloniki, Thessaloniki 54124, Greece
e-mail: ajkap@civil.auth.gr

M. Garevski, A. Ansal (eds.), *Earthquake Engineering in Europe*, Geotechnical, Geological, and Earthquake Engineering 17, DOI 10.1007/978-90-481-9544-2_11,

In Section 11.3, a number of selected case-studies are presented, involving reinforced concrete buildings designed to a number of the aforementioned procedures. The different designs are compared in terms of economy, i.e. required quantity of materials and estimated labour costs, and also in terms of easiness to apply.

In Section 11.4, an overview of the currently available procedures for seismic assessment is presented, and the seismic performance of the different designs (of Section 11.3) is assessed using state-of-the-art methods, involving inelastic analysis of the static and/or dynamic type, and verification of different performance indices (local and global). On the basis of the assessment, each method is evaluated and critically discussed.

Finally, in Section 11.5, some general conclusions are drawn, regarding the feasibility of using new procedures that aim at a better control of the seismic performance of buildings under different levels of seismic loading.

11.2 Seismic Design of Buildings

Currently, the two leading seismic codes worldwide, are arguably Eurocode 8 (CEN, 2004), the prevailing code in Europe (and some other countries in the world), and the International Building Code (International Conference of Building Officials, 2009), which has recently replaced the long-established previous codes, such as the *Uniform Building Code* (International Conference of Building Officials, 1997) in North and Central America (and other parts of the world). It is noted here that, as far as seismic design actions are concerned, the IBC generally adopts the ASCE 7 standard (American Society of Civil Engineers, 2006). Most of the international codes share essentially the same principles and design procedures (notably the "capacity" philosophy that aims at the development of a favourable ductile plastic mechanism), although differences in some of their provisions do exist and (for the same design assumptions) the designs resulting from each code are not the same.

Critical overviews of these codes can be found in a number of publications, including a recent one by the writer (Kappos, 2009), and they will not be repeated herein. The emphasis in the remainder of this section will be on performance-based design, which can be thought of as an explicit design for more than one *limit state* (or *performance objective*, in US terminology).

Performance can be monitored in a number of ways, but it is clear that parameters that are directly related to damage (Kappos, 1997a), such as member deformation or interstorey drift, are preferred choices. For a number of reasons, the best-known procedure that falls within this category, is the so-called displacement-based design (DBD), whose roots can be traced in a paper by Moehle (1992), but its full development and extensive calibration were carried out by Priestley (1993) and Priestley and Kowalsky (2000), who recently produced an entire book (Priestley et al., 2007) describing all aspects of the methodology (for both buildings and bridges). A number of different displacement-based methods are described in a comprehensive state of the art report by the *fib* Task Group on Displacement-based Design and Assessment (*fib* Task Group 7.2, 2003). An interesting categorisation is made therein, assigning the various proposals to three categories, namely:

- Deformation-Calculation Based (*DCB*),
- Iterative Deformation-Specification Based (*IDSB*), and
- Direct Deformation-Specification Based (*DDSB*).

The first category of methods (DCB) involve calculation of the expected maximum displacement for an already designed structural system; detailing is then provided such that the displacement capacity of the building and its components exceeds the calculated maximum displacement. The second category, IDSB methods, are similar to the DCB in that they involve analysis of an already designed system to evaluate the expected maximum displacement. However, unlike the DCB methods, a target displacement is selected, and as a result, changes are made to the structural system such that the calculated displacements are kept below the specified limit; hence the iterative nature of the process. The last category (DDSB) includes the aforementioned method developed by Priestley and Kowalsky and utilizes as a starting point a pre-defined target displacement. The design of the structure then progresses in a direct manner whereby the end result is the required strength, and hence stiffness, to reach the target displacement under the design level earthquake.

Another way to classify design methods is with respect to the earthquake input and the type of analysis used; hence, the input may consist of either the well-known acceleration response spectrum of current codes, or a displacement spectrum (a key component of direct DBD methods), or a suite of ground motions (accelerograms). Analysis can be (equivalent) static, or dynamic modal, or response-history ("time-history"). Various performance-based design methods are presented and discussed in (*fib* Task Group 7.2, 2003).

As will be shown in the following, displacement, and in particular interstorey drift in buildings, albeit valuable as a damage parameter (hence appropriate for PBD) is not always fully adequate for practical design. Structures such as dual frame-wall systems which are the prevalent structural system used for mid-rise and high-rise R/C buildings, are often not sensitive to drift, while in a number of actual buildings ensuring that a target interstorey drift develops during the design earthquake does not necessarily mean that deformations of the individual members are also equal (or even close) to the values envisaged by design. For these and other reasons, adoption of DBD methods by practising engineers is still far from a reality, and attempts to include such methods as an alternative procedure in design codes are accompanied by requirements for verification of the design resulting from DBD through nonlinear analysis (SEAOC Ad Hoc Committee, 1999); this is, of course, a rigorous way to design a structure, but also a very time-consuming one if realistic multistorey and/or extensive in-plan buildings are involved. A recent attempt to develop a method based directly on both displacement and member deformation is that by Kappos et al. (2007) and Kappos and Stefanidou (2010). As will be shown in Section 11.2.2, this method that evolved from a DCB procedure to a direct deformation-based one, ensures that ductility requirements in the individual members (rather than storey drift only) are reasonably close to those targeted by design. In the writer's opinion these two methods (Priestley et al. and Kappos et al.) represent two viable alternatives to the currently used code procedures, and deserve some attention; hence they will be critically presented and reviewed in the following two sections. The focus in the presentation of both methods will be on critically identifying their advantages as well as their weaknesses and limitations.

11.2.1 The Direct Displacement-Based Approach

It should be noted that the method presented in the following is that by Priestley (1993), Priestley and Kowalsky (2000), Priestley et al. (2007) and not the somewhat simplified version of the method included as an Appendix in the SEAOC 1999 document (SEAOC Ad Hoc Committee, 1999). It should be recalled, though, that (unlike Priestley et al.) SEAOC explicitly requires a verification of the initial DB design through nonlinear static (pushover) analysis.

11.2.1.1 Step 1: Target Displacement Pattern and Equivalent SDOF System

A key feature of displacement-based procedures is the definition of the target displacement of the structure to be designed. Unlike current code procedures wherein not only the overall geometry of the building but also the member stiffnesses have to be fully defined prior to the definition of the design seismic action, in direct DBD only the overall geometry and the structural system of the building are selected, while the stiffness of the constituent members (beams, columns, and walls, if present) will be defined at a later stage with a view to corresponding to the selected target displacement. The procedure commonly adopted in DBD methods is to transform the actual (model of the) building into an equivalent single-degree-of-freedom (SDOF) system (Fig. 11.1a), an idea that is by no means new; originally it can be found in the book by Biggs (1964), while most of the subsequent structural dynamics textbooks present the topic of a "generalised SDOF system" based on an assumed displaced shape (e.g. the fundamental mode shape) of the corresponding MDOF system. The shape to be used for the SDOF system in DBD should be as close as possible to the prevailing mode shape of the building in the direction considered, dully accounting for inelasticity effects, since the method relies on developing ductile behaviour of members. Priestley and Kowalsky (2000) have studied the displacement profiles of typical structural systems and suggested "standard shapes" that can be used in defining the equivalent SDOF, as described in the following. These are valuable proposals but one should keep in mind that using them in actual three-dimensional buildings (particularly asymmetric ones) is far from straightforward, while good results are not always guaranteed. The developers of the method recognise this, but argue that final results are not particularly sensitive to the accuracy of the assumed displacement pattern The alternative to using the SDOF approach is, clearly, to use inelastic analysis (see next section), something that, in the writer's opinion, is not beyond the realm of design practice anymore.

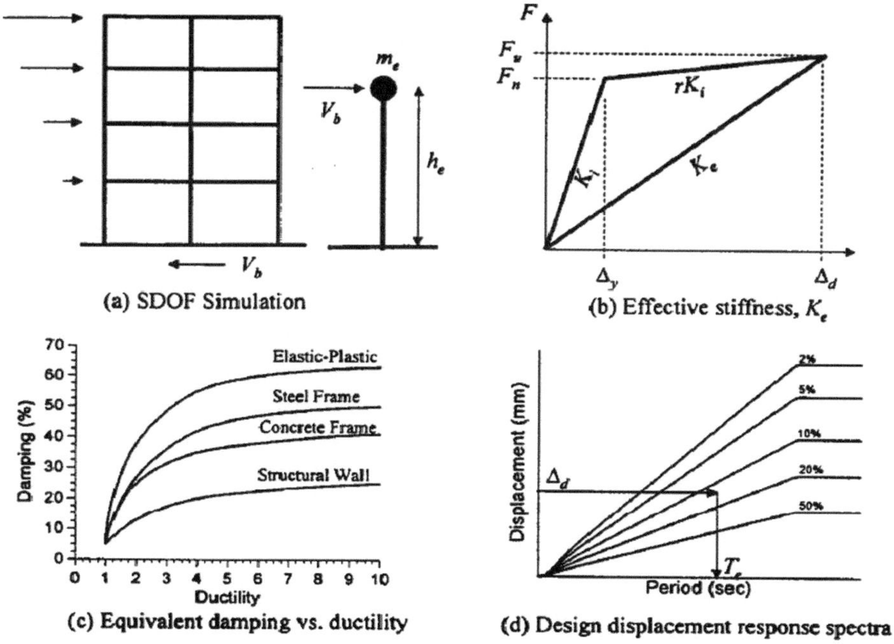

Fig. 11.1 Key aspects of the DBD procedure (Priestley and Kowalsky, 2000)

The equivalent SDOF design displacement is a function of the target displacement profile for the building. For buildings with structural system consisting of *frames*, the target displacement profile (Δ_i) as a function of number of stories, n, building height h_n, distance of storey i from the base h_i, and target interstorey drift ratio, θ_d, is given by the following relationships

$$\text{for } n < 4: \ \Delta_i = \theta_d h_i \tag{1a}$$

$$4 < n < 20: \ \Delta_i = \theta_d h_i \left(1 - \frac{0.5(n-4)h_i}{16h_n}\right) \tag{1b}$$

$$n \geq 20 \quad \Delta_i = \theta_d h_i (1 - 0.5\, h_i/h_n) \tag{1c}$$

Values for θ_d can be obtained from limitations on member ductility, or from code-specified drift limits.

For structural *walls*, the target displacement profile is given by

$$\Delta_i = \Delta_{ei} + \Delta_{pi} = \frac{2}{3}\varepsilon_y \frac{h_i^2}{\ell_w}\left(1.5 - \frac{h_i}{2h_n}\right) + \left(\theta_d - \frac{\varepsilon_y h_n}{\ell_w}\right)\left(h_i - \frac{\ell_p}{2}\right) \tag{2}$$

where ε_y is the strain at yield of the reinforcement, ℓ_w the wall length and ℓ_p the plastic hinge length (given from empirical formulae, see Priestley and Kowalsky, 2000; *fib* Task Group 7.2, 2003).

Dual systems, consisting of walls and frames, represent a major challenge for direct DBD since they are more complex than systems consisting of frames or walls (only), hence less amenable to being reduced to equivalent SDOF systems. Therefore, it is no surprise that until a few years ago (*fib* Task Group 7.2, 2003) they were not covered by the method; very recently, though, the work of Sullivan et al. (American Society of Civil Engineers, 2006), building on some concepts previously suggested by Paulay (2002), has made feasible the application of DBD to (at least a class of) dual systems. Due to space limitations and to the fact that the case-studies presented here involve R/C frame structures, DBD of dual systems will not be further addressed herein; details of the procedure can be found in Sullivan et al. (2006) and Priestley et al. (2007), while a critical summary is given in Kappos (2009).

Having established an appropriate displacement profile, the *target displacement* for the equivalent SDOF system is obtained (in all cases) from

$$\Delta_d = \sum_{i=1}^{n} \left(m_i \Delta_i^2\right) / \sum_{i=1}^{n} \left(m_i \Delta_i\right) \tag{3}$$

which considers equivalence in work between the MDOF and SDOF system.

The effective mass, m_e, of the SDOF system represents the first inelastic mode participating mass and is obtained from

$$m_e = \sum_{i=1}^{n} \left(m_i \Delta_i\right) / \Delta_d \tag{4}$$

Typically, m_e is about 70% of the total building mass.

11.2.1.2 Step 2: Estimation of Effective Damping of SDOF System

Another key feature of the DBD method is that the design displacement spectra are not inelastic spectra, but rather elastic spectra for viscous damping ratios consistent with the expected level of inelasticity, in other words hysteretic damping (resulting from inelastic response at the plastic hinges) is expressed as equivalent viscous damping; this is also a long-established practice, especially in seismic isolation design. It is beyond the scope of this work to discuss advantages and disadvantages of using over-damped elastic spectra in lieu of inelastic spectra; it is simply noted that Chopra and Goel (2001) have suggested a DBD method, similar to that by Priestley and Kowalsky (2000) and Priestley et al. (2007) but involving inelastic spectra, and also introducing acceptable member plastic rotation directly as a design parameter.

The effective damping, ξ_e, can be obtained as a function of the ductility requirement Δ_d/Δ_y, where Δ_d is taken from Step 1 and Δ_y is the system displacement at yield (Fig. 11.1b) which, at this stage, can only be estimated from empirical relationships; for instance, for R/C *frame* systems

$$\Delta_y = 0.5\varepsilon_y \left(\frac{\ell_b}{h_b}\right)(0.6h_n) \tag{5}$$

where ℓ_b and h_b are the span and depth of the beam, and the rest of the symbols are as defined previously. Clearly, in the usual case of frames with unequal beam spans and depths some average value has to be introduced in (5) and this is a typical indication of the difficulties involved in estimating global response quantities of buildings (that depend on a large number of parameters) from only a few selected quantities. Note also that in a realistic 3D building, design displacements Δ_d, Δ_y have to be estimated for at least two mutually orthogonal axes of the building (provided, of course, that such axes can be appropriately defined). Furthermore, if *walls* of unequal length are included in the system (also a very common case in practical design) Δ_y, which can be estimated from

$$\Delta_y = \frac{2.0\varepsilon_y}{3\ell_w}(0.7h_n)^2 \tag{6}$$

can differ substantially in each wall. In this case Priestley and Kowalsky (2000) recommend weighing damping in proportion to the force resisted by each wall, i.e.

$$\xi_e = \sum_{j=1}^{m}(V_j\xi_j) \,/\, \sum_{j=1}^{m} V_j \tag{7a}$$

For walls of equal height and thickness (and length ℓ_{wj}), Eq. (7a) can be expressed as

$$\xi_e = \sum_{j=1}^{m}\left(\ell_{wj}^2\,\xi_j\right) \,/\, \sum_{j=1}^{m} \ell_{wj}^2 \tag{7b}$$

Typical ξ_e curves as functions of ductility only are given in Fig. 11.1c for different types of structural systems; note that substantially lower ξ_e values are recommended in the recent book by Priestley et al. (2007), based on recent research by the group. The systems mentioned in Fig. 11.1c are supposed to be the parts of the actual system that dissipate the earthquake energy (through plastic hinging), hence in the common case that different sub-systems are involved in energy dissipation some averaging is again required.

11.2.1.3 Step 3: Calculate Design Base Shear

With the design displacement Δ_d determined (Step 1) and the damping estimated from the expected ductility demand (Step 2), the effective period T_e at maximum displacement response can be read from a set of design displacement spectra

(Fig. 11.1d). Representing the structure as an equivalent SDOF oscillator, the effective stiffness K_e at maximum response displacement can be found by inverting the equation for natural period of an SDOF oscillator i.e.

$$T_e = 2\pi\sqrt{\frac{m_e}{K_e}} \tag{8}$$

$$K_e = \frac{4\pi^2 m_e}{T_e^2} \tag{9}$$

where m_e is the effective mass.

The design base shear at maximum response is then derived on the basis of Fig. 11.1b (assuming for simplicity $F_n = F_u = V_b$, i.e. approximating the bilinear F-Δ diagram as elastic-perfectly-plastic)

$$V_b = K_e \Delta_d \tag{10}$$

This is the core of the DBD approach, and its key difference from the ("force-based") Code procedure, since the stiffness of the structure is not defined a-priori, but is determined during the design process in such a way that a target displacement (which is the initially selected design variable) is reached.

There are several problems associated with this crucial stage (Step 3), for instance the appropriate displacement spectrum to be used, and the characteristics of the selected structural system that could render it not controlled by drift, which is not the same as saying that it is not sensitive to seismic damage (e.g. extensive yielding of some regions). As will become clear from the case-studies in Section 11.3, the DBD approach is a promising procedure for drift-controlled structural systems, a typical example being frames (and under certain conditions wall systems without strong frames), situated in seismically active zones. The writer (among several others) believes that DBD is generally a poor choice for "inherently very stiff" systems (such as dual systems with large and/or numerous reinforced concrete walls), as well as for all structural systems if they are situated in zones with relatively low (or even "moderate") seismic activity; in the latter case it is quite common to find that target displacements selected on the basis of typical drift values (say, 2–3%) are well above the horizontal plateau of the displacement spectra of Fig. 11.1d, hence DBD cannot be applied, unless the target displacement is lowered by adopting conservative drift limits.

11.2.1.4 Step 4: Lateral Force Analysis

The base shear derived in Step 3 can be distributed along the height of the building, for structural analysis to be performed; a distribution based on the displacement profile Δ_i is used

$$F_i = V_b\,(m_i\Delta_i) \,/\, \sum_{i=1}^{n} (m_i\Delta_i) \tag{11}$$

It is important to recognise that since the outcome of the previous steps is a base shear (generally different in each direction of a 3D building), only a static analysis can be subsequently carried out. Clearly, higher-mode effects (e.g. in tall buildings) cannot be properly captured, unless more sophisticated distributions than that suggested by Eq. (11) are used. In the recent book by Priestley et al. (2007) the well-known code distribution with 10% of V_b acting at the top of the building is proposed, but it is clear that such simplified distributions cannot always provide the same result as a proper dynamic (modal) analysis.

In order to determine the design moments at potential plastic hinge locations, the lateral force analysis of the structure under the forces resulting from the aforementioned distribution should be based on member stiffnesses representative of conditions at maximum displacement response. This is an essential component of the *substitute structure* approach (Shibata and Sozen, 1976), which forms the theoretical basis of the DBD procedure adopting the secant stiffness (Fig. 11.1b). For cantilever wall buildings, this can be simplified to distribution of the forces between walls in proportion to ℓ_w^2, and the walls separately analysed.

For *frame* buildings, the member stiffness should reflect the effective stiffness at maximum response, rather than the elastic cracked-section stiffness I_{cr} (or stiffness at first yield) usually adopted for force-based analysis. With a weak beam – strong column design, beam members will be subjected to inelastic actions, and the appropriate stiffness will be

$$I_{\text{b}} = I_{\text{cr}}/\mu_{\text{b}} \tag{12}$$

where μ_{b} is the expected beam displacement ductility demand. Analyses have shown (*fib* Task Group 7.2, 2003) that member forces are not particularly sensitive to the level of stiffness assumed, thus it is acceptable to assume $\mu_{\text{b}} = \mu_{\text{s}}$, the frame design ductility.

Since the columns will be protected against inelastic action by capacity design procedures, their stiffness should be I_{cr}, with no reduction for ductility. An exception exists for the ground floor column, where plastic hinges will normally be expected at the base level, but not at first floor level. Priestley and Kowalsky (2000) suggest an ad-hoc procedure for dealing with such columns, based on introducing a hinge at the base of the column and pre-selecting the point of contraflexure at 60% the column height above the base.

11.2.1.5 Step 5: Design of Structural Members

Based on the results of the lateral force analysis, design of structural members can be carried out in such a way that the latter obtain a strength consistent with the demand from the lateral force analysis at the chosen design limit state, in a fashion similar to the familiar procedure used in current codes. For instance, in R/C buildings, flexural reinforcement for the structural members is proportioned at this stage. If displacement-based design is performed at the life-safety limit state, then reinforcement is proportioned such that the ultimate flexural capacity of plastic hinges equals the moment demands from the lateral force analysis. Conversely, if design is performed at the yield limit state, then reinforcement is proportioned such that the yield moment capacity of the plastic hinges equals the moment demand from the lateral force analysis. It should be pointed out, though, that commonly available design aids (tables, charts) provide only factored flexural capacities based on (conservative) values of strain in the reinforcement and concrete; hence differentiating between yield moment and actual flexural capacity of plastic hinges in the design requires developing new design aids. In the book by Priestley et al. (2007) moment – curvature ($M - \phi$) analysis is suggested in lieu of design aids; one should recall, though, that $M - \phi$ analysis can only be carried out if section reinforcement is known, hence iteration is necessary for designing a section.

11.2.1.6 Step 6: Detailing of Structural Members

Based on the limit state under consideration, plastic hinges are detailed to sustain the required deformation demand, which was specified at the beginning of the procedure (Step 1). Capacity design principles are employed to ensure that the chosen mechanism can be developed (e.g. strong column – weak beam). This step is important, but both material-dependent and similar to that used by modern codes, and will not be further dealt with herein.

11.2.2 The Direct Deformation-Based Approach

In earlier versions (Kappos, 1997b; Kappos and Manafpour, 2001; Kappos and Panagopoulos, 2004) of this method inelastic deformations were included as a design verification, not as a design parameter. To overcome this weakness, a direct deformation-based design method was sought, maintaining the key features of the aforementioned performance-based procedure. Moreover, while the application of earlier versions of the method was restricted to regular buildings, the method is applied here (Section 11.3.2) to multistorey irregular buildings with setbacks, noting that response-history analysis based procedures appear better suited to irregular structural systems (Sullivan et al., 2003). The steps involved in the proposed "direct deformation-based design method" are described in the following.

11.2.2.1 Step 1: Flexural Design of Plastic Hinge Zones Based on Serviceability Criteria

The purpose of this step is the establishment of a basic level of strength in the structure that would ensure that the structure remains serviceable ("immediate occupancy" requirement in ASCE Standard 41-06; ASCE/SEI, 2007) after an earthquake having a high probability of exceedance (usually taken as 50%/50 years). The verifications include specific limits for member ductility factors and plastic hinge rotations of critical members (see Step 4) and the corresponding demands are estimated from inelastic analysis of a reduced inelastic model of the structure (described in Step 3). Hence, an initial analysis is required, which would provide the strength of the members (energy dissipation zones) that will respond inelastically during the serviceability verification; this analysis constitutes Step 1 and is a vital part of the procedure.

The design of selected dissipation zones like the beam ends and the bases of ground storey columns, is carried out using conventional elastic analysis. The strength of these zones is estimated taking into consideration the range within which the inelastic deformations should fall, which corresponds to the degree of damage allowed for the selected performance level. The procedure proposed in the following leads to reaching the permissible values of inelastic deformations (expressed through rotational ductility factors), since the latter are directly related to the reduction of element forces corresponding to elastic behaviour. This is a critical feature, not included in previous versions of the method that simply included a serviceability check, the result of which typically was that most members either remained elastic or were well below the allowable deformation limits (Kappos and Panagopoulos, 2004).

To reach the aforementioned goal, element forces and rotations are first obtained from the results of an elastic analysis. Design for flexure is carried out in terms of design values, using commonly available design aids. On the other hand, serviceability checks are based on the results of inelastic analysis, for which mean values are commonly adopted; furthermore, several members are expected to posses some overstrength with respect to the design moments used in their dimensioning, due to detailing requirements, i.e. rounding (upwards) of required reinforcement areas and use of minimum reinforcement specified by codes. For these two reasons, the initial elastic analysis should be carried out for an appropriate fraction ν_o of the earthquake level associated with the serviceability performance level (50%/50 years); the suggested (Kappos and Stefanidou, 2010) value is $\nu_o = 2/3$.

Subsequently, elastic rotations (θ_{el}) are related to the corresponding inelastic ones (θ_{inel}), using an empirical procedure (like that proposed in Panagiotakos and Fardis, 2001). Referring to Fig. 11.2, having defined the target rotational ductility factor (μ_θ) and the maximum inelastic rotation, θ_{inel} (this is the total chord rotation, not the plastic one), from the θ_{el} found in the elastic analysis, the yield rotation (θ_y) is calculated for every structural member. For simplicity of the procedure one could assume first that M-θ response is elastic-perfectly plastic and second that the slope of the elastic and the elastoplastic M-θ diagram is the same. However, a more accurate procedure, depicted in Fig. 11.2, can be used, recognising that the relationship of element forces (moments) to rotations (M-θ) is dependent on the loading history (which is non-proportional). Moments and rotations due to permanent loading (gravity and reduced live loads) are first applied and held constant, and any decrease of the elastic forces (M_{el}) should refer to the seismic loading that is applied after the permanent one. Then the corresponding yield moment (M_y) can be computed, as

$$M_y = M_g + aM_E \tag{13}$$

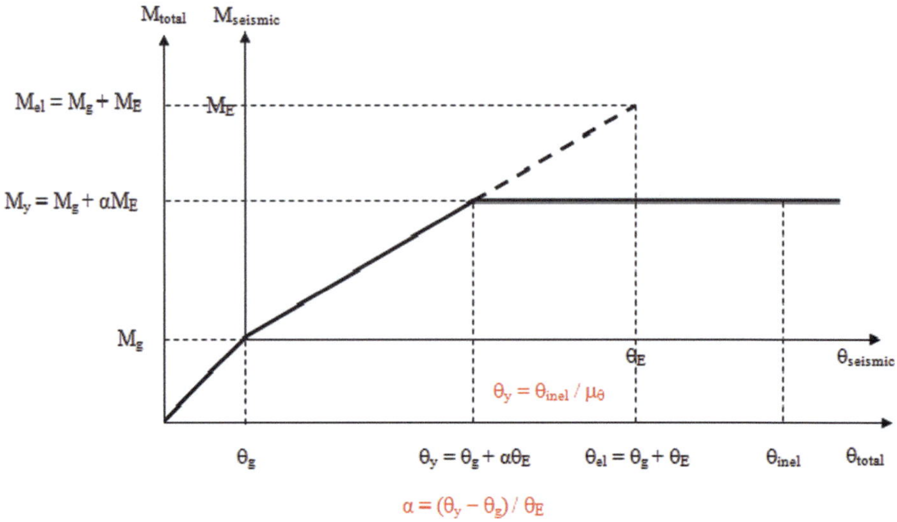

Fig. 11.2 Definition of the slope of M-θ_{inel} diagram and of $a\theta_E$ for beams

where the reduction factor a (that is the same for moments and rotations) is (Kappos and Stefanidou, 2010)

$$\alpha = \frac{\theta_y - \theta_g}{\theta_{E,el}} \tag{14}$$

The various rotation values are explained in Fig. 11.2.

According to the aforementioned procedure, the reduced design forces are computed for every beam element, and they are directly related to the target rotational ductility selected for the serviceability performance level. The longitudinal reinforcement demand for the beams is calculated using standard flexural design procedures and compared to the minimum requirements according to code provisions. In case the longitudinal reinforcement demands are found to be less than the minimum requirements, reduction of cross-sections is in order (reduction of stiffness), otherwise deformations for the considered performance level will be less than the allowable ones; clearly, this stage involves striking a balance between economy and performance.

11.2.2.2 Step 2: Selection of Seismic Actions

The response-history analyses necessary for seismic design according to the proposed method require the definition of appropriately selected input seismic motions. The accelerogram set used for the analysis should include a pair of components for every seismic motion and it is recommended that it be selected based on the results of a seismic hazard analysis ("deaggregation" phase, wherein M and R for the site in consideration are determined). Hence the selected input seismic motions should conform to certain criteria concerning magnitude (e.g. $Ms = 6.0 \sim 6.5$), and epicentral distance (e.g. $R = 10 \sim 25$ km), and also peak ground acceleration (PGA > ~0.1 g).

The earthquake motions used for design, should be properly scaled in order to correspond to the level associated with the limit state examined ("serviceability" limit state for the design of energy dissipation zones, and "life safety" for the other members). Several scaling procedures have been explored (Kappos et al., 2007) and the one adopted by EC8-Part 2 (CEN, 2005a) is used here, duly tailored to the needs of the performance-based design method.

11.2.2.3 Step 3: Set-Up of the Partially Inelastic Model

During this step a partially inelastic model (PIM) of the structure is set up, where the beams and the base of ground storey columns (and walls, if present) are modelled as yielding elements, with their strength based on the reinforcement calculated for reduced element forces according to the inelastic deformations allowed for the serviceability limit state (Step 1). In the same model, the remaining columns (and walls) are modelled as elastic members. Details regarding the PIM can be found in Kappos (1997b), Kappos and Manafpour (2001), and Kappos and Panagopoulos (2004).

11.2.2.4 Step 4: Serviceability Verifications

The usage of inelastic dynamic response-history analysis in the PIM, involves a set of recorded motions scaled to the intensity associated with the serviceability requirement (e.g. 50%/50 years). The verifications include specific limits for maximum drifts and plastic hinge rotations of critical members; recommended interstorey drift values range from 0.2 to 0.5% the storey height, while permissible plastic hinge rotations vary between 0.001 and 0.005 rad for columns and about 0.005 rad for beams (Kappos and Manafpour, 2001). The purpose of this step, apart from checking the inelastic performance of the structural system, is the verification that the required rotational ductility factor (μ_θ) of beams and bases of ground storey columns is consistent with the values considered during the design. Hence, this step is basically an assessment (or verification) of the seismic response of the structure for the "serviceability" limit state; in principle, it can be skipped if adequate calibration of the method is carried out in the future.

Since inelastic dynamic analysis is used in order to check the seismic response of the structure for the aforementioned performance level, mean values of material strength are considered (f_{cm} and f_{ym} for concrete and steel respectively).

11.2.2.5 Step 5: Design of Longitudinal Reinforcement in Columns (and Walls) for the "Life Safety" Limit State

The design of members (such as columns at locations other than the base of the structure) considered elastic in setting up the PIM, is based on the results of inelastic response-history analyses of the aforementioned model for each of the selected sets of input motions properly scaled to the intensity of the earthquake associated with the "life safety" requirement (probability of exceedance 10%/50 years). Simultaneous values of M_1, M_2, N are considered (biaxial bending and axial force), while the design is based on the most critical combinations. Consideration of mean values of material strength during the design leads to an overestimation of the longitudinal reinforcement of columns (Kappos et al., 2007). Since the input to the columns directly depends on the strength of the adjoining beams (designed to form plastic hinges) and the latter's yield moments are based on the mean value of steel strength (f_{ym}), then design column moments are over-estimated by the ratio f_{ym}/f_{yd} (equal to 1.26), which is deemed as over-conservative. The specific performance objective to be satisfied is that for the considered seismic action (10%/50 years) columns should not yield (except at the base), and mean values of column yield moments are used for this verification; hence the 1.26 factor is redundant. Since design for biaxial bending was carried out using commonly available design aids (based on f_{cd},f_{yd}) it was more convenient to use design values of material strength in the dynamic analysis of the PIM as well as in the design of the columns.

11.2.2.6 Step 6: Design for Shear

To account for the less ductile nature of this mode of failure, shear forces should correspond to seismic actions corresponding to the 2%/50 years earthquake (associated with the "collapse prevention" performance level). However, to simplify the design procedure, design and detailing for shear can be carried out using shear forces calculated from inelastic response-history analysis for the seismic action associated with the "life safety" performance level, and implicitly relate them to those corresponding to the 2%/50 years earthquake through appropriately selected magnification factors (γ_v); recommended γ_v factors (Kappos and Panagopoulos, 2004) for beams and columns are equal to 1.20 and 1.15 respectively.

11.2.2.7 Step 7: Detailing for Confinement, Anchorages and Lap Splices

Detailing of all members for confinement, anchorages and lap splices, is carried out with due consideration of the level of inelasticity expected in each member. Structural members where the development of extended inelastic performance is anticipated (bases of ground storey columns or walls), are detailed according to the provisions of EC8 (CEN, 2004) concerning ductility class "Medium" ("DCM"), while others where inelastic performance is expected to be restricted (columns of upper storeys) are detailed according to the provisions for ductility class "Low" ("DCL").

11.3 Case-Studies of Application of Different Methodologies

Two case studies are presented in the following, involving 4-storey and 10-storey R/C buildings, designed to different procedures, but for the same reference earthquake (same response spectrum). The 4-storey building (Kappos, 2009) is designed to the direct DBD method described in Section 11.2.1, the DBD procedure adopted by SEAOC (SEAOC Ad Hoc Committee, 1999) (referred to only very briefly in Section 11.2.1), and to a current Code procedure (the Greek Seismic Code, which is very similar to Eurocode 8). The 10-storey building (Kappos and Stefanidou, 2010) is designed to the deformation-based method described in Section 11.2.2 (with two alternative selections of member geometry) and to the Eurocode 8 provisions (for both DCM and DCH). Both reference buildings share two important structural features: First, their lateral-load resisting system consists entirely of (moment-resisting) frames, hence they represent cases where displacements are normally expected to be an issue; examples of designs involving walls can be found in *fib* Task Group 7.2 (2003) and Sullivan et al. (2003) for most of the DBD methods, while applications of the direct DBD method to dual structures can be found in Sullivan et al. (2006). Second, the buildings studied are structures with irregularities in plan and/or in elevation; this is the type of structures that challenges most the PBD/DBD methods, which involve more design quantities than normal code-type methods, some of which are difficult to estimate properly in irregular structures.

11.3.1 Four-Storey Building with Irregularity in Plan

The configuration of the 4-storey reinforced concrete building is shown in Fig. 11.3; the large re-entrant corner automatically classifies the building as irregular in plan according to code provisions. The building is designed for two α_g values, 0.24 and 0.36 g (Zones II and III of the Greek Seismic Code), for site conditions B (firm soil). The materials used were C20/25 concrete (characteristic cylinder strength $f_{ck} = 20$ MPa) and S500 s steel ($f_{yk} = 500$ MPa).

The following alternative design and analysis procedures were implemented

- Equivalent static method according to the Greek Code (similar to EC8)
- Dynamic response spectrum method according to
- Direct displacement-based design according to the method of Priestley et al. (2007), described in Section 11.2.1 of this paper
- Direct displacement-based design according to Appendix I-Part B of SEAOC 1999 (SEAOC Ad Hoc Committee, 1999).

11.3.1.1 Discussion of Different Design Aspects

All analyses were carried out using the commercial software ETABS (Computers and Structures Inc., 2005). In the Code design cracked section stiffnesses were assumed (50%EI_g for beams and EI_g for columns, as per EAK). The first three natural periods of the building were found to be $T_1 = 1.08$ s, $T_2 = 0.80$ s, and $T_3 = 0.76$ s.

In applying the SEAOC procedure (SEAOC Ad Hoc Committee, 1999), the building was designed for structural performance level 2 (SP2) for an earthquake

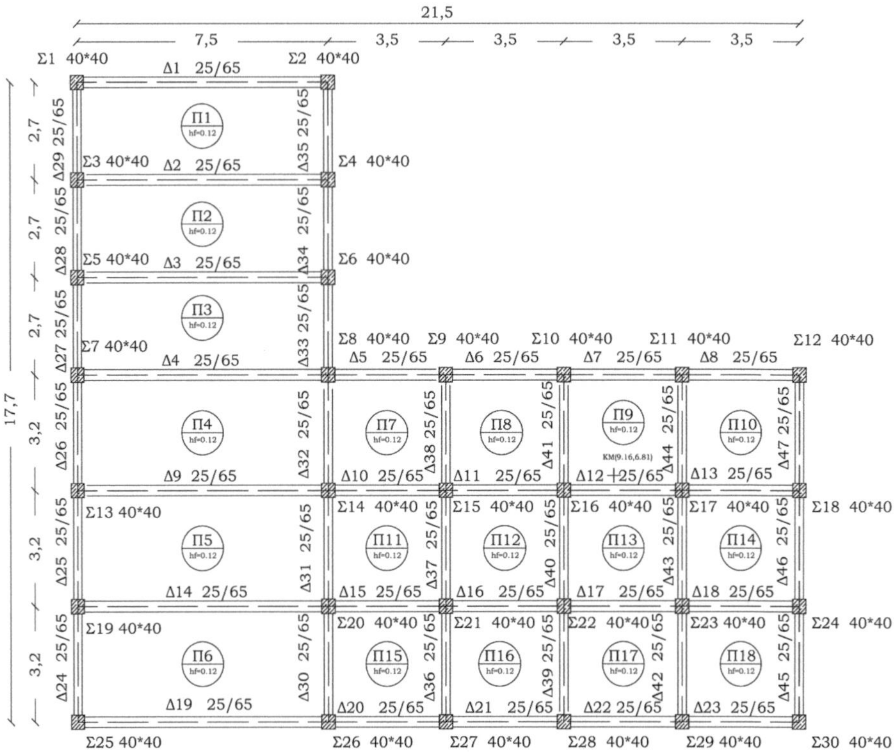

Fig. 11.3 Typical storey plan and member dimensions (in cm) of the 4-storey building

with return period of 72 years and for performance level 3 for a return period of 475 years (the same as that used for the Code design); the return period of 72 years results in PGA's of 0.17 and 0.30 g, for 475 years values of 0.24 and 0.36 g, respectively, using attenuation relationships from hazard studies in Greece. For frame structures the drifts recommended by SEAOC for SP2 and SP3 are 1.5 and 3%, respectively. However, if the Eurocode 8 design spectrum for displacements (S_d) is used (wherein the plateau starts at 2 s) the SEAOC recommended drifts result in displacement values that are above the plateau values of the spectrum (see also comments on Step 3, in Section 11.2.1). Hence, for the DBD to be applied, the SEAOC drifts had to be reduced to 0.75 and 1.30% for SP2, and 1.40 and 1.85% for SP3, for α_g equal to 0.24 and 0.36 g, respectively. Note that for the medium seismic hazard Zone II ($\alpha_g = 0.24$ g) the code-recommended drifts had to be reduced by 50% or more, a clear indication of the irrelevance of DBD procedures in low and medium seismic hazard zones. To be fair with all methods, one should note that the parameters adopted for the design displacement spectrum, in particular the corner period T_D (beginning of horizontal branch), have a major influence on the feasibility of DBD; if instead of $T_D = 2$ s (the EC8-adopted value), one assumes $T_D = 4$ s (the SEAOC-adopted value), the resulting design displacements are much closer to those corresponding to the recommended drifts.

In applying the Priestley et al. procedure (Priestley et al., 2007; Priestley and Kowalsky, 2000), the building was designed for serviceability and for damage limitation limit states, for return periods of 92 and 475 years, respectively. The design drifts had again to be reduced in order not to exceed the maximum values from the displacement spectra; values of 0.75 and 1.00% for Zone II, and 1.00 and 1.40% for Zone III (first value in each case is the serviceability value). It is worth noting that in this method, due to the difference in the return period adopted for the higher performance level (serviceability), the drastic reduction in the recommended value was for the lower performance level (damage limitation), for which Priestley et al. recommend a drift of 2.5%. It is also worth mentioning that while for Zone II (0.24 g) the critical base shear resulted from the serviceability requirement, for Zone III (0.36 g) the critical base shear was that from the damage limitation limit state; hence, it is not a-priori known which limit state is the most critical, and multiple limit states have to be checked, which is a key feature of PBD.

11.3.1.2 Evaluation of Different Designs

The "economics" of each design method can be inferred from comparisons such as those shown in Fig. 11.4, where the reinforcement required for flexure (longitudinal bars) is shown for the four different designs (the static and dynamic analysis based designs to the EAK Code are shown as separate cases). Several interesting trends are revealed from these comparisons: First, that the economy of DBD procedures depends on the seismic zone wherein the design is made; for the medium seismicity zone II, both DBD procedures result in more reinforcement than the reference Code procedure (the dynamic one, which is required for irregular buildings), whereas for the (relatively) high seismicity zone III the DBD procedures, especially the one by Priestley et al., result in less flexural reinforcement than the Code. As anticipated, the Code procedure based on static analysis was more conservative than the dynamic analysis based, and resulted in more reinforcement, regardless of seismic zone.

In the writer's opinion one of the most important conclusions from this case-study (and other similar ones) is that one should be very careful when comparing different design methods. The comparison of cost of materials (mainly of reinforcement, if R/C structures are addressed) should be properly made; referring again to the charts of Fig. 11.4, if one considers that the DBD methods are interrelated with static analysis procedures, hence they are compared with the static analysis based design of the Code, then the DBD methods are more economical. However, the reference method of current codes (such as the Eurocode or the Greek EAK) is the dynamic one, and indeed for most of the irregular structures (particularly those in medium and high seismicity zones) their use is compulsory; hence a more appropriate comparison should be between the second chart in Fig. 11.4 and the two on its right, in which case Code design appears to be more economical than DBD in zone II (medium seismic hazard) and less economical in zone III (high seismic hazard). In all cases, though, differences in the cost of reinforcement are not very large, particularly if one considers it as a fraction of the total cost of the building (which makes perfect sense in a practical design context).

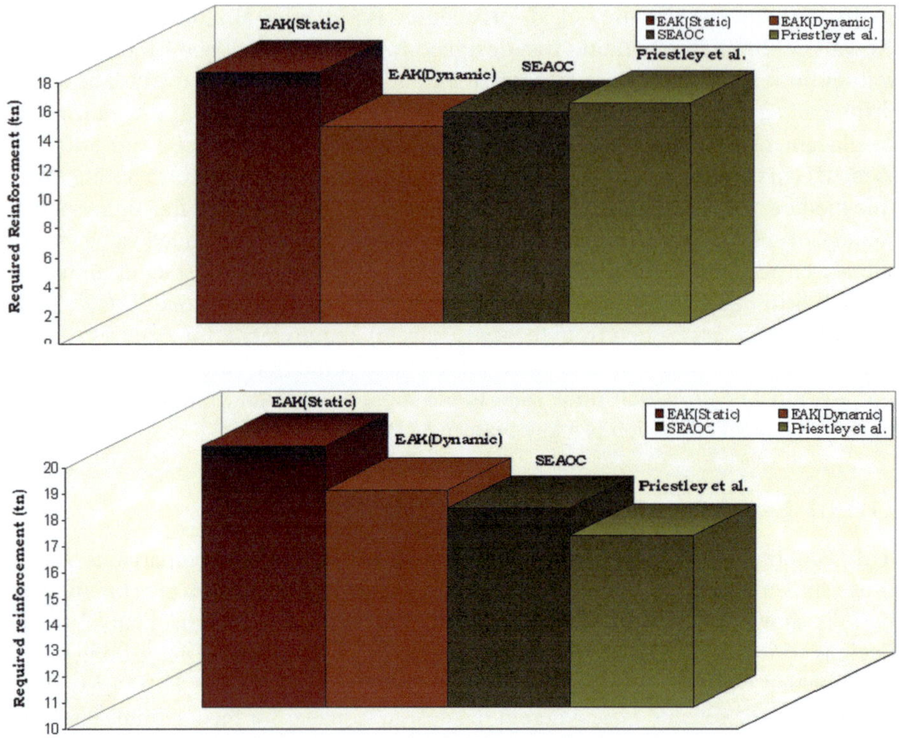

Fig. 11.4 Required flexural reinforcement for the 4-storey building: Zone II (*above*) and Zone III (*below*)

11.3.2 Ten-Storey Building with Irregularity in Plan and Elevation

The geometry of the ten-storey R/C building with setbacks at the two upper storeys, having a 3D frame structural system is shown in Fig. 11.5.

11.3.2.1 Discussion of Different Design Aspects

The building was first designed according to the provisions of EC8 (CEN, 2004) for ductility classes "M" and "H", and then redesigned to the performance/deformation-based procedure described in Section 11.3.3. The design ground acceleration was taken equal to 0.24 g, while ground conditions were assumed to be type "B" according to EC8 classification. The materials used for design were concrete class C25/30 and steel S500. The structure is classified as irregular in both directions according to the provisions of EC8, which has repercussions on the behaviour factor q and the type of analysis to be used for design. The q-factors for the DCH and DCM structures, were found equal to 4.14 and 2.76, respectively. The method of analysis used was the response spectrum method, since the equivalent static method is not

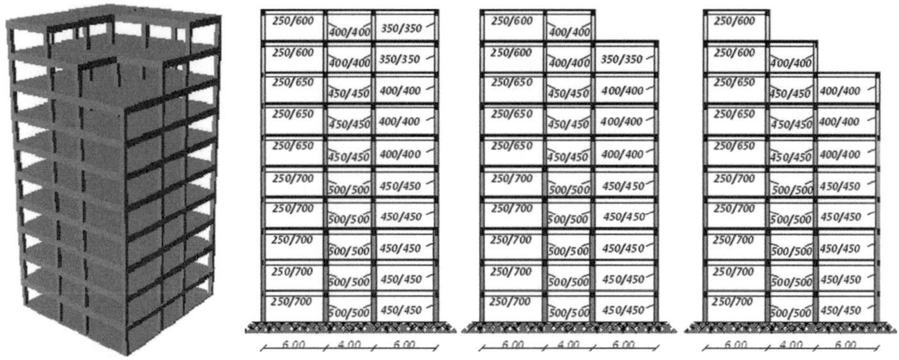

Fig. 11.5 Three-dimensional view (*left*) and geometry of typical frames of 10-storey building

allowed in the case of irregular buildings. The rigidity of structural members was taken equal to $0.5EI_g$ for all members, as prescribed in EC8.

In applying the direct deformation-based method, both elastic and inelastic analyses of the structure were carried out using the software package *Ruaumoko* 3D (Carr, 2004); modelling of members' inelastic performance was done by means of a spread plasticity model and bilinear elastoplastic hysteresis rule. The effective rigidity was taken equal to 50% the gross section rigidity (EI_g) for T-beams and for columns (same as in EC8). For the dynamic response-history analyses, a set of six pairs of actually recorded motions was selected from the European Database (Ambraseys et al., 2000) and a synthetic record was added to form the final set of 7 records. All input motions were scaled to the intensity of the design spectrum (the same used for EC8 design), and pairs of horizontal components were applied simultaneously in each horizontal direction of the structure.

The resulting longitudinal reinforcement demands were found to be generally less than the minimum Eurocode requirements. This hinted to the need for re-dimensioning the cross sections initially selected for the structural members (especially beams). Therefore, the proposed design method was additionally applied to a second structure ("Building 2") having the same geometry as Building 1 depicted in Fig. 11.5 and properly reduced cross sections (details are given in Kappos and Stefanidou, 2010). Design according to the provisions of EC8 was applied mainly with a view to comparing the required reinforcement to the one resulting from the proposed design procedure, and providing a basis for evaluating the performance of complex structures designed to different methods.

11.3.2.2 Evaluation of Different Designs

Comparison of the longitudinal reinforcement requirements for the building having the *same* cross-sections as the one designed to EC8 provisions, showed that the proposed design method leads to more economical design as far as beams and the bases of ground storey columns are concerned. Differences are even more marked

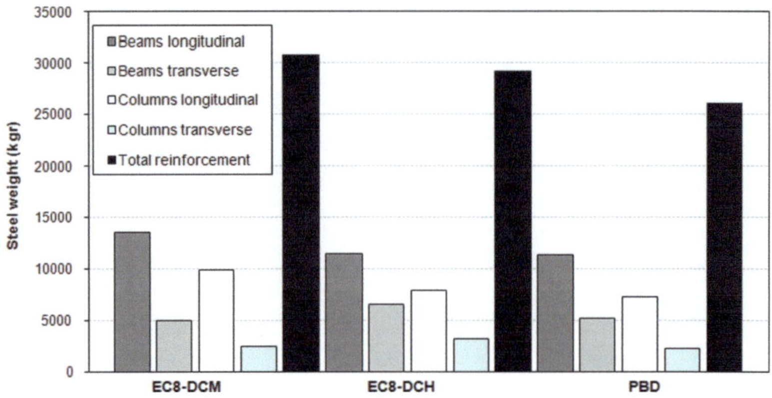

Fig. 11.6 Required amount of steel in beams and columns for code design (EC8) and PBD

in the case of "Building 2" that was designed for reduced beam cross-sections. The quantity of steel required in each member type is shown in Fig. 11.6 for the three different designs; it is clear that the application of the PBD method led to lower total reinforcement demands, the more important difference being in the transverse reinforcement in columns, which also implies easier detailing on site.

11.4 Assessment of Different Designs

11.4.1 Overview of Available Assessment Procedures

A variety of analytical procedures are currently available for the seismic assessment of structures. It has long been recognised that a proper assessment can be carried out only if the post-elastic response of the structure is captured in the analysis, hence revealing the actual plastic mechanism that will develop under a given level of earthquake action. It is well-known that this mechanism is hardly ever an "ideal" one ("beam mechanism" or "column sidesway mechanism", as described in current codes) and, indeed, in older and/or poorly designed structures this mechanism can be an unfavourable one, involving concentration of ductility demands in one (or a few) storeys. Therefore, leading code-type documents for seismic assessment, such as ASCE 41-06 (ASCE/SEI, 2007) and Eurocode 8 – Part 3 (CEN, 2005b) recommend and, under specific conditions (such as the presence of irregularities), impose the use of inelastic analysis methods. Both types of inelastic analysis are allowed, but the static (pushover) method is presented in more detail in documents related to seismic assessment, particularly the American ones, such as ASCE/SEI (2007). This is clearly done under the presumption that inelastic static analysis is simpler to apply in practice, which may or may not be true if the limitations of the method are dully accounted for. More specifically, irregular structural configurations are quite common in both "old" and new buildings, and irregular structures are typically affected by higher modes and/or by changes in their dynamic characteristics in the post-elastic range of their response to seismic actions. Typical examples are high-rise buildings, and also medium-rise buildings with setbacks in the upper storeys, wherein consideration of at least the second mode (in each direction of a 3D building) is mandatory, and the formation of a weak (or "soft") storey mechanism in the ground storey (very common in older buildings with masonry infills discontinued at that level, the so-called "pilotis" buildings), which drastically changes the fundamental mode from an essentially "triangular" to a "uniform" one. Consideration of multiple loading pattern in pushover analysis (as prescribed in ASCE/SEI, 2007; CEN, 2004) and several other codes) is a mixed blessing, in the sense that higher mode effects can still be missed (especially in the upper part of the building), whereas basing the final assessment on an "envelope" of the action effects derived from each pattern is very often over-conservative. Therefore, use of inelastic dynamic (response-history) analysis is in many respects an appropriate choice and, with the currently available tools like ETABS Nonlinear (Computers and Structures Inc., 2005) and Ruaumoko (Carr, 2004) it is also a feasible one. A broader discussion of the "pros" and "cons" of the aforementioned procedures and the analytical tools for their implementation can be found in a previous paper by the writer (Kappos, 2000).

The seismic performance of the buildings designed in Section 11.3 is assessed in the following sections using both inelastic analysis procedures, i.e. static (pushover) and dynamic (response-history); hence, the case-studies also serve for furnishing a good idea of the possibilities of current assessment procedures and the parameters that can (and should) be checked in each case. It is noted that rather than using code-prescribed values, assessment is based herein on state-of-the-art methods for estimating the local (plastic rotation) and global (interstorey drift) capacity of R/C buildings. Another aspect that is treated here in a rather detailed way is the assumption regarding the stiffness of R/C members outside the plastic hinge region, which is a critical one in inelastic analysis (Kappos, 2000).

11.4.2 Assessment of the Building Designed to the Displacement-Based Procedure

In this case-study (Section 11.3.1) pushover analysis was used for all designs; recall that for the SEAOC (SEAOC Ad Hoc Committee, 1999) design, this is a compulsory final step of the method. In this analysis two different assumptions were used for member stiffnesses, one using the conventional values (percentages of EI_g) recommended by the codes used, and one using the secant stiffness at yield (M_y/φ_y) calculated from detailed moment-curvature analysis of all critical sections. Inelastic response of members was modelled using the familiar point-hinge model, in the version implemented in ETABS Nonlinear (Computers and Structures Inc., 2005). The spectra used for design were also used for estimating target displacements in pushover analysis (for each earthquake level considered); both the ASCE

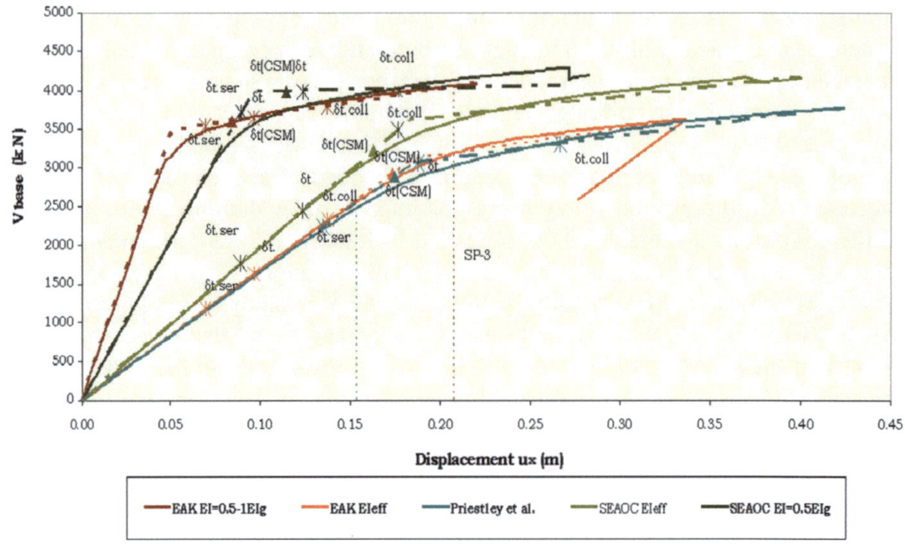

Fig. 11.7 Pushover curves for buildings designed to different procedures (zone II, 475 years)

41-06 (ASCE/SEI, 2007) "coefficient method" and the "capacity-demand spectra" approach (Fajfar, 1999) were used for calculating target displacement.

The pushover curves resulting for the designs carried out for exactly the same seismic action (spectrum for 475 years earthquake) and for zone II (0.24 g) are shown in Fig. 11.7; similar trends were observed for zone III design. A bilinear approximation to each curve is also shown in the figure, which also includes target displacements calculated in each case: $\delta_{\mathrm{t.ser}}$, for the "serviceability earthquake" (50%/50 years), δ_{t} for the "damage limitation earthquake" (10%/50 years), $\delta_{\mathrm{t,coll}}$ for the "no-collapse earthquake" (2%/50 years); $\delta_{\mathrm{t,CSM}}$ is the target displacement for the 10%/50 years earthquake, estimated using the capacity spectra approach. Differences between the calculated δ_{t} and $\delta_{\mathrm{t,CSM}}$ were less than 15% in all cases.

A first remark regarding the curves in Fig. 11.7 is that, as expected, the stiffness assumed has a substantial effect on the initial stiffness of the building; the stiffest one is the EAK-designed structure modelled with $EI_{\mathrm{ef}} = 0.5 \div 1.0EI_{\mathrm{g}}$, then the SEAOC structure with $EI_{\mathrm{ef}} = 0.5EI_{\mathrm{g}}$, then the SEAOC structure with $EI_{\mathrm{ef}} = M_y/\varphi_y$, then the EAK structure with $EI_{\mathrm{ef}} = M_y/\varphi_y$, and finally the Priestley et al. structure with $EI_{\mathrm{ef}} = M_y/\varphi_y$. Clearly, no meaningful comparison between methods can be made if different stiffness assumptions are adopted in each case; moreover the result of the assessment might be different depending on the modelling assumptions. For all three designs (EAK, SEAOC, Priestley) when $EI_{\mathrm{ef}} = M_y/\varphi_y$ is assumed, the displacement corresponding to the 10%/50 years earthquake (the usual design earthquake in current codes) is within the elastic branch of the bilinear curve, hence little inelasticity is expected in the building. For the "no-collapse earthquake" (2%/50 years) all designs are safe (regardless of stiffness assumption) since all buildings remain well within their ductility capacity. Nevertheless, the displacements predicted from pushover analysis for the DBD structures are slightly larger than those considered at the design stage. The overstrength ratio V_y/V_d was 1.85 or 1.66 for the EAK design (depending on the stiffness assumption), 1.58 or 1.44 for the SEAOC design, and 1.42 for the Priestley et al. design; hence for a common assumption $EI_{ef} = M_y/\varphi_y$, the Code design is more conservative in terms of strength, which is not surprising, while the overstrength in the two versions of the DBD method is very similar.

Clearly, the seismic reliability of each design is a major criterion for judging the appropriateness of each design method. Pushover analysis of all designs in this case-study, using currently available advanced analysis tools, has shown that the performance requirements in each method (checked either explicitly or implicitly during the design) are met by the "end product". All designs remained essentially within the elastic range for the serviceability-related earthquake and all designs were well within their ductility capacities even when subjected to about twice the intensity of the "design earthquake" (2%/50 years event, as opposed to the 10%/50 years event explicitly considered in design). Hence, from the safety point of view, there does not appear to be any real merit in revising the current code provisions and switching to DBD; in fact it appears that in most cases the overstrength margins (which are a measure of the safety of the building against earthquakes substantially stronger than the design one) are higher in the current code-designed structures.

The conclusion is then that any possible advantages of the DBD methods should be traced in the direction of economy, i.e. to potentially save material by avoiding over-conservatism in design. This is a tricky issue, though, and certainly more case-studies are required before any definite trends are identified; it is worth recalling that in the comprehensive (albeit involving "academic" structures) study by Sullivan et al. (2003), there were instances wherein the base shear resulting from the DBD method was higher than that resulting from other procedures.

Finally, a trend which appears to be very clear is that, at least at this stage of development, any potential use of DBD should be confined to high seismic hazard areas (design PGA of about 0.3 g or higher), whereas it is almost irrelevant in zones with design PGA's of less than about 0.2 g.

11.4.3 Assessment of the Building Designed to the Deformation-Based Procedure

The seismic performance of the alternative designs of Section 11.3.2 was assessed by carrying out inelastic response-history analysis of fully inelastic models of the 3D R/C buildings (as opposed to the partially inelastic model used in design). A total of 8 pairs of ground motion records were used (an extra pair was added to those used for design, and scaling factors were all adjusted accordingly in the new set). Verifications regarding interstorey drifts and plastic rotations were carried out for different levels of seismic action (50%/50 years, 10%/50 years and 2%/50 years), related to serviceability, life safety and collapse prevention objectives. Additional

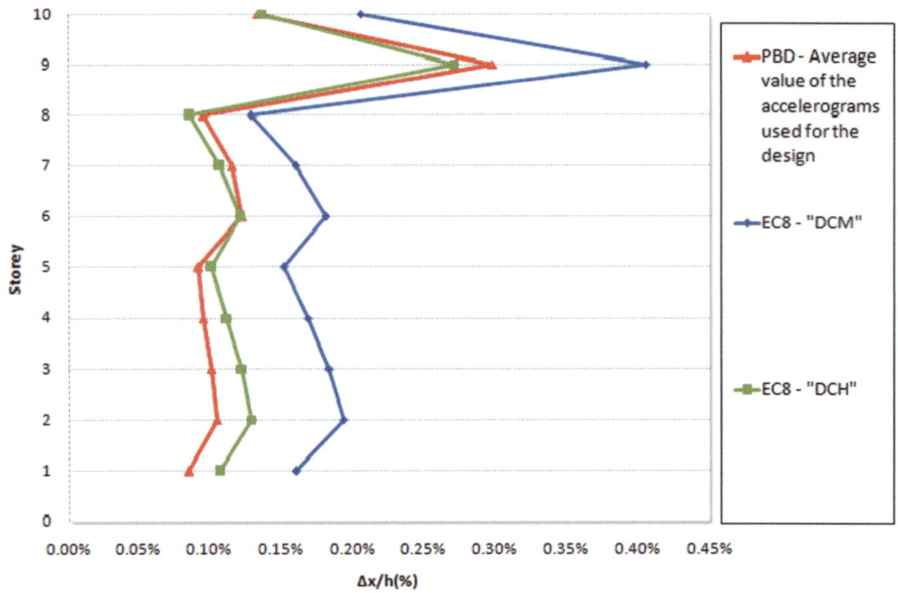

Fig. 11.8 Serviceability verification: Building 1 – Interstorey drifts in x-direction

to the set of analyses based on stiffness assumptions corresponding to moderate levels of inelasticity ($EI_{\text{ef}} = 0.5EI_g$), extra analyses were carried out, where the secant stiffness of the cracked section at yield, $EI_{\text{ef}} = M_y/\varphi_y$, was used for all R/C members.

From the drifts at the serviceability-related earthquake shown in Fig. 11.8, it is clear that the seismic performance of both the EC8 designs and the building designed for target deformations having the same cross-sections was very satisfactory. Moreover, the maximum value (average of 8 motions) of interstorey drift ratio, was equal to 0.32% for the PBD Building 1 (recorded at the 9th storey, i.e. at the set-back), and increased to only 0.35% when a number of cross-sections were reduced ("Building 2"). As far as the development of plastic hinge rotations is concerned, the values obtained from the results of inelastic response-history analysis are significantly lower than the adopted serviceability limits (maximum value equal to about 0.002 and 0.003 for buildings 1 and 2, respectively).

From the several results of the performance assessment of the alternative designs of the irregular 10-storey building for the various levels of earthquake intensity, reported in detail in Kappos and Stefanidou (2010), which showed that both the EC8-designed buildings and those designed to the PBD satisfied the "life safety" criteria for the 10%/50 years event and the "collapse prevention" criteria for the 2%/50 years event, a potentially critical situation is shown in Fig. 11.9. It refers to the case that the 8 pairs of records were scaled to the intensity of the 2%/50 years earthquake and all R/C members were modelled with the reduced stiffness ($EI_{\text{ef}} = M_y/\varphi_y$), i.e. lower than those used for design; furthermore, the results are for Building 2 (reduced cross-sections), hence this is expected to be a critical case.

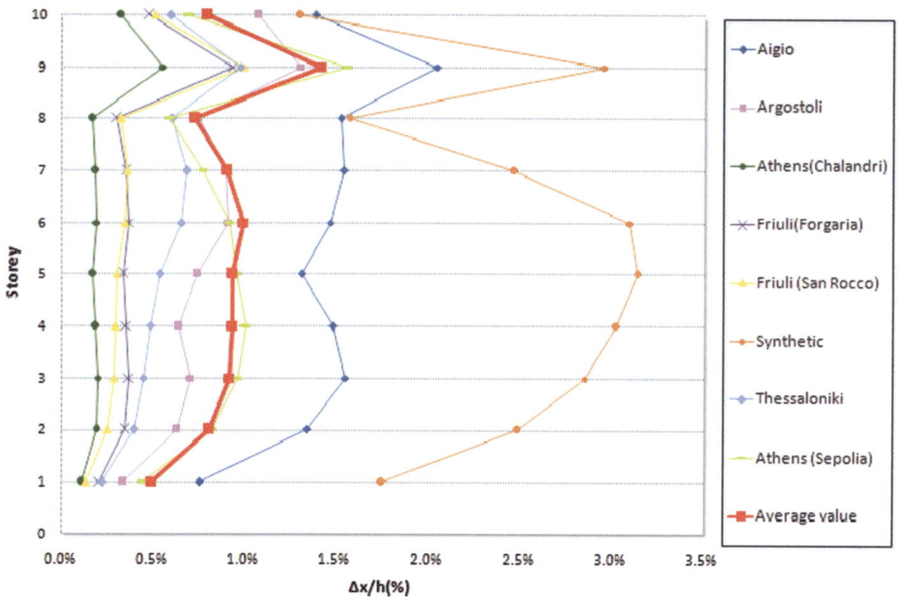

Fig. 11.9 Interstorey drifts for the "collapse prevention" performance level, $EI_{ef} = M_y/\varphi_y$

It is noted in Fig. 11.9 that even in this extreme case the maximum drift value (average of 8 records) is equal to 1.4% for Building 2 (and 1.3% for Building 1, not shown in Fig. 11.9), values that fall well below the drift capacity of R/C frame structures estimated on the basis of a large number of test results (Dymiotis et al., 1999). It is noted that analysis results should be interpreted on the basis of the average of the calculated values of each response-history analysis set, since the scaling procedure was based on the consideration of a mean spectrum. As depicted in Fig. 11.9, some analysis results (typically the ones concerning the synthetic ground motion in the set) can lead to an overestimation of interstorey drift values.

Furthermore, regarding the plastic hinges developed, the corresponding rotations were quite low in all cases, while the values of column plastic hinge rotations are very low compared to those in beams (Kappos and Stefanidou, 2010); hence a ductile failure mechanism is ensured for all limit states considered.

The deformation-based procedure is characterised by greater complexity compared to the current code procedures, but the results of applying this method to the design of irregular structures were encouraging. Since the deformation-based method accounts for the design according to the inelastic deformations anticipated for every performance level, basically the ductility of each member, the cross-sections required for the specific performance can be defined. Eventually, by designing according to the deformation-based design method, economy is obtained (in comparison to Code design, see Fig. 11.6), concerning not only the cross-sections used but also the reinforcement requirements (especially the transverse reinforcement of columns). It should be noted, however, that these and other assessment exercises have clearly shown that Code-designed (e.g. according to the EC8 DC "M" and "H") buildings also perform very satisfactorily for several earthquake levels.

Therefore, as already noted in Section 11.4.2 (referring to the direct DBD), possible advantages of the PBD methods should be traced mainly in the direction of economy, i.e. to potentially save material by avoiding over-conservatism in design; nevertheless, better control of seismic performance at different earthquake intensities might also be a critical issue, especially in some important buildings.

11.5 Closing Remarks

It was attempted to provide here an overview and discussion of the various seismic design procedures available for buildings, with emphasis mainly on whether new proposals for improved design methods (such as the direct displacement-based and deformation-based design procedures presented herein) could be useful within the frame of a "new generation" of codes. As far as the performance of structures designed to current codes is concerned, the answer is straightforward: Far from being perfect (whatever this might mean in the context of practical design), current codes like Eurocode 8 and the American Code IBC lead to designing sound structures with ample margins of safety against collapse, and in this respect they are, indeed, adequate. One can argue that sometimes current codes tend to be over-conservative and/or to result in building members that are difficult to detail on-site, but others could argue that earthquakes keep surprising us, in the sense that ground motions stronger than those recorded in the past keep being recorded, hence the extra safety margins apparently provided by current codes should not be reduced. It is perhaps worth noting here that the final version of Eurocode 8 generally results in less amount of reinforcement than earlier versions of this Code (like the ENV one, see detailed presentation and examples in Penelis and Kappos, 1997), in contrast to what happened until recently, i.e. that new seismic codes generally led to more stringent requirements and increased the cost of building. Interestingly, comparative studies (Athanassiadou et al., 2003) have shown that the more economic design resulting from the final EC8 does not lead to any noticeable reduction in safety margins.

The second question, i.e. whether new performance-based design proposals could or should be incorporated in future seismic codes, is more difficult to answer in a definitive way. Based on the (undoubtedly limited) available evidence, it appears that there are two main issues wherein new proposals can "entice" code developers: better damage control for a number of different earthquake intensities (in particular those lower that the commonly used single design earthquake with 10%/50 years probability of exceedance), and, of course, economy. As far as damage control is concerned, the writer's opinion is that the direct deformation-control method (Section 11.2.2) is better suited for inclusion in future codes, not only for "format" reasons (i.e. that it can be incorporated in existing codes by revising them, rather than by, essentially, completely replacing them), but also because, as already pointed out herein, displacement-based methods, even when applied to structural systems for which they were properly calibrated, do not always guarantee that local inelastic deformations will be within the acceptable limits, since checking of these deformations is not part of the procedure. It is clear, nevertheless, that explicitly checking these local deformations requires more refined and costly types of analysis than the simple equivalent static approach put forward by the DBD developers. In principle, only inelastic analysis can offer a viable alternative here, and for several types of buildings this analysis should be dynamic (response history) rather than static. Moreover, in many cases, analysis should account not only for inelastic member response but also for (nonlinear) soil-structure interaction (SSI) effects, a crucial issue that has not been raised here due to space limitations. Just as an indication of its importance, one could note that both the effective period (Eq. 8) and the effective damping of the system (e.g. the SDOF system forming the basis of the direct DBD method) can be strongly affected by SSI and by radiation damping, i.e. the damping resulting from the scattered wave energy from the foundations. Of course, as one keeps refining the analysis, the latter is made more complex and difficult to apply in a design office context (and within the stringent time schedules that usually apply). Seen from a slightly different perspective, the key difference in the interesting new proposals reviewed here is in the level of approximation, since the goal is common in both of them, i.e. control of damage. The direct DBD procedure assumes that the (generally complex) real building can be properly reduced to an SDOF system based on a reasonable (inelastic) displacement pattern, whereas the direct deformation-based procedure arrives at the inelastic displacement pattern and the associated local deformations through inelastic analysis, albeit of a reduced inelastic model. Nevertheless, for the latter method to be direct, rather than iterative (which would increase substantially the cost of analysis), it has to introduce an approximation in the way inelastic rotations are estimated from elastic ones (in Step 1).

Last and not least, the issue of economy has to be addressed, which is arguably the one most difficult to tackle in a comprehensive way. The available evidence is certainly too limited for drawing conclusions of general validity. Moreover, it should be emphasised that the economy of the final design does not depend solely on the way seismic action is defined and the analysis method used (e.g. code-type or PBD), but on several other issues that have not been studied systematically so far. For instance, comparisons among "old" and "new" procedures are in most studies carried out for 2D building models, hence the influence of important design assumptions such as torsion and accidental eccentricity effects, and combination rules for multi-component earthquake input, have not been properly addressed. Some pilot studies within the writer's research group have indicated that the final action effects (moments, shears) can be influenced more by the way torsion and accidental eccentricity are taken into account, than by whether the base shear was determined using the Code procedure or the DBD approach. Furthermore, as clearly illustrated by the case study presented in Section 11.3.1, answers to the economy question depend strongly on the code method (static or dynamic) to which the results of PBD procedures are compared. In view of these remarks, the only definitive conclusion regarding the issue of economy is that additional and, especially, more systematic and comprehensive, studies are required to compare the final products resulting from each procedure, wherein these products should be realistic, 3D buildings like those that one finds in the real world (as opposed to academic studies).

Acknowledgments A number of the author's students have made significant contributions to some of the studies summarised herein. The contributions of S. Stefanidou, S. Papista, and G. Panagopoulos, graduate students at the Aristotle University of Thessaloniki, and A. Manafpour, former graduate student at Imperial College, London, are particularly acknowledged.

References

Ambraseys N, Smit P, Berardi R, Rinaldis D, Cotton F, Berge C (2000) Dissemination of European strong-motion data. CD-ROM collection. European Commission, DGXII, Science, Research and Development, Bruxelles

American Society of Civil Engineers (2006) Minimum design loads for buildings and other structures. ASCE/SEI 7-05, Reston, VA

ASCE/SEI (2007) Seismic rehabilitation of existing buildings – ASCE standard 41-06. American Society of Civil Engineers, Reston, VA

Athanassiadou CJ, Kappos AJ, Ziakos K (2003) Seismic performance of multistorey r/c buildings designed to the new Eurocode 8 (prEN-1998-1). fib 2003 Symposium: Concrete structures in seismic regions (Athens), CD Proceedings, paper no. 018

Biggs JM (1964) Structural dynamics. McGraw-Hill, New York, NY

Carr A (2004) RUAUMOKO, manuals, "vol. 1 theory and user guide to associated programs, vol. 3 user manual for the 3-dimensional version". University of Canterbury, New Zealand

CEN (2004) Eurocode 8: Design of structures for earthquake resistance – Part 1: General rules, seismic actions and rules for buildings (EN 1998-1: 2004). CEN, Brussels

CEN (2005a) Eurocode 8: Design provisions of structures for earthquake resistance – Part 2: Bridges (EN1998-2:2005). CEN, Brussels

CEN (2005b) Eurocode 8: Design of structures for earthquake resistance – Part 3: Assessment and retrofitting of buildings (EN 1998-3:2005). CEN, Brussels

Chopra AK, Goel RK (2001) Direct displacement-based design: Use of inelastic vs. elastic design spectra. Earthquake Spectra 17(1):47–65

Computers and Structures Inc. (2005) ETABS nonlinear v.9.1.4 "extended 3D analysis of building systems". Program Manuals, Berkeley, CA

Dymiotis C, Kappos AJ, Chryssanthopoulos MC (1999) Seismic reliability of R/C frames with uncertain drift and member capacity. J Struct Eng ASCE 125(9):1038–1047

Fajfar P (1999) Capacity spectrum method based on inelastic demand spectra. Earthquake Eng Struct Dyn 28(9):979–993

fib Task Group 7.2 (2003) Displacement-based seismic design of reinforced concrete buildings. *fib* Bull. 25, Lausanne

International Conference of Building Officials (1997) Uniform building code – 1997 edition, vol 2. Structural Engineering Design Provisions, Whittier, CA

International Conference of Building Officials (2009) International Building Code/Building Officials and Code Administrators International, Country Club Hills, IL; Whittier, CA; and Southern Building Code Congress International, Inc., Birmingham, AL

Kappos AJ (1997a) Seismic damage indices for R/C buildings: Evaluation of concepts and procedures. Prog Struct Eng Mater 1(1):78–87

Kappos AJ (1997b) Partial inelastic analysis procedure for optimum capacity design of buildings. In: Proceedings of the international workshop on seismic design methodologies for the next generation of codes (Bled, Slovenia, June 1997), Balkema, pp 229–240

Kappos AJ (2000) Feasibility of using advanced analytical tools in the seismic design of R/C structures. In: Proceedings of the G. Penelis international symposium on concrete and masonry structures, Ziti editions, Thessaloniki, pp 47–60

Kappos AJ (2009) Design of earthquake resistant buildings. In: Invited lecture, international conference on earthquake engineering, Banja Luka, 26–28 October 2009, pp 147–184

Kappos AJ, Goutzika E, Stefanidou S (2007) An improved performance-based seismic design method for 3D R/C buildings using inelastic dynamic analysis. In: Conference on computational methods in structural dynamics and earthquake engineering (COMPDYN), Rethymno, Greece, June, paper no. 1375

Kappos AJ, Manafpour A (2001) Seismic design of R/C buildings with the aid of advanced analytical techniques. Eng Struct 23(4):319–332

Kappos AJ, Panagopoulos G (2004) Performance-based seismic design of 3D R/C buildings using inelastic static and dynamic analysis procedures. ISET J Earthquake Technol 41(1):141–158

Kappos AJ, Stefanidou S (2010) A deformation-based seismic design method for 3D R/C irregular buildings using inelastic dynamic analysis. Bull Earthquake Eng 8(4):875–895

Moehle JP (1992) Displacement-based design of RC structures subjected to earthquakes. Earthquake Spectra 8(3):403–428

Panagiotakos TB, Fardis MN (2001) A displacement-based seismic design procedure for R/C buildings and comparison with EC8. Earthquake Eng Struct Dyn 30:1439–1462

Paulay T (2002) A displacement-focused seismic design of mixed building systems. Earthquake Spectra 18(4):689–718

Penelis GG, Kappos AJ (1997) Earthquake-resistant concrete structures. E&FN SPON, London

Priestley MJN (1993) Myths and fallacies in earthquake engineering—Conflicts between design and reality. In: Proceedings of the Tom Paulay symposium—Recent developments in lateral force transfer in buildings, ACI SP-157, pp 229–252

Priestley MJN, Calvi GM, Kowalsky MJ (2007) Displacement-based seismic design of structures. IUSS Press, Pavia

Priestley MJN, Kowalsky MJ (2000) Direct displacement-based design of concrete buildings. Bull N Z Natl Soc Earthquake Eng 33(4):421–444

SEAOC Ad Hoc Committee (1999) Tentative guidelines for performance-based seismic engineering. App. I of: Recommended lateral force requirements and Commentary, SEAOC, Sacramento, CA

Shibata A, Sozen M (1976) Substitute structure method for seismic design in reinforced concrete. J Str Div ASCE 102(1):1–18

Sullivan TJ, Calvi GM, Priestley MJN, Kowalski MJ (2003) The limitations and performances of different displacement based design methods. J Earthquake Eng 7(1):201–244

Sullivan TJ, Priestley MJN, Calvi GM (2006) Direct displacement-based design of frame-wall structures. J Earthquake Eng 10(1):91–124

Chapter 12
Performance-Based Design of Tall Reinforced Concrete Core Wall Buildings

John W. Wallace

Abstract Reinforced concrete (RC) walls are commonly used as the primary lateral-force-resisting system for tall buildings, although for buildings over 49 m (160 ft), IBC 2006 requires use of a dual system. Use of nonlinear response history analysis (NRHA) coupled with peer-review has become a common way to assess the expected performance of tall buildings at various hazard levels to avoid the use of a backup Special Moment Frame for tall buildings employing structural walls. Modeling of the load versus deformation behavior of reinforced concrete walls and coupling beams is essential to accurately predict important response quantities for NRHA. It also has become important to assess the impact of the floor diaphragms, gravity framing system, and foundation system on the expected performance, as well as to compare the expected performance of code-compliant and performance-based designed buildings to assess the merits of using a performance-based design approach. Given this critical need, an overview of modeling approaches used for RC core wall systems is reviewed to assess the ability of common modeling approaches to accurately predict both global and local responses. Application of fragility relations within a performance-based framework is reviewed for selected components and analytical studies are used to address system level issues such as the impact of slab coupling on gravity column axial loads and higher mode impacts on wall moment and shear demands. Based on the results, recommendations for performance-based design are made and research needs are identified.

12.1 Introduction

Reinforced concrete (RC) structural walls are effective for resisting lateral loads imposed by wind and earthquakes as they provide substantial strength and stiffness and limit the deformations resulting from strong earthquake ground shaking. Use of

---

J.W. Wallace (✉)
NEES@UCLA Laboratory, Department of Civil and Environmental Engineering, University of California, Los Angeles, CA 90095-1593, USA
e-mail: wallacej@ucla.edu

M. Garevski, A. Ansal (eds.), *Earthquake Engineering in Europe*, Geotechnical, Geological, and Earthquake Engineering 17, DOI 10.1007/978-90-481-9544-2_12,

structural wall(s) alone to resist lateral loads is not allowed for buildings over 49 m (160 ft) in ASCE 7-05; this limit has been bypassed by employing a code section that allows use of any system that can be shown to have equivalent performance to allowed systems. This alternative approach requires use on nonlinear response history analysis (NRHA) and peer-review by a panel of experts, and thus, has only become possible in recent years as the tools for conducting nonlinear response history analysis have improved. Use of a slab-column gravity frame has emerged as one of the preferred gravity systems for tall buildings.

Application of NRHA requires use of an analytical model that reasonably represents the hysteretic response of the primary lateral force resisting elements (including the foundation), as well as the interaction between the wall and other structural and non-structural (gravity) members. For tall buildings, use of a relatively simple model is required to reduce computer run times; therefore, it is important to balance model simplicity with the ability of the model to reliably predict inelastic responses both at the global and local levels. Despite the availability of prior research, which is considerable, the scale and complexity of the overall system requires variation in model and material parameters to assess the sensitivity of the computed responses and to ensure adequate safety against collapse.

Primary lateral-force-resisting elements of core walls include wall segments and coupling beams, typically supported on a mat foundation that also supports the gravity framing system. The lateral and gravity systems are tied together by a floor diaphragm; however, code provisions for new buildings require that earthquake demands be resisted entirely by the lateral-force-resisting system. Thus, considerable importance is placed on accurate modeling of the wall segments and coupling beams that make of the primary components of the lateral-force-resisting system. Given that design requirements are focused on promoting flexural yielding, use of appropriate flexural stiffness values is particularly important as it impacts lateral drift estimates and the degree of coupling between various wall segments comprising the core wall.

Slab-column frames, with their limited forming, low story heights, and open floor plan, are efficient gravity systems for tall core wall buildings. The slab-column frame is typically designed to resist only gravity loads; however, the ability of the slab-column gravity frame to maintain support for gravity loads under the lateral deformations imposed on it by the lateral-force resisting system must be checked in current codes. The primary objectives of this "deformation compatibility" check are to verify that slab-column punching failures will not occur for SLE and MCE shaking levels, as well as to assess the need to place slab shear reinforcement adjacent to the column to enhance slab shear strength. Design requirements for these checks are included in ACI 318-08 S21.13.4. For tall core wall systems, considerable coupling is likely to exist between the core wall and gravity framing. Therefore, studies have been conducted to assess the impact of this coupling on the overall system response and design (Salas, 2008). Detailing of the slab-wall connection also is an important design consideration, as the rotation of the core wall can impose relatively large rotation demands on the slab at the slab-wall interface (Klemencic et al., 2006). Slip forming of the core wall is common to reduce construction time; requiring special attention to slab shear and moment transfer at the slab-wall interface.

Recent guidelines (e.g., LATBSDC, 2008) recommend a two-level design, where a Service Level Earthquake (SLE) is used to assess damage in relatively frequent earthquakes (e.g., 43 year return periods; 50%/30 year) and Maximum Considered Earthquake (MCE) is used to assess the potential for collapse in very rare earthquakes (2,475 year return period; 2%/50 year). Use of either a linear model with acceptance based on evaluation of demand-to-capacity ratios (D/C ratios) or a nonlinear model subjected to base accelerations (minimum of three ground motions) with acceptance based on element demands (e.g., rotations, strains) are recommended for SLE checks. For MCE shaking, nonlinear models subjected to seven pairs of ground motion acceleration are common with acceptance based on average responses, although in some cases, more stringent acceptance criteria have been adopted (e.g., median plus one standard deviation response values).

The preceding paragraphs provide an overview of several important issues associated with analysis and design of tall reinforced concrete core wall buildings. A more detailed discussion, including a review of relevant recent research and specific recommendations, are presented in the following sections.

12.2 Wall Modeling

Orakcal and Wallace (2006) present a comprehensive study on the ability of nonlinear modeling approaches to capture the cyclic response of relatively slender reinforced concrete walls for combined bending and axial load. A MVLE model, which is conceptually the same as the fiber model approaches that are embedded in some commercially available computer programs (e.g., PERFORM 3D), was employed in their study for isolated walls subjected to reversed, cyclic loading. The overall modeling process involves: (1) subdividing the wall cross section into unconfined concrete fibers, confined concrete fibers, and reinforcement fibers, (2) selecting appropriate material relations, (3) subdividing the wall into a specified number of elements (components) over the wall height, (4) defining appropriate boundary conditions, and (5) imposing a prescribed load/displacement history. Some of the results of their study are shown in Fig. 12.1 for a test of a 12 ft tall wall with a 4 in. by 48 in. cross section subjected to constant axial load and reversed cyclic lateral displacements at the top of the wall. The test walls were approximately one-fourth scale models of prototype walls proportioned using the 1991 Uniform Building Code (Thomsen and Wallace, 1995, 2004).

It is noted that Orakcal and Wallace (2006) reduced the test data into lateral force versus deformation relations for flexure and shear. As well, contributions from foundation rotation or slip between the test wall foundation and the strong floor were removed. Several important observations can be gleaned from their results. The effective linear stiffness to the yield point is very close to the $0.5EI_g$ value commonly used for design (Fig. 12.1a) and that the wall lateral load capacity computed

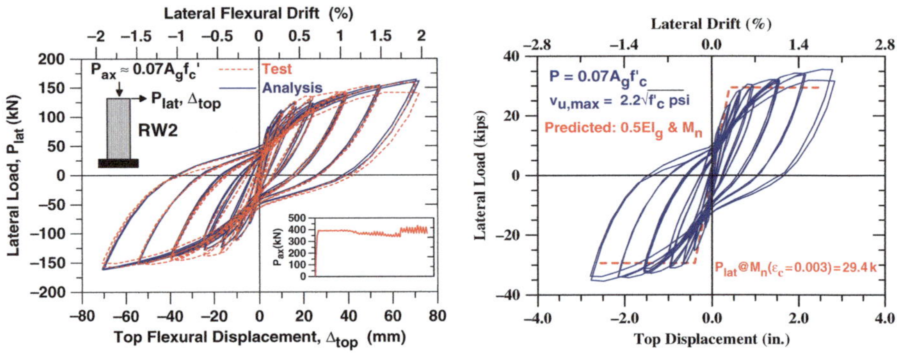

(a) (b)

Fig. 12.1 Test results for specimen RW2. (**a**) Model results (Orakcal and Wallace, 2006). (**b**) Bilinear fit for $0.5EI_g$ and M_n

using the nominal moment capacity at the wall critical section located at the wall base for as-tested material properties is slightly less than the maximum lateral load achieved during the test (Fig. 12.1b; Thomsen and Wallace, 2004). It is noted that, when a fiber element model is used, selection of effective flexural stiffness values is not possible, since the effective stiffness is "automatically" determined based on the selected material relations, level of axial load, and the current state (including history for nonlinear response history analysis).

The results presented indicate that cyclic material relations for concrete and reinforcing steel can be selected to produce overall load versus deformation responses which are generally consistent with test results for a wide range of responses. Orakcal and Wallace (2006) report that model and test results for first story displacements and rotations, where inelastic deformations dominate over elastic deformations, compare very favorably. Results for wall curvature and average wall strain over a 23 mm (9 in.) gauge length at the base of the wall presented in Figs. 12.2 and 12.3, respectively, reveal that tensile strains are well represented with the model; however, model compressive strains substantially under estimate the peak compressive strains measured for several tests. In general, for the relatively slender wall tests ($h_w l_w = M_u/(V_u l_w) = 3$), peak measured compressive strains were about twice the model predicted strains. Given these results, the maximum compressive strains derived from analytical models available in commonly used computer programs are likely to underestimate compressive strains. Preliminary analytical studies have indicated that one reason for this discrepancy may be interaction that occurs between flexural and shear behavior (Wallace, 2007); however, models that account for interaction are not available in commonly used commercial computer programs. The shear stress levels for the tests reviewed were in the range of $0.17 - 0.5\sqrt{f_c'}$ MPa, whereas design shear stress levels for tall buildings are likely to approach $0.67\sqrt{f_c'}$ MPa, the ACI 318 nominal design limit, indicating a potential for greater discrepancies between model and actual compressive strains for typical

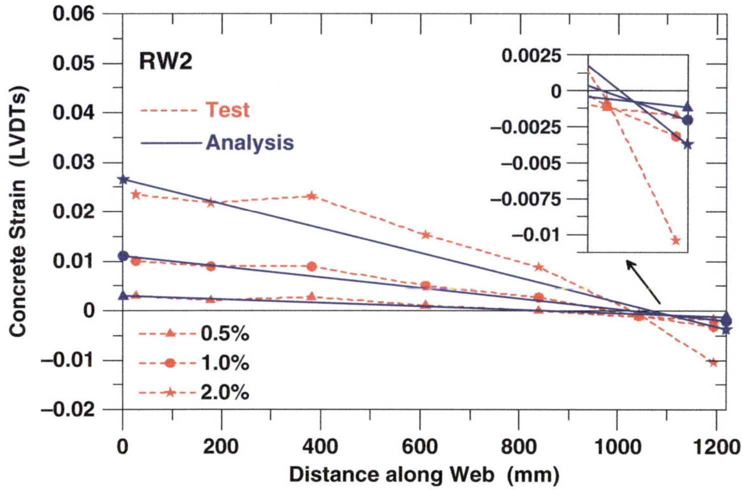

Fig. 12.2 Wall curvature profiles

tall core wall buildings. Given these findings, acceptance criteria for spalling and required transverse reinforcement at wall boundaries should be selected carefully. Wallace (2007) recommends doubling the compressive strain obtained from analytical results and using a limiting compressive strain of 0.004 to assess the potential for concrete spalling (or reducing the strain limit of 0.004 to 0.002).

The results presented in Figs. 12.1, 12.2, and 12.3 represent nonlinear flexural behavior. In cases where nonlinear flexural responses occur, linear shear behavior is typically assumed, i.e., flexural behavior and shear behavior are uncoupled. It is

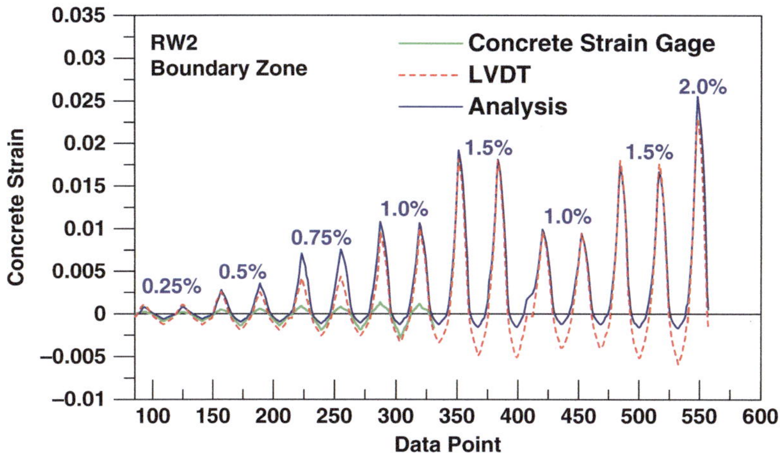

Fig. 12.3 Wall average strain at critical section

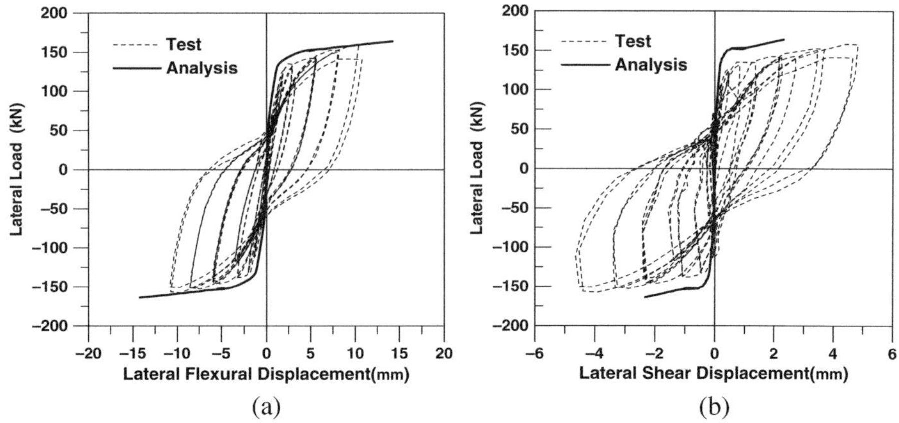

Fig. 12.4 Load displacement relations: (**a**) flexure, (**b**) shear

apparent from the experimental results presented in Fig. 12.4 for first story deformations that significant inelastic shear deformations initiate at the same applied lateral load as inelastic flexural deformations, that is, the flexural and shear responses are coupled. The results presented in Fig. 12.4 are for a wall with nominal shear capacity of approximately 300 kN, which is about twice the maximum shear applied to the wall; therefore, for an uncoupled model, linear shear behavior would be assumed. The analysis results presented in Fig. 12.4 are for a coupled model for monotonic material behavior (Massone, 2006) and reveal that a coupled model can reproduce observed test results reasonably well. However, models that account for coupled flexure-shear behavior for *cyclic* loading are not yet available in commercially available computer programs as development of coupled cyclic material models remains a significant research challenge.

Given these limitations, use of a simplified (or approximate) modeling approach is suggested that reasonably captures the observed experimental trends for displacement responses of walls governed by flexure with modest shear stresses (i.e., $v_u \leq 0.5\sqrt{f_c'}$ MPa); additional studies are needed to address higher shear stress levels (the issue of under-estimating concrete compressive strain remains). Flexural responses are modeled as noted above and a nonlinear (translational) shear spring is used to capture shear behavior (Fig. 12.5a). It is noted that flexural behavior and shear behavior are uncoupled in this model. The relation used to model the nonlinear shear behavior is similar to that given in ASCE 41-06 (2007), except the cracking level is taken as 0.5 times the shear force required to reach the yield moment at the wall base. The initial stiffness of the shear spring is defined as $0.4E_c$ up to the point where shear cracking occurs; the post-yield stiffness is arbitrarily reduced to $0.1E_c$ to produce a good match in between the overall load versus top displacement relations (Fig. 12.5b) i.e., to account for nonlinear shear deformations (Gogus Wallace, 2010).

Model sensitivity studies reported by Orakcal et al. (2004) indicate that lateral load versus lateral top displacement relations are quite insensitive to the number of

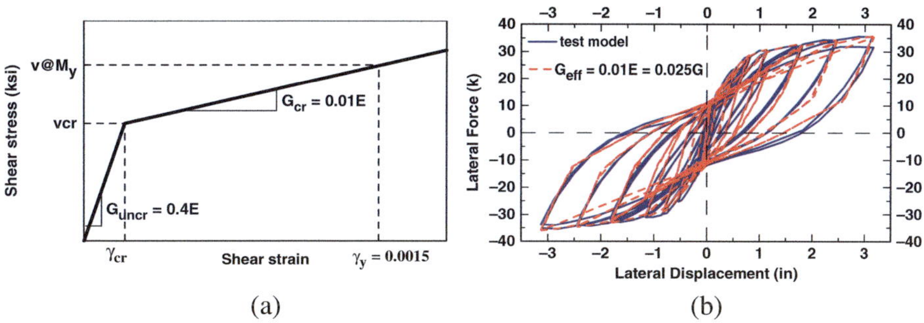

Fig. 12.5 Nonlinear wall modeling – combined flexure and shear behavior. (**a**) Nonlinear shear spring relation. (**b**) Model and test results – Wall RW2

material fibers and number of elements used, that is, mesh and element refinements do not markedly improve the response prediction presented in Fig. 12.5b. This result is encouraging in that a coarse mesh can be used to assess drift responses for tall buildings, leading to reduced computer run times. However, the study also revealed that use of a course mesh is likely to modestly underestimate the peak strains for the material fibers (by about 30%). Subsequent studies by Salas (2008) indicated that this underestimation of peak model compressive strains is mitigated if the element length used approximately corresponds to the expected plastic hinge length and modest strain hardening is assumed for the reinforcement (3–5% post-yield). Therefore, for analysis of tall core wall buildings, it is important to select element sizes and material relations with these issues in mind.

12.3 Coupling Beams

Requirements for transverse reinforcement for diagonally-reinforcement coupling beams with clear length to total depth less than four were introduced into ACI 318-95 S21.7.7. The objective of the requirement is to confine the concrete strut and to suppress reinforcement buckling; however, placement of the transverse reinforcement around the diagonal bar bundles is difficult where the diagonal groups intersect at the beam mid-span (particularly for shallow beams) as well as at the beam-wall interface due to interference with the wall boundary vertical reinforcement. ACI 318-08 S21.9.7 introduced an alternative detailing option, where transverse reinforcement is placed around the entire beam cross section, i.e., no transverse reinforcement is provided directly around the diagonal bar bundles. Nonlinear modeling of coupling beams has received increased attention as the use of performance-based design for tall core wall buildings has become more common. Of particular interest is the selection of the effective secant bending stiffness at yield $E_c I_{eff}$ and the allowable plastic rotation prior to significant lateral strength degradation. The value used for coupling beam bending stiffness has a significant impact on the system behavior. Test results that helped support the code change in the required transverse reinforcement for diagonally-reinforced coupling beams are reviewed in the following paragraphs. The test results also are used to evaluate modeling approaches for coupling beams. More detail is available in the report by Naish et al. (2009).

The test beam prototypes were based on two common tall building configurations for residential and office construction. Typical wall openings and story heights produce coupling beams with aspect ratios of approximately 2.4 for residential buildings and 3.33 for office buildings. A coupling beam with cross-section dimensions of 24 in. × 30 in. and 24 in. × 36 in. reinforced with two bundles of 8-#11 (d_b=35.8 mm) diagonal bars is common for residential and office construction, respectively. Due to geometric and strength constraints of an existing reaction frame, tests were conducted on one-half scale replicas of the prototype beams. Thus the test specimens were 12 in. × 15 in. or 0.3 m × 0.38 m (CB24F and CB24D) and 12 in. × 18 in. or 0.3 m × 0.46 m (CB33F and CB33D) with two bundles of 6-#7 (d_b=22.2 mm) diagonal bars, for the residential and office beams, respectively (Fig. 12.6).

Beams with transverse reinforcement provided around the bundles of diagonal bars (referred to as "Diagonal confinement") were designed according to ACI 318-05 S21.7.7.4, whereas beams with transverse reinforcement provided around the entire beam cross section (referred to as "Full section confinement") were designed according to ACI 318-08 S21.9.7.4(d). Due to maximum spacing requirements, the volumetric ratios of transverse reinforcement provided in both the prototype and test beams exceed that calculated using the requirement for columns (ACI 318-08 S21.6.4.4). Three test specimens with aspect ratio of 2.4 were constructed with 4 in.-thick slabs. One specimen (CB24F-RC) contained a slab reinforced with #3 (d_b = 9.5 mm) bars @12 in. spacing, on the top and bottom in the transverse direction, and on the top only in the longitudinal direction, without post-tensioning strands. Two specimens (CB24F-PT and CB24F-1/2-PT) both contained a similar reinforced-concrete slab, but also were reinforced with 3/8 in. (d_b = 9.5 mm) 7-wire strands post-tensioned to apply 1.0 MPa (150 psi) to the slab in the longitudinal direction. Specimen geometries and material properties are summarized in Table 12.1. Further details can be found in (Naish et al., 2009).

The test specimens were each placed in a vertical position with end blocks simulating wall boundary zones at each end, and tested using the setup shown in Fig. 12.7. The lateral load was applied via a horizontal actuator. Two vertical hydraulic actuators were used to ensure zero rotation at the top of the specimen,

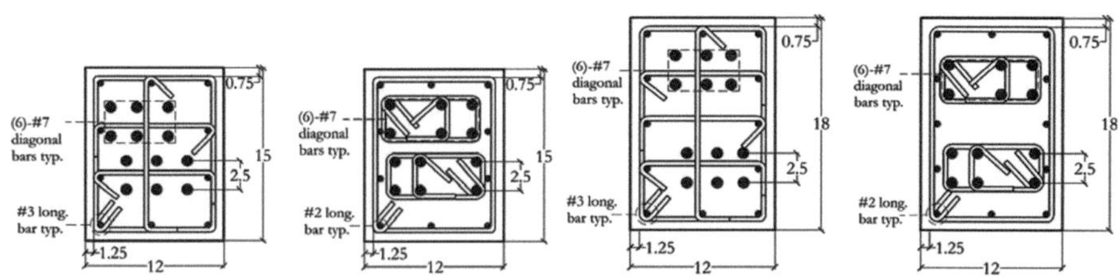

Fig. 12.6 From *left to right*, beam cross-section for CB24F, CB24D, CB33F, and CB33D. A 4 in. slab is included on the *top* of CB24F-RC, PT, ½-PT (1 in. = 25.4 mm)

Table 12.1 Test matrix and material properties

Beam	l_n/h type	α[°]	Transverse reinforcement		f'_c[psi]	f_y[psi]	f_u[psi]	Description
			Full section	Diagonals				
CB24F		15.7	#3 @ 3 in.	NA	6,850			Full section confinement ACI 318-08
CB24D			#2 @ 2.5 in.	#3 @ 2.5 in.	6,850			Diagonal confinement ACI 318-05
CB24F-RC	2.4 residential		#3 @ 3 in.	NA	7,305			Full section conf. w/ RC slab ACI 318-08
CB24F-PT			#3 @ 3 in.	NA	7,242	70,000	90,000	Full section conf. w/ PT slab ACI 318-08
CB24F-1/2-PT			#3 @ 6 in.	NA	6,990			Full section conf. (reduced) w/ PT slab ACI 318-08
CB33F	3.33 office	12.3	#3 @ 3 in.	NA	6,850			Full section confinement ACI 318-08
CB33D			#2 @ 2.5 in.	#3 @ 2.5 in.	6,850			Diagonal confinement ACI 318-05

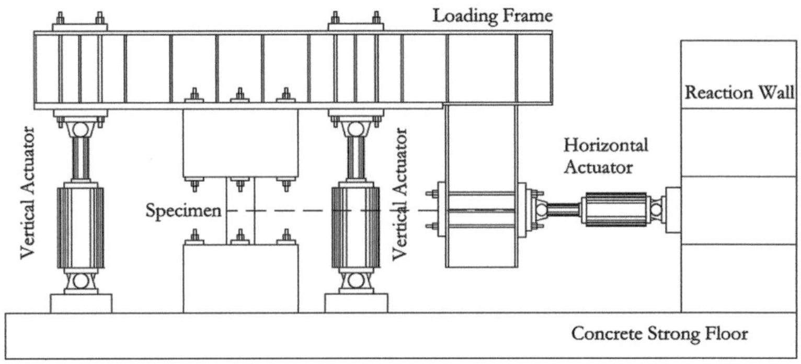

Fig. 12.7 Laboratory test setup

while maintaining constant (zero) axial force in the beam. Load-control testing was performed at 0.125, 0.25, 0.50, and 0.75 V_y, where $V_y = 2M_y/l_n$ to ensure that the load-displacement behavior prior to yield was captured. Beyond 0.75 V_y, displacement-control was used in increments of percent chord rotation (θ), defined as the relative lateral displacement over the clear span of the beam (Δ) divided by the beam clear span (l_n).

12.3.1 Experimental Results

Load-deformation responses of CB24F and CB24D are very similar over the full range of applied rotations (Fig. 12.8a). Notably, both beams achieve large rotation (˜8%) without significant degradation in the lateral load carrying capacity, and the beams achieve shear strengths of 1.25 and 1.17 times the ACI nominal strength (Table 12.2). Figure 12.8b plots load vs. rotation relations for the 3.33 aspect ratio beams with full section confinement (CB33F) vs. diagonal confinement (CB33D). Similar to the 2.4 aspect ratio beams, Fig. 12.8b reveals that the beams have similar strength, stiffness, and deformation characteristics. The test results presented in Fig. 12.8 indicate that the full section confinement option of ACI 318-08 provides equivalent, if not improved performance, compared to confinement around the diagonals per ACI 318-05.

The transverse reinforcement used for CB24F-1/2-PT was one-half that used for CB24F-PT to assess the impact of using less than the code-required transverse reinforcement given that the requirements of S21.6.4 are based on column requirements. Figure 12.9 plots load-deformation responses and reveals similar loading and unloading relations up to 3% total rotation, which approximately corresponds to the Collapse Prevention limit state per ASCE 41-06. At higher rotations ($\theta \geq 4\%$), modest strength degradation is observed for CB24F-1/2-PT, whereas the strength of CB24F-PT continues to increase slightly; however, both beams achieve rotations of ˜8% before significant lateral strength degradation ($< 0.8\,V_{ave}$). V_{ave} is defined as the average shear force resisted by the beam between the yield point and the onset of significant lateral strength degradation.

Table 12.2 Summary of calculated and experimental coupling beam parameters

Beam	M_n^+ [in-k]	M_n^- [in-k]	$V@M_n$ [k]	$\frac{V@M_n}{\sqrt{f'_c}A_{cv}}$	$V_n(ACI)$ [k]	$\frac{V_n(ACI)}{\sqrt{f'_c}A_{cv}}$	V_{ave} [k]	$\frac{V_{ave}}{\sqrt{f'_c}A_{cv}}$	V_y [k]	Δ_y [in.]	V_{max} [k]	$\Delta@V_{max}$ [in.]
CB24F	2,850	2,850	158.3	10.65	136.3	9.15	154.9	10.40	121.3	0.360	171.0	1.08
CB24D	2,850	2,850	158.3	10.65	136.3	9.15	150.7	10.12	128.8	0.363	159.2	2.16
CB24F-RC	2,890 3,550[a]	2,890 3,350[a]	160.6 191.7[a]	10.45 12.50[a]	136.3	8.87	181.0	11.77	147.2	0.362	190.8	2.16
CB24F-PT	3,160 3,960[a]	3,160 3,625[a]	175.6 210.7[a]	11.45 13.75[a]	136.3	8.90	198.9	12.98	163.2	0.361	211.8	2.16
CB24F-1/2-PT	3,145 3,940[a]	3,145 3,610[a]	174.7 209.7[a]	11.61 13.90[a]	136.3	9.06	182.4	12.12	158.1	0.365	189.6	1.08
CB33F	3,615	3,615	120.5	6.77	107.8	6.03	118.3	6.62	107.7	0.600	124.0	1.80
CB33D	3,615	3,615	120.5	6.77	107.8	6.03	114.7	6.42	95.94	0.601	120.6	3.60

[a]Calculations that consider the impact of the slab

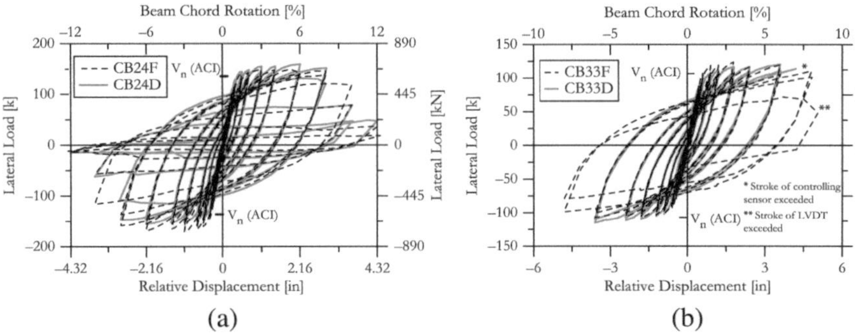

Fig. 12.8 Cyclic load-deformation: (**a**) CB24F vs. CB24D; (**b**) CB33F vs. CB33D

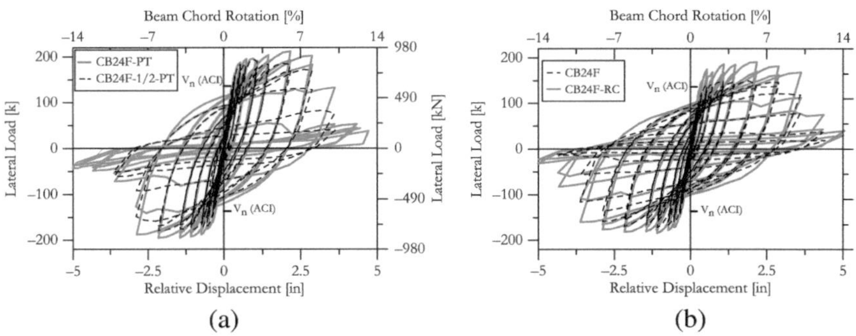

Fig. 12.9 Cyclic load-deformation relations. (**a**) CB24F-PT vs. CB24F-1/2-PT. (**b**) CB24F vs. CB24F-RC

The results indicate that the one-half scale coupling beams tested with ACI 318-08 detailing are capable of achieving total rotations exceeding 8%, whereas ASCE 41 limits plastic rotation to 3% without strength degradation and 5% with a modest strength degradation of 20%. The potential influence of scale factor on the test results, which is an important consideration, is discussed later. The test results indicate that there is little difference in load-deformation response between CB24F-PT and CB24F-1/2-PT; therefore, the potential to reduce the quantity of required transverse reinforcement exists, but requires further study since only one beam test was conducted.

Four beams with aspect ratio of 2.4 were tested to systematically assess the impact of a slab on the load-deformation responses. CB24F did not include a slab, whereas CB24F-RC included an RC slab, and CB24F-PT and CB24F-1/2-PT included PT slabs (with 1.0 MPa or 150 psi of prestress). Fig. 12.9b, which directly compares the load-displacement responses of CB24F vs. CB24F-RC, reveals that the slab increases shear strength by 17% (155–181 k); however, this strength increase can be taken into account by considering the increase in nominal moment strength due to the presence of the slab, i.e. slab concrete in compression at the beam-wall interface at one end, and slab tension reinforcement at the beam-wall interface at the other end. The results, summarized in Table 12.2, indicate that the higher test shear strength observed is primarily due to the increase in nominal moment capacity when a slab is present.

12.3.2 Modeling

Elastic analysis approaches require estimation of effective elastic bending and shear stiffness values. Approaches commonly used to determine effective (secant) stiffness values of coupling beams at yield are summarized and compared to test results in Table 12.3. Of the various approaches, only the approach in ASCE 41-06 addresses the impact of slip/extension deformations on the effective yield stiffness. The contribution of slip/extension to the yield rotation was estimated for the test beams using the approach recommended by Alsiwat and Saatcioglu (1992), where the crack width that develops at the beam-wall interface depends on bar slip and bar extension (strain). As noted in ASCE 41-06 Supplement #1, the impact of slip/extension on effective bending stiffness can be accounted for using two approaches, one which uses a moment-curvature to define the effective bending stiffness along with a bond-slip model that accounts for the added flexibility due to bond-slip, or an alternative model where the effective bending stiffness determined from a moment-curvature analysis is reduced to account for the added flexibility due to bond-slip. These two approaches are assessed using the test results to develop a simple recommendation.

As previously stated, the tests were conducted at one-half scale; therefore, it is important to understand the potential impact of scale on the effective yield stiffness as well as the overall load-deformation behavior. The relative contribution of flexural deformations (curvature) and slip/extension to the yield rotation of the test beams at full scale (i.e. prototype beams) is assessed using the same approach as noted in the previous paragraph for the one-half scale beams. The study is extended to consider coupling beam aspect ratios beyond those tested, by varying the beam length. Results are reported in Fig. 12.10, where the effective yield rotation is plotted against beam aspect ratio (l_n /h) for various scale factors. For a given aspect ratio, slip rotation at yield is significantly impacted by scale, with a 35–40% reduction for beams at one-half versus full scale. The effective bending stiffness at yield for the

Table 12.3 Summary of effective secant stiffness at yield – Test and recommended values

	Test results	FEMA 356	ASCE 41	ASCE 41 S1, w/slip hinge	NZS-3101 95 (μ=1)
EI_{eff} [% EI_g]	14.0 12.5[a]	50.0	30.0	16.5 13.0[a]	50.0
θ_y[% *drift*]	0.70 1.00[a]	0.23	0.39	0.75 0.95[a]	0.23

[a]Modifications for 1/2-scale

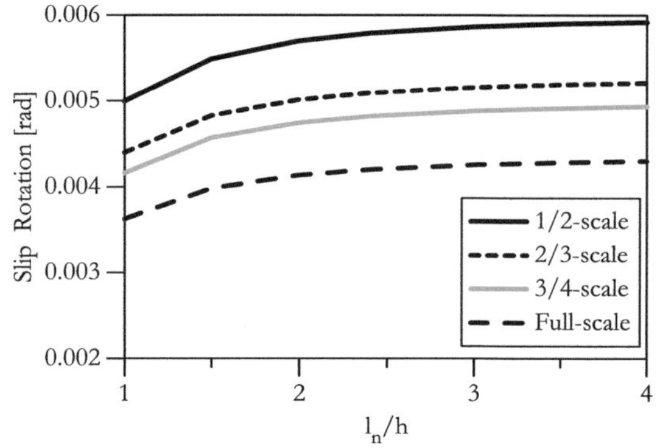

Fig. 12.10 Yield rotation due to slip/extension

one-half scale tests of $0.12\,E_cI_g$ increases to $0.14\,E_cI_g$ for the full-scale prototypes due to the reduction in the relative contribution of slip rotation.

Linearized backbone relations for normalized shear strength versus rotation are plotted in Fig. 12.11 as dotted lines for the three configurations of beams tested. The backbone relations that are modified to represent full-scale beams are also plotted in Fig. 12.11, as discussed in the prior paragraphs. For configurations with multiple tests, an average relation is plotted. Backbone relations modified to represent full-scale beams indicate that the total rotations at yield, strength degradation, and residual strength are reduced to 0.70, 6.0, and 9.0%, respectively (from 1.0, 8.0, and 12.0%). ASCE 41-06 with Supplement #1 modeling parameters also are plotted on Fig. 12.11. Relative to ASCE 41-06, the relations derived for the full-scale beams have a lower effective yield stiffness ($0.14E_cI_g/0.3E_cI_g\ =\ 0.47$) and substantially greater deformation capacity (5.3%/3.0% = 1.77). It is reasonable to use a plastic rotation value of 5.0% with no strength degradation, with moderate residual strength (0.3 V_n) up to a plastic rotation of 7.0%, compared to the ASCE 41-06 residual strength ratio of 0.8 at a plastic rotation value of 5.0%. It is noted that the ASCE 41-06 relation applies to all diagonally-reinforced coupling beams, including beams with aspect ratios significantly less than the values of 2.4 and 3.33

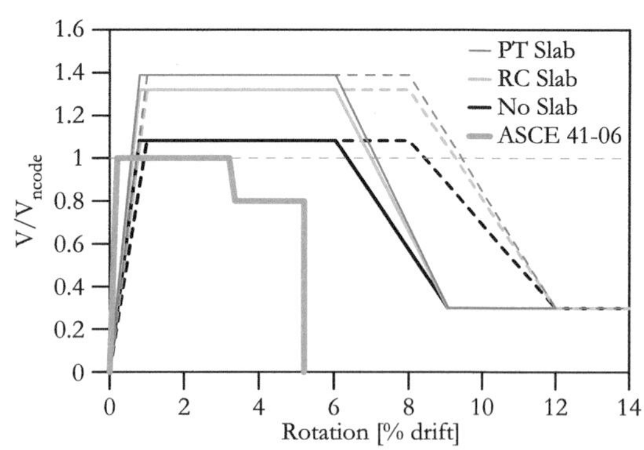

Fig. 12.11 Modified backbone relations (test results – *dashed lines*)

investigated in this test program. Results presented in Fig. 12.11 apply for the beam aspect ratios tested (2.4 and 3.33), as well as to beams between these ratios. It is reasonable to assume these values can be extrapolated modestly to apply to beams with $2.0 < l_n/h < 4.0$.

Based on the backbone and effective stiffness relations discussed above, nonlinear modeling approaches commonly used by practicing engineers were investigated to assess how well they were able to represent the measured test results. Two models were considered, one utilizing a rotational spring at the ends of the beam to account for both nonlinear flexural and slip/extension deformations (M_n hinge) and one utilizing a nonlinear shear spring at beam mid-span to account for both shear and slip/extension deformations (V_n hinge). The M_n-hinge model (Fig. 12.12a) consists of an elastic beam cross-section with $E_cI_{eff} = 0.5E_cI_g$, elastic-rotation springs (hinges) at each beam-end to simulate the effects of slip/extension deformations, and rigid plastic rotational springs (hinges) at each beam-end to simulate the effects of nonlinear deformations. The stiffness of the slip/extension hinges were defined using the Alsiwat and Saatcioglu model discussed above, whereas the nonlinear flexural hinges are modeled using the backbone relations derived from test results (Fig. 12.11, excluding the elastic portion). The V_n-hinge model (Fig. 12.12b) also consists of an elastic beam cross-section and slip/extension hinges. However, instead of using flexural hinges at the beam ends, a shear force versus displacement hinge (spring) is used at the beam mid-span to simulate the effects of nonlinear deformations. The shear hinge properties are defined using the backbone relations derived from the test results (Fig. 12.11).

Figure 12.12 shows cyclic load-deformation plots for the two models and the test results for CB24F. Both models accurately capture the overall load-displacement response of the member; however, the M_n-hinge model (Fig. 12.12c) captures the

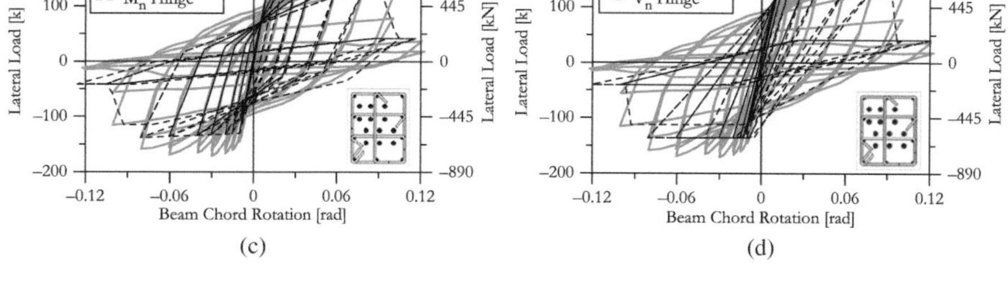

Fig. 12.12 Cyclic load-deformation modeling results ($ln/h = 2.4$). (**a**) M_n-hinge model. (**b**) V_n-hinge model. (**c**) CB24F vs. moment hinge model. (**d**) CB24F vs. shear hinge model

unloading characteristics better than the V_n-hinge model (Fig. 12.12d), due to the fact that unloading stiffness modeling parameters, which help to adjust the slope of the unloading curve, are available for the flexural hinges in the commercial computer program used, but not for the shear hinges. Therefore, depending on the computer program used, similar modeling studies should be conducted to calibrate available model parameters with test results.

12.4 Shear Modeling

The nominal shear strength of walls is typically defined using ACI 318-08 provisions as:

$$V_n = A_{cv}\left(\alpha_c\sqrt{f_c'} + \rho_t f_y\right) \tag{1}$$

where the coefficient α_c is 3.0 for $h_w/h_w \leq 1.5$, is 2.0 for $h_w/l_w \geq 2.0$, and varies linearly between 3.0 and 2.0 for h_w/l_w between 1.5 and 2.0. In this equation, A_{cv} represents the cross-sectional web area of a wall, ρ_t is transverse reinforcement ratio, f_y is the yield strength of transverse reinforcement, and f_c' is the compressive strength of concrete. The variation of α_c for h_w/l_w (height-to-length) ratios between 1.5 and 2.0 accounts for the observed strength increase for low-aspect ratio walls. An upper limit on nominal shear strength is set at $V_n = A_{cv}\left(10\sqrt{f_c'\psi}\right)$ for a single wall, the same limit used for beams (ACI-ASCE Committee 426, 1973), and $V_n = A_{cv}\left(8\sqrt{f_c'}\right)$ for walls sharing lateral load. Test data were reviewed by Cardenas et al. (1973) as part of an ACI 318-71 code background paper to show that the limit of $V_n = A_{cv}\left(10\sqrt{f_c'}\right)$ was satisfactory for design.

Wallace (1998) evaluates wall shear strength for concrete strengths exceeding approximately 70 MPa (10 ksi) using results reported by Kabeyasawa et al. (1998) on 37 tests of with concrete strengths between approximately 70 and 100 MPa and shear-span-to-depth ratio ($M_u/V_u l_w$) between 0.6 and 2.0. The assessment indicates that the ratio of the maximum shear force obtained in the test to the ACI 318-95 nominal shear strength (V_{test}/V_n) was 1.38 with a standard deviation of 0.34, indicating that the ACI shear strength provides close to a lower-bound estimate of wall shear strength. Similar results are reported by Wood (1990) and Orakcal et al. (2009) for concrete compressive strengths from approximately 15–40 MPa. Based these studies, a median shear strength of the tests is approximately $V_{test} = 1.5\ V_{n,ACI}$.

For walls controlled by shear, the relation suggested by Elwood et al. (2007) for ASCE 41-06 Supplement #1 (with shear strength of $1.5\ V_{n,ACI}$) is recommended until more test data are available to assess the deformation capacity of walls with detailing consistent with current codes are available. For walls controlled by flexure, relatively sparse data exist to judge whether shear strength should be degraded as the

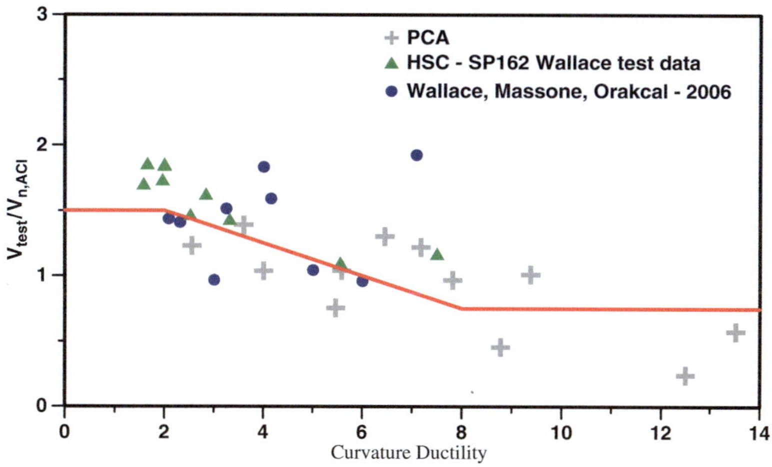

Fig. 12.13 Wall shear strength as impacted by flexural ductility

wall is subjected to increasing nonlinear deformations. Results from various tests are summarized in Fig. 12.13 (Gogus and Wallace, 2010). The data clearly show that wall shear strength degrades with increasing ductility. Oesterle et al. (1984) suggest that the reduction in drift capacity is related to increased contribution of inelastic shear deformations leading to web crushing failures. The relation suggested is approximately a median value.

12.5 Capacity Design

The sensitivity of core wall system responses, such as core wall moment, shear, and lateral displacement over the building height, and diaphragm transfer forces are likely impacted by modeling parameters. A case study of a single building is undertaken to demonstrate the potential impact of model parameters on response parameters (Fig. 12.14). The structural system consists of a core wall tower with a multi-level podium with perimeter walls. To facilitate discussion, the core wall is divided into three regions, an assumed hinge zone at the base of the core wall above the podium levels, and the core wall above and below the hinge zone. A fiber wall model with specified material models is used to capture P-M behavior, whereas shear responses are modeled with a bilinear spring. Two models are considered, one with nonlinear fiber elements in the hinge zone and elastic elements above, and one that uses nonlinear elements over the full height of the building. Nonlinear dynamic analyses were conducted for simultaneous application of the North–South, East–West, and vertical ground motion records for the 1994 Northridge earthquake Beverly Hills-14145 Mulholland (USC-90013 station) and the 1964 Niigata, Japan record from 701 B1F SMAC-A station (Salas, 2008).

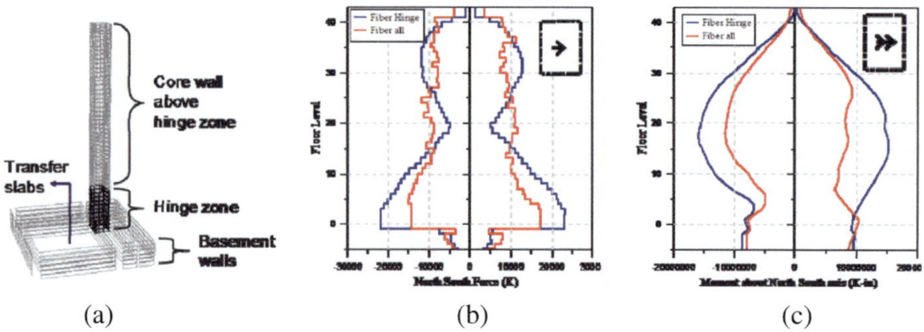

Fig. 12.14 (**a**) Case study building, (**b**) Wall Shear, (**c**) Wall moment (1994 Northridge record)

The difference in the wall shear force distributions presented in Fig. 12.14 indicate that allowing yield at upper stories has a substantial impact, as shear forces are substantially reduced, especially within the hinge region. Moments at upper levels also are substantially reduced. Examination of core wall strains indicated that the reduction was a result of modest yielding of reinforcement at upper levels ($< 2\varepsilon_y$), typically spread out over several stories. If elastic behavior is assumed at upper stories, spurious results are produced as higher modes produce large moment demands at upper levels, sometimes much larger than at the wall base.

Capacity design concepts can be applied to other response parameters and elements, such as wall shear and floor diaphragms. Wall shear demands determined from nonlinear models are most often computed as the average value obtained from the model subjected to seven pairs of horizontal ground motions (versus the maximum value from three pairs of ground motions). The nonlinear models are typically use expected material properties and a capacity reduction factor $\phi = 1.0$ (LATBSDC, 2008) using the nominal shear strength of ACI 318-08 (as opposed to 1.5 V_n shown in Fig. 12.13). It is questionable whether this approach provides adequate safety against shear failure, especially within the hinge region where nonlinear flexural deformations reduce shear strength (Fig. 12.13). Given this issue, for some tall building peer-review projects, a shear demand of the mean plus one standard deviation has been required, although shear capacity at low flexural deformation demands also appears to be higher than indicated by the ACI equation. Further study of this issue is underway.

Capacity design of floor diaphragms also is needed to ensure a proper load path at typical tower floor levels as well as at the podium level (Fig. 12.14a). The diaphragm openings and large transfer forces at podium levels between core walls and perimeter basement walls can significantly complicate the design process. The magnitude of the podium level slab transfer forces can be significantly impacted by the assumed stiffness of the diaphragm (Salas, 2008) and large deformations can result given long load path even if slab reinforcement provided to transfer the force does not yield.

12.6 Slab-Column Frames

Modeling of slab-column frames, which are commonly used as gravity systems for tall core wall structures, involves assigning appropriate stiffness and strength values, and consideration of punching failures. ASCE/SEI 41 Supplement #1 provisions provide useful information on modeling slab-column frame systems (Elwood et al., 2007). Therefore, the focus of the material presented in the following paragraphs is to assess the potential impact of slab-column gravity frames on the performance of tall core wall buildings as coupling between the core wall and the gravity frame, via the slab, will impact lateral story deformations and axial force demand on the gravity column.

Consider the plan view of a core-wall with a gravity slab-column frame system shown in Fig. 12.15 (Salas, 2008). A simple model is used to account for coupling between the core wall and the gravity column (Fig 12.14b) to minimize the computational effort given the general objectives of this study. The slab is modeled using an equivalent beam whose properties are determined using the effective beam width model (Allen and Darvall, 1977; Hwang and Moehle, 2000; Kang and Wallace, 2005) recommended in ASCE/SEI 41 Supplement #1. Four equivalent columns are used to represent the behavior of the gravity columns. The stiffness values of the equivalent slab-beams are determined as shown in Fig. 12.16, where two slab effective widths are used to model a span with different elastic effective widths at each end. For example, beams B1 and B2 are used as shown in Fig. 12.14, B1 with the effective width determined by the ratios of c_1/l_1 and c_2/l_2, and B2 with the effective width is equal to l_2 given the core wall spans the entire width. The two beams meet at a nodal point at the center of the span based on the approach recommended by Hwang and Moehle (2000).

The model shown in Fig. 12.15b is developed to provide an equivalent beam based on the properties determined for Beams B1–B4, as well as the columns.

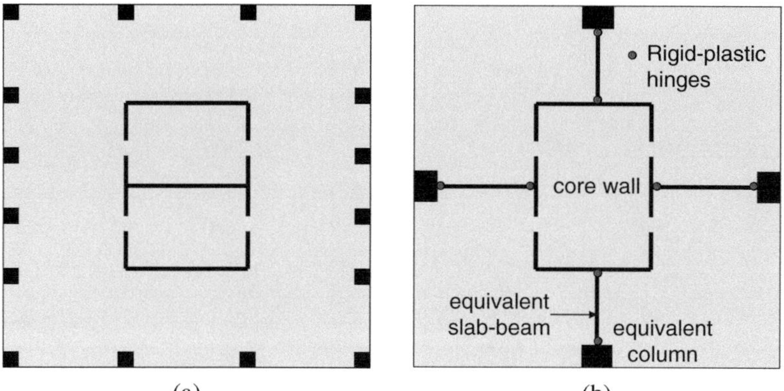

Fig. 12.15 Slab column layout and slab column modeling. (**a**) Floor plan. (**b**) Simple model

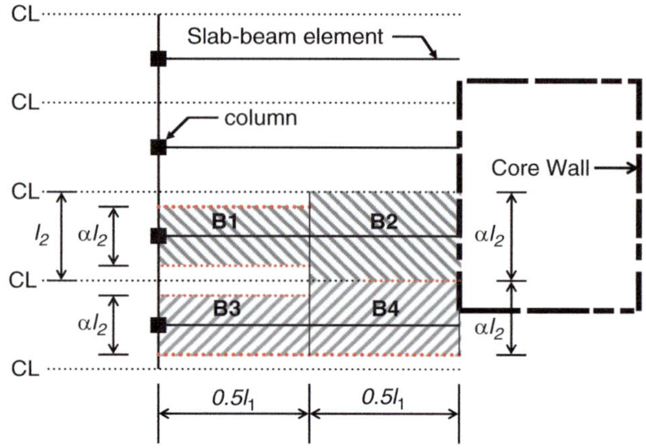

Fig. 12.16 Application of effective width model to core wall

Note that effective EI values determined for the slab are multiplied by a factor (β) to account for cracking (Elwood et al., 2007). The yield moments for positive and negative bending for slab-beams B1–B4 are determined based on properly anchored slab flexural reinforcement. For example, for slab-beam B2, the positive and negative yield moments at the slab-beam end that frames into the core wall would be based on (developed) reinforcement within the entire width l_2, whereas the positive and negative yield moments for the slab-beam end that frames into the column would typically be based on (developed) reinforcement within the column strip. It is noted that the spans for this model are relatively large given the slab is post-tensioned, which reduces the effectiveness of the coupling.

The envelopes of gravity column axial stress over the height of the study building are presented in Fig. 12.17 for two different cases: (a) 1.2 DL + 1.6 LL, and (b) 1.0 DL + EQ (for a single ground motion record); the drop in axial stress at levels 9

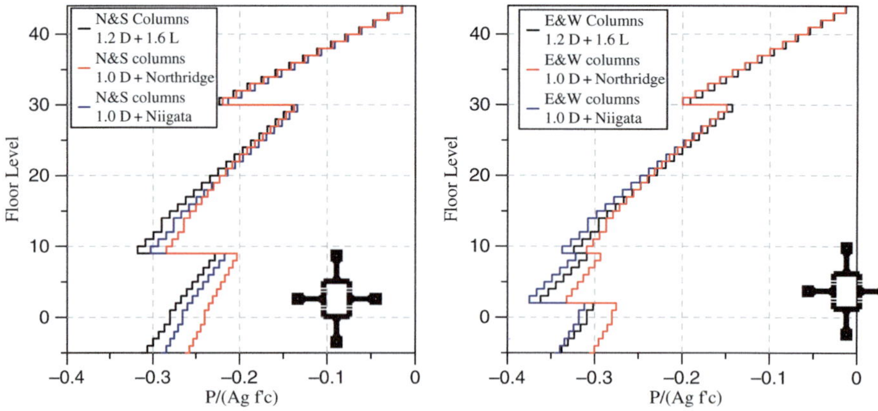

Fig. 12.17 Variation in gravity column axial load due to slab coupling

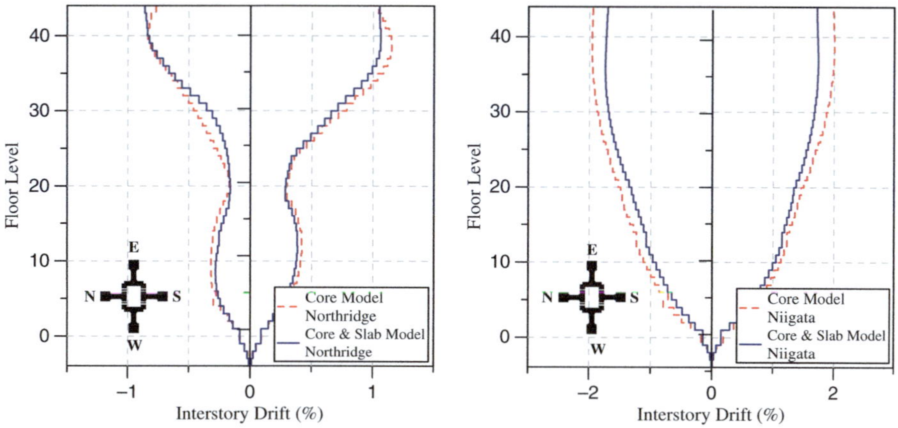

Fig. 12.18 Variations in story lateral displacements

and 30 occur due to changes in the column cross-sections at these levels. For this building and the given ground motion, the variation in the peak column axial stress for the two load cases is insignificant. The results for this case study suggest that the variation in column axial load due to slab shear forces developed under lateral loading do not produce significant increases relative to the pure gravity load case. For slabs with more longitudinal reinforcement and shorter spans, a greater variation in column axial load would be expected. Figure 12.18 plots the differences between lateral story displacements for the two models and reveals that the coupling, in this case, has a relatively minor impact.

12.7 Slab-Wall Connections

As previously noted, post-tensioned slab-column frames are commonly used to support gravity loads in core wall construction. Given that wall strain gradients can be quite large (e.g. see Fig. 12.2), the wall can impose relatively large rotations on the slab, particularly at the slab-wall interface where the core wall is in tension (axial growth in tension can be substantial). When subjected to these large rotation demands, the slab-wall connection must be capable of transferring gravity loads to the core wall. To speed up construction, slip-forming is sometimes used for core wall buildings, i.e., the core wall is cast prior to the floor slab, creating a potential weak connection at the slab-wall interface. One approach that has been used to accomplish this connection is shown in Fig. 12.18, where post-tensioning strands stop short of the wall interface and are lapped to top and bottom reinforcement connected to the wall via mechanical couplers. Shear keys are typically provided at the slab-wall interface.

Two full-scale tests were undertaken by Klemencic et al. (2006) to investigate the behavior of slab-to-wall connections (Fig. 12.19) subjected to reverse cyclic

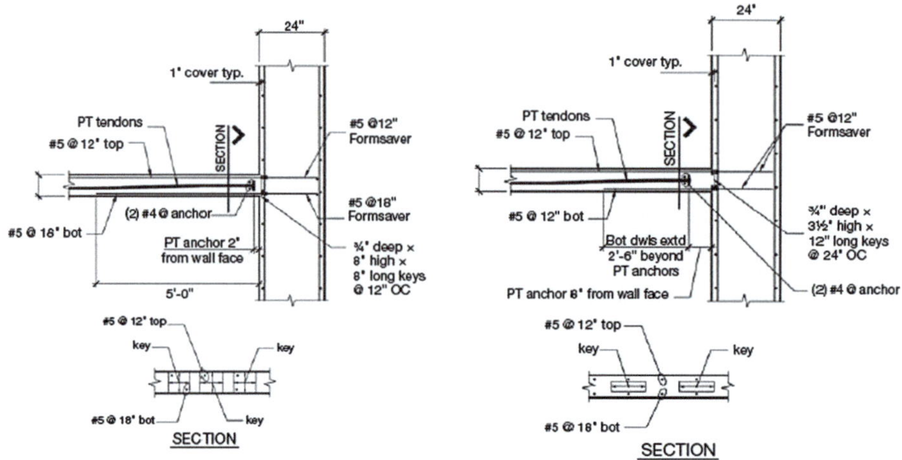

Fig. 12.19 Slab-to-wall connection details (Klemencic et al., 2006)

loading to demonstrate that the connection could sustain gravity loads for interstory drift levels up to 3% (collapse prevention performance at MCE demands). Other goals of the tests were to assess the impact of lateral drift on the degree of cracking in the connection region, the influence of varying the location of the anchor on the unbonded post tensioning cables, and the behavior of the mechanical couples at slab-wall interface. The two specimens, with common architectural dimensions (Figs. 12.19 and 12.20) were subjected to constant gravity load and then increasing lateral deformation.

The displacement history applied to the slab-wall connection involved applying negative peak drift values equal to twice the positive peak drift values to account for the impact of wall growth on the tension face on the rotation demand. Essentially elastic behavior was observed up to a peak drift ratio of 0.85, with significant

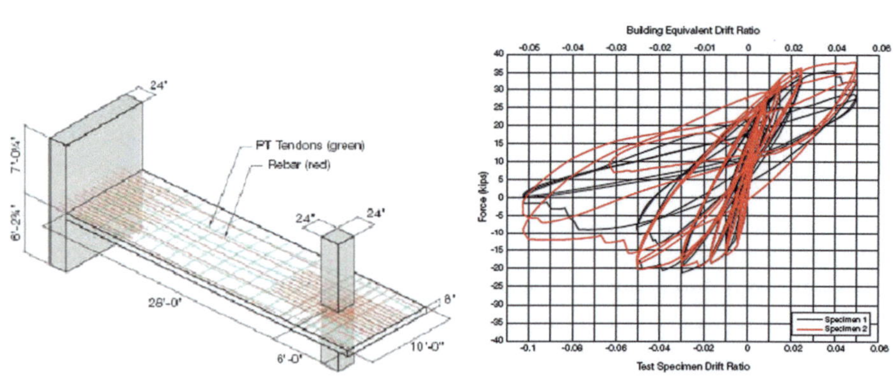

(a) (b)

Fig. 12.20 (**a**) Overall test geometry; (**b**) Force-deformation response (Klemencic et al., 2006) yielding at a drift ratio of approximately 1.0% (Fig. 12.20b). Lateral strength degradation initiated on the first cycle to 2.5 and 5.0% lateral drift for specimens 1 and 2 (Fig. 12.20b), respectively, due to pullout of top dowels (Specimen 1) and buckling (Specimen 2); however, both specimens were subjected to multiple drift cycles at 2.5 and 5% drift without collapse (loss of gravity load support). The biggest difference in performance between the two tests was the degree of cracking at the slab-wall interface. For Specimen 1, where the anchor for the post-tensioning tendons was placed only 2 in. from the wall face, large cracks were observed between the anchor and the wall. In Specimen 2, with the anchor 8 in. from the wall face, cracks were more distributed and narrower. It is noted that Specimen 2 also used equal amounts of top and bottom slab bonded slab reinforcement at the slab-wall interface, which also likely led to modest improvements in the observed behavior.

12.8 Instrumentation for Seismic Monitoring

Tall building construction also provides a unique opportunity to employ monitoring equipment to measure structural responses for a variety of conditions (ambient, high-level wind, and earthquake). Ideally, a broad spectrum of sensor types capably of measuring floor accelerations, wind pressures, average concrete strains, rebar strains, and rotations should be employed. In addition to a broad spectrum of sensors, key attributes of a robust monitoring system include: rapid deployment, energy efficiency, event detection, robust analog-to-digital conversion, local storage, redundant time synchronization, multi-hop wireless data transport, and remote sensor and network health monitoring. Recent developments in all of these areas reveal that robust structural health monitoring is likely to emerge over the coming years. Therefore, careful consideration should be given to increased use of sensors in existing and planned buildings. In general, more sensors are needed than are often employed in buildings, that is, only one triaxial accelerometer at the base, a mid-level, and the roof (e.g., instrumentation required by the City of Los Angeles, 2002).

Given the complexity and geometry of tall buildings, laboratory studies, which are hindered by scale, materials, and appropriate boundary conditions, are unlikely to provide definitive results for a variety of important modeling issues. For a given instrumented building, the details of the embedded sensor network design should be model-driven, i.e., sensor types and locations determined based on response quantities obtained from 3D dynamic finite element models (FEM) subjected to a suite of site-specific ground motions. In a concrete core wall system, response quantities of interest might be average core wall concrete strains within the plastic hinge (yielding) region and rotations imposed on coupling beams (or slab-wall connections). Other modeling and design issues could also be targeted, such as so-called podium effects and appropriate ground motion building inputs at subterranean levels (Stewart, 2007). Given the uncertainty associated with the response of structural systems to earthquake ground motions, a probabilistic distribution of response quantities of interest (e.g., interstory displacements, coupling beam deformations) should be determined for the structural model subjected to the suite of ground motions and the sensor layout should target specific regions versus a single response quantity.

The City of Los Angeles requires building instrumentation (accelerometers) be installed at the base, mid-level, and roof to obtain a building permit for all buildings over ten stories as well as for buildings over 6 stories with an aggregate floor area exceeding 60,000 ft^2 (LA Building Code §1635, 2002). The owner is required to maintain the instrumentation in working order; the City of LA has an extensive program for monitoring the equipment currently installed in approximately 400 buildings. Currently, data collected by the required accelerometers are not archived and are not readily available for use either for rapid post-event assessment or by researchers to improve our ability to model buildings. Clearly, there are buildings where the measurement of interstory drift (moment frame) or average concrete strain (base of a shear wall system) might produce more useful and meaningful data than acceleration data alone. The instrumentation requirements for the City of Los Angeles were recently updated to address these issues and also require more detailed sensor layouts for tall buildings designed using NRHA.

12.9 Engineering Demand Parameters and Fragility Relations

Over the last several years, considerable effort has been expended on development of relations that describe damage states and consequences for various Engineering Demand Parameters (EDPs). A majority of this effort has been organized by the ATC 58 project. Several of the relations developed are presented the following paragraphs.

Example relationships for slab-column connections and coupling beams are shown in Fig. 12.21. It is relatively easy to develop relations for specific points for backbone relations reported in the literature, such as points where yielding and significant strength degradation are observed; however, within a performance-based design framework it is more informative if relationships are developed address the

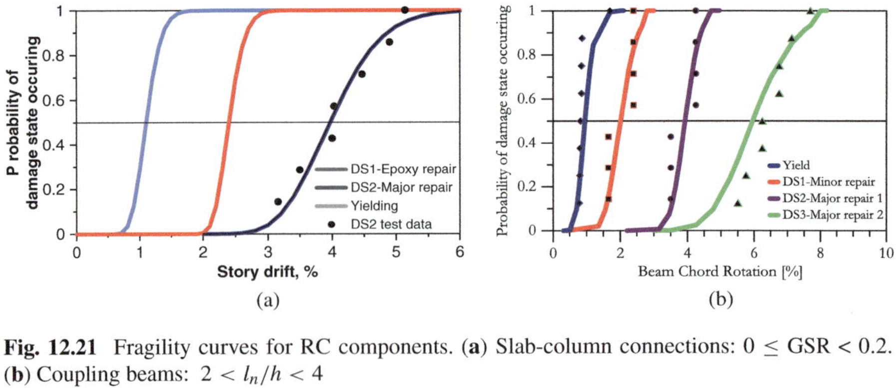

Fig. 12.21 Fragility curves for RC components. (**a**) Slab-column connections: $0 \leq \text{GSR} < 0.2$. (**b**) Coupling beams: $2 < l_n/h < 4$

degree of repair required and the associated cost (the consequence of the damage). For example, minor repair of damaged reinforced concrete structural elements is more likely to be related to the degree of concrete spalling and residual crack widths (versus crack widths at peak load). Unfortunately, this type of information is rarely reported, even for tests conducted in recent years. Thus, developing relationships between EDPs and repair states is often subjective.

A fragility relation for reinforced concrete slab-column connections without shear reinforcement is shown in Fig. 12.21a, since this is one of the more common conditions reported in the literature. Relationships are shown for yielding and punching failure, which are typically reported in the literature but may need to be modified to ensure consistent treatment of data. At yield, repair is unlikely to be required; therefore, in this example, a minor repair state was subjectively defined at an interstory displacement (EDP) of twice the displacement associated with the yield point. Although this definition is subjective, it is probably reasonable since residual crack widths requiring epoxy injection repair are likely to exist once significant yielding of slab reinforcement has occurred. As well, for slab-column connections, larger crack widths are likely at the slab-column interface, where nonlinear deformations are larger than those represented in the overall load versus deformation relation. Test results are then reviewed to determine likely crack lengths that would require epoxy injection. Major repair, for this case, is associated with punching failure where concrete would have to be chipped away, reinforcement removed, and new reinforcement spliced to existing bars. Major repair is likely to require shoring and possible jacking to return the slab to its original position since punching failures are typically accompanied by a drop in the slab elevation. A similar approach was used to develop a rough relation for reinforced coupling beams (Fig. 12.21b). Given the complex geometry of reinforced concrete core walls, strain is the most likely EDP, with minor repair associated with spalling (e.g., compressive strain of 0.002) and residual cracks at the wall boundary and within the wall web. The degree of repair for at wall boundaries and the wall web are likely to be related to the wall tensile strain and the web shear stress, respectively. Major repair is more likely to be associated with buckling and fracture of flexural reinforcement for slender walls (low shear stress) and web crushing (high shear stress).

12.10 Performance Assessment

As part of the PEER Center Tall Buildings Initiative (TBI), a project is underway to quantify differences in performance of tall buildings designed using performance-based versus code-prescriptive design approaches for three building types (RC core wall, RC dual system, Steel BRB). Models of the buildings are being subjected to 15 pairs of ground motions to obtain EDPs at five different hazard levels (25, 43, 475, 2,475, and 4,975 year return periods). The EDPs are being used to assess likely damage and repair costs, included costs for contents and non-structural components. Comprehensive studies like this are essential to help assess and quantify the potential benefits associated with use of performance-based design approaches as well as to help identify shortcomings in our modeling approaches and to prioritize research needs.

12.11 Conclusions

An overview of some important issues associated with performance-based design of tall reinforced concrete core wall buildings was presented. Based on this review, the following observations and conclusions are noted.

Existing commercially available computer programs that incorporate fiber models (or similar models) are capable of reproducing lateral-load versus top displacement relations measured from moderate-scale, relatively slender walls subjected to constant axial load and cyclic lateral displacements. Model element heights used within the potential plastic hinge region should be selected to be approximately equal to the anticipated plastic hinge length and a modest reinforcement strain hardening ratio of 3–5% should be used to avoid potential problems associated with concentration of inelastic deformations within a single element of short height. Sufficient elements and fibers should be used to ensure the strain distribution along the cross section is adequately represented; however, even with these steps, current models underestimate the peak compressive strains measured in a limited number of tests by a factor of about two. Coupling between nonlinear flexural and shear deformations appears to be one factor that could explain this observed discrepancy. To account for shear deformations in walls with moderate stress levels ($< 0.5\sqrt{f_c' MPa}$), the post-crack shear stiffness of shear springs can be reduced to approximately $0.01E_c$.

For stout walls, it is important to consider the impact of concrete cracking on the lateral stiffness; therefore, a backbone relation which defines cracking and yielding points is recommended, similar to the one incorporated into ASCE 41-06 Supplement #1. The ASCE 41-06 relation captures the load-deformation response of lightly-reinforced wall segments with very low axial load reasonably well; however, for new walls with modest axial load levels $\left(P = 0.05A_g f_c'\right)$, the model underestimates the peak strength and overestimates the yield displacement based on a review of very limited test data. A yield displacement at yield of 0.001–0.002 is more realistic. Additional test data are needed to assess the impact of axial load on wall shear strength and the yield point.

An alternative detailing approach for coupling beams was introduced in ACI 318-08 to reduce congestion and reduce construction time. Tests results indicate that the alternative detailing approach, which uses transverse reinforcement around the entire beam section, produces beams with strength and deformation capacity equal to or better than beams with transverse reinforcement only enclosing the diagonal reinforcement. Test results, adjusted for the impact of scale, indicate total rotation capacities of approximately 0.06 for coupling beam geometry and quantities of reinforcement typically used in tall buildings. Commonly employed modeling approaches for coupling beams were able to reproduce the observed test results very closely provide an effective elastic stiffness of roughly $0.15E_cI_g$ was used.

The reduced stiffness was primarily a result of slip/extension deformations at the beam-wall interface.

Use of capacity design to promote flexural yielding only at the base (or just above the podium level) is not realistic for taller buildings due to the impact of higher modes. In this limited study, modest nonlinear behavior at upper levels ($\varepsilon_s < 2\varepsilon_y$) led to significant reductions in wall moments and shears. The results indicate that nonlinear modeling over the full height of the core wall is required to accurate assess core wall demands. Large transfer forces at podium levels can lead to large deformations for longer transfer lengths, even if the reinforcement is designed to remain elastic. The potential impact of these deformations on the integrity of the transfer slabs should be considered.

Modeling of the complete lateral and gravity systems for nonlinear response history analysis may be appropriate provided computer run times are not excessive (and continue to improve dramatically). For post-tensioned slab-column gravity frames, incorporating the gravity system in the model for one case study building had only minor impacts on the computed lateral displacements and column axial stress levels were similar to those for gravity loads alone. However, for shorter spans or higher reinforcing ratios, more significant variations are likely. Incorporating the gravity system into the model allows rotation and drift demands to be assessed at slab-column and slab-wall connections directly.

Given the complex issues that arise for tall buildings and the expense associated with laboratory testing for such large structures (even at reduced scale), an aggressive program to incorporate building instrumentation is needed. A program is being initiated in Los Angeles to start to address this need.

Comprehensive studies are needed to systematically assess the relative merits of using performance-based design approaches, particularly for tall buildings. The PEER Center Tall Buildings Initiative is currently conducting a limited study on three tall buildings to provide insight into this issue.

Acknowledgements The work presented in this paper was supported by various sources, including the National Science Foundation, the Charles Pankow Foundation, the Applied technology Council (Projects ATC-58, -72, and -76), and the PEER Center Tall Buildings Initiative with support from the California Seismic Safety Commission. The results presented represent the work of numerous students in recent years, including: Dr. Leonardo Massone, now at the University of Chile, Dr. Kutay Orakcal, now at Bogazici University, Turkey, Marisol Salas, MSCE UCLA 2008, and David Naish and Aysegul Gogus, both currently Ph.D. students at UCLA. The author also has benefited from numerous interactions with PEER Center researchers, and in particular, Prof. Jack Moehle at UC Berkeley and Mr. Ron Klemencic at Magnusson Klemencic Associates in Seattle. Any opinions, findings, and conclusions or recommendations expressed in this paper are those of the author and do not necessarily reflect those of the supporting organization or other people acknowledged herein.

References

ACI 318-05 (2005) Building code requirements for structural concrete (ACI 318-05) and commentary (ACI 318R-05), American Concrete Institute, Farmington Hills, Michigan

ACI 318-08 (expected 2008) Building code requirements for structural concrete (ACI 318-08) and commentary (ACI 318R-08), American Concrete Institute, Farmington Hills, Michigan

Allen FH, Darvall P (1977) Lateral load equivalent frame. ACI J, Proc 74(7):294–299

Alsiwat J, Saatcioglu M (1992) Reinforcement anchorage slip under monotonic loading. J Struct Eng, ASCE 118(9):2421–2438

ASCE (2007) Seismic rehabilitation of existing buildings (ASCE/SEI 41-06, Including Supplement #1), ASCE, Reston, VA

Cardenas AE, Hanson JM, Corley WG, Hognestad E (1973) Design provisions for shearwalls. ACI J, Proc 70(3):221–230 [PCA test]

Elwood KJ, Matamoros AB, Wallace JW, Lehman DE, Heintz JA, Mitchell AD, Moore MA, Valley MT, Lowes LN, Comartin CD, Moehle JP (2007) Update to ASCE/SEI 41 concrete provisions. Earthquake Spectra 23(3):493–523

Gogus A, Wallace JW (2010) ATC 76-4: Trial Application of Reinforced Concrete Structural Walls, ATC Project 76–4, Applied Technology Council (under review)

Hwang S, Moehle JP (2000) Models for laterally loaded slab-column freames. ACI Struct J 97(2):345–353

IBC: International Building Code (2006) IBC-2006, International Code Council

Kabeyasawa T, Hiraishi H (1998) Tests and analysis of high-strength reinforced concrete shear walls in Japan (ACI Special Publication, SP-176), American Concrete Institute, Farmington Hills, MI, pp 281–310

Kang THK, Wallace JW (2005) Dynamic responses of flat plate systems with shear reinforcement. ACI Struct J 102(5):763–773

Klemencic R, Fry JA, Hurtado G, Moehle, JP (2006) Performance of post-tensioned slab-core walls connections. PTI J 2:7–23

LATBSDC (2008) An Alternative Procedure for Seismic Analysis and design of Tall Buildings Located in the Los Angeles Region: A Consensus Document – 2008 Edition, Los Angeles Tall Buildings Structural Design Council, April 2008, 32 pp

Los Angeles Building Code, §1635, 2002

Naish D, Fry JA, Klemencic R, Wallace JW (2009) Experimental evaluation and analytical modeling of ACI-318/05/08 reinforced concrete coupling beams subjected to reversed cyclic loading. Report SGEL 2009/06, Department of Civil and Environmental Engineering, University of California, Los Angeles, CA, Aug 2009, 109 pp

Oesterle RG, Aristizabal-Ochoa JD, Shiu KN, Corley WG (1984) Web crushing of reinforced concrete structural walls. ACI J, Proc 81(3):231–241 [PCA test]

OpenSees—open system for earthquake engineering simulation. Pacific Earthquake Engineering Research Center, University of California, Berkeley, CA. http://opensees.berkeley.edu/OpenSees/developer.html

Orakcal K, Conte JP, Wallace JW (2004) Flexural modeling of reinforced concrete walls – model attributes. ACI Struct J 101(5):688–698

Orakcal K, Massone LM, Wallace JW (2009) Shear strength of lightly reinforced wall piers and spandrels. ACI Struct J 106(4):455–465

Orakcal K, Wallace JW (2006) Flexural modeling of reinforced concrete walls – model calibration. ACI Struct J 103(2):196–206

Perform V4 (2006) Computer and Structures Inc., Perform 3-D, Nonlinear analysis and performance assessment for 3D structures, Version 4, Aug 2006

Salas MC (2008) Modeling of tall reinforced concrete wall buildings. MSCE thesis, Department of Civil and Environmental Engineering, University of California, Los Angeles, CA, May 2008, 84 pp

Stewart JP (2007) Input motions for buildings with embedment. Proceedings, Los Angeles Tall Buildings Structural Design Council, Annual Meeting, May 2007

Thomsen JH IV, Wallace JW (1995) Displacement-based design of RC structural walls: experimental studies of walls with rectangular and T-shaped cross sections. Report CU/CEE-95/06, Department of Civil and Environmental Engineering, Clarkson University, Potsdam, NY

Thomsen JH IV, Wallace JW (2004) Displacement-based design of slender rc structural walls – experimental verification. J Struct Eng, ASCE 130(4):618–630

Wallace JW (1998) Behavior and design of high-strength RC walls. ACI Struct J, SP-176, American Concrete Institute, Farmington Hills, MI, pp 259–279
Wallace JW (2007) Modeling issues for tall reinforced concrete core wall buildings. The structural design of tall and special buildings, vol 16. Wiley, New York, pp 615–632
Wood SL (1990) Shear strength of low-rise reinforced concrete walls. ACI Struct J 87(1):99–107

Part IV
Earthquake Resistant Engineering Structures

Chapter 13
Open Issues in the Seismic Design and Assessment of Bridges

Paolo E. Pinto and Paolo Franchin

Abstract The chapter presents an overview of recent research on the seismic assessment/retrofit and design of bridges, focussing on some of the aspects which are still not adequately covered in the codes. These are: The level of protection to be provided when upgrading an existing bridge, and in particular whether this should be differentiated between new designs and retrofit of existing bridges; The appropriate methods of analysis and modelling, with emphasis on the scope of nonlinear static methods and to the problems related to the selection of the input for dynamic analysis; Soil-foundation-structure interaction and non uniform support input, representing two controversial issues that may be mature for an inclusion in routine bridge analysis.

13.1 Introduction

The first realization of the seismic vulnerability of bridges coincides with the destructive events that struck California and Japan in the early 1970s of the last century. Vulnerability of both older and quite recent bridges (some severely affected bridges in the San Fernando event where just built) was exposed. The events spurred both emergency interventions to increase the protection of existing bridges, and a wave of research on seismic design and assessment of bridges. The efforts on the US side resulted in a series of progressively more encompassing documents: ATC (1983), FHWA (1995) and FHWA-MCEER (2006). The parallel evolution in Japan is summarized in Unjoh et al. (2000).

A possible reason for the delayed awakening of Europe to the problem is the absence of events with similar effects on bridges during the same period. Starting from the 1990s, however, a considerable amount of activity on the subject has been carried out, e.g. with the PREC8 project (Calvi and Pinto, 1996), which is reflected

---

P.E. Pinto (✉)
Department of Structural and Geotechnical Engineering, University of Roma "La Sapienza", 00197 Rome, Italy
e-mail: paolo.pinto@uniroma1.it

M. Garevski, A. Ansal (eds.), *Earthquake Engineering in Europe*, Geotechnical, Geological, and Earthquake Engineering 17, DOI 10.1007/978-90-481-9544-2_13,

both in the code for new bridges (Eurocode 8 Part 2, CEN 2005), as well as in national guidance documents on assessment and retrofit in Italy (Pinto et al., 2009) and Greece (TEE, 2007). The continuing interest and the state of advancement of European research on this topic is also demonstrated by the work carried out within the International Federation of Structural Concrete (*fib*) with Task Group 7.4 (*fib*, 2007) and the European Association of Earthquake Engineering (EAEE) with Task Group 11 (EAEE, 2010). Beside, most of the highway bridge stock in Europe was built in structural concrete after WWII, and represents now an ageing infrastructure. The interest in seismic upgrade is thus far from theoretical since many major arteries, summing up to several hundreds of kilometres, are currently undergoing functional upgrade to increase their traffic capacity. Seismic assessment and retrofit is carried out concurrently.

As a result of the described evolution, current efforts on the assessment and design of bridges are left to focus on more advanced issues, which initially were regarded as of secondary importance. This chapter concentrates on some of these issues, which are still awaiting a satisfactory solution from ongoing research:

- The level of protection to be provided when upgrading an existing bridge. Should it be allowed to be lower than that for new bridges, as often advocated for existing buildings? What are effective structural limit-state definitions that correlate well with functional performance measures, such as residual traffic capacity?
- Methods of analysis and modelling issues. The current trend is towards almost exclusive use of nonlinear methods. What is the lower level of sophistication considered reliable/effective? What is the scope for nonlinear static procedures? What is the current capability of modelling deformation and strength capacities for typical bridge elements?
- Soil-foundation-structure interaction. The relevance of the phenomenon has been controversial since the very early studies, and the question is not yet settled. Is it still a real problem awaiting a general answer, or possibly recognising that the answer is case-dependent, we have already the tools to include it routinely in the analysis?
- Non uniform support input. Another question difficult to deal with. The only notion that the motion exciting a plan-extended structure such as a bridge can be identical at all supports defies common sense. Hard data, on the other hand, are likely beyond reach. This appears to be another phenomenon whose relevance can only be stated after actually including it in the analysis and on a case-by-case basis. Are there physically sound simplified approaches to its treatment?

13.2 Level of Protection

By *level of protection* it is meant the choice of an adequate performance objective, i.e. the association of a performance level with the intensity of the seismic action.

There is a growing consensus that bridges, at least those on main arteries, should retain their full traffic capacity even after major events. This is justified, on one hand, by the critical importance of the transportation infrastructure in facing an emergency, and, on the other, by the fact that a high level of protection comes, for a new bridge, at a very low extra cost. This argument plays in favour of design choices which make only reduced, and well controlled, recourse to structural ductility, a goal that the global force-reduction factor q cannot properly achieve. On the other hand, Eurocode 8 explicitly provides for an alternative design strategy called *limited ductile behaviour*, which exempts from both the onerous capacity design provisions and the detailing for ductility. Further, the widespread adoption of seismic isolation greatly facilitates the achievement of such performances.

Existing bridges require a more articulate discussion. Starting from the point of view that looks at a bridge as part of an infrastructure, at parity of strategic importance there is no apparent logic in advocating lower performance requirements for existing bridges.

In most cases, however, these bridges are far from possessing the strength required for a substantially elastic response, and often they lack also the ductility sufficient to compensate. Even in the case that available ductility is enough to avoid collapse, the corresponding level of damage is often not compatible with continued traffic capacity. The latter question touches an aspect where research is still open. Actually, attempts to relate residual traffic capacity to damage states expressed in terms of structural responses can be found in the literature (Mackie and Stojadinovic, 2005) but they are admittedly still far from a definitive stage. The solution is to limit considerably the ductility demand. This can be done either be increasing the strength, a strategy that may easily turn out to be economically unfeasible for its effect on the foundations, or by seismic isolation. The latter is the solution of choice when the substructures are kept and the deck is replaced with a new one.

13.3 Methods of Analysis and Modelling

The analysis of an existing bridge for the purpose of checking its seismic capacity is a task requiring in general more accurate analytical tools and engineering expertise than for the design of new ones. From such an analysis one expects unequivocal indications on whether a given bridge is actually in need of retrofit or not and, if the answer is positive, what and where are the deficiencies to be remedied and what is the appropriate extent of the intervention. The economic relevance of a realistic diagnosis calls for the use of "adequately sophisticated" analytical tools (Priestley et al., 1996). This statement nowadays translates into the requirement of a generalized use of nonlinear methods of analysis, as implicitly recognised in recent documents such as the comprehensive one being prepared within TG11 of EAEE (2010).

Nonlinear analyses can be either static or dynamic. Though they share with dynamic ones the burden of nonlinear modelling, static methods have the appeal of avoiding a description of the hysteretic behaviour and the often debated issue of records selection. On the other hand, as it is discussed in the following, they readily reach their limits as soon as the geometry of the bridge is not rather simple, when several mode contribute to the dynamic response, the supports are founded on different soils and/or the interaction with the soil-foundation system is significant. Hence, notwithstanding their apparent appeal, their scope might be narrower than for buildings.

13.3.1 Nonlinear Static Methods

All such methods belong to two categories: single-mode methods and multi-mode methods. The single-mode method is reliably applicable when a fundamental mode governs the response of the bridge (with a participating mass larger than, say, 80%). It consists of the application of an invariant pattern of forces in general proportional to the fundamental mode (or its approximation) to derive the so-called capacity-curve, which is idealized to obtain an equivalent nonlinear SDOF. The variants differ in the second step, where the response of this SDOF is determined (demand). The version included in the Eurocode, with reference to buildings, is the N2 method (Fajfar, 2000), where an inelastic displacement spectrum is employed. The application of this method to bridges requires slight modifications that are now well-known (EAEE, 2010).

When the bridge length increases the response of different portions of the bridge are contributed by different mode shapes. Multi-modal proposals have been developed in large number, both with invariant and adaptive force patterns. Comparative studies have already been undertaken and have shown that differences are not significant and there is no single method whose accuracy is systematically higher, the choice of a method over another being largely a matter of taste (Pinho et al., 2007). The current capability of one of the most recent proposals (Paraskeva and Kappos, 2009), which includes a refinement over the so-called Modal Pushover Analysis (MPA) by Goel and Chopra (2002), is demonstrated in Figs. 13.1 and 13.2 (from EAEE, 2010).

Figure 13.1 shows the numerical model of a continuous 12-span curve concrete bridge on the recently completed Egnatia highway in Greece. Piers are rectangular hollow-core RC members, and their heights varies between 11 and 27 m. The deck, fixed over piers P4–P8 and free over the remaining lateral ones, is a prestressed concrete box-girder.

Figure 13.2 shows the comparison of the lateral deck displacement as obtained from conventional single-mode static analysis (SPA), with both uniform and modal force patterns, from the multi-modal method (MPA) and from time-history analysis (RHA). The figure shows the comparison for two intensity levels. The inadequacy of the single-mode methods is apparent, especially for the higher intensity. The MPA method, however, does not exhibit a constant approximation over all piers.

13.3.2 Nonlinear Dynamic Method

In spite of the routine statements on its being the ultimate reference method for accurate response determination, nonlinear time-history analysis (NLTHA) appears

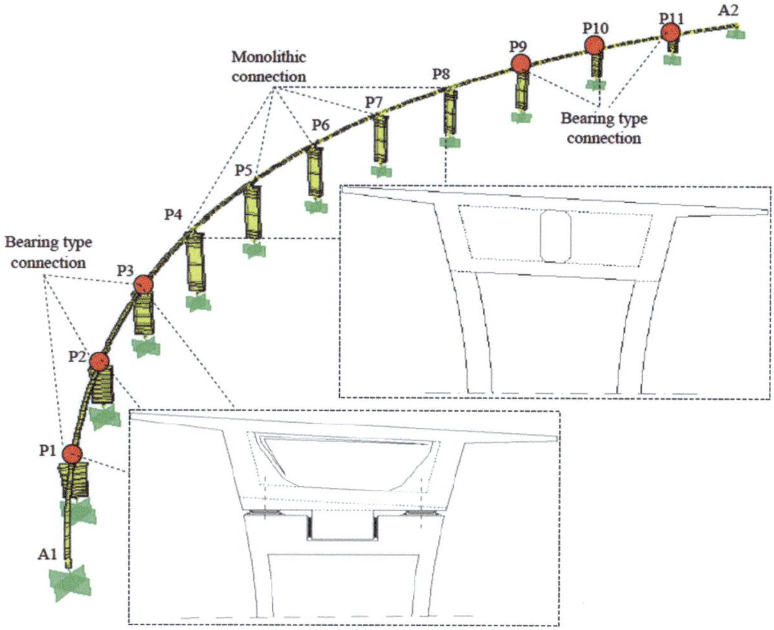

Fig. 13.1 Krystallopigi bridge model

to be still perceived as a too complicated and inaccessible tool for practicing engineers. This is compounded by the treatment of the topic to be found in codes and guidelines, where the method is usually allocated not more than about one page.

On the other hand, ongoing earthquake engineering research is heavily relying on nonlinear dynamic analysis, and to its systematic use are due the most important advancements in the implementation of the performance-based philosophy. Further, commercial codes have experienced considerable progress and are now sufficiently reliable for nonlinear response history analysis of bridge structures.

It is important that the present dichotomy be removed by providing adequate education and more detailed guidance in normative documents, to match the available tools.

Taking as a reference Eurocode 8 Part 2 on CEN (2005), the major part of the (few) indications regarding NLTHA, apart generic requirements concerning the modelling capabilities, is devoted to specifying the minimum number of signals to be used as well as the conditions for spectrum-compatibility.

EC8-2 in Section 3.2.3 puts forward two principles (marked P) for the selection of input signals for NLTHA.

The first one states that natural recordings should be preferably used, in minimum number of three pairs of horizontal components, and that they should be selected from events with magnitude, distance and mechanism consistent with those defining the design seismic action (specified in terms of a uniform-hazard elastic response spectrum). This principle can be interpreted as suggesting that the records should be

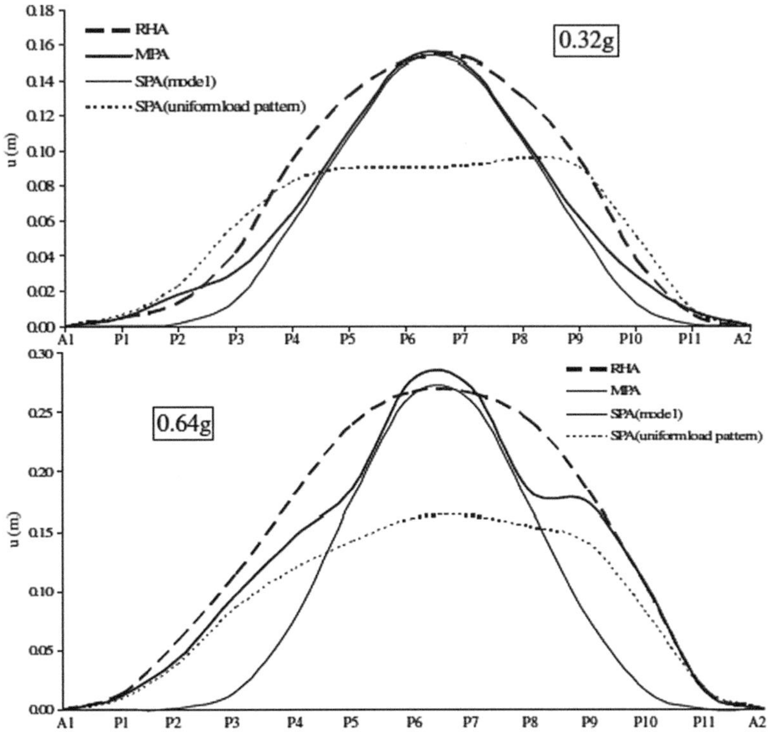

Fig. 13.2 Results of the proposed improved MPA procedure versus time-history and single-mode static procedures

selected matching the outcome of seismic hazard de-aggregation, i.e. sets of magnitude, distance and epsilon values (the modes), each set conditional on an average return period.

The second principle states that the selected records should be compatible with the design spectrum. The consistency is checked in terms of the average of the SRSS spectra of the two horizontal components, and records can be scaled in order to match the target spectrum.

In the first place it can be observed that the number of three pairs is definitely insufficient to describe the ground motion variability and to yield representative values for the design actions. Even the other common number of seven pairs, specified later as the minimum number of recordings to take the average values of the response quantities as design actions, may not be adequate, as shown by recent research (Cornell, 2005).

It can be also observed how asking for records to be selected according to a de-aggregation procedure, and to match at the same time a uniform hazard spectrum (UHS) is not entirely consistent. Recent literature has shown how the influence on the spectral shape of magnitude and distance, as compared with that of ε (i.e. the distance of an ordinate of the spectrum of one recorded motion from the target one, expressed in units of the standard deviation of the attenuation law) is relatively minor (Baker and Cornell, 2006). As a consequence records should better be selected to match the spectrum shape conditional on the dominating ε for the average return period of interest (expressions are available for this spectrum, usually called *conditional mean spectrum* and denoted by CMS-ε). The CMS-ε is usually lower than the UHS at the same return period, as illustrated in Fig. 13.3. As an intermediate, conservative step towards an improved selection procedure, some progress could already be achieved if the code were to specify intensity-dependent spectral shapes that reflect the influence of the varying M,R,ε with increasing return period. Incidentally this happens to be the case for last seismic code in Italy (MI, 2008), where local UHS spectra are specified for 11 return periods on a fine geographic grid of about 5 km side (an example is given in Fig. 13.4).

The envisaged selection procedure is actually feasible due to both the number of large data bases of recorded motions now freely accessible, and to the existence of ad-hoc software that facilitate the selection by interfacing with such data bases (Iervolino et al., 2009).

It is important to underline how all of the above considerations, however, are relevant only as long as the seismic excitation can be considered as a rigid (uniform) motion under the bridge supports, and no significant near fault effects are expected (including directivity and a strong vertical action).

Research on the latter aspect, and in particular on ways to include near-fault effects in the design seismic action, has not progressed enough to provide quantitative indications. EC8-2 states only that if an active fault is closer than 10 km from the site near-source effects should be included via site-specific spectra.

Spatial variability of the ground motion, on the other hand, is a more mature topic and EC8-2 caters for this phenomenon with an approximate procedure. The

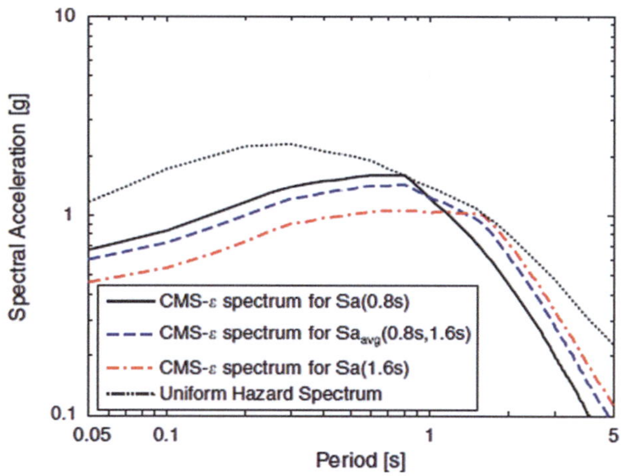

Fig. 13.3 Conditional mean spectrum for the de-aggregation ε value of S_a at two different vibration periods, together with the UHS at the same average return period of 2,500 years (from Baker and Cornell, 2006)

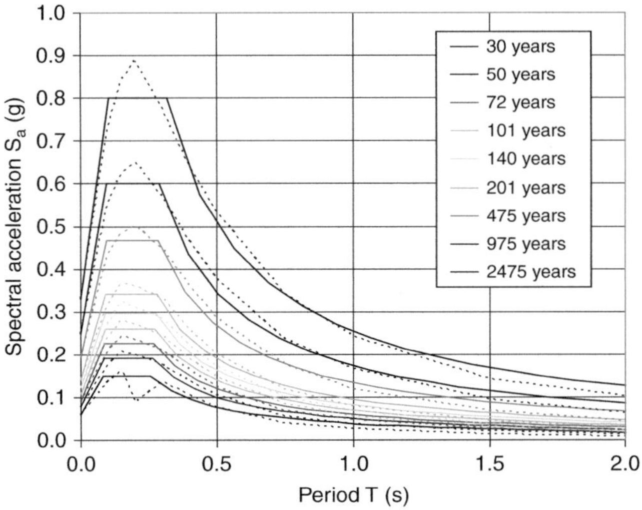

Fig. 13.4 Sample of the intensity-dependent UHS for the site of Barberino del Mugello, Italy. The *dotted lines* are the original UHS as provided by the Italian National Institute for Geophysics and Vulcanology (INGV), while the *solid lines* are the corresponding code approximation (EC8 standard shape fit to UHS)

topic is dealt with in detail in a following section. As long as NLTHA is concerned, however, it can be anticipated that there seem not to be much room at present for the use of natural recordings.

13.3.3 Modelling

With the exception of few special typologies, such as large arch and cable-supported bridges, ordinary bridges are amenable of relatively simple modelling. Seismic resistance is provided essentially by bearings, joints, piers and their foundations, while the deck, normally engaged in transverse bending, has a flexural strength more than adequate to cope in the elastic range with the demands.

In particular modelling is even simpler when the design objective is to keep response essentially in the linear range.

In the case of existing bridges, accurate modelling of defective response mechanisms becomes a necessity for fine calibration of the upgrade design. Defective mechanisms arise due to one or more of the following causes: (a) inadequate strength for fixed bearings or displacement capacity of movable ones (including loss of support), (b) inadequate shear strength or flexural strength/ductility, e.g. due to insufficient lap-splicing, (c) inadequate foundation system.

As far as bearings are concerned, fixed ones are normally modelled as constraints assuming infinite strength, either verifying a posteriori the force demand, or simply assuming that they will be replaced with strong enough new elements, while moving bearings are modelled as rollers with infinite displacement capacity, checking that the resulting displacements do not exceed the seating length. It is now possible to use concentrated hysteretic elements to model the dissipation occurring in these elements, even if this is not commonly done.

The situation is less favourable for what concerns piers and their foundations. Modelling of the nonlinear flexural response of piers can follow two alternative strategies. The first one is based on the fiber-discretisation of the cross-section, the second one on the use of so-called plastic hinges with pre-defined moment-curvature or moment-rotation laws, possibly dependent on the axial force level. Both approaches can be enhanced to model degrading mechanisms, though this is usually more easily accomplished with the latter. To model degradation within a fibre-section approach one must include bar slippage and buckling, as well as degrading concrete behaviour. This is often quite consequential on the convergence properties and speed of the algorithm. On the other hand, even plastic hinge models can pose problems. Piers can have hollow-core, multi-cell, polygonal cross-section: evaluation of their ultimate deformation capacity at present can only be undertaken employing formulas based on the yield and ultimate curvature, and on plastic hinge length (e.g. Eurocode 8 Part 2, Annex E). Both curvatures (yield and ultimate) are of subjective evaluation for sections with distributed reinforcement, and hinge length has no precise physical definition.

The situation is made more complex for shear-sensitive piers and in particular in zones where shear and bending have a strong interaction, such as at the base of squat piers. A large body of literature is available on the topic, though the more rigorous solutions are still confined to the realm of research. A recent comprehensive review can be found in Ceresa et al. (2007).

When the interaction is less significant and with practical application in mind, there are approximate ways to account for the contribution of nonlinear shear response to the overall response. One such way is to use a fiber-section model or a plastic-hinge model for the flexural behaviour and couple it (enforcing equilibrium) with a 1D hysteretic shear force-deformation law. This can be done e.g. with the *section Aggregator* feature in OpenSEES (McKenna et al., 2007). Figure 13.5

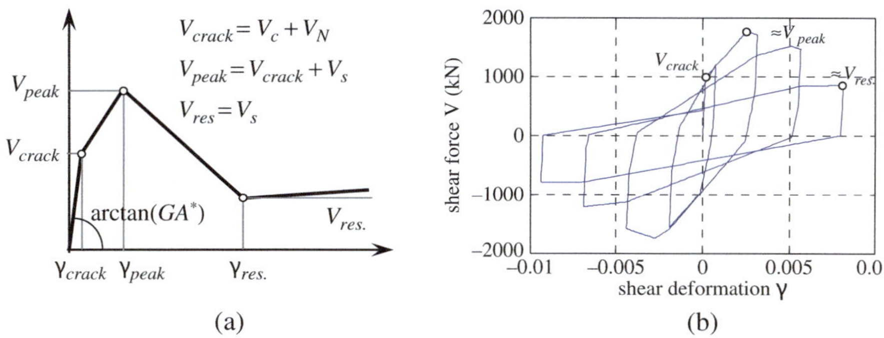

Fig. 13.5 Shear force-deformation law adopted in (Franchin and Pinto, 2009): (**a**) monotonic envelope (**b**) sample of cyclic degradation. The terms V_c, V_N and V_s represent the strength contributions of concrete, axial force and shear reinforcement, respectively

shows an example from (Franchin and Pinto, 2009). The problem in this case is that of establishing the parameters of the envelope and the hysteretic rule. In particular, the definition of the shear strength poses the same difficulty encountered in verification, i.e. that of choosing amongst the several proposals available (Kowalsky and Priestley, 2000; Biskinis et al., 2003; Sezen and Mohele, 2004; Sezen, 2008).

Coming now to the foundations, the standard performance requirement has always been that of practically linear response with negligible residual strains. This view is starting being challenged (Gerolymos, 2009; Pecker, 2006) since recent research has shown how advantageous can be even a moderate, controlled amount of inelasticity in the soil-foundation system for the response of the structure. Modeling the inelastic flexibility and energy dissipation associated with the soil-foundation system appears as a resource both for design and, most importantly, for assessment of existing bridges.

13.4 Soil-Foundation-Structure Interaction

As anticipated, the relevance of the SSI phenomenon has been controversial since the very early studies and, to some extent, it still is at the present time. In recognising that its importance is case-dependent, the main question becomes whether the tools are now available for its inclusion as a regular feature in bridge analysis.

Several proposals are available in the literature to evaluate the response of soil-foundation-structure systems. They can be lumped in two main classes:

- The *global* approaches, where the model encompasses soil, foundation and structure, and that can be either full three-dimensional as, e.g., in Elgamal et al. (2008), or mixed 2D and 1D as in Klar (2003), or finally 1D, with a Winkler-type modeling, as in El Naggar and Novak (1996).
- The *sub-structuring* approach (Dobry and Gazetas, 1988; Makris and Gazetas, 1991, 1992; Mylonakis et al., 1997), where the soil-foundation system and the structural system are analyzed separately.

It is fair to say that the state of the art in dealing with the phenomenon in rigorous ways, taking into account the whole soil-foundation-structure system with a refined constitutive model for the soil medium and the soil-foundation interface (global 3D or mixed 2D/1D models) is still not ready for practice. Discrete models where the soil and the soil-structure interface are idealized as shear beams and Winkler springs, respectively, are relatively inexpensive and can easily include inelasticity (e.g. Badoni and Makris, 1996).

By far the most widely used approach, however, is the sub-structuring one, where inelasticity can be included in terms of effective (secant) properties. The approximation involved can be considered acceptable since the amount of inelasticity is in any case limited. As it is well known the approach makes use of superposition and consists of the separate evaluation of the soil-foundation system and structural system responses. The first system is analyzed with the two-fold objective of establishing the modified input motion to the structure (due to kinematic interaction) and of determining the dynamic impedance to be used at the structure base. The second system, consisting of the structure, flexibly connected to the support, is then analyzed under the modified motion. The scheme employed for this second analysis is shown with reference to a portion of a viaduct in Fig. 13.6. Notice that the motion is applied to the structure in the form of forces at the degrees of freedom of the pier base nodes (Luco, 1982; Mylonakis, 1995). The force to be applied at the foundation joint of the i-th pier, in each horizontal direction, is given by:

$$f_i = m_i^* \ddot{u}_i + c_i \dot{u}_i + k_i u_i$$

where u_i is the input ground motion (a component in each horizontal direction), as modified by the presence of the foundation, and k_i, c_i and m_i^* are the components of the dynamic impedance.

The modified input motion in each direction, at each support, as well as the dynamic impedance to be applied at each pier base can be obtained as the solution of two algebraic problems as shown in (Makris and Gazetas, 1991, 1992). This solution, however, can be directly employed for frequency domain analysis only. For the time-domain, direct integration analysis the dynamic impedance needs to be complemented by fictitious mass terms to approximately reproduce the frequency-dependence of the dynamic stiffness in the period-range of interest (Wolf, 1991).

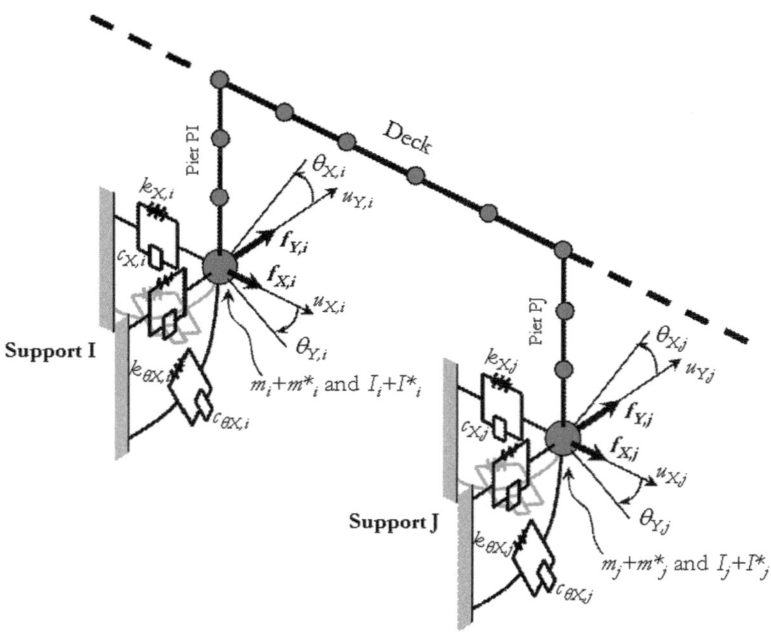

Fig. 13.6 Sub-structuring approach applied to a generic viaduct. Components of the impedance matrix at the piers bases

The kinematic modification of the input motion is usually important only for large embedded foundation shapes, such as e.g. caissons, where the main effect is that of averaging (reducing) the translational components and introducing rotational components. For piled foundations, apart from the case of very large stiffness contrast between soil and pile group, usually the effect is much reduced and limited to the highest frequency range. For shallow foundations of ordinary dimensions the alteration is negligible.

For the purpose of practical application a number of simplified expressions for the impedance terms are available. For caissons foundations the approach in (Gerolymos and Gazetas, 2006) can be used. For piled foundations one particularly handy, recent set of expressions is the one provided in Taherzadeh et al. (2009).

Without any pretense of drawing conclusions of general applicability on the phenomenon, but with the only goal of showing a practical application to an actual design case, in the following some details are given from an example in (Pinto and Franchin, 2010).

The example refers to the functional and seismic upgrade of a more than 2 km long viaduct which, in its present configuration, consists of a series of simply supported spans of equal length of about 34 m. In the designed intervention the existing piers are kept, in some cases reinforced with concrete jackets, and the decks are replaced with a continuous composite dual-girder one, which is subdivided into four portions by three expansion joints. The existing foundations, resting on an alluvial sediment of coarse grained, loose soil, are on a variable number of large diameter piles (1.2 m diameter), ranging from six to eight, depending on pier height, with length between 22 and 25 m. The intervention design includes also an important upgrade of the seismic capacity of the piled foundations for all piers with the addition of a large number of 25 m long micro-piles and the casting of a topping on the foundation mat to achieve a better connection.

The analyses described in the following refer to the first portion, up to pier P7, for a total length of 274 m. Figure 13.7 shows the present pier-deck configuration and the designed intervention. In the analyzed portion the viaduct crosses a riverbed and the intervention includes the demolition of some of the existing piers to increase the span lengths and minimize scouring risk by moving supports outside the riverbed (the span sequence is 22+34+42+60+60+42+34).

The seismic design of the intervention relies on piers ductility. The deck is supported at each pier on unidirectional longitudinal flat sliding bearings, which feature a shock-transmitter. Under the seismic action all piers are engaged in both longitudinal and transverse direction. The deck is connected with an elastic restrainer at the new abutments. Taller piers in the third and fourth portions of the viaduct, are strengthened with an RC jacket to increase their total diameter to 3.5 m. This is also the diameter of the new piers in the riverbed.

The properties of the soil along the bridge axis cannot be attributed to a single soil category. The looser soil in the riverbed portion, with about 150 m/s average shear wave velocity, is a category D soil, while the remaining supports stand on slightly better C soil (about 300 m/s).

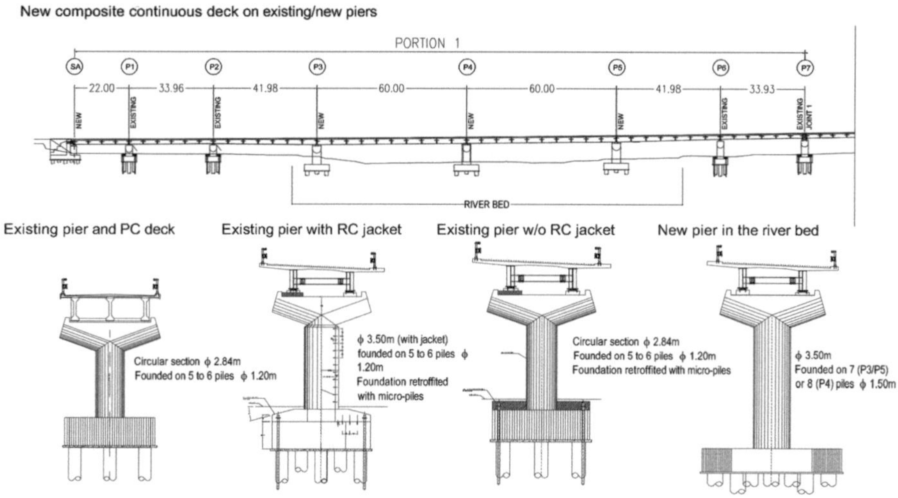

Fig. 13.7 The "Salso" viaduct in South-Western Sicily: longitudinal profile of the upgraded viaduct, analyzed portion (*top*); existing pier and simply supported prestressed concrete deck, retrofitted and new piers with the new continuous steel-concrete deck (*bottom*)

The viaduct is analyzed by means of nonlinear time history analysis (with a plain fiber-model within OpenSEES), employing artificial records compatible with the soil D spectrum, in accordance with EC8-2 prescription of using the spectrum corresponding to the worst soil category when carrying out the analysis without considering the variability of the soil profile along the bridge. According to design practice, seven two-components records have been generated, and for verification purposes the demand is assumed equal to the average of the seven maxima. The considered response quantities are the piers shear forces and chord rotations in two orthogonal vertical planes. The D/C ratios in the two planes are then combined to yield a single index according to (piers have circular cross-section):

$$\rho_\theta = \max_t \sqrt{\left(\theta_T(t)/\theta_{uT}(t)\right)^2 + \left(\theta_L(t)/\theta_{uL}(t)\right)^2} \cong \frac{\max_t \sqrt{(\theta_T(t))^2 + (\theta_L(t))^2}}{\theta_u} \leq 1 \tag{1}$$

$$\rho_V = \max_t \sqrt{\left(V_T(t)/V_{uT}(t)\right)^2 + \left(V_L(t)/V_{uL}(t)\right)^2} \cong \frac{\max_t \sqrt{(V_T(t))^2 + (V_L(t))^2}}{V_u} \leq 1 \tag{2}$$

Figure 13.8 shows the main mode shapes of the fixed-base model of the bridge.

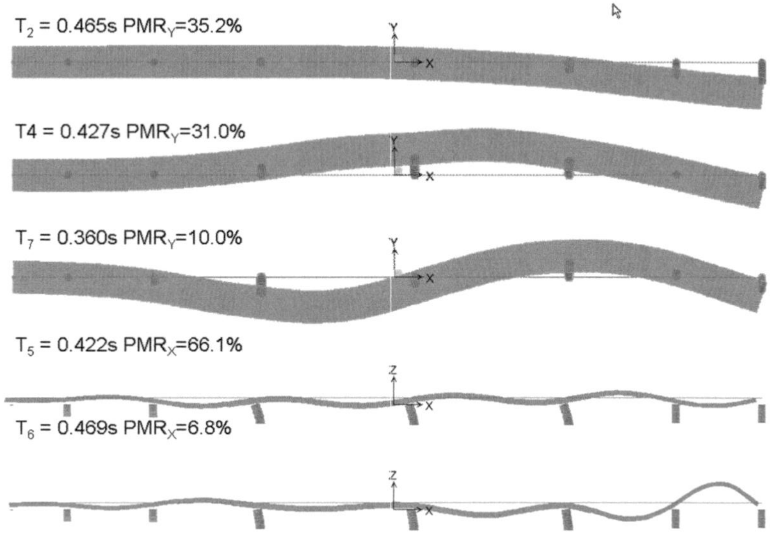

Fig. 13.8 Mode shapes and periods for the main modes of the fixed-base model of the bridge

Table 13.1 Impedance terms at each support

		Abutment	P1, P2, P6 and P7	P3	P4 and P5
V_{s30}	m/s	300	300	300	150
k_h	kN/m	6.30E+06	3.50E+06	4.90E+06	1.36E+06
c_h	kNs/m	9.28E+04	2.91E+04	5.77E+04	3.43E+04
m^*	kNs^2/m	81.11	12.23	31.17	54.28
k_θ	kNm/rad	6.21E+08	1.61E+08	3.72E+08	1.07E+08
c_θ	kNms/rad	9.14E+06	1.36E+06	4.84E+06	2.59E+06
I^*	kNm^3	43,686	4,810	20,101	23,089

Table 13.1 reports the impedance terms employed at each support, determined according to the formulas in Taherzadeh et al. (2009). The resulting variation of the modal properties are shown in Table 13.2, where the periods of corresponding modes of the fixed-base and compliant-base models are compared: the evaluated impedances lead to an increase of the vibration periods in the order of 20–30% (the effect is minor on the mode shapes, not shown).

A sample of the results is illustrated in Fig. 13.9 which shows the pier top displacements histories in the longitudinal and transversal directions, for both the fixed-base model and the compliant-base one. It is noted that period lengthening is more than compensated by the increased amount of global damping, contributed by the radiated energy. Quantitative figures for all piers averaged over all signals are reported later, in Table 13.3, together with corresponding variations due to non-uniform excitations (see next section). The variations with respect to the reference fixed-base/uniform excitation case greatly differ from pier to pier and for response

Table 13.2 Comparison of vibration periods for the main modes of the fixed-base and the compliant-base models

Dir.	Mode	Tfixed (s)	Mode	Tcompliant (s)
T	3	0.405	2	0.485
T	5	0.376	5	0.420
T	8	0.318	7	0.350
L	4	0.399	3	0.480
L	6	0.353	4	0.450
L	7	0.327	6	0.365

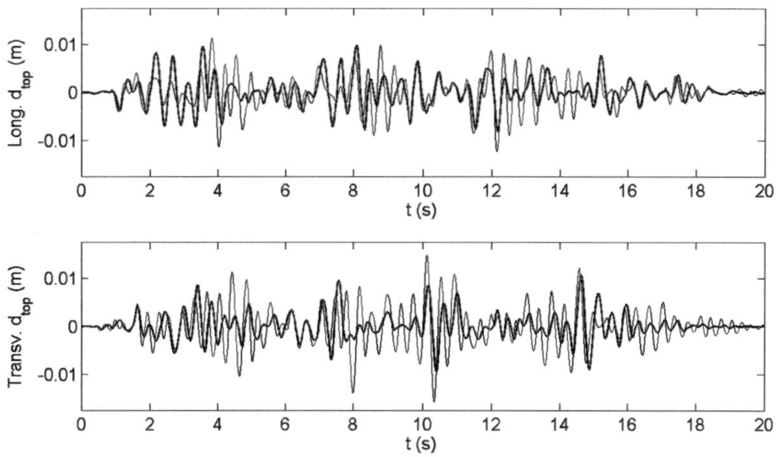

Fig. 13.9 Analysis with SSI: pier P4 top displacement relative to the base for the uniform-excitation/fixed (*grey*) and the uniform-excitation/compliant (*black*) cases, in the longitudinal (*top*) and transversal (*bottom*) direction

quantities. They do not exceed 25%, with the exception of pier P7 (50%), which is the tallest one and is the location of the deck joint, with the corresponding lower restraint by the deck.

13.5 Non Uniform Support Input

The reason why knowledgeable bridge engineers have systematically used identical excitations at the base of the piers, independently of their distance and of the length of the bridge, can be explained, apart from the simplification of the analysis, by the notion that by so doing the dynamic response is conservatively estimated. This approach, however, neglects the physical facts that the ground motion has a propagatory character and that the propagation occurs in a non-homogeneous medium with properties that can never be fully modelled as deterministic. Both of these aspects have led a number of investigators to propose (stochastic) models of the ground motion field over an extent of space, whereby the difference in motion between two points is described in terms of decrease of correlation between the field values at the two points. One such model, quite well known, is that in (Der Kiureghian, 1996). The same author had gone further in proposing also an extension of the classical response-spectrum method to be used for the case of non uniform excitation, which employed multiple displacement spectra (Der Kiureghian and Neuenhofer, 1992). This, however, is inherently limited to linear structures.

A number of numerical researches (e.g. Sextos et al., 2003a, b; Lupoi et al., 2005) have explored the sensitivity of several bridge configurations to a wide range of parameters regulating the ground motion field model. Two rather firm conclusions emerged from these studies:

- in cases where the soil properties are approximately uniform, it cannot be anticipated whether the effect of a differential input at the supports will be beneficial or detrimental, but in any case it will generally be not substantial. This result must also be seen in light of the fundamental uncertainty affecting the parameters of the coherency models.
- the phenomenon becomes relevant when soil conditions under the supports cannot be considered approximately uniform. Indeed, according to EC8-2 consideration of spatial variability of the ground motion is mandatory whenever soil conditions at supports cannot be attributed to the same soil category. The approximate method proposed by EC8-2 to deal with the problem, however, appears neither to match the results of more accurate methods, nor to be conservative.

The above considerations lead to a tentative formulation of a relatively straightforward method to deal with the spatial variability phenomenon. The proposal represents a considerable simplification with respect to the rigorous method for nonlinear time-history analysis of structures subjected to differential input-motions. This latter requires the generation of samples of correlated motions at the supports from one of the cited stochastic models. This operation can only be carried out with an ad-hoc specialistic software and requires as input the values of highly uncertain parameters.

The simplified approach consists instead in performing non-linear time-history analysis with input consisting in different, independently generated ground motion samples, each one compatible with the local support soil conditions. The basis for this proposal is the relatively minor influence of the cross-correlation terms on the response, when soil conditions differs under the piers, as shown for instance in (Monti and Pinto, 1998). This approach would only require the standard tools for artificial spectrum-compatible signal generation currently used in practice.

It is of a certain interest to observe that recorded ground motions cannot presently be used to solve the problem at hand.

Incidentally, one more comment on this issue. As it is well known, soil categories almost ubiquitously adopted by international codes, with modest variations, are the result of drastic lumping of results obtained for a variety of soil strata, and that the attribution of a site to a category is based on the much debated global parameter V_{s30}, which is simply an average and just over 30 m, a depth that may not always be sufficient and representative to characterise the site. This said, it would seem not irrational to propose that independently generated motions be used even for the case of an homogeneous soil category at the supports.

One last remark is needed about the procedure to be employed for the generation of the input histories. The response of the bridge to differential inputs can be shown to be the sum of a dynamic portion, and a pseudo-static one. The latter is the response to the slowly varying differential displacements at the supports. These relative displacements are proportional to the absolute ground displacement, hence the importance that the generated histories describe accurately the *actual* displacement spectra at large periods (up to ~10 s). Recent results (Faccioli and Villani, 2009) based on high-quality digital records showing how these long-period displacements may be larger than currently specified in the EC8-1 (informative Annex A) become particularly relevant for bridge design/assessment.

A case in point to test the significance of the differential input for the response of a bridge is given by the same viaduct examined in the previous section. As it was already mentioned, the viaduct crosses a river-bed and the soil stiffness on the two sides is larger (about 300 m/s vs 150 m/s of average shear wave velocity).

According to the idea put forward above, the bridge is subjected to NLTHA with independently generated ground motion samples (three components at each support) compatible with spectra for soils C (abutment and piers P1, P2, P3, P6 and P7) and D (P4 and P5). Seven suites of ground motions have been employed.

Figure 13.10 shows the response for one of the analyses, with reference to the displacement components on top of pier P4, which is at the transition between the two soil categories, and thus is expected to experience larger effects of the differential input. The time-histories show that the response is considerably larger

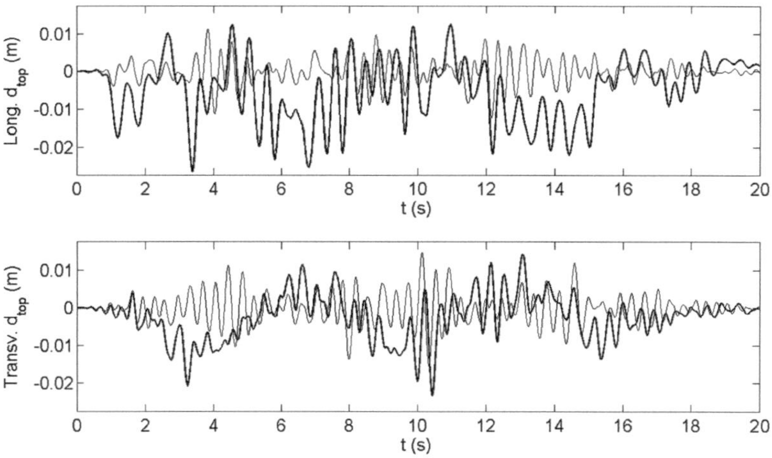

Fig. 13.10 Analysis under differential support motion: pier P4 top displacement relative to the base for the uniform-excitation/fixed-base (*grey*) and the differential-excitation/fixed-base (*black*) cases, in the longitudinal (*top*) and transversal (*bottom*) direction

Table 13.3 Comparison of vibration periods for the main modes of the fixed-base and the compliant-base models

	$\Delta\rho\theta$		$\Delta\rho v$	
	MSE (%)	SSI (%)	MSE (%)	SSI (%)
P1	288	−10	53	−4
P2	415	−6	43	−2
P3	1, 167	−17	45	−16
P4	93	−1	35	−5
P5	47	−11	20	−7
P6	313	−25	2	−18
P7	−11	−51	−5	−34

than in the reference case. Also, it appears clearly how the dynamic response is superimposed to a longer period pseudo-static response to the slowly-varying differential displacements.

Quantitatively, the overall influence of the phenomenon is summarized in Table 13.3, in terms of average (over the seven motion suites) percent difference of chord-rotation and shear-force D/C ratios with respect to the uniform-excitation/fixed-base reference case (which, as already mentioned, is analyzed with uniform soil D). But for pier P7, the variation is always detrimental and may reach one order of magnitude.

Of course the above results refer to a single case and they cannot have any pretence of generality. They fit well, however, in the more general picture that comes out of a large number of numerical explorations carried out by the authors as well as other researchers. The presented case can be considered on the severe side, the results would have been less unfavourable were the same difference between soil categories related to stiffer soils (e.g. A vs B).

As a final comment the phenomenon should be carefully considered for the design of isolated bridges. The demand increments highlighted by the previous example would directly transfer to a larger displacement capacity requirement for the isolation devices.

References

ATC, Applied Technology Council (1983) Seismic retrofitting manual for highway bridges, Report ATC6-2

Badoni D, Makris N (1996) Nonlinear response of single piles under lateral inertial and seismic loads. Soil Dyn Earthquake Eng 15:29–43

Baker J, Cornell CA (2006) Spectral shape, epsilon and record selection. Earthquake Eng Struct Dyn 35:1077–1095

Biskinis D, Roupakias G, Fardis MN (2003) Cyclic deformation capacity of shear-critical RC elements. In: Proceedings of the fib 2003 symposium: concrete structures in seismic regions, Athens, Greece

Calvi GM, Pinto PE (1996) Experimental and numerical investigation on the seismic response of bridges and recommendations for code provisions. ECOEST-PREC8 Report No 4, LNEC, Lisbon

CEN (2005) European committee for standardization: Eurocode 8 design of structures for earthquake resistance Part 2. Bridges, Brussels, Belgium

Ceresa P, Petrini L, Pinho R (2007) Flexure-shear fiber beam-column elements for modeling frame structures under seismic loading – state of the art. J Earthquake Eng 11:46–88

Cornell CA (2005) On earthquake record selection for nonlinear dynamic analysis. In: Proceedings of the Luis Esteva symposium, Mexico City, Mexico

Der Kiureghian A (1996) A coherency model for spatially varying ground motions earthquake. Earthquake Eng Struct Dyn 25:99–111

Der Kiureghian A, Neuenhofer A (1992) Response spectrum method for multi-support seismic excitations. Earthquake Eng Struct Dyn 21:713–740

Dobry R, Gazetas G (1988) Simple method for dynamic stiffness and damping of floating pile groups. Géotechnique 38:557–574

EAEE, European Association of Earthquake Engineering (2010) Task Group 11: inelastic methods for seismic design and assessment of bridges (Draft)

El Naggar MH, Novak M (1996) Non linear analysis for dynamic lateral pile response. Soil Dyn Earthquake Eng 15:233–244

Elgamal A, Yan L, Yang Z, Conte JP (2008) Three-dimensional seismic response of humboldt bay Bridge-Foundation-Ground System. J Struct Eng 134:1165–1176

Faccioli E, Villani M (2009) Seismic hazard mapping for Italy in terms of broadband displacement response spectra. Earthquake Spectra 25:515

Fajfar P (2000) A nonlinear analysis method for performance-based seismic design. Earthquake Spectra 16:573–592

FHWA, Federal Highway Administration (1995) Seismic retrofitting manual for highway bridges, Publ. No. FHWA-RD-94 052

FHWA-MCEER, Federal Highway Administration and Multi-disciplinary Center for Earthquake Engineering Research (2006) Seismic retrofitting manual for highway structures. Part 1 – Bridges, FHWA-HRT-06-032

fib, International Federation of Structural Concrete (2007) Seismic bridge design and retrofit – structural solutions, Bulletin 39

Franchin P, Pinto PE (2009) Allowing traffic over mainshock-damaged bridges. J Earthquake Eng 13:585–599

Gerolymos N (2009) Seismic soil-structure interaction: New Approaches in Performance Based design of Foundations, Earthquake Engineering by the Beach Workshop, July 2–4, 2009, Capri, Italy

Gerolymos N, Gazetas G (2006) Development of a Winkler model for static and dynamic response of caisson foundations with soil and interface nonlinearities. Soil Dyn Earthquake Eng 26:363–376

Goel RK, Chopra AK (2002) A modal pushover analysis procedure for estimating seismic demands for buildings. Earthquake Eng Struct Dyn 31:561–52

Iervolino I, Galasso C, Cosenza E (2009) REXEL: computer aided record selection for code-based seismic structural analysis. Bull Earthquake Eng. 8:339–362. doi 10.1007/s10518-009-9146-1

Klar A (2003) Model studies of seismic behaviour of piles in sands. PhD thesis, Technion, Israel Institute of Technology, Haifa, Israel

Kowalsky M, Priestley MJN (2000) Improved analytical model for shear strength of circular reinforced concrete columns in seismic regions. ACI Struct J 97:388–396

Luco JE (1982) Linear soil-structure interaction: a review. Earthquake ground motion effects struct. ASME, AMD 53:41–57

Lupoi A, Franchin P, Monti G, Pinto PE (2005) Seismic design of bridges accounting for spatial variability of ground motion. Earthquake Eng Struct Dyn 34:327–348

Mackie KR, Stojadinovic B (2005) Fragility basis for California highway overpass bridge seismic decision making. Technical Report 2005-02, Pacific Earthquake Engineering Research Center, University of California, Berkeley, CA

Makris N, Gazetas G (1991) Dynamic pile-soil-pile interaction. Part I: Analysis of axial vibration. Earthquake Eng Struct Dyn 20:115–132

Makris N, Gazetas G (1992) Dynamic pile-soil-pile interaction. Part II: Lateral and seismic response. Earthquake Eng Struct Dyn 21:145–162

McKenna F, Fenves GL, Scott MH (2007) OpenSees: open system for earthquake engineering simulation. http://opensees.berkeley.edu , Pacific Earthquake Engineering Research Center, University of California, Berkeley, CA

MI, Ministero delle Infrastrutture (2008) Decreto Ministeriale del 14/1/2008 recante Nuove norme tecniche per le costruzioni (Decree 14/1/2008, New technical norms for constructions)

Monti G, Pinto PE (1998) Effects of multi-support excitation on isolated bridges. In: Proceedings of the US–Italy workshop on seismic protective systems for bridges, Technical Report MCEER-98-0015, 225–247

Mylonakis G (1995) Contributions to static and seismic analysis of piles and pile-supported bridge piers. Ph.D. dissertation, State University of New York, Buffalo, NY

Mylonakis G, Nikolaou A, Gazetas G (1997) Soil-Pile-Bridge seismic interaction: kinematic and inertial effects, Part I: Soft soil. Earthquake Eng Struct Dyn 26:337–359

Paraskeva Th, Kappos AJ (2009) Further development of a multimodal pushover analysis procedure for seismic assessment of bridges. Earthquake Eng Struct Dyn 39:211–222

Pecker A (2006) Enhanced seismic design of shallow foundations: example of the rion-antirion bridge. In: Proceedings of the 4th Athenian lecture on geotechnical engineering, Athens

Pinho R, Casarotti C, Monteiro R (2007) An adaptive capacity spectrum method and other nonlinear static procedures applied to the seismic assessment of bridges. In: Proceedings of the 1st US–Italy workshop on seismic design and assessment of bridges, Pavia, Italy

Pinto PE, Franchin P (2010) Issues in the upgrade of Italian highway structures. J Earthquake Eng 14

Pinto PE, Franchin P, Lupoi A (2009) Seismic assessment and retrofit of existing bridges (in Italian). IUSS Press, Pavia

Priestley MJN, Seible F, Calvi GM (1996) Seismic design and retrofit of bridges. Wiley, New York, NY

Sextos AG, Kappos AJ, Pitilakis KD (2003b) Inelastic dynamic analysis of RC bridges accounting for spatial variability of ground motion, site effects and soil–structure interaction phenomena. Part 2: Parametric study. Earthquake Eng Struct Dyn 32:629–652

Sextos AG, Pitilakis KD, Kappos AJ (2003a) Inelastic dynamic analysis of RC bridges accounting for spatial variability of ground motion, site effects and soil–structure interaction phenomena. Part 1: Methodology and analytical tools. Earthquake Eng Struct Dyn 32:607–628

Sezen H (2008) Shear deformation model for reinforced concrete columns. Struct Eng Mech 28:39–52

Sezen H, Moehle JP (2004) Shear strength model for lightly reinforced concrete columns. J Struct Eng 130:1692–1703

Taherzadeh R, Clouteau D, Cottereau R (2009) Simple formulas for the dynamic stiffness of pile groups. Earthquake Eng Struct Dyn 38:1665–1685

TEE, Technical chamber of Greece (2007) National greek retrofit code (draft version)

Unjoh S, Terayama T, Adachi Y, Hoshikuma J (2000) Seismic retrofit of existing highway bridges in Japan. Cement Concr Compos 22:1–16

Wolf JP (1991) Consistent lumped-parameter models for unbonded soil: physical representation. Earthquake Eng Struct Dyn 20:12–32

Chapter 14
Recent Developments on Structural Health Monitoring and Data Analyses

Erdal Şafak, Eser Çaktı, and Yavuz Kaya

Abstract The term "Structural Health Monitoring (SHM)" refers to continuous monitoring of a structure in order to track the changes in its dynamic characteristics and detect damage. In Civil/Structural Engineering, the majority of SHM applications are directed towards studying the response and damage from natural hazards, such as earthquakes and strong winds. The monitoring typically involves measuring continuously the vibrations of the structure by acceleration sensors. Some recent applications have also included GPS sensors, which provide superior accuracy for measuring displacements. Although a significant number of structures are now installed with SHM systems, the utilization of data for practical applications are still lacking. Some of the new findings resulting from SHM include the significant influence of environment on structural frequencies and damping, strong dependency of damping on amplitude and frequency, exponential decay in modal damping values with increasing building height, and the prevalence of 3D modes and non-proportional damping. A critical need in SHM is the simple tools and techniques for real-time data analysis and interpretation. Since data come continuously, the analysis cannot be done in batch mode; it should be done in real-time. This chapter summarizes the latest developments in SHM, with emphasis on data analysis and damage detection. The topics discussed include real-time analysis techniques, noise reduction in ambient vibration data, utilization of wave propagation approach as an alternative to spectral analysis, inadequacy of modal parameters for damage detection, applications of *Seismic Interferometry* for data analysis, and identification and damage detection for historical structures.

E. Şafak (✉)
Kandilli Observatory and Earthquake Research Institute, Boğaziçi University, 38684 Istanbul, Turkey
e-mail: erdal.safak@boun.edu.tr

M. Garevski, A. Ansal (eds.), *Earthquake Engineering in Europe*, Geotechnical, Geological, and Earthquake Engineering 17, DOI 10.1007/978-90-481-9544-2_14,

14.1 Introduction

Structural Health Monitoring (SHM) involves continuous monitoring of the dynamic characteristics of a structure by digital instruments (i.e., sensors and recorders). The monitoring is typically done by recording the vibrations of the structure continuously by acceleration sensors. The main objective in SHM is to track the changes in the characteristics of the structural system in order to detect and locate damage. In addition, SHM is also used for the following objectives:

- Determine in-situ dynamic characteristics of the structure.
- Develop analytical models calibrated with recorded data.
- Check the design and analysis methods used.
- Improve structural design codes.
- Develop new retrofit and strengthening techniques.
- Predict behavior for future extreme loads.
- Develop instantaneous damage distribution and loss maps.

Since the monitoring is done continuously and in real time, the data processing and analysis should also be done in real time.

In general, extreme loads (e.g., a large earthquake) do not occur frequently. Therefore, most of the data collected by a SHM system are the vibrations of the structure caused by ambient forces, such as wind, traffic loads, and micro tremors. For most structures, ambient vibration data are sufficient to identify the dynamic properties for linear behaviour. They include modal properties (such as natural frequencies, damping ratios, and mode shapes), torsion, and soil-structure interaction.

Low-amplitude vibration data generated by ambient forces or small excitations provide a means to predict behaviour under large excitations. For example, data collected from a small earthquake can be used to predict the behaviour of the structure for a future large earthquake. This typically involves the following steps:

1. Develop a linear analytical model of the structure based on the vibration data generated by the small earthquake.
2. Estimate ground input for the large earthquake by extrapolating the recorded ground input from the small earthquake.
3. Estimate the response to the large earthquake by using the analytical model and allowing nonlinear behaviour.

The extrapolation of ground input from small to large earthquake can be done in the time domain or in the frequency domain. In the time domain, a large earthquake is assumed as a sum of small earthquakes superimposed with a time shift as schematically shown in Fig. 14.1. In the frequency domain, the relationship between the FAS (Fourier Amplitude Spectra) of small and large earthquakes can be approximated as shown in Fig. 14.2. For frequencies lower than the corner frequency of the large earthquake, the scaling is constant and is proportional to the ratio of seismic

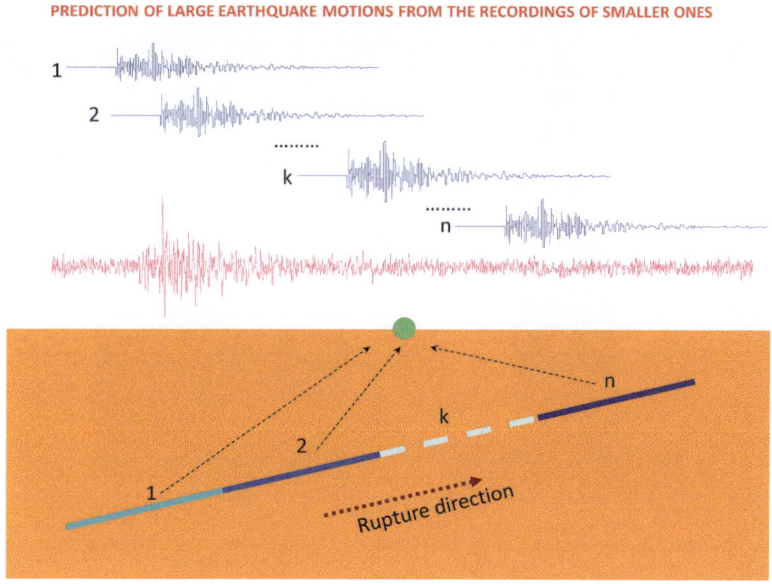

Fig. 14.1 Time-domain relationship between small and large earthquakes

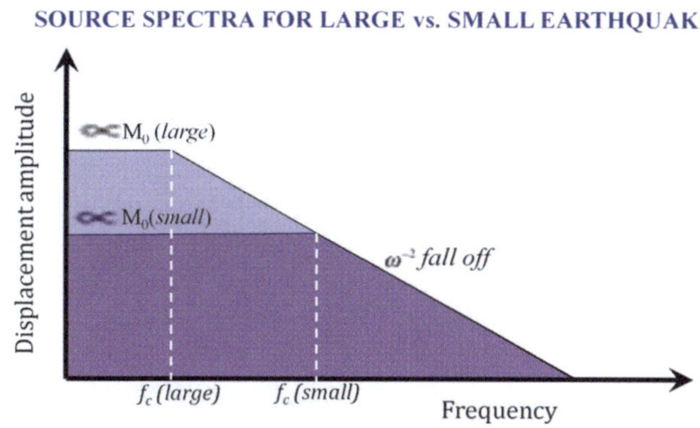

Fig. 14.2 Frequency-domain relationship between small and large earthquakes

moments. For frequencies higher than the corner frequency of the small earthquake, the scaling is one (i.e., there is no difference between the large and the small earthquakes). For frequencies in between, the scaling is linear in the log-log plot.

As the amount of data from instrumented structures are increasing, it is now possible to find sufficient number of structures that have multiple sets of data under different levels of excitations. Such data would allow studying the correlations of modal characteristics with vibration amplitudes for different structural categories. Figure 14.3 shows schematically how these correlations would look like for natural frequency and damping. Using such curves, one can easily extrapolate

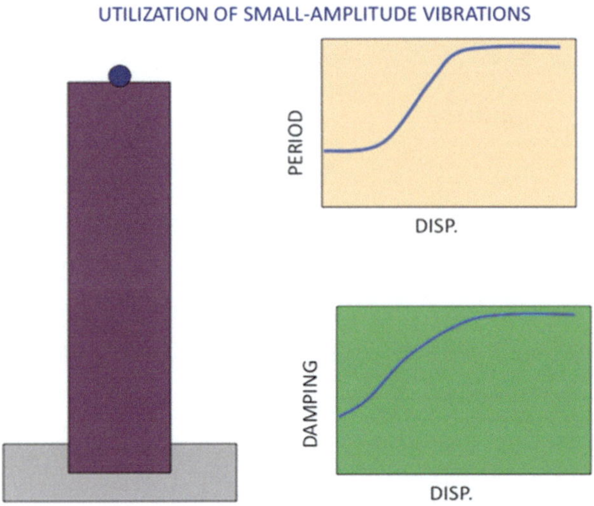

Fig. 14.3 Schematic representation of the changes in frequency and damping with vibration amplitude

modal characteristics calculated from low-amplitude motions to those expected for high-amplitude motions.

14.2 Damage Detection Based on Natural Frequencies

The natural frequency of a structure is the fundamental parameter defining its dynamic response. For earthquake loads, the natural frequency and the damping are the only parameters needed to describe the response of the structure. Therefore, it is natural to use the change in natural frequency as a damage indicator. Damage detection typically involves analyses of acceleration response data from a damaging event to see if there are any changes in the structure's natural frequencies. However, the dynamic response of a damaged structure is nonlinear and in most cases hysteretic, as schematically shown in Fig. 14.4a. The stiffness, and consequently the natural frequencies, rapidly change during the damaging vibrations and are hard to track for short-duration, transient loads such as earthquakes. Moreover, data from earthquakes have shown that even though a structure is damaged, the stiffness before and after the damage may not be that much different, as characterized by the hysteretic force-deformation curve in Fig. 14.4b.

Natural frequencies of a structure can also change due to soil-structure interaction and environmental factors, such as temperature, rain, wind, etc. without any damage. By studying a 2-year long continuous data from the Millikan Library building at Caltech, Clinton (2004) has shown that the building's natural frequency can change significantly due to environmental factors. He has found a strong correlation between the changes in the natural frequency and the rainfall, because the building has significant soil-structure interaction in the form of rocking motions. The rainfall has affected the stiffness of the soil around the foundation.

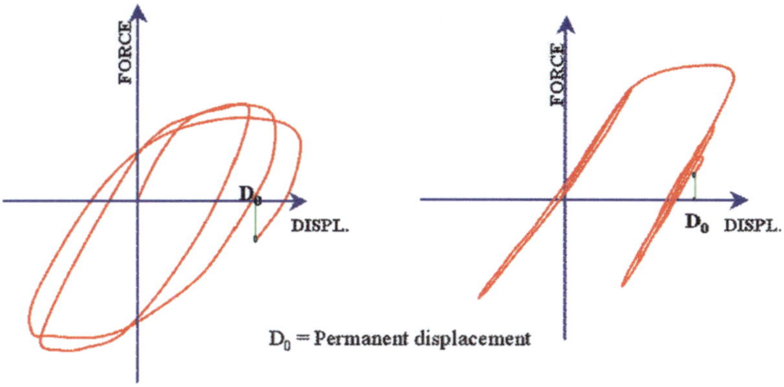

Fig. 14.4 Hysteretic force-deformation curves for damaged structures

In some buildings, although it is damaged, no changes in the frequency can be observed from the records. This was the case for a 7-story, reinforced concrete hotel building in Van Nuys, California, which suffered significant damage to the fourth floor columns during the 1994 Northridge earthquake. More on the building and the damage can be found in Trifunac et al. (1999). The building had records from several earthquakes, including the Northridge earthquake. Figure 14.5 gives a comparison

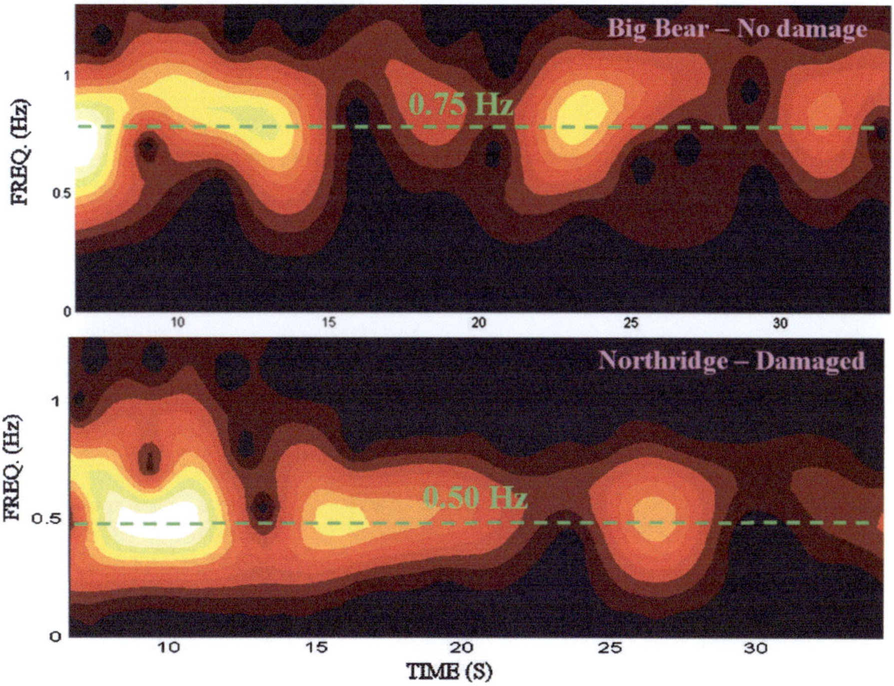

Fig. 14.5 Time variations of the fundamental frequency of a 7-story RC building during a non-damaging and damaging earthquakes

of the time variation of the building's first modal frequency calculated from the recorded top-floor accelerations during the two earthquakes, the 1992 Big Bear, California earthquake, which did not cause any damage, and the 1994 Northridge, California earthquake, which caused extensive damage. No distinct changes in the frequency during the course of the earthquakes can be detected from either set of data, although there was major damage from the Northridge earthquake. However, the Northridge frequency is 50% lower than the Big Bear frequency, suggesting that the damage probably occurred very early and instantaneously (i.e., in a brittle fashion) during the Northridge earthquake.

Multiple sets of earthquake records (http://nsmp.wr.usgs.gov/) from a 40-story steel building in Los Angeles have shown that small nonlinearities, which are always present in buildings, and the variations in damping can also cause changes in the observed frequencies. The foundation-to-roof transfer functions of the building for six different earthquakes and ambient vibrations are plotted in Fig. 14.6. The reason for using the transfer functions, rather than Fourier spectra, is that the transfer functions are independent of soil-structure interaction effects (Safak, 1995). Only the frequencies near the fundamental frequency are shown in the figure. The figure confirms that there are significant shifts in the fundamental frequency, although the building did not suffer any damage.

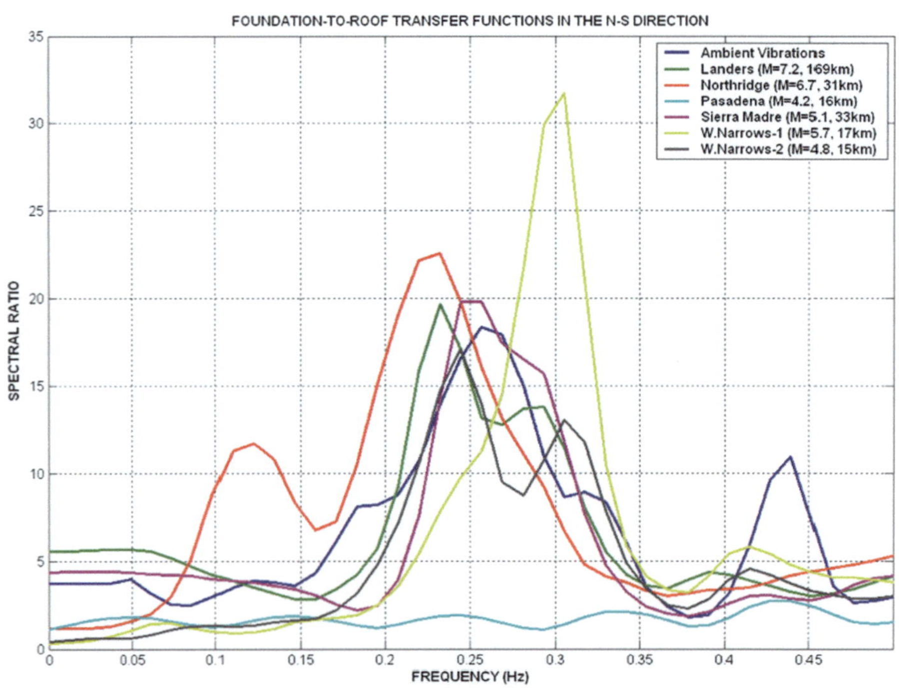

Fig. 14.6 Foundation-to-roof transfer functions of a 40-story steel high-rise building during six earthquakes and ambient vibrations

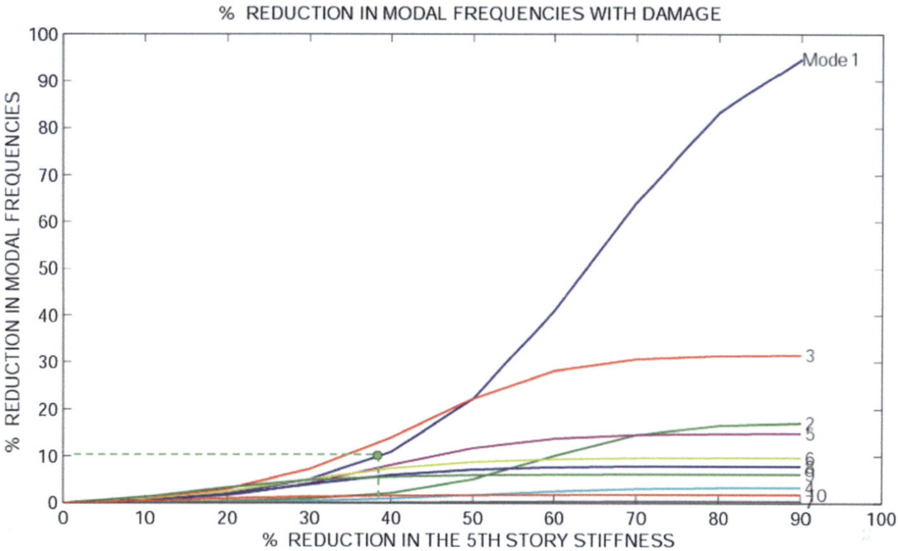

Fig. 14.7 Changes in the modal frequencies of a 10-story building due to reductions in the 5th story stiffness

Analytical studies also confirm the unreliability of using frequency changes for damage detection. An example is presented in Fig. 14.7, which shows the changes in the natural frequencies of a 10-story building for a gradual reduction in the stiffness of the fifth story. The curves show the percent reduction on each natural frequency against the percent reduction in the fifth story stiffness. As the figure indicates, in order to see a 10% reduction in the fundamental frequency we need about 40% reduction in the fifth floor stiffness. Any damage due to such a large reduction in stiffness would be visible to naked eye.

14.3 Damage Detection Based on Permanent Deformations

As the examples in the preceding section clearly show, changes in natural frequencies are not always a reliable indicator for damage. As stated earlier, the structure goes through nonlinear, hysteresis-type force-deformation loops when the damage takes place. An important characteristic of hysteretic behaviour is that the structure does not return to its original configuration when the excitation stops. In other words, the structure shows permanent deformations, such as permanent displacements and/or permanent rotations, after the earthquake. Unlike the trigger-based monitoring, the continuous monitoring can detect permanent deformations. This is accomplished by comparing pre- and post-earthquake ambient records. Analyses of pre- and post-earthquake records, along with the earthquake records, provide a more reliable approach to damage detection.

The earthquake-induced damage in a structure can be detected by continuous monitoring, based on the following two criteria: (1) the dynamic characteristics of the structure change during the earthquake, and (2) the structure exhibits permanent deformations after the earthquake. In terms of signal properties, these criteria correspond to the following: (1) the spectral characteristics of the signal (namely, the frequency content and damping) change during the earthquake and (2) the mean values of the signal before and after the earthquake are different (i.e., the post-earthquake portion of the signal shows permanent DC offset). Most of today's instrumented structures use acceleration sensors. Displacements and rotations are calculated by the integration of accelerations. Accelerations are not the best quantity to measure when trying to detect permanent (i.e., static) displacements and rotations. Such deformations can best be measured by special sensors, such as GPS sensors and tiltmeters. The latest GPS sensors are able to detect displacements as small as 1.0 mm. Once the presence of permanent displacements and rotations are confirmed, the question becomes whether they represent damage or not. Statistical hypothesis tests can be used to make such decisions (e.g., Lehmann, 1959). These concepts are summarized schematically in Figs. 14.8 and 14.9.

The change in the spectral characteristics of the signal can be detected by using a large number of tools available in the literature for time-varying signal analysis (e.g., Durbin, 1959; Burg, 1968; Griffiths, 1977; Widrow and Stearns, 1985; Brammer and Siffling, 1989). Applications of such tools to real-time vibration data from structures are outlined in Safak (1991, 2004). Adaptive filters and Kalman filters are more

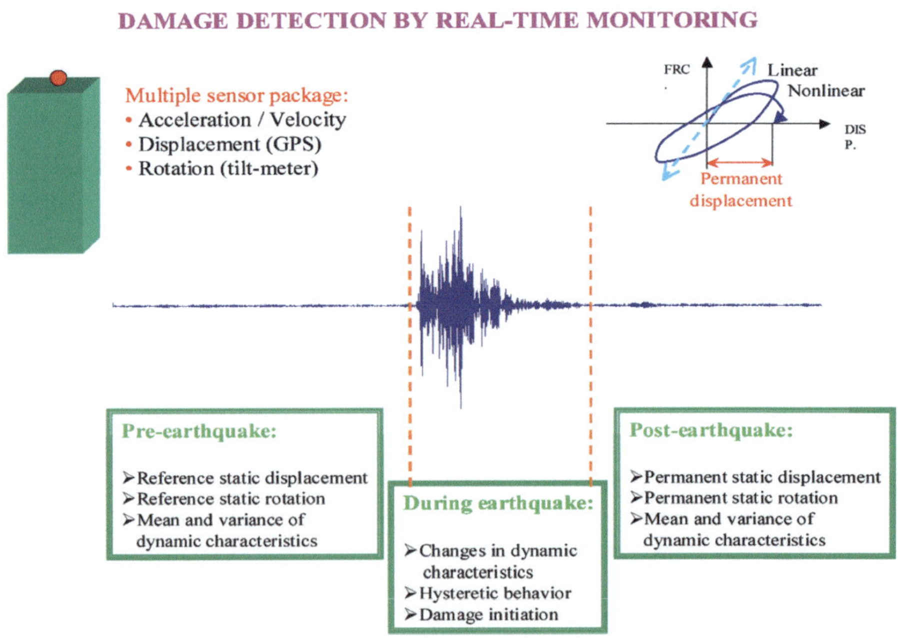

Fig. 14.8 Components of damage detection by continuous monitoring

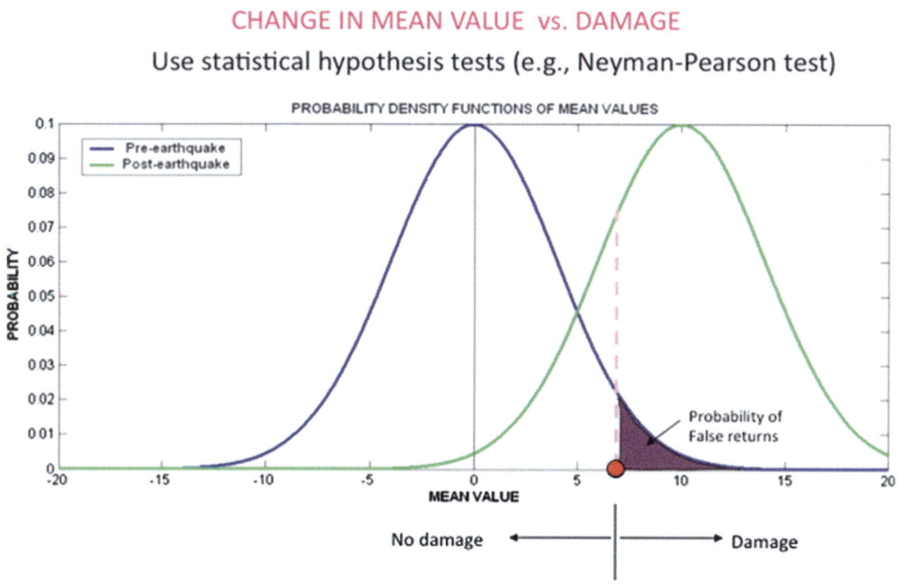

Fig. 14.9 Statistical tests for damage detection

appropriate for real-time data because they can be applied in real-time. Instead of monitoring the changes in the frequencies and damping of the structure, it is easier and faster to monitor the changes in the parameters of such filters to detect changes. One caution, the change in signal characteristics would occur much faster during a damaging earthquake than that during ambient vibrations. Therefore, in an automated system, it is advisable to have two parallel adaptive identifications, one with a longer time window and the other with a shorter time window. The longer-window identification detects slow changes during the ambient vibrations and is more appropriate for the low signal-to-noise signals, whereas the shorter-window identification detects the sudden changes during the earthquake and is more appropriate for high signal-to-noise signals.

14.4 Damage Detection Based on Wave Propagation

The vibrations of structures under dynamic loads can be considered as a wave propagation problem. For multi-story buildings, for example, the vibrations can be characterized in terms of wave propagation parameters; namely, wave velocities, attenuation of wave amplitudes, and the wave reflections and transmission coefficients (Safak, 1999). Recorded earthquake motions from instrumented structures clearly show the propagation of seismic waves. For example, Fig. 14.10 shows a 17-story steel-frame building instrumented with four accelerometers at every floor, and its recorded accelerations during a small earthquake (Kohler et al., 2005). If

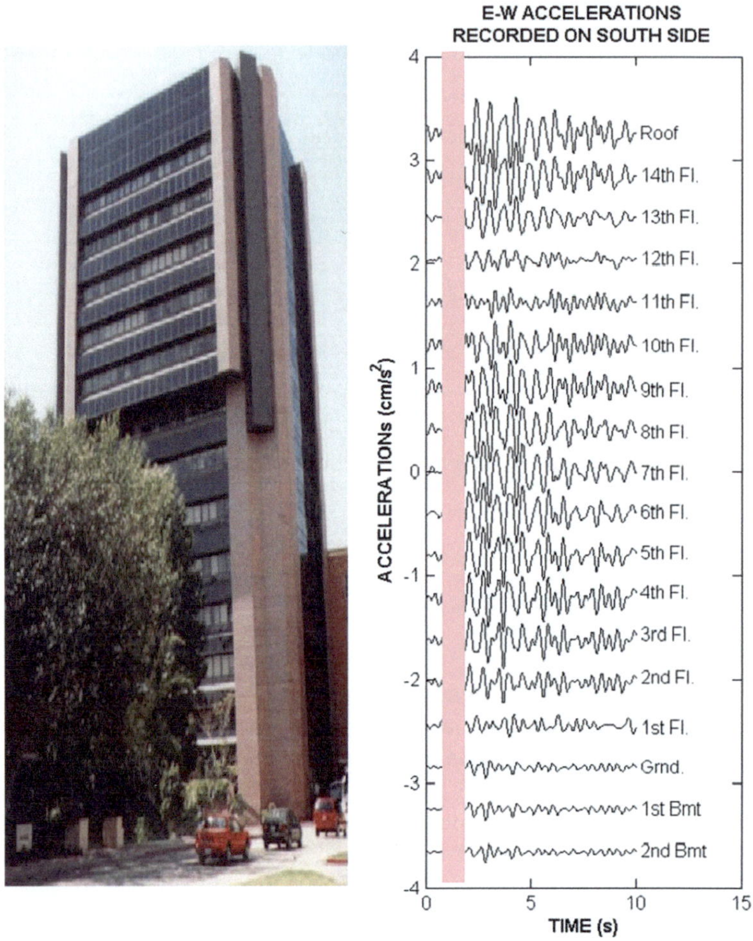

Fig. 14.10 Recorded floor accelerations in a 17-story steel building during an earthquake

we take a closer look to a 1-s long segment, as marked in Fig. 14.10 and shown in Fig. 14.11, the propagation of waves becomes very clear. The horizontal axes in Fig. 14.11 denote the time and the floor level, and the vertical axis is the accelerations deconvolved by the recorded ground accelerations. The accelerations are color-coded based on their amplitudes. As the figure shows, the incoming seismic waves travel upward in the building reaching to the roof in about 0.4 s. They are then reflected by the free surface on the roof, propagating downward, and again reflected back upwards by the ground. This up and down bouncing of the waves in the building is what causes the vibrations, and they last until the earthquake stops and the vibrations are damped out.

For system identification and damage detection, it has been shown that, when compared to modal parameters, the wave propagation parameters are more reliable and robust, and also more sensitive to damage (Safak, 1998). For historical

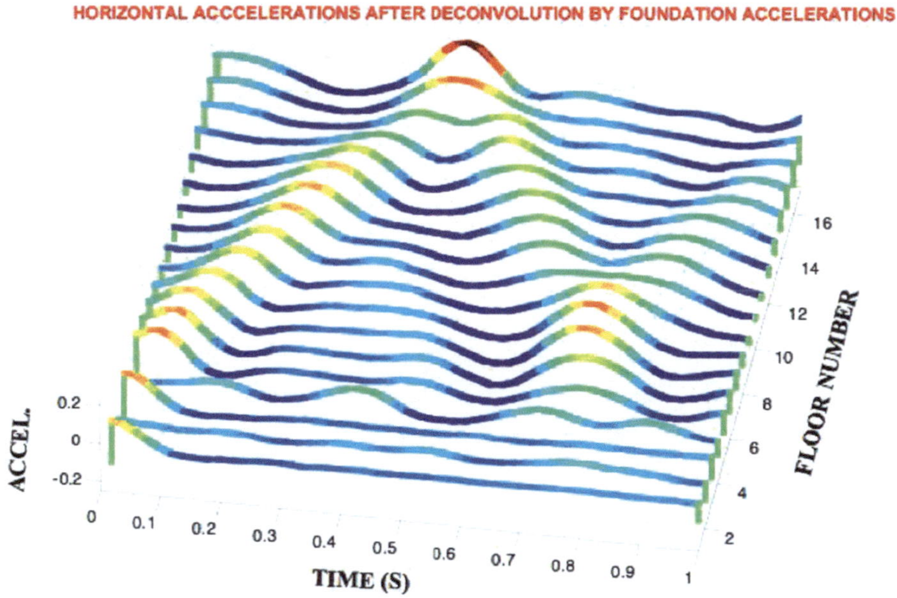

Fig. 14.11 Propagation of seismic waves during a 1-s interval in the 17-story building

structures, the utilization of wave propagation approach for system identification and damage detection is particularly convenient because, in most cases, due to their age, geometry, construction material, and the structural system historical structures do not meet the requirements of the classical modal analysis, such as elasticity, linearity, mass and/or stiffness proportional damping.

For earthquake induced waves, another wave propagation approach is the *Seismic Interferometry* (Snieder and Safak, 2006). Seismic Interferometry is based on the correlation of synchronized records collected from different locations. In structures, this correlation can be shown to lead to the Green's functions that account for the wave propagation between different receivers in the structure. The properties of the waves can be investigated without knowing the seismic input that generated the waves. The travel times of the waves, and their reflections and transmissions at floor levels provide a critical insight into the characteristics of the structure.

A critical step in using wave propagation approach for identification and damage detection is the accurate calculation of wave travel times. This first requires high-quality and high-sampling recording. The two standard approaches to calculate wave travel times between two recording points have been to use the time differences between characteristic peaks in the signals, or to determine the time lag where the cross-correlation of the signals has a maximum. These methods are acceptable for non-dispersive, non-attenuating media, where the waveforms do not change their shape as they travel. In structures, the waves attenuate due to damping. The attenuation changes the shape (i.e., the phase) of the waves. In other words, the phase shifts in two records are caused by the combined effects of wave travel times, plus the phase distortions due to damping. This is shown schematically in Fig. 14.12.

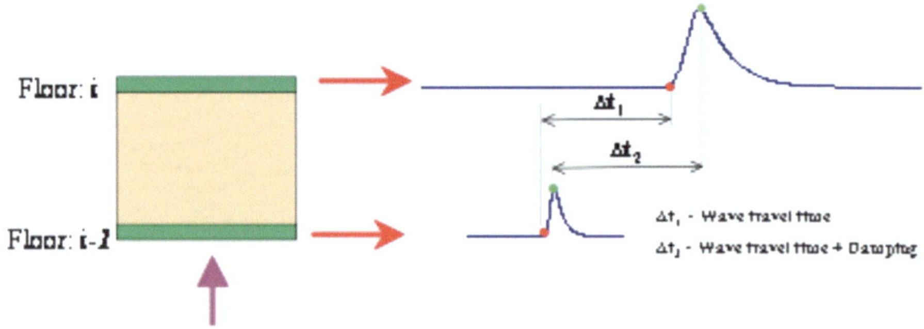

Fig. 14.12 Schematic representation of the effect of damping on the signal's phase

A rigorous theoretical analysis of wave dispersion in an attenuating medium can be found in Aki and Richards (1980).

It is possible to eliminate the phase shifts introduced by damping on the calculated wave travel times. This can be accomplished by using the envelope functions of the signals instead of the signals themselves to calculate the time shifts. The time shifts from the envelope functions can be calculated by observing the time delay between two specific phases, or the peaks of cross-correlation functions. The envelope functions of the signals are calculated by taking their Hilbert transforms and calculating the corresponding analytic functions.

It can be shown that for narrow-band signals that are propagating in a frequency-dispersive medium, the phase and group velocities are such that the peaks of the signal and the envelope do not coincide. Envelope functions are not affected by the dispersive properties of the medium (Bendat and Piersol, 1985). This property of envelope functions provide a convenient tool to remove the phase shifts due to damping, and calculate wave travel times more accurately. More on calculating wave velocities in structures can be found in Safak et al. (2009).

As an example, we calculate the wave travel times in the main pillars of a 1,500 year-old historical structure, the Hagia Sophia Museum, in Istanbul, Turkey. The structure is permanently instrumented with 12 acceleration sensors, collecting data continuously in real time (Durukal et al., 2003). Figure 14.13 shows the locations of ground sensors and the sensors at the top of the four main pillars of the structure, and the accelerations recorded (the sensors on the arches are not shown in the figure). These records are used to calculate the wave travel times in the pillars. The height of the pillars from the ground level to the bottom of the arches are approximately 23 m.

To calculate the wave travel times, we first band-pass filter the recorded accelerations around a narrow frequency band centered at the dominant frequency of the structure, and increase the sampling rate of the records to 1000 sps by using interpolation. The original sampling rate in the records (100 sps) is not sufficient for the accurate calculation of wave travel times in 23-m high pillars. Next, we determine the envelope of the filtered accelerations by using the Hilbert transforms and the corresponding analytic functions. Filtered ground versus pillar-top accelerations and their envelopes are shown for each pillar in Fig. 14.14.

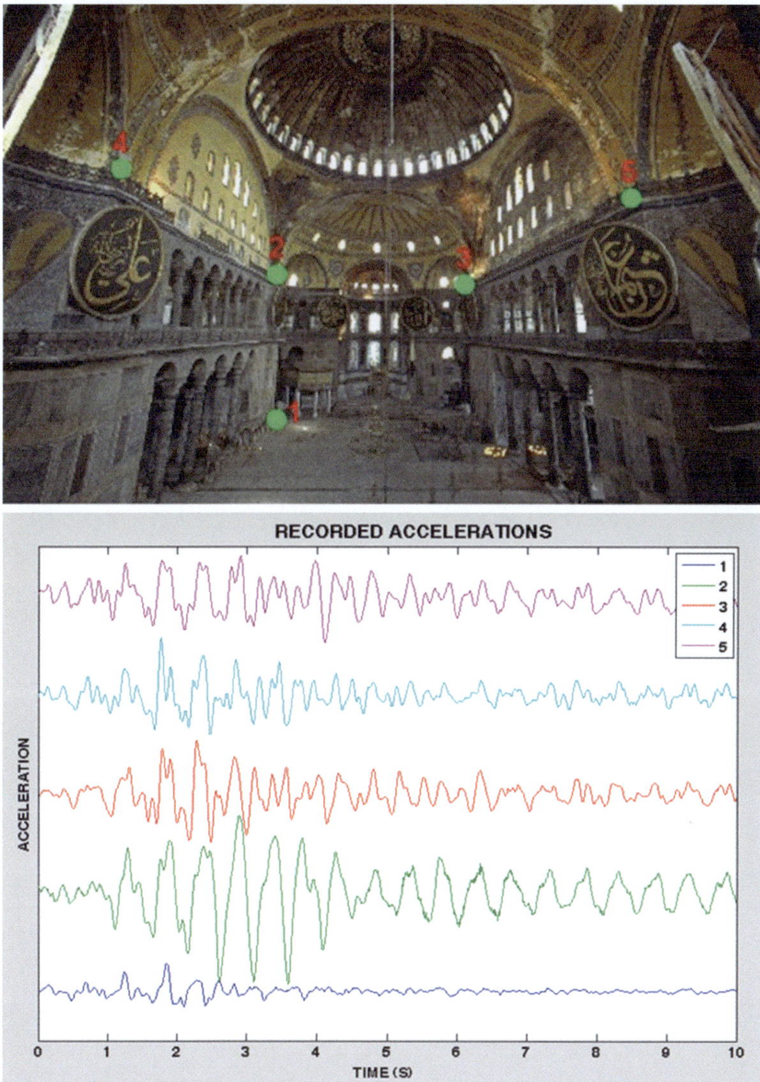

Fig. 14.13 Locations of sensors and recorded accelerations in the Hagia Sophia Museum

We then calculate ground to pillar-top wave travel times twice, first by using the filtered signals, and next by using their envelopes. The calculated wave travel times are also given in Fig. 14.14. As expected, the wave travel times calculated from the envelope functions are smaller, and represent the actual travel times. The wave travel times calculated from the filtered accelerations are larger. The difference in the calculated wave travel times represents the phase shifts due to damping. The relationship between the phase shifts and the corresponding damping coefficients can be found in Safak (1999). The wave travel time for the first pillar is significantly

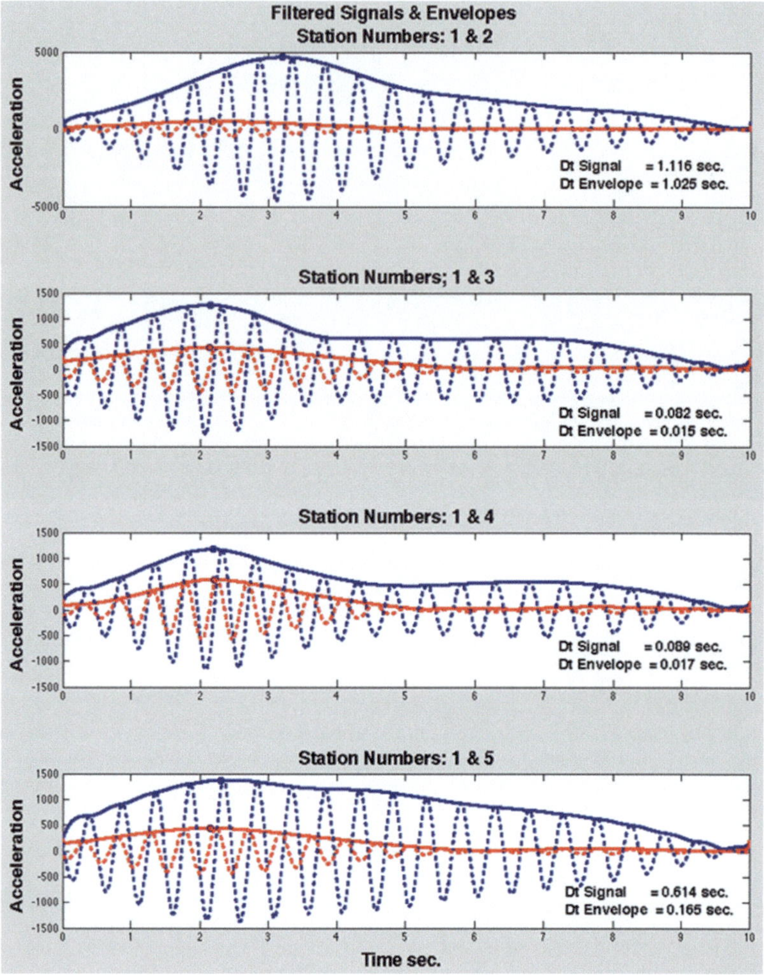

Fig. 14.14 Wave travel times calculated from the filtered signals and their envelopes (the difference represents the effect of damping)

larger than the others because of a known damage in that pillar. The calculated wave travel times correspond to wave velocities of approximately 23 m/s for the first pillar, and 435 m/s for the other pillars. More detail on the study can be found in Safak et al. (2009).

14.5 Minimizing Effects of Noise in Spectral Analysis

Fourier spectral analysis has been the standard method to analyze vibration data from structures. When used for SHM data, the main source of errors in spectral analysis is the noise in the records and the time-varying characteristics of the signals under transient loads. Noise alters the amplitudes and the frequency content of Fourier spectra, and introduces spurious resonant peaks. For a typical structure, the vibrations recorded by a SHM system are almost entirely composed of ambient vibrations generated by ambient loads, such as wind, traffic, and micro tremors. Therefore, the signals are fairly noisy with very low SNR (signal-to-noise ratio). Unless special processing techniques are used to reduce the effects of noise, Fourier spectral analysis can give misleading results, particularly for records from stiff structures.

Three techniques are described below to improve the accuracy of Fourier spectral analysis.

14.5.1 Segmentation and Averaging

Assume that the recorded signal, $x(t)$, is the sum of actual (i.e., noise free) signal, $s(t)$, plus zero-mean Gaussian white noise, $n(t)$:

$$x(t) = s(t) + n(t) \text{ with } n(t) = \mathrm{N}\,[0, \sigma] \tag{1}$$

The discrete Fourier expansions of $s(t)$ and $x(t)$ can be written as

$$\begin{aligned} s(t) &= \sum_{k=1}^{N/2+1} a_k \cdot \cos(2\pi f_k t) + \sum_{k=1}^{N/2+1} b_k \cdot \sin(2\pi f_k t) \\ x(t) &= \sum_{k=1}^{N/2+1} \hat{a}_k \cdot \cos(2\pi f_k t) + \sum_{k=1}^{N/2+1} \hat{b}_k \cdot \sin(2\pi f_k t) \end{aligned} \tag{2}$$

For zero-mean Gaussian $n(t)$, we can calculate the statistical properties of the Fourier coefficients $\hat{a}_k$ and $\hat{b}_k$ of $x(t)$, and show the following

$$\begin{aligned} &\mathrm{Mean}[\hat{a}_k] = a_k \text{ and } \mathrm{Mean}[\hat{b}_k] = b_k \\ &\mathrm{Variance}[\hat{a}_k] = \tfrac{2\sigma^2}{N} \text{ and } \mathrm{Variance}[\hat{b}_k] = \tfrac{2\sigma^2}{N} \end{aligned} \tag{3}$$

where σ^2 is the variance of $n(t)$ and N denotes the number of points in the record. The first equation confirms that the mean values of the Fourier coefficients of the noisy signal, $x(t)$, are equal to those of the noise-free signal, $s(t)$. The second equation shows that the variances of the Fourier coefficients of $x(t)$ are inversely proportional to the record length; that is, the longer the record length the smaller the variance of Fourier spectrum (i.e., the more accurate the results). This observation suggests that we should consider very long signals when calculating the Fourier spectra of ambient data, provided that the signal characteristics remain stationary.

If the stationarity condition is not met, the alternative would be to divide the signal into equal-length stationary segments, and calculate the Fourier spectrum as the average of the Fourier spectra of these segments.

14.5.2 Selection of Optimal Smoothing Windows

A widely used technique to reduce the influence of noise in Fourier spectra is to apply smoothing windows. There are no straightforward rules on selecting smoothing windows. Too short smoothing windows may not provide sufficient noise reduction, whereas too long smoothing windows may eliminate some of the real peaks. A simple technique for selecting the optimal smoothing window length is suggested in (Safak, 1997). It involves plotting the area under the squared Fourier amplitude spectrum with increasing window length. The plot shows a decaying curve with increasing window length. Initially, the decay is very fast, but becomes much slower as the window length increases. If it is assumed that the noise-free Fourier amplitude spectrum is a smooth function of frequency, it can be shown that the window length where the rate of decay in the curve changes from fast to slow corresponds to the optimal window length. The procedure for finding this point in the curve is shown schematically in Fig. 14.15.

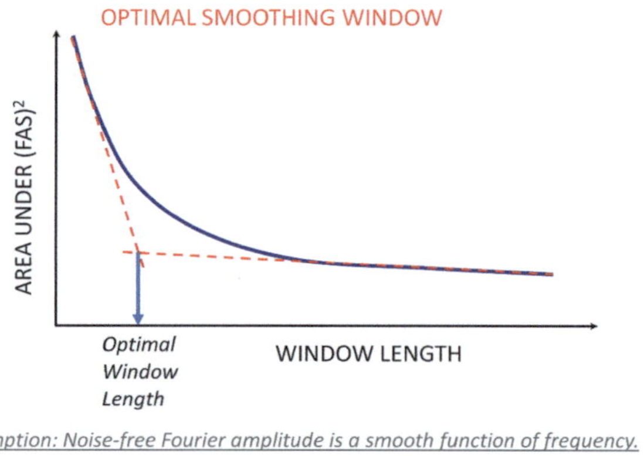

Fig. 14.15 Estimation of optimal smoothing window

14.5.3 Least-Squares Estimation of Fourier Spectra

The discrete Fourier expansion of the noise-free signal, $s(t)$, is given by the first expression in Eq. (2). For a given signal length, N, and sampling interval, Δt, the discrete frequencies, f_k, of the Fourier spectra are calculated as

$$f_k = \frac{k}{N \cdot \Delta t} \text{ where } k = 1, \cdots, (N/2 + 1) \tag{4}$$

Therefore, all the sine and cosine terms in the Fourier expansion of $s(t)$ are known. The unknowns are the Fourier coefficients, a_k and b_k. Instead of determining a_k and b_k by standard Fast Fourier transforms, we can calculate them by minimizing the error, V, between the noise-free signal, $s(t)$, and the recorded signal, $x(t)$ by using the following equations:

$$\begin{aligned} V &= \sum_{t=1}^{N} [x(t) - s(t)]^2 \quad \text{where} \\ s(t) &= \sum_{k=1}^{N/2+1} a_k \cdot \cos(2\pi f_k t) + \sum_{k=1}^{N/2+1} b_k \cdot \sin(2\pi f_k t) \\ \min_{a_k, b_k}(V) \quad &\rightarrow \quad \frac{\partial V}{\partial a_k} = 0 \quad \text{and} \quad \frac{\partial V}{\partial b_k} = 0 \end{aligned} \tag{5}$$

The minimization results in a linear set of equations for a_k and b_k, which can easily be solved by matrix inversion. The calculated a_k and b_k represent the least-squares estimate of the Fourier coefficients of the noise-free signal. Figure 14.16 shows a microtremor, and the corresponding standard and least-square Fourier

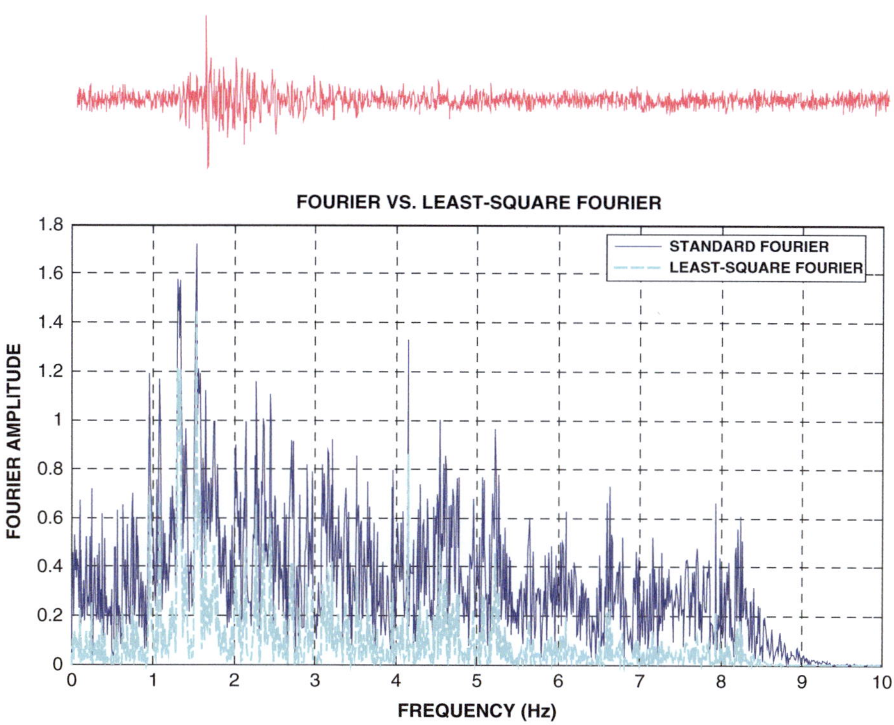

Fig. 14.16 Comparison of standard and least-squares Fourier amplitude spectra of a record

amplitude spectra. The least-squares Fourier spectra have smaller amplitudes outside the dominant frequency band (i.e., 1.0–2.0 Hz). The reduction is due to the minimization of noise amplitudes in those regions.

14.6 Statistical Signal Processing

Statistical signal processing accounts for the randomness of the noise in the records, and tries to remove it by using the statistical properties of the signals. The data from SHM systems are mostly stationary (i.e., its temporal and frequency characteristics do not change significantly with time). The SHM signals are also infinitely long with low SNR. These properties make statistical signal processing tools very appropriate for the analysis of SHM data.

Some of the simple statistical signal processing tools are presented below.

14.6.1 Autocorrelation Functions and Optimal Filters

The auto-correlation function $R(\tau)$ of a signal $x(t)$ is defined by the following equation:

$$R(\tau) = \frac{1}{N}\sum_{t=1}^{N} x(t) \cdot x(t-\tau) \tag{6}$$

For stationary signals, such as ambient ground noise, the auto-correlation function depends only on the time lag τ. It can be shown that the expected auto-correlation function of a sinusoid buried in noise has the same frequency as the sinusoid. That is

$$x(t) = A \cdot \cos(\omega t) + n(t) \;\rightarrow\; \mathrm{E}[R(\tau)] = \frac{A^2}{2} \cdot \cos(\omega t) \tag{7}$$

where E[ ] denotes the expected value. In other words, taking the autocorrelation does not change the frequency content of the signal. It can also be shown that the autocorrelation improves the SNR, i.e., the SNR in the autocorrelation of a signal is higher than that of the original signal. This is because the autocorrelation operation amplifies the amplitudes of any periodic components in the data. Therefore, when calculating Fourier spectra of ambient noise, it is advantageous to use the autocorrelation functions of the records instead of the original records. The Fourier spectrum of the autocorrelation function is commonly known as the power spectral density function.

A concept directly related to autocorrelation functions is the optimal filtering. Optimal filtering aims to remove noise by searching correlated (i.e., periodic) components in the record. We assume that the periodic components in the record correspond to the actual (i.e., noise free) signal, and the remaining components are considered to be the noise. A characteristic of a periodic signal is that its value at any given time can be written as a linear combination of its past values. Therefore, if we were able to separate the record, $x(t)$, into its periodic and random components, we can express $x(t)$ as

$$x(t) = \sum_{k=1}^{m} a_k \cdot x(t-k) + n(t) \tag{8}$$

where the first term on the right hand side is the periodic component. Assume that we know all the values of $x(t)$ up to time step $(t-1)$ and want to predict the value at the next time step, t. Since the mean value of $n(t)$ is zero, the most likely value, $\hat{x}(t)$, of $x(t)$ would be

$$\hat{x}(t) = \sum_{k=1}^{m} a_k \cdot x(t-k) \tag{9}$$

The difference between the predicted and the recorded values of $x(t)$ is the error in our estimation. We can select the coefficients a_k in Eq. (9) such that the estimation error, V, is minimum. That is,

$$\min_a(V) = \left[x(t) - \sum_{k=1}^{m} a_k \cdot x(t-k)\right]^2 \quad \rightarrow \quad \frac{\partial V}{\partial a_k} = 0 \tag{10}$$

Equation (10) results in a set of linear equations to determine the coefficients a_k. These coefficients define the filter to remove noise from the signal. We calculate the noise-free signal by filtering the record using Eq. (9).

The procedure presented above describes the basic idea in optimal filtering. There are numerous variations of the procedure suggested in the literature with their unique names such as Wiener filtering, Recursive Least Squares, Least Mean Squares, Durbin Algorithm, Burg Algorithm, and Yule-Walker Algorithm. More detail on these methods can be found in textbooks on optimal filtering and linear estimation (e.g., Kailath et al., 2000).

14.6.2 Eigenvalues of Autocorrelation Matrix

Another set of powerful tools to separate signal from the noise can be developed based on the eigenvalues and eigenvectors of the autocorrelation matrix. The autocorrelation matrix, Q, is defined by the following equation

$$Q = \begin{pmatrix} R(0) & \dots & R(M) \\ \vdots & \ddots & \vdots \\ R(-M) & \dots & R(0) \end{pmatrix} \tag{11}$$

where

$$R(\tau) = \frac{1}{N}\sum_{t=1}^{N} x(t) \cdot x(t-\tau) \text{ and } \tau = -M, \cdots, 0, \cdots, M;$$

Q is a $(M+1) \times (M+1)$ dimensional matrix that has $(M+1)$ eigenvalues and eigenvectors. The well known Karhunen–Loeve expansion states that a stationary signal can be represented in terms of the eigenvectors of its autocorrelation matrix (Karhunen, 1947; Loeve, 1978). That is

$$x(t) = \sum_{i=0}^{M} c_i \cdot q_i(t) \tag{12}$$

where $q_i(t)$ denotes the ith eigenvector and c_i is a constant. It can also be shown that the eigenvalues that correspond to the correlated (i.e., periodic) components of the record are much larger than those that correspond to the uncorrelated (i.e., noise) components in the record. Therefore, the eigenvalues and eigenvectors of the correlation matrix can be used to separate the noise from the signal.

An important assumption made in the derivation of Eq. (11) is that $x(t)$ is a stationary signal. In other words, the temporal and frequency characteristics of $x(t)$ does not change significantly with time, and therefore the autocorrelation function R is the function of the time lag only between the two components. The assumption of stationarity is appropriate for vibrations under ambient forces and wind loads, but not for vibrations under transient loads such as earthquakes or blast loads.

There are several filtering methods based on this approach, such as Pisarenko Harmonic Decomposition, Multiple Signal Classification (MUSIC), and Rotational Invariance Techniques (ESPRIT). Details of these methods can be found in advanced textbooks on signal processing (e.g., Moon and Stirling, 2000). Figure 14.17 shows a microteremor and the corresponding standard, auto-correlation based (Burg), and eigen-based (MUSIC) Fourier amplitude spectra. There is a significant reduction in noise effects by the Burg and MUSIC algorithms.

A key parameter that needs to be selected in the optimal filtering and the eigenvalue approach is the filter order, m. A filter with too small m does not accurately represent the signal, whereas a filter with too large m may try to represent noise as well as the signal. There several criteria available in the literature to select m (see, Soderstrom, 1987). A simpler and more straightforward selection can be made by plotting the variation of $V = \sum \varepsilon^2(t)$ with m (Safak, 2004). This sum typically shows a fast drop with increasing m, and then level off. The m value where the sum starts to level off can be taken as the optimal filter order.

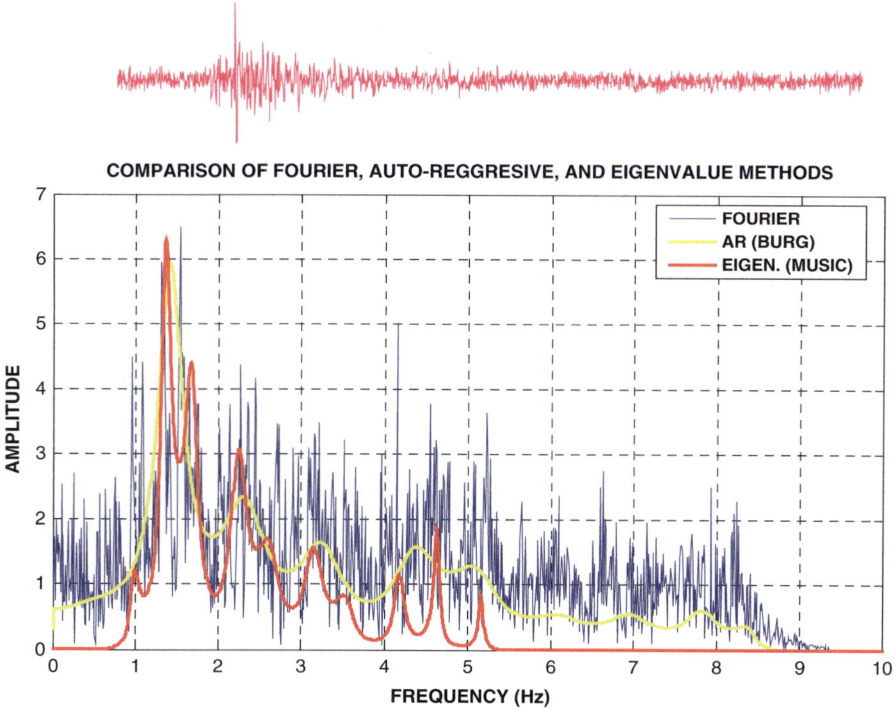

Fig. 14.17 Comparison of Fourier, Burg, and MUSIC spectral estimates

14.7 Tracking Time Variations of Signal Properties

Continuous monitoring requires continuous and automated data processing and analysis. The methods that are used for analysis should be able to adapt and account for any changes in signal characteristics. The simplest and most straightforward approach to analyze continuous data is the block-data approach. In this method, the records are handled in blocks of specified length. Each block is processed and analyzed as soon as it is full, and while the data for the next block are being acquired. More efficient ways to analyze continuous data can be developed by utilizing running time windows. Running windows are in essence weighting functions that emphasize recent data, while gradually deemphasizing past data. The windows ensure that any property calculated from data contains measurements that are relevant to the current state of the structure.

The two widely used weighting functions are exponentially decaying windows and sliding rectangular windows. The exponentially decaying window is defined as

$$w(t,i) = \frac{1-\lambda}{1-\lambda^t} \cdot \lambda^{t-i} \qquad \text{with } i = 0, 1, 2, \cdots, t \qquad \text{and } \sum_{i=1}^{t} w(t,i) = 1 \quad (13)$$

where λ is known as the forgetting factor with $0.0 < \lambda < 1.0$. The window applies exponentially decaying weights to past data points. The memory time constant, N_0, of the window is defined as the number of sampling points over which the characteristics of the structure can be assumed to remain constant. It can be approximated as:

$$N_0 \approx \frac{1}{1-\lambda} \tag{14}$$

This equation provides a simple criterion for the selection of λ. Typically, $\lambda = 0.900$–0.999.

Sliding rectangular windows consider only a limited number of past data points with equal weights. It is defined as:

$$w(t,i) = \frac{1}{L} \text{ for } t-L \leq i \leq t \qquad \text{with } i = 0,1,2,\cdots,t \qquad \text{and} \sum_{i=1}^{t} w(t,i) = 1 \tag{15}$$

where L is the window length and the memory time constant.

As an example, consider the tracking of the mean value of a signal. The mean value in a vibration record represents the static component of the structure's response. Normally, the mean value of vibration records should be zero. In real-time monitoring, the mean value may fluctuate around zero due to various factors, such as imperfections in sensors, environmental factors (e.g., wind, rain, etc.), natural changes in structure (e.g., structural modifications, change in loads, aging, etc.), or permanent damage after an extreme event (e.g., inelastic deformations during an earthquake). Therefore, it is important that the changes in mean value are tracked accurately.

We can consider the calculation of mean value of a signal as a problem of a weighted least-squares fit of a constant to the record by minimizing the following error function:

$$\varepsilon(t) = \sum_{i=1}^{t} w(t,i) \cdot [x(i) - m(t)]^2 \tag{16}$$

where $x(i)$ is the signal, $w(t,i)$ is the weighting function, and $m(t)$ is the mean value. For the weighting functions described above, the minimization gives the following expressions for the mean value at time t

$$\begin{aligned} &\text{For exponentially decaying window: } m(t) = \frac{1-\lambda}{1-\lambda^t} \cdot \sum_{i=1}^{t} \lambda^{t-i} x(i) \\ &\text{For sliding regtangular window: } m(t) = \frac{1}{L} \cdot \sum_{i=t-L+1}^{t} x(i) \end{aligned} \tag{17}$$

Above expressions can be put into a recursive form such that the mean value at time t is calculated from the mean value at time t–1. The recursive forms are (Safak, 2004)

$$\begin{aligned}&\text{For exponentially decaying window: } m(t) = \lambda \cdot m(t-1) + (1-\lambda) \cdot x(i)\\&\text{For sliding regtangular window: } m(t) = m(t-1) + \frac{1}{L} \cdot [y(t) - y(t-L)]\end{aligned} \tag{18}$$

Computationally, the recursive forms are more appropriate for real-time data.

Similarly, we can derive the following equations for the mean-square value, $s(t)$, of the signal (Safak, 2004):

$$\begin{aligned}&\text{For exponentially decaying window: } \mathrm{s}(t) = \lambda \cdot s(t-1) + (1-\lambda) \cdot x^2(i)\\&\text{For sliding regtangular window: } \mathrm{s}(t) = s(t-1) + \frac{1}{L} \cdot [y^2(t) - y^2(t-L)]\end{aligned} \tag{19}$$

Similar recursive expressions can be developed for spectral analysis and system identification (Ljung and Soderstrom, 1983; Safak, 1988; Safak, 1989a; Safak, 1989b; Ljung, 1999).

14.8 Conclusions

Structural Health Monitoring (SHM) involves continuous monitoring of the dynamic characteristics of a structure by digital instruments. The main objective in SHM is to track the changes in the structure's dynamic characteristics and detect damage. The current data from SHM systems show that the natural frequencies of the structure are not always a reliable indicator of damage. Two additional parameters that can be used for damage detection are the permanent change in the geometry of the structure and the changes in the characteristics of propagating waves within the structure.

Data from SHM systems are mainly ambient vibration data, which typically have very low signal-to-noise ratios (SNR), particularly for stiff structures. Standard Fourier-based spectral analysis approach is not always reliable for noisy data. Statistical signal processing tools, such as autocorrelation analysis, optimal filters, and eigenvalue-based spectral analysis are more appropriate.

Also, since SHM data are continuous and recorded in real-time, the data processing and analysis should ideally be done in real time. Adaptive filtering and system identification techniques provide tools to analyze SHM data in real time.

References

Aki K, Richards PG (1980) Quantitative seismology: theory and Methods, vol I. W.H. Freeman and Company, San Francisco, CA

Bendat JS, Piersol AG (1985) Random data: analysis and measurement procedures. Wiley, New York, NY

Brammer K, Siffling G (1989) Kalman-bucy filters. Artech House, Norwood, MA
Burg JP (1968) A new analysis technique for time-series data. NATO Advanced Study Institute on Signal Processing, Enschede, The Netherlands
Clinton J (2004) www.ce.caltech.edu/~jclinton/thesis.html
Durbin J (1959) Efficient estimators of parameters in moving average models. Biometrica 46: 306–316
Durukal E, Cimilli S, Erdik M (2003) Dynamic response of two historical monuments in Istanbul deduced from the recordings of Kocaeli and Duzce earthquakes. Bull Seism Soc Am 93(2): 694–712
Griffiths LJ (1977) A continuously adaptive filter implemented as a lattice structure. In: Proceedings of the IEEE International Conference on Acoustics, Speech and Signal Processing, Hartford, CT, pp 683–686
Kailath T, Sayed AH, Hassibi B (2000) Linear estimation. Prentice Hall, Upper River Saddle, NJ
Karhunen KK (1947) Über lineare methoden in der wahrscheinlichkeitsrechnung. Ann Acad Sci Fennicae Ser A I Math-Phys 37:1–79
Kohler MD, Davis PM, Safak E (2005) Earthquake and ambient vibration monitoring of the steel frame UCLA factor building. Earthquake Spectra 21(3):715–736
Lehmann EL (1959) Testing statistical hypothese. Wiley, New York, NY
Ljung L (1999) System identification: theory for the user. Prentice Hall, Upper Saddle River, NJ
Ljung L, Soderstrom T (1983) Theory and practice of recursive identification. MIT Press, Cambridge
Loève M (1978) Probability theory, vol II, 4th edn. Graduate texts in mathematics, vol 46, Springer, New York, NY
Moon TK, Stirling WC (2000) Mathematical methods and algorithms for signal processing. Prentice Hall, Upper River Saddle, NJ
Safak E (1988) Analysis of recordings in structural engineering: adaptive filtering, prediction, and control. Open-File Report 88–647, US Geological Survey, Menlo Park, CA
Safak E (1989a) Adaptive modeling, identification, and control of dynamic structural systems: Part I – Theory. J Eng Mech, ASCE 115(11):2386–2405
Safak E (1989b) Adaptive modeling, identification, and control of dynamic structural systems: Part II – Applications. J Eng Mech, ASCE 115(11):2406–2426
Safak E (1991) Identification of linear structures using discrete-time filters. J Struct Eng, ASCE 117(10):3046–3085
Safak E (1995) Detection and identification of soil-structure interaction in buildings from vibration recordings. J Struct Eng, ASCE 121(5):899–906
Safak E (1997) Models and methods to characterize site amplification from a pair of records. Earthquake Spectra, EERI 13(1):97–129
Safak E (1998) Detection of seismic damage in multi-story buildings by using wave-propagation analysis. In: Proceedings (in CD-ROM) of the 6th US national conference on earthquake engineering, Seattle, Washington, DC, May 31–June 4, 1998. Elsevier Science Inc, New York, NY
Safak E (1999) Wave propagation formulation of seismic response of multi-story buildings. J Struct Eng, ASCE 125(4):426–437
Safak E (2004) Analysis of real-time data from instrumented structures. In: Proceedings (CD ROM) of the 13th world conference on earthquake engineering, Vancouver, BC, Canada, 1–6 August 2004 (Paper No 2193)
Safak E, Cakti E, Kaya Y (2009) Seismic wave velocities in historical structures: a new parameter for identification and damage detection. In: Proceedings of (IOMAC 2009) the 3rd international modal analysis conference, Ancona, Italy, 4–6 May 2009
Snieder R, Safak E (2006) Extracting the building response using interferometric imaging: theory and application to the Millikan Library, Pasadena, CA. Bull Seism Soc Am 96:586–598
Soderstrom T (1987) Model structure determination. In: Singh M (ed) Encyclopedia of systems and control. Pergamon Press, Elmsford, NY

Trifunac MD, Ivanovic SS, Todorovska MI (1999) Instrumented 7-storey reinforced concrete building in Van Nuys, California: description of Damage from the 1994 Northridge earthquake and strong motion data. Report CE 99–02, Department of Civil Engineering, University of Southern California, Los Angeles, CA

Widrow B, Stearns SD (1985) Adaptive signal processing. Prentice-Hall, Englewood Cliffs, NJ

Part V
New Techniques and Technologies

Chapter 15
Large Scale Testing

Achievements and Future Needs

Artur Pinto

Abstract A brief summary of the European research involving Large-scale testing is given. Present and near future opportunities to further advance knowledge on earthquake behaviour and performance of structures are depicted. Performance-Based Seismic Design (PBSD) and Risk Assessment are recalled in view of the definition of structural testing procedures and protocols and also in view of the type of tests required. The importance of non-structural components is highlighted. Some tests performed at ELSA in support of the European Design Code (Eurocode 8) and on assessment and retrofit of existing structures are summarized. As examples of more advanced testing technique, Pseudo-dynamic tests with nonlinear substructuring carried out on bridges, are illustrated. The contribution and role of Large-scale testing to the challenging development and implementation of PBSD are addressed.

15.1 Introduction

Contrarily to other research fields, in structural/earthquake testing, the term "Large-scale testing" refers to the use of Large-scale physical models and not to the number of models tested. Recourse to large/full scale is necessary as similitude does not extend to nonlinear behaviour. In addition, extrapolation of complex phenomena, such as bond-slip in reinforced concrete structures, in the nonlinear range is not straightforward. This implies use of realist concrete mixes, rebar diameters, etc. Moreover, large-scale testing is often associated with a system (assemblage/structural physical model) in contrast with element testing (beam, column).

Earthquake testing has always played a central role in the development of earthquake engineering (EE) research and practice. There are primary aspects related to validation of modelling and analysis procedures, together with aspects related to

A. Pinto (✉)
ELSA, IPSC, Joint Research Centre, Ispra, 21020 VA, Italy
e-mail: artur.pinto@jrc.ec.europa.eu

M. Garevski, A. Ansal (eds.), *Earthquake Engineering in Europe*, Geotechnical, Geological, and Earthquake Engineering 17, DOI 10.1007/978-90-481-9544-2_15,

structural innovation (new materials, assemblages, etc.), which require the adoption of laboratory experimentation.

In Europe there is a specific case of intensive use of experimental facilities and associated numerical exploitation of experimental results for the calibration of the European Standards for design (Eurocode 8 for seismic design), whose enforcement is foreseen for 2010 is many European countries. Design according to the Eurocodes must be accepted, for public works, in all European Union Member States (Eurocodes, 2010).

Contrarily to most of the existing codes worldwide, Eurocodes are new codes, not built on any specific existing code, and embody many innovations, including a clear statement on performance requirements and compliance criteria (see details in Fardis, 2004). There was therefore a need to check performance of structures designed to Eurocodes and to check capacities and limit-state requirements. In fact, since the beginning of the 1990s a large experimental research work has been carried out in Europe, at the European Laboratory for Structural Assessment (ELSA) of the Joint Research Centre (JRC) and at many other laboratories and universities equipped with shaking-tables and other testing facilities.

With the project SERIES (Seismic Engineering Research Infrastructures for European Synergies), financed by the European Commission, the earthquake engineering research community and industry is offered a new opportunity to access to Large-scale research infrastructures and develop their experimental research. Thus, European researchers have the opportunity to further contribute to new Eurocode developments and to the preparation of its revision foreseen by 2015.

In what concerns testing, there are several aspects to take into account, as component, assemblage and structure testing should be thoroughly considered. Regarding "structure testing", there is a need to define appropriate testing protocols, which include loading type, intensity and test sequence, together with any variables representative and relevant to the control of performance, on the basis of realistic loading conditions for different test levels. Intensity, sequence and number of tests represent a compromise between an ideally refined response/capacity evaluation and the need to limit the number of sequential tests on the same model causing unrealistic cumulative damage. This requires a close interaction between various actors, namely experimentalists and analysts.

In order to meet the requirements of PBSD there is also a need for experimental facilities capable to handle complex structures and systems, including 3D earthquake response, asynchronous input motions, to understand real effects of phenomena like soil-structure interaction (SSI) and to combine physical and numerical testing online and offline in a sort of "real-virtual testing environment" where local and global, point and field digital measuring and visualization systems and corresponding processing can provide detailed information on demands and on the corresponding consequences, namely type and evolution of physical damage. A more risk oriented design approach will require a better quantification (estimation) of structural and non-structural damage and corresponding losses. This requires testing models (test specimens) integrating also non-structural components.

There will be the need to improve test capacities and enhance the characteristics of the testing facilities and probably to design and construct new research infrastructure, which can better satisfy the present and future needs. The EFAST project, a design study financed by the European Commission, addresses the challenge of a new European advanced facility for earthquake and dynamic testing.

This chapter is based in two previous publications on related topics (Pinto et al., 2004a, 2006). It summarizes the recent "history" of the European projects involving large-scale seismic testing underlying the impact on the Eurocode and the role in strengthen the European earthquake engineering research community. Then, a short reference to the integrated project SERIES is made, in what concerns new opportunities for experimentation. The first results from the EFAST design study are given. Technical and scientific aspects related to large-scale testing are discussed in view of the new challenges of PBSE and risk-based design.

15.2 The Co-operation between European Large-Scale Earthquake Testing Facilities

In Europe, the earthquake engineering laboratories initiated an enlarged collaboration at the beginning of the 1990s, as a direct consequence of financial support from the European Commission within the various Framework Programmes. Five major shaking tables were involved, together with the ELSA reaction wall facility at JRC, Ispra. As stated by (Severn, 2000), as a result of providing access to researchers from Member States, a major step forward was made in the fidelity and accuracy with which these six facilities could be used, leading to significantly enhanced performance. This major advance has put European earthquake engineering infrastructures at international level and has opened up several new research areas for experimental study. Collaboration between the infrastructures group and three successive Research Networks has allowed significant progress to be made towards the validation of many aspects of Eurocode 8 and the mitigation of seismic risk.

15.2.1 Past Co-operation

The consortium of the European Large Scale-Facilities in Earthquake Engineering (ECOEST), coordinated by Prof. Roy Severn, from the University of Bristol, UK, developed several activities and addressed several technical and scientific topics, namely: (1) Shaking table standardization studies and research; (2) Access to shaking tables for external researchers from all European countries; (3) Access by others in ongoing research at the ECOEST laboratories; (4) Research programmes initiated by users; (5) Training in shaking table procedures.

It is noted that the first large-scale earthquake tests on structures design with the Eurocodes (Eurocode 8) were performed in the framework of ECOEST with an enlarged research network of the earthquake engineering community (PREC8 Research Network, coordinated by Prof. G.M. Calvi from the University of

Pavia) acting as external user(s) of the facilities. The PREC8 network involved 18 European research groups working on the following 4 broad themes, chosen because of their importance to the new European Seismic Design Code, namely: (1) Reinforced Concrete Highway Bridges (ISMES, Bristol, JRC-Ispra); (2) Reinforced Concrete Frames (Athens); (3) Infilled Frames (Athens, ISMES, Bristol, LNEC); (4) Geotechnical structures (Bristol).

Given the success of the co-operation and collaboration between research infrastructures, external researchers (Universities and Laboratories) and the outcome of the research activity and its relevance for European standards in process of development and approval by the European countries, the consortium was further financed to continue its activities, addressing now the new topics, as part of the programme of the research network ICONS – Innovative Concepts for New and Existing Structures, with emphasis on: (1) Seismic Actions; (2) Assessment, strengthening and repair; (3) Innovative design concepts; (4) Composite structures; (5) Shear-wall structures.

In the context of the fifth Framework Programme of the European Commission, the consortium was further expanded (ECOLEADER) and the earthquake engineering research concentrated on the topics of the Network – Safety Assessment for Earthquake Risk Reduction (SAFEER) research programme, namely: (1) Characterization of seismic hazard; (2) Assessment and design in low seismicity regions; (3) Strategies/techniques for risk reduction; (4) Risk assessment systems.

Together with the unique reference tests performed in the consortium facilities, there was research to improve functioning and control of the testing facilities and to further advance testing methods and techniques. The following projects were developed subsequently: (1) CESTADS – Control Enhancement of Shaking Tables from Analogue to Digital systems; (2) FUDIDCOEEF – Further Developments in Dynamic control of Earthquake Engineering Facilities; and, (3) NEFOREE – New fields of research in earthquake engineering experimentation. These projects allowed for important enhancement of the European testing facilities with substantial progress on: (1) Control of shaking tables (Minimal Control Synthesis (MCS) algorithm, developed by Stoten and Gomez, 1999); (2) Controlled testing in the non-linear range of material behaviour; (3) Substructuring on shaking tables; (4) Multiple support input motions; (5) The effects on response of spurious motion; (6) Continuous Pseudo-Dynamic Testing (CPSD).

As a platform for enlarged discussion between facilities, their users and external community, the Access and Research projects were accompanied by Coordination Actions consisting of thematic workshops, organized in different countries, and by technical reports responding to the requirements for dissemination of the scientific results and outreach of the scientific and technical communities and academia as well as policy makers. The report series ECOEST-PREC8 (1996), ECOEST2-ICONS (2001) and CASCADE (2005) [see also Taucer, 2005; Taucer and Franchioni, 2005], constitute a set of 27 technical reports. They were written and edited by several European researchers and published by the National Laboratory for Civil Engineering, Lisbon, constituting important milestones of the pioneering

European cooperative research on earthquake engineering, involving experimental facilities, their users and the research community.

Details of the experimental research carried out at the European laboratories can be found in (Severn, 2000) and in the report series referred to above.

15.2.2 Present Opportunities: The SERIES Project

After a short "interregnum" of the European Commission funding, the Research Infrastructures in Earthquake Engineering and Structural Dynamics have setup a new and enlarged consortium coordinated by Prof. M. Fardis, from the University of Patras, Greece, the SERIES project (Seismic Engineering Research Infrastructures for European Synergies). The project was approved by the Commission (FP7 – Integrating Activity – Combination of Collaborative Project and Coordination and Support Action) and financed for a period of 4 years starting at the beginning of 2009.

As stated in its Document of Work, SERIES aims at bridging the two gaps of RTD in experimental earthquake engineering and structural dynamics: (a) between Europe and the US or Japan; and, (b) between European countries. It will do so by integrating the entire European RTD community in earthquake engineering via:

1. A concerted program of Networking Activities, fostering a sustainable culture of co-operation among all research infrastructures and teams active in European earthquake engineering, including:
   - A distributed database of test results, pooling data from the beneficiary research infrastructures and others, accessible and maintained by a virtual research community after the project's end;
   - Tele-presence and geographically distributed concurrent testing at the research infrastructures;
   - Standards, protocols and criteria for qualification of RTD infrastructures in earthquake engineering;
   - Enhancement of human resources by training new users and beneficiary technical/research personnel in courses on good practices in operation and use of research infrastructures.
2. Coordinated Transnational in-person Access of Users to a world class portfolio combining:
   - EU's four largest earthquake Shaking Tables, each one with diverse capabilities: (a) the TAMARIS laboratory of CEA/Saclay (FR), (b) the EUCENTRE/TREES Lab in Pavia (IT), (c) LNEC in Lisbon (PT) and (d) the Bristol University Earthquake and Large Structures Laboratory (UK);
   - EU's largest Reaction Wall and Pseudo-Dynamic testing facility (ELSA) at the JRC, Ispra, and unique Centrifuge Test facilities at (a) LCPC in Nantes (FR) and (b) Cambridge University (UK).

3. Joint innovative Research toward new fundamental technologies and techniques promoting efficient and joint use of the research infrastructures, in three areas where the beneficiaries excel at world level:
   - Concepts, technical requirements and prototyping for new-generation electrodynamic actuators (including coupling with hydraulic ones) for high-performance, enhanced-quality real-time testing;
   - New instrumentation and sensor techniques for improved sensing and test control. Dedicated software for data collection, processing and communication, serving current needs for model calibration and interpretation of structural response. Use of data assimilation and model updating to develop virtual models of the equipment-specimen system, in combination with recent advances in control, to reduce calibration pre-tests, optimise instrumentation and improve the quality results;
   - New capabilities and techniques for experimental study of soil-structure-interaction and seismic wave propagation phenomena, currently insufficiently covered by experimental research infrastructures at world level.

More details on the SERIES project can be found in (Fardis, 2009) and in the project website (http://www.series.upatras.gr/), which gives also detailed information on the procedures to apply for Transnational Access to the Research Infrastructures. Researchers (individual researchers or research groups), technical community and industry have a unique opportunity to develop their ideas and to check them under realistic testing conditions in one the world-class European experimental facilities.

15.2.3 New Research Infrastructures in Europe: The EFAST Project

EFAST (European Facility for Advanced Seismic Testing) is an FP7 European collaborative project, coordinated by Dr. Ionnis Politopoulos, from CEA, France (http://efast.eknowrisk.eu/EFast/). It deals with the study of all aspects regarding the design (design-study) of a major testing facility in Europe that would complement and collaborate with the existing ones. Actually, earthquake risk is a major civil protection issue which is related to mitigation measures for protecting citizens, infrastructures, property and the human cultural heritage. Seismic testing plays a key role for the better understanding of physical phenomena, the validation and improvement of analysis and design methods and for the qualification of sensitive equipment. Therefore the availability of high level experimental facilities is essential to meet the objectives of earthquake mitigation.

A required first step for a design study for a new advanced research infrastructure is to collect, understand and classify the future needs, in terms of research. Then, there is the development of suitable experimental technologies to address the new problems and finally the preliminary design of the research infrastructure should be carried out eventually in parallel with some demonstration activity.

The first round-table of the EFAST project had the contribution of several experts from Europe and abroad. The following points were taken from the needs identified for the near future (Marazzi and Molina, 2009):

- A better knowledge of the behaviour of flat-slab buildings, pre-stressed framed structures, masonry structures, structures with masonry infills, cultural heritage buildings and bridges is needed;
- Further harmonization of the Eurocodes through reduction of the number of Nationally Determined Parameters (NDPs). Further studies regarding assessment and retrofitting of buildings and bridges are needed;
- Realistic tests of Soil-(Foundation)-Structure Interaction (S(F)SI) systems;
- Qualification of protection devices;
- Better integration of experimental and numerical simulations underlying the importance of networking and complementarities;
- New measurement technologies should be used allowing field measurements;
- Meaningful probability risk assessment requires knowledge on actual margins of the structures have to be estimated. To this end tests should be performed up to collapse or resulting in a relevant significant damage. This implies that the new facility should have the capability to reproduce high intensity excitations (high acceleration, velocity and displacement);
- Easy share the data: new “Informatics Technologies” (IT) should be adopted.

A second round-table on Testing Methods and Technology has agreed on the following points (Marazzi and Molina, 2009):

- A key feature of the future testing facility must be its versatility (e.g. capability for applying multi-axial loading and for substructuring testing (including fast and real-time substructuring), possibility extension and upgrading);
- Strong coupling between the experimental and the numerical aspects. Software harmonization should be promoted;
- Information, dissemination and collaboration must be stressed;
- Besides the main testing facility, some dedicated Testing Facilities (e.g. MATS – Multi-Axial Testing System for testing of non structural components) should be considered.

More details on the EFAST project and round-tables can be found in (Marazzi and Molina, 2009) and in the project website (http://efast.eknowrisk.eu/EFast/).

15.3 Examples of Large-Scale Testing

15.3.1 Objectives of Structural Testing

Experimental verification of the performance of structures subjected to earthquake input motions can be made through either shaking table (dynamic) tests or reaction-wall (pseudo-dynamic) tests. However, if strain rate effects are important and condensation to a reduced number of test DOFs is not realistic, dynamic testing should be sought. On the other side, if large-full scale models should be considered, pseudo-dynamic (PSD) testing becomes the appropriate solution because complex nonlinear phenomena are often accurately simulated only at full or large model scales. Furthermore, expansion of the time scale makes up for much more handy tests, in that the tests can be stopped at any critical event and be re-started if necessary. Furthermore, PSD testing allows hybrid (physical and numerical) online simulation of large structures and systems to be carried out by substructuring techniques already familiar to analysts.

The basic objectives of earthquake testing of structures can be summarized as:

- to check the accuracy of numerical models and to adjust/calibrate model parameters (modelling of single components may not capture the behaviour of a complete structural system);
- to check structural performance for different input motion intensities (compare: demand, control variables and damage descriptions with capacity, limit state characterization and, ultimately, to reach collapse of the structure, which is normally associated with: (1) severe degradation of the structural properties often not accurately simulated by the analytical models, and/or (2) brittle failure modes not captured by the models);
- to build confidence and trust on the performance of new structural solutions, new design methods (e.g.: new design codes) and innovative materials, as well as to provide evidence on good or bad performance (demonstration).

A test campaign normally involves a series of phases as described in Table 15.1. However, there is no standard procedure to conduct a test campaign. It should be tailored to the research/demonstration/qualification scope and objectives.

Table 15.1 Full-scale seismic tests: stages and corresponding description

Stage	Description
A0	Define scope and objectives of the experimental campaign
A	Define a test specimen representative of a class of structures
B	Subject test specimen to EQ ground motions with specific intensities, I1, I2, I3, ..., corresponding to characteristic lifetime exceedence probabilities (e.g. 50, 10 and 2%) and achieve collapse stage (Ultimate Capacity)
C	Record demands, in terms of deformation (e.g. drifts) and corresponding damage description
D	Carry out engineering quantification of damage (damage model, damage indices), taking into account the problem of cumulative damage resulting from sequential tests
E	Carry out calibration of damage cost functions relating drifts and/or damage indices with repair costs
F	Compare performances with corresponding performance objectives
G	Identify implications on modelling, design methods and procedures, redefinition of performance objectives

The minimal scope of structural seismic tests would be to check the performance of a model when subjected to the loading considered in its design and to check also its ultimate capacity in order to evaluate safety margins. In fact, the present limit-state based design codes explicitly consider one or two limit-states (safety and serviceability) and implicitly assume that the structure should be able to withstand (without collapse but with important/severe damage) earthquake intensities much higher than the design ones, which is achieved through capacity design (preferential-stable dissipation mechanisms) and requirements on ductility capacity. Explicit quantification of the seismic intensities associated to limit states other than safety is not given, nor is required performance to be checked. Therefore, one relies on prescriptive design procedures and on intended performances, which require verification and/or calibration. This has been the main scope of most of the tests performed at ELSA on structures designed according to the Eurocodes. Building and bridge models were tested and the results were used by the European research community and code-makers, to calibrate models, to refine some parts of the code (e.g. ductility classes, behaviour factors), to introduce new design rules (e.g. structures with infill panels) and analysis methods, to introduce new materials (e.g. composite structures) and to introduce new technologies (base-isolation and distributed passive dissipation systems).

A few examples of tests performed at ELSA in support of Eurocode 8 are given in this chapter. One is concerned with new structures and the corresponding tests were carried out for earthquake intensities corresponding to serviceability life-safety and ultimate capacity. The second example is concerned with the assessment of existing structures, for which a test protocol tailored for life-safety and for ultimate capacity was adopted. The third example addresses the issue of torsion (in-plan irregularity) and the last one constitutes a pioneering example of pseudo-dynamic testing with non-linear substructuring applied to bridges.

15.3.2 Testing of a 4-Storey RC Structure Designed to the Eurocodes

The first experiments performed at ELSA in support of the European Codes consisted on a series of tests on a full-scale 4-storey RC frame building designed according to Eurocodes 2 and 8 (see Fig. 15.1). This was the first "Eurocode structure", built and seismically tested for two different earthquake input motion intensities corresponding to serviceability and life-safety limit states. The structure was subsequently subjected to a displacement controlled cyclic test up to collapse in order to check its ultimate capacity. Earthquake intensities corresponding to 40 and 150% of the "design-earthquake" (DE) were used in the PSD tests. Illustrative results are given in Figs. 15.1 and 15.2. Detailed description of the research programme, test results and analysis can be found elsewhere (Negro et al., 1996). It is however important to note that the low-level test caused only minor cracking in the structure and apparent low damage was sustained in the high-level test, with cracks

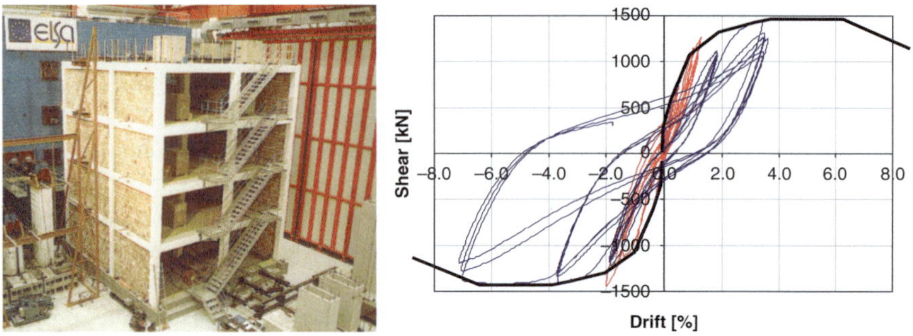

(a) (b)

Fig. 15.1 R/C structure tested at ELSA: (**a**) Infilled frame configuration, (**b**) Bare frame 1st storey shear-drift diagrams

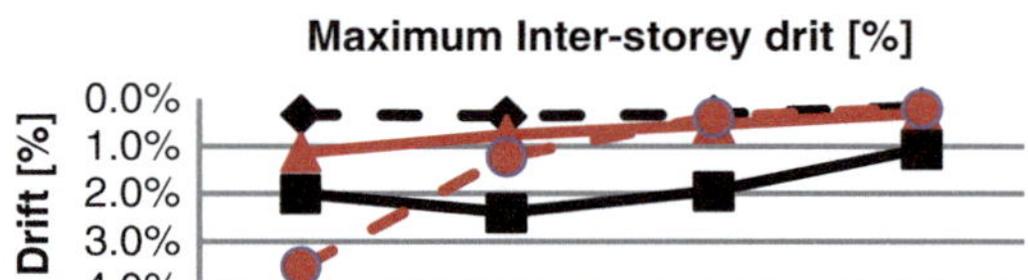

	Storey 1	Storey 2	Storey 3	Storey 4
Bare frame 0.4DE	0.3%	0.4%	0.3%	0.2%
Unif Infill 1.5DE	1.1%	0.8%	0.5%	0.3%
Bare frame 1.5DE	2.0%	2.4%	2.0%	1.1%
Soft-storey Infill	3.5%	1.2%	0.4%	0.2%

Fig. 15.2 R/C structure tested at ELSA: drift profiles for three earthquake tests: bare frame 0.4DE (0.4×design earthquake), bare frame 1.5DE, Unif Infill (uniformly infilled) 1.5DE and Soft-storey Infill (open first storey) 1.5DE

remaining open only in the critical parts of the beams. Yielding of rebars took place in the beams plastic hinge zones and at the base of the ground floor columns, but neither spalling of concrete (only slight indication of spalling at the base of the 1st storey columns) nor buckling of rebars was observed.

Before the final cyclic collapse test, two additional pseudo-dynamic tests were carried out: one with infill panels uniformly distributed along the height (see Fig. 15.1) and another one with infills at the all but the ground storey. Maximum inter-storey drift profiles for these tests are given in Fig. 15.2. The final cyclic test on the bare frame was performed with imposed top displacement and inverted triangular force distribution. Figure 15.1 shows also the first storey shear-drift diagrams for the tests on the bare frame structure, confirming that the structure experienced a maximum drift of 7% without significant loss of load carrying capacity.

15.3.3 Assessment and Retrofit of Existing RC Frame Structures

A series of pseudo-dynamic tests on two full-scale models of a 4-storey R/C frame (Fig. 15.3) representative of existing structures designed without specific seismic

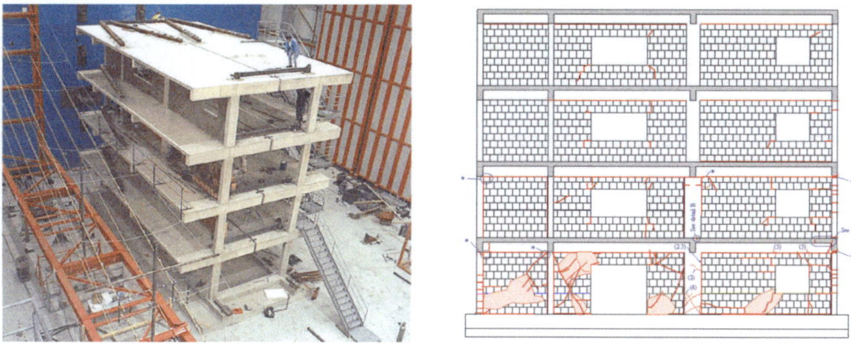

(a) (b)

Fig. 15.3 Full-scale models of an existing RC frame structure: (**a**) Test set-up at ELSA, (**b**) damage patterns for the infilled frame configuration after the 975-yrp pseudo-dynamic test

resisting characteristics (common practice of the 1960s in South European countries) were carried out at ELSA. Four testing campaigns were performed aiming at: (1) Vulnerability assessment of a bare frame; (2) Assessment of a selective retrofit solution; (3) Earthquake assessment of an identical frame with infill masonry walls; (4) Assessment of shotcrete retrofitting of the infill panels.

Contrarily to the strategy adopted in the tests described in the last section, which aimed at the verification of serviceability and life-safety limit-states and check of the ultimate capacity, the tests on the model representing existing structures were focused on the behaviour and performance for input motions corresponding to the design actions of new structures as well as on the assessment of their ultimate capacity. Therefore, an input motion corresponding to 475 yrp was adopted for the first test on the bare frame. The second test aimed at reaching ultimate capacity of the frames and was carried out with an input motion intensity corresponding to 975 yrp. The tests on the retrofitted structure and on the infilled frame structure adopted the same input intensities in order to allow for direct comparison with the original configuration. A subsequent PSD test with an intensity corresponding to 2,000 yrp was carried out. Illustrative results are given in Fig. 15.3 (damage patterns in the infill panels) and Fig. 15.4 (storey shear-drift diagrams for bare and infilled frames) whereas a detailed analysis of the test results can be found elsewhere (Pinto et al., 2002). Figure 15.5 compares the maximum inter-storey drift for the 475 and 975 yrp PSD tests on the infilled frame with reference drift values corresponding to damage limit states.

15.3.4 3D Tests on a Torsionally Unbalanced Structure

A substantial improvement of the testing capabilities has been obtained by the commissioning of a bi-directional PSD implementation.

In the framework of the research activities of ELSA, PSD testing of a real-size plan-wise irregular 3-storey frame structure was carried out as the core of

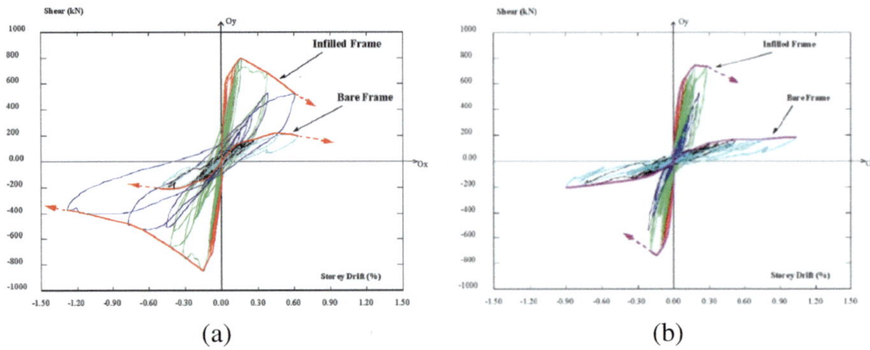

Fig. 15.4 Model of an existing RC frame structure: storey shear-drift diagrams and envelope curves for the bare and infilled frames: (**a**) first-storey, (**b**) second-storey

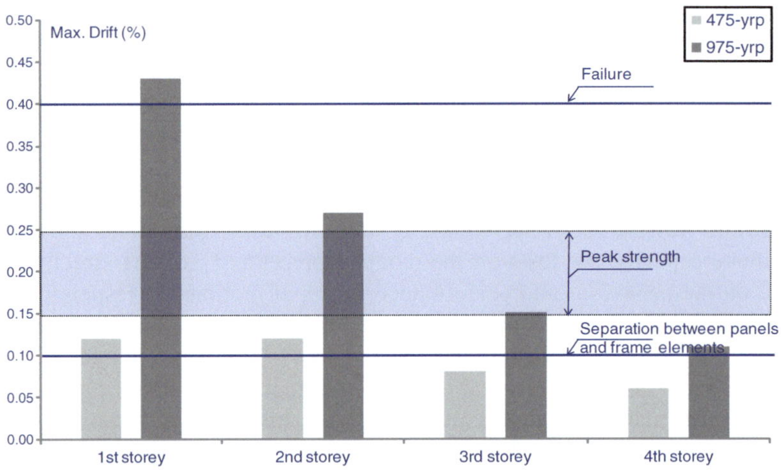

Fig. 15.5 Maximum inter-storey drift for the 475 and 975 yrp PSD tests: comparison with drift values corresponding to damage limit states

the research project SPEAR (Seismic PErformance Assessment and Rehabilitation of existing buildings). The project was specifically aimed at throwing light onto the behaviour of existing old RC frame buildings lacking seismic provisions. A balanced combination of numerical and experimental activities was considered, including a series of full-scale PSD tests on a torsionally unbalanced 3-storey RC frame structure, representing a common configuration of housing units in most earthquake-prone areas of Europe. The experimental phase focused on a real-size specimen (see Fig. 15.6). The first tests were carried out on the structure in its original "as built" configuration. Following these tests, a light (i.e., member-level) retrofitting intervention (FRP wrapping of columns to improve ductility) was carried out. A new round of tests was performed on the retrofitted configuration, so that the

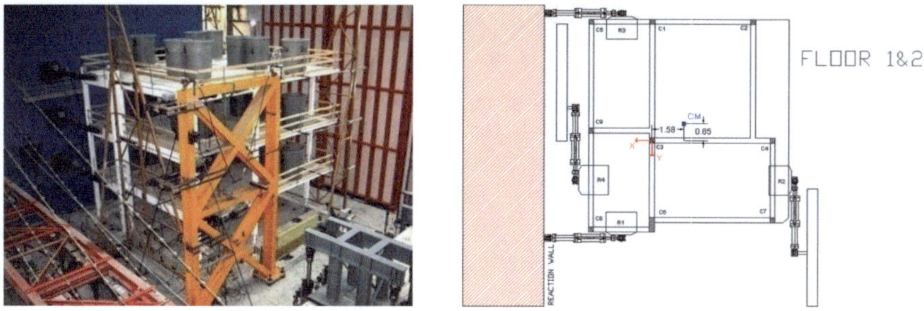

(a) (b)

Fig. 15.6 3D tests on an full-scale model of a torsionally unbalanced RC structure: (**a**) View of the model at the testing site; (**b**) Actuators layout and location of floor centre of mass (CM)

effectiveness of currently available guidelines for the design of retrofitting interventions will be judged. Finally, the damage inflicted by the second round of tests was repaired and the structure more heavily retrofitted, by means of interventions aimed at improving the global structural configuration.

The bi-directionality of the PSD test, consisting in the simultaneous application of the longitudinal and the transverse component of the earthquake to the structure (see Fig. 15.6), introduces a higher degree of complexity, from both the analytical and technical points of view, with respect to usual unidirectional PSD testing. In fact, three DOFs per storey need to be taken into account: two translations and one rotation along the vertical axis, as opposed to the single degree of freedom per storey that is usually taken into account in unidirectional PSD testing. Four actuators per storey were connected to the structure, three of which were strictly necessary. The structure was subjected to two tests (with PGA of 0.15 and 0.20 g). Each test with one accelerogram in each direction. Illustrative results from the second test (0.20 g) are given in Fig. 15.7, which shows rather different column drift histories resulting from the induced torsion of the building. Detailed analysis of the test results and test set-up can be found elsewhere (Negro et al., 2004).

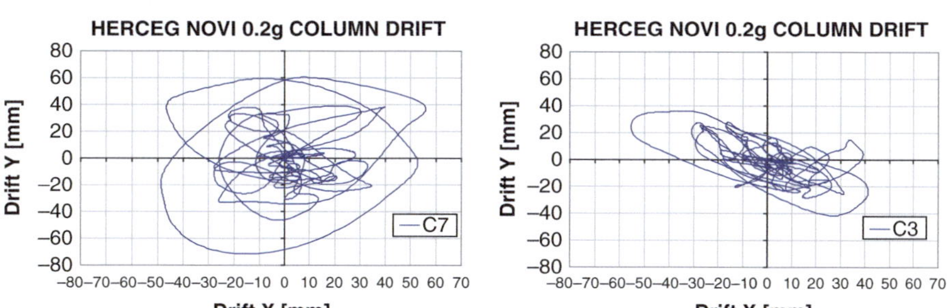

(a) (b)

Fig. 15.7 Second storey X and Y direction drifts for two columns: (**a**) external-corner column, (**b**) internal-‘central’ column

15.3.5 *Pseudo-Dynamic Testing of Bridges with Non-linear Substructuring*

Another major advancement in testing techniques was obtained by the development of non-linear substructuring techniques in PSD testing.

ELSA already had experience on PSD testing with substructuring in application to bridges. PSD testing with linear substructuring was successfully applied to bridges at ELSA in the mid 1990s (Pegon and Pinto, 2000). A series of PSD tests were performed on regular and irregular bridges designed according to the Eurocodes, with three piers of the model-bridge built and physically tested and the deck simulated numerically with linear FEM. The test campaign comprised also isolation solutions to tackle the irregularity problem and addressed the issue of asynchronous input motion (Pinto, 1996). An extension of this technique is the use of non-linear models for the numerical parts of the structure – "non-linear substructuring".

Non-linear substructuring was for the first time applied at ELSA (Pinto et al., 2004) allowing the assessment of the performance of a six-pier bridge to be made with physical testing of two piers and on-line simulation of the remaining piers (non-linear numerical models) and deck (linear numerical model). A schematic representation of the test set-up is shown in Fig. 15.8. The bridge was tested for three input motion intensities corresponding to probabilities of exceedance of 50, 10 and 2%, in 75 years (tests: 0.4×NE, 1.0×NE and 2.0×NE). Recorded values of the maximum top displacement of the piers are shown in Fig. 15.9. Detailed description and analysis of the results can be found elsewhere (Pinto et al., 2004).

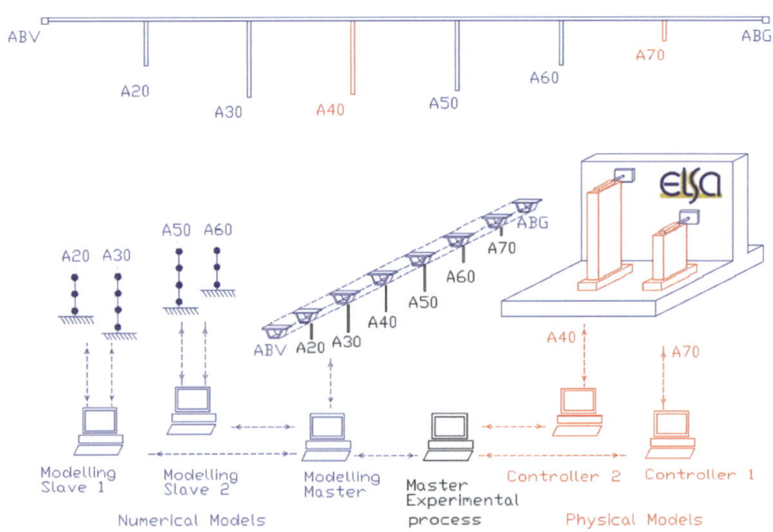

Fig. 15.8 PSD testing with substructuring – application to bridges: test setup

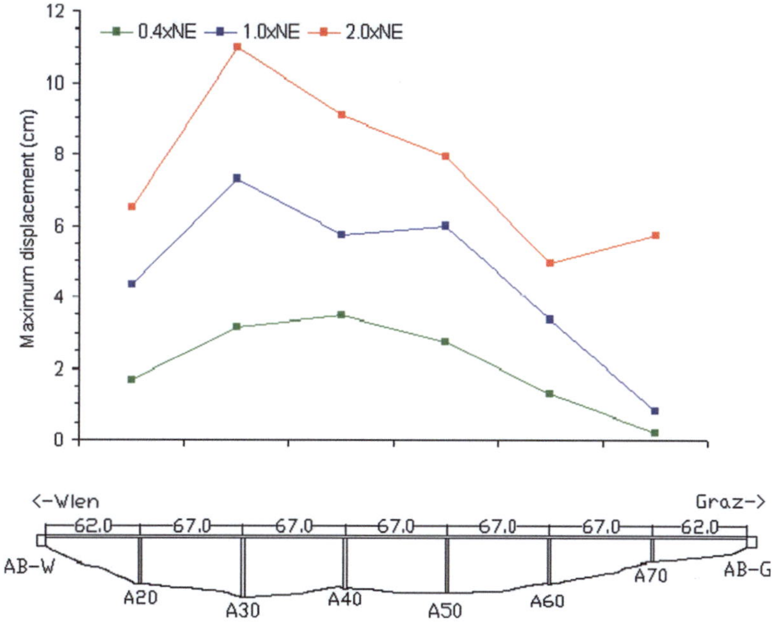

Fig. 15.9 PSD testing with substructuring: Bridge EQ demands for three testing intensities (NE-Nominal (Design) Earthquake)

15.4 Large-Scale Testing: Needs and Opportunities

15.4.1 New Concepts and Concerns in Seismic Design

Seismic design has experienced a substantial evolution in the last 50 years achieving the fundamental objective of life safety and accepting/incorporating solutions and technologies that enable critical facilities to remain operational after major seismic events. The present seismic design codes state clear objectives in terms of life safety (strength and ductility requirements), which can be mostly achieved, and state also objectives in terms of damage control that are typically checked indirectly. This means that damage control checks are derived from demands based on the values calculated from ultimate limit states.

As economical aspects are also becoming overriding objectives in our societies, measurable consequences of earthquakes, such as structural and non-structural damage (e.g. repair costs) in earthquake events, as well as other economical consequences (e.g. loss of operation/revenue) and "non-measurable" consequences, such as social impacts (quality of life), should be considered in the planning and design of our infrastructures, living and production facilities. As a matter of fact, the economic losses resulting from the last major events in US and Japan can be considered as the motivation for PBEE, which is deemed to provide an appropriate platform to achieve safer and more economic constructions.

The conceptual frameworks proposed in the USA for PBEE (Krawinkler, 1999), such as Vision 2000, can be considered as a step forward on a more rational seismic design and assessment/redesign of engineered facilities. In fact, explicit consideration of multi-level performance objectives together with specific seismic intensities leads to a more controllable/predictable seismic performance (see Fig. 15.10). This represents a significant improvement relatively to the single-level explicit approach of current design codes because it requires explicit consideration and check of key performance objectives and it conveys it clearly to the designer that a structure is likely to be subjected to different seismic intensities during its life, including severe ones with low probability of occurrence.

Experience from recent earthquakes in Economically More Developed Countries (EMDC) indicate that there is increased expectation from modern societies concerning performance of structures. It is claimed that also ordinary structures should remain operational after rare events, which implies a shift in the multi-level design procedure as illustrated in Fig. 15.10 (Pampanin, 2009).

However, this multi-performance approach still embeds a prescriptive concept, in the sense that the association of a series of performance objectives with specific input levels does not leave space to differentiated choices and might not satisfy the requirements and expectations of different stakeholders (the general public, owners, lenders, insurers, businesses and government). It is believed that decisions regarding acceptable earthquake risk should be left to the stakeholders and the scientific/technical communities should focus on the issues related to calculation of these risks and associated costs.

It is however advocated that a risk-based approach should be followed for seismic design. It should include prescriptive performance objectives related to safety as

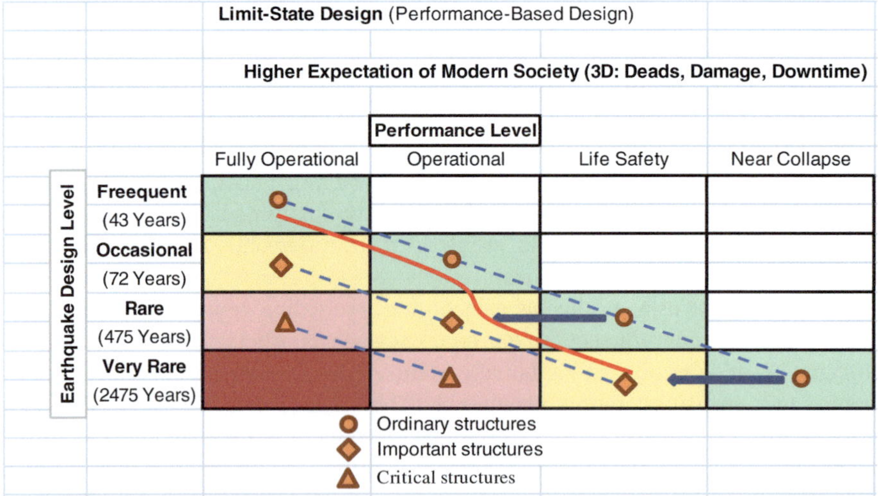

Fig. 15.10 Vision 2000 matrix (multi-level performance objectives with corresponding demands) and increased performance expected by modern societies

well as to other relevant macroeconomic minimum requirements (stakeholders: state and authorities) and leave the economic aspects on the other stakeholders, who are deemed to focus on the mitigation of the adverse economic consequences (Hadjian, 2002). This mixed approach (minimum requirement performance – optimum-risk based) imposes minimum safety levels as well as minimum social adverse consequences and brings seismic design to a new level, where safety, design optimization (allowed trade-off between different performance levels) and innovation can coexist.

15.4.2 Earthquake Expected Losses: Illustrative Example

Reliability analysis and risk assessment of structures can be carried out following well-established methodologies. Difficulties may be encountered in system reliability approaches, for which correlation between different failure mechanisms exists as well as in the quantification of demand and capacity dispersions and loss (cost) functions. An application of reliability and risk assessment tools and methodologies to structures designed according to the Eurocodes was made by (Pinto, 1998). A case studied in that research work is herein revisited to underline a few important aspects relevant to risk assessment and also to the definition of appropriate earthquake testing protocols.

The four-storey reinforced concrete frame building presented in previous section is the subject of this example. The structure was modelled numerically and was assumed to be in a High-seismicity zone in Europe with a hazard compatible with its design seismic action. Response simulations (non-linear models under earthquake input motions) were obtained for several input intensities (each using five artificially generated accelerograms). The response curve was approximated by an analytical function (average values of the simulation results) and a constant c.o.v. of 25% was assumed for the sectional lognormal distribution of the response. Performance curves were obtained for a few different cases (using the same number of "experimental" points but distributed differently along the intensity ranges to approximate the response curves) and subsequently Annualized Earthquake Losses (AEL) were derived (see Fig. 15.11). It was concluded that the approximation of the structural response curves represents a key component of the risk assessment process, with very significant implications on the values of the expected earthquake losses. Approximation should be based on well-distributed "experimental points" covering low, medium and high input intensities.

There is another important aspect to take into consideration in the risk assessment process, which is concerned with the contribution of the damage ranges to the total expected losses. Figure 15.11 shows the partial contribution of the damage ranges (0–0.05, 0.05–0.1, 0.1–0.15, ...) to the total expected losses. It is noted that damage states in the vicinity of 0.1 are predominantly contributing to the repair and economic losses whereas "human losses" are practically constant for all damage ranges other than for the damage values lower than 0.1, for which they are very limited.

The key concluding note is that reduction of economic losses is effective in the zones corresponding to low–low/medium damage indices, which can be addressed

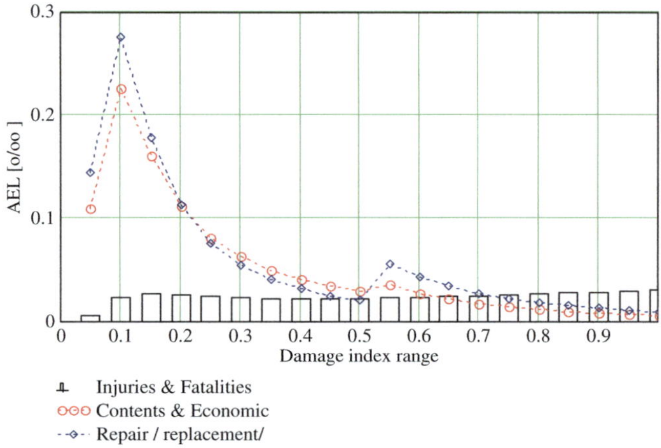

Fig. 15.11 Contribution of damage ranges to total expected AEL

Fig. 15.12 Typical Investments in building construction (after Taghavi and Miranda, 2003)

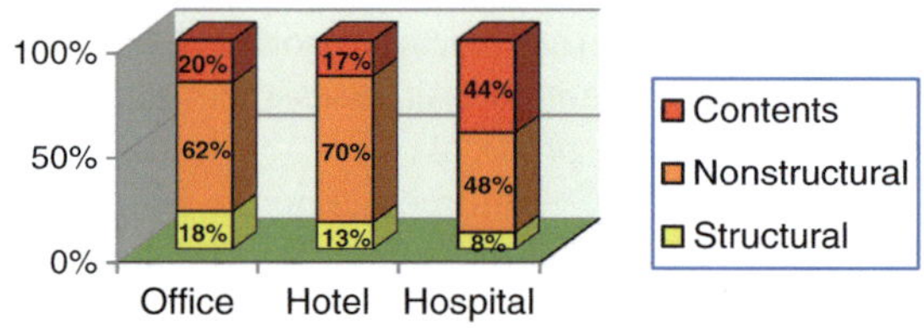

by the reduction of damage-inducing demands (e.g. drifts) corresponding to low-medium input intensities with high probability of occurrence.

It should be noted that the above mentioned calculations were performed assuming that the structural part represents one third (1/3) of the total investment in the buildings construction. The fact that non-structural components and contents tend increasingly to dominate building investment costs, beyond the classic 2/3 (as illustrated in Fig. 15.12), further underlines the need to reduce damage-inducing demands for low-medium earthquake input intensities with high probability of occurrence.

15.4.3 Further Development of the Eurocodes

The process of transformation of the Eurocodes into European Standards was completed in 2007 and their national implementation is progressing aiming at their "enforcement" by 2010 in most European countries (Eurocodes, 2010). In order to undertake a period of stability, no major revision is foreseen within the next 5 years. However, the rapid progress in seismic design and the request for more risk-oriented design procedures call for the preparation of an eventual revision after that period (preparing the next generation of the Eurocodes). As illustrated in Fig. 15.13,

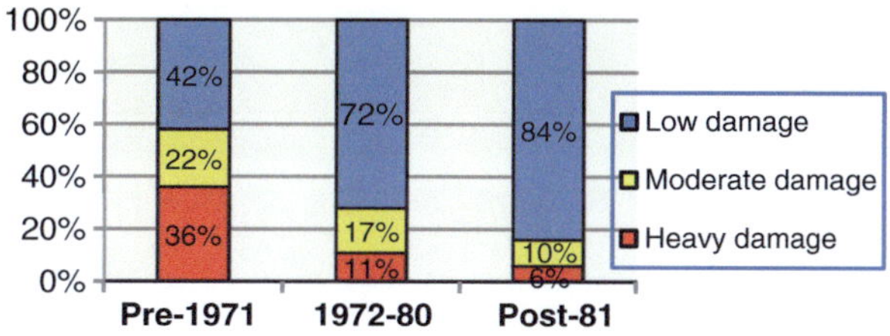

Fig. 15.13 The role of the Design code: performance of buildings constructed in Kobe by Ohbayashi Corporation (Pinto, 1998)

past experience on updating of the design codes represents always progress on live protection and reduction of economic loss).

The present version of Eurocode 8 (EN1998-1, 2004) adopts a two-level seismic design procedure with two explicit performance objectives, namely: (1) Protection of life under a rare seismic action, by preventing collapse of the structure or parts of it and ensuring structural integrity and residual load capacity; (2) Limited property loss in a frequent earthquake, via limitation of structural and non-structural damage. The hazard levels corresponding to these limit-states verification is left to the National Authorities (through the Nationally Determined Parameters, NDPs).

There are two aspects to take into consideration: one is related to the number of limit-states requiring explicit verification, which might be increased in light of a more refined performance assessment/verification and also due to the increased weight of the non-structural parts in the building investments. The other aspect is related with the expected variations of the NDPs among the European countries. They should represent a rational choice and if possible there should be progress towards harmonization. All these aspects require research and development, including experimental evidence.

In addition to this evolution of concepts, other research needs in support of Eurocode 8 were already identified (Carvalho, 2009), (Fardis, 2007), (Pinto et al., 2007) and (Spence et al., 2005). Aiming at to concentrating on the topics not covered in the present version of EN 1998 and to resolve the aspects concerned with safety, the following research topics were identified: (1) Harmonized European Seismic map; (2) Provisions for the design of irregular-in-plan buildings; (3) Primary vs secondary seismic elements: Elaboration of the implications and re-evaluation of the concept; (4) Seismic design rules for flat slab systems; (5) Seismic design rules for prestressed concrete elements and systems; (6) Design rules for masonry buildings; (7) Seismic assessment and retrofitting (emphasis to masonry-infilled frame buildings); (8) Seismic design of the structure-foundation-soil system; (9) Seismic protection of sensitive or valuable equipment and artefacts.

Concerning the issue of Seismic assessment and retrofitting (emphasis to masonry-infilled frame buildings) it should be noted that, in addition to safety, there are other emerging requirements for buildings, concerned with energy saving and sustainability. The picture shown in Fig. 15.14 illustrates a case were conflicting requirements co-exist: in one side there is the infill-panel full-confinement safety requirement, on the other side there is the energy efficiency/saving requirement, which requires full-involvement in order to avoid thermal bridging.

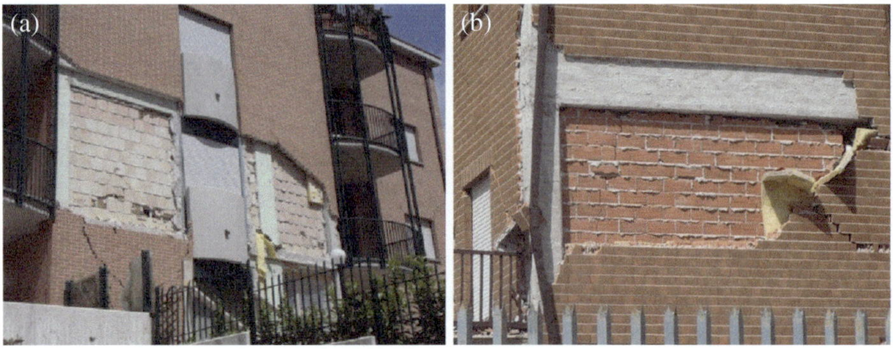

Fig. 15.14 L'Aquila, 2009: safety/damage control vs energy efficiency/saving: (**a**) building 1; (**b**) building 2

15.4.4 Harmonization of Experimental and Analytical Simulations

The Bled Workshop, "Performance-based Seismic Design Concepts and Implementation" (Fajfar and Krawinkler, 2004; Fardis, 2004), included a working group session on Harmonization of Experimental and Analytical Simulations, which discussed and agreed on a set of conclusions and recommendations, which should be placed in a paper on Large-scale testing. The Working group offered the following conclusions and recommendations:

Advanced experimental facilities have become available worldwide, for example: NEES, E-Defense, JRC, and NCREE. Experiments on complex structural systems at larger scales become more practicable. They provide great opportunities for more accurate characterization of various limit states of structures and ultimately for accelerated acceptance of PBEE. New experimental facilities, techniques, and devices require new approaches to research and development. The following specific recommendations are along these lines:

- Testing procedures. Experimentation should cover the full range of behaviour from damage initiation to collapse. Test structures should contain non-structural and content systems to the extent feasible. In simplified test configurations, much attention needs to be paid to simulation of boundary conditions. Field testing should be encouraged to provide realistic performance data. A great need exists to develop testing protocols, including interaction between testing and analysis, peer review of procedures, careful selection of input motion, and specialized protocols for testing of non-structural components and for material testing. Advanced instrumentation should be developed (including high-resolution image processing) for comprehensive documentation of damage data. All experimental data should be documented, archived, and shared publicly after verification, taking into account intellectual rights.
- Analytical prediction of behaviour until collapse. Improved approaches need to be developed to simulate collapse and behaviour of non-structural systems, and for constitutive modelling of new and existing materials. Computer analysis programs should emphasize user-friendliness and should be developed through partnerships of researchers and practicing engineers with software companies.

- Distributed simulations. The benefits obtained from geographically distributed simulation should be clearly advocated, including the identification of systems that necessitate distributed simulation and cannot be dealt with otherwise. To raise public awareness, news media should be utilized to inform the general public, including the technical community and policy makers, of major distributed simulation efforts and to encourage tele-observation of experimental activities.

15.5 Conclusion

A summary of the large-scale testing activity carried out at ELSA in the framework of the European large-scale-facilities joint projects was given and the new opportunities for testing under the SERIES project were highlighted.

The new multi-limit-state design procedures were revisited and discussed in light of the increased importance of non-structural components in building costs.

In what concerns experimental activity, focus has been placed on the issue of the definition of test protocols aiming at the assessment of the structure performance for different levels of the input motion. A performance-oriented test protocol would require several tests for different levels of the input motion. However, this may be unrealistic because sequential tests on the same structure would lead to unrealistic damage accumulation. A test protocol considering input motions corresponding to serviceability and life-safety limit-states with a subsequent test to derive ultimate capacity appears to be the most appropriate. It is underlined that a "serviceability test" is indispensable to calibrate loss functions required in performance and risk-based design and assessment approaches. Non-structural elements must be also included in the test models.

New experimental facilities and test methods able to perform complex tests and to combine numerical and physical simulation should play an important role in the clarification of open issues in the design and assessment of structures, such as structural irregularity, SSI, variability and type of input motions and in the study of complex systems such as structures with dissipation devices, which may address the problem of excessive expected economic losses. Eventually, a new European Research Infrastructure subject of the EFAST Design Study should respond to these requirements. Performance and risk-based design shall benefit from the creation of a comprehensive database of experimental results, which is presently addressed in the framework of the SERIES project.

In view of the preparation of the next generation of the Eurocodes, it is expected that a strong activity will develop in the next 5 years comprising both numerical and experimental research.

New measuring/recording systems, such as digital video systems to record response and damage during the tests will also provide better information and evidence on local and global damage evolution and allow better damage descriptions.

Assuming that performance based design is achieved, accepted and implemented, practice will move from prescriptive to performance-based codes widening the possibilities for creativity and innovation but also transferring more responsibility to the designer, to the owner and to other players in the process. Anticipating that recourse to testing and testing/simulation will be necessary, in particular for innovative solutions, it will be required to agree on a "qualification procedure", focused on standard testing protocols, which provide realistic/reliable performance evaluation.

Acknowledgements Research results presented in this work were obtained from tests carried out in the framework of projects co-financed by the European Commission under R&D Framework Programmes. Credits are due to the following colleagues for the co-ordination and participation in the experimental campaigns: G. Magonette, J. Molina, P. Negro, P. Pegon, F. Taucer, H. Varum, E. Mola, G. Tsionis. The author thanks also H. Varum for the revision of the manuscript.

References

Carvalho EC (2009) Pre- and co-normative research needs for Eurocode 8. In: Proceedings of the SAFECAST workshop, JRC, Ispra, 5–7 October

CASCADE (2005) Series reports, LNEC, Lisbon

ECOEST-PREC8 (1996) Series reports, LNEC, Lisbon

ECOEST2-ICONS (2001) Series reports, LNEC, Lisbon

EFAST (2009) Design Study of a European facility for advanced seismic testing. http://efast.eknowrisk.eu/EFast/

EN1998-1 (2004) Eurocode 8: Design of structures for earthquake resistance – Part 1: General rules, seismic actions and rules for buildings. CEN-European Committee for Standardization Eurocodes (2010) Eurocodes – Building the Future. The European Commission website on the Eurocodes. http://eurocodes.jrc.ec.europa.eu

Fajfar P, Krawinkler H (2004) Performance-based seismic design concepts and implementation. In: Proceedings of an international workshop, Bled, Slovenia, June 28–July 1. PEER report 2004/05, Berkeley, CA, USA, ISBN 0-9762060-0-5

Fardis MN (2004) A European perspective to performance-based seismic design, assessment and retrofitting. In: Proceedings of the international workshop, Bled, Slovenia, June 28–July 1. PEER report 2004/05, Berkeley, CA, USA, ISBN 0-9762060-0-5

Fardis MN (2007) Pre- and co-normative research needs for Eurocode 8. In: Pre-normative research needs to achieve improved design guidelines for seismic protection in the EU. JRC scientific and technical report, EUR 22858 EN, ISSN 1018-5593

Fardis MN (2009) SERIES project overview. In: Proceedings of the workshop on opportunities for users to access European research infrastructures in earthquake engineering, Iasi, Romania, 13 July

Hadjian AS (2002) A general framework for risk-consistent seismic design. Eng Struct Dyn 31:601–626

Krawinkler H (1999) Challenges and progress in performance-based earthquake engineering. In: Proceedings of the international seminar on seismic engineering for tomorrow – in honour of Professor Hiroshi Akiyama, Tokyo, Japan, 26 November

Marazzi F, Molina F (2009) First EFAST workshop challenges, needs and open questions. JRC scientific and technical report, EUR 23822 EN, ISSN 1018-5593

Negro P, Pinto AV, Verzeletti G, Magonette GE (1996) PSD tests on a four-story R/C building designed according to Eurocodes. J of Struct Eng 122(11):1409–1417

Negro P, Mola E, Molina F, Magonette GE (2004) Full-scale PSD testing of a torsionally unbalanced three-storey non-seismic RC frame. In: Proceedings of the 13th WCEE, Vancouver, Canada

Pampanin S (2009) Emerging solutions for damage-resisting precast concrete buildings: an update on New Zealand's practice and R&D. In: Proceedings of the SAFECAST workshop, JRC, Ispra, 5–7 October

Pegon P, Pinto AV (2000) Pseudo-dynamic testing with substructuring at the ELSA laboratory. Earthquake Eng Struct Dyn 29:905–925

Pinto AV (ed) (1996) Pseudo-dynamic and shaking table tests on RC bridges. ECOEST-PREC8 report No. 5, LNEC, Lisbon

Pinto AV (1998) Earthquake performance of structures: behavioural, safety and economical aspects. Special Publication N. I.98.111. Phd thesis, European Commission, Joint research centre, Ispra, Italy

Pinto AV, Varum H, Molina F (2002) Experimental assessment and retrofit of full-scale models of existing RC frames. In: Proceedings of the 12th European conference on earthquake engineering, London, Elsevier Science Ltd

Pinto AV, Negro P, Taucer F (2004a) Full-scale laboratory testing: strategies and procedures to meet the needs of PBEE. In: Proceedings of an international workshop, Bled, Slovenia, June 28–July 1. PEER report 2004/05, Berkeley, CA, USA, ISBN 0-9762060-0-5

Pinto AV, Pegon P, Magonette GE, Tsionis G (2004) Pseudo-dynamic testing of bridges using non-linear substructuring. Earthquake Eng Struct Dyn 33:1125–1146

Pinto AV, Pegon P, Taucer F (2006) Shaking table facilities and testing for advancement of earthquake engineering: international cooperation, experiences, values, chances. In: Proceedings of the 1st European conference on earthquake engineering and seismology, Geneva, Switzerland, 3–8 September

Pinto AV, Taucer F, Dimova S (2007) Pre-normative research needs to achieve improved design guidelines for seismic protection in the EU. JRC scientific and technical report, EUR 22858 EN, ISSN 1018-5593

SERIES, Seismic Engineering Research Infrastructures for European Synergies. http://www.series.upatras.gr/

Severn RT (2000) Earthquake engineering research infrastructures. In: Proceedings of the workshop – mitigation of the seismic risk: support to recently affected European countries, Belgirate 2000. JRC, Ispra, Italy

Spence R, Lopes M, Bisch Ph, Plumier A, Dolce M (2005) Earthquake risk reduction in the European Union: proposals for a European earthquake risk reduction programme – a discussion document, EAEE

Stoten DP, Gomez EG (1999) MCS adaptive control on shaking tables using retrofit strategies. IASTED conference on control, Banff, Canada

Taghavi S, Miranda E (2003) Response assessment of non-structural building elements. PEER report 2003/05, Berkeley, CA, USA

Taucer F (2005) Recent advances and future needs in experimental earthquake engineering. CASCADE series report No. 7, LNEC, Lisbon, ISBN 972-49-1971-4

Taucer F, Franchioni G (2005) Directory of European facilities for seismic and dynamic tests in support of industry. CASCADE series report No. 6, LNEC, Lisbon, ISBN 972-49-1970-6

Chapter 16
The Contribution of Shaking Tables to Early Developments in Earthquake Engineering

R.T. Severn

Abstract Using examples, the thesis developed here is that shaking tables were essential to the progress made in earthquake engineering during the period 1900–1980. This period covers that from the very first shaking tables at the beginning of the twentieth century and ends just before the rapid advances in computing and control engineering made such major changes in shaking table performance capabilities that a separate paper would be required to record them. Not surprisingly, progress was linked to major earthquakes, but corroboration of existing theories or new design methods based on theoretical advances also played a part, as did the specific practical needs of the construction industry. Examples of the first of these spurs to activity are foundation issues highlighted by the 1906 Californian earthquake and similar events in 1964 in Alaska and Niigata in Japan in the same year; also, elevated water tanks by the 1933 event in California, and the general range of structures by the 1923 Tokyo (Kanto)earthquake which devastated the city. Relating to shaking table developments due to theoretical advances, the 1933 "added-mass" analysis by Westergaard for the effect of a rigid wall vibrating against a body of water (as in a dam), prompted several shaking table studies to test the validity of practical design issues, as did Biot's 1943 introduction of spectral curves to represent earthquake input in a simplified way.

16.1 Introduction

Lord Rayleigh's two-volume work "The Theory of Sound" published in 1877–1878, established the fundamental principles of vibrations in the elastic range of behaviour, but the application of these principles to problems occurring in any industry required a great many simplifying assumptions before they were able to produce results, and these very simplifications made the results of doubtful value unless they could be corroborated by practical experiences of some sort. This was

R.T. Severn (✉)
Earthquake Engineering Research Centre, University of Bristol, Bristol, UK
e-mail: R.T.Severn@bristol.ac.uk

M. Garevski, A. Ansal (eds.), *Earthquake Engineering in Europe*, Geotechnical, Geological, and Earthquake Engineering 17, DOI 10.1007/978-90-481-9544-2_16,

particularly so in the construction industry. Basic parameters such as natural frequencies and modal shapes might be obtainable, but dynamic response – even to harmonic inputs – could not always be obtained with adequate accuracy, and the prediction of structural failure, or any kind of inelastic behaviour, was certainly beyond theoretical reach.

These shortcomings of theoretical methods were doubly valid for earthquake engineering at the end of the nineteenth century because no firm knowledge was available about the actual inputs which they caused structures to suffer. Indeed, it was thought at the time that the principal effects of an earthquake were due to an initial shock, with scant regard for any vibration which followed, this being due to the crude seismographs available at that time, which were not devised to record such vibrations.

In the absence of applicable theoretical methods, engineers turned to experiments, and these were thought to have originated in Japan in 1890. Muir-Wood (1988) reproduces a drawing of a uni-directional table built by Milne and Omeri in which a hand-driven wheel was connected by an eccentric crank to a trolley running on rails. It is unfortunate that no details are available of any tests which used this table.

For this present chapter the 1906 Californian earthquake becomes the starting point, and it is remarkable that the first detailed account of the use of a shaking table was concerned with a practical problem in foundation engineering, a discipline which even today presents great difficulties in shaking table laboratories. This Californian beginning has determined that the structure of this chapter will progress through civil engineering topics, rather than through developments of shaking tables themselves, which has been presented in Severn (to be published) The foundation studies will therefore be followed by research on structure/fluid interaction – particularly relating to dams and fluid-containing structures – followed by bridges and then general structures. It is not to be assumed of course that shaking tables have been used exclusively for studies relating to the construction industry during the period covered by this chapter, but the lack of published evidence is most likely due to the commercial value of the results obtained; this is certainly true for the period since 1985.

16.2 Soil Mechanics and Foundation Engineering

16.2.1 Studies on Dry and Wet Sand by F.J. Rogers

Direct observation of the effects of the 1906 Californian earthquake indicated greater destruction in structures built on soft ground than to those built on firmer foundations. Because this was contrary to general expectations, F.J. Rogers of Stanford University (Rogers, 1906) began a series of simple tests which consisted of upending buckets of sand of varying degrees of wetness on to a table which he shook using an eccentric crank attached to an electric motor.

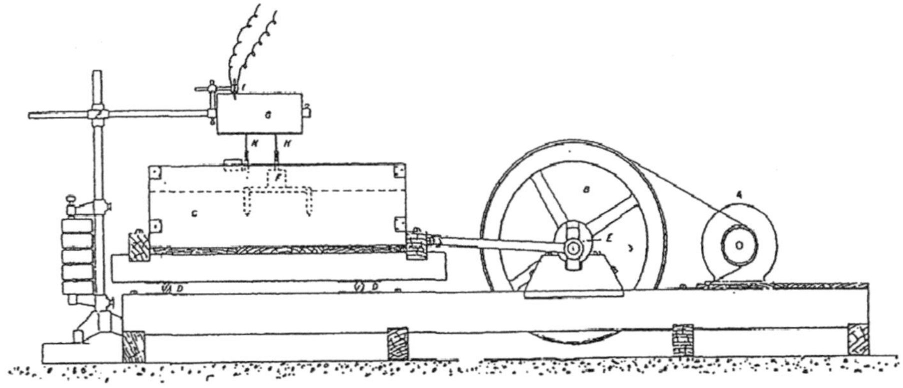

Fig. 16.1 The shaking table designed and used by F.J. Rogers for soil dynamics studies

Though simply qualitative – indicating that the degree of wetness affected the behaviour of the soil – the results were of sufficient interest to cause the Californian State Earthquake Investigation Commission to promote the building of the shaking table which is shown in Fig. 16.1. Rogers accepted that earthquake motion was exceedingly complex, rather naively commenting

> . . .but it was not thought worthwhile to imitate this complexity but rather to confine the shaking motion to simple to-and-fro horizontal motion in one direction.

A box of dimensions $100 \times 86 \times 30$ cm, which Rogers referred to as a "car", was fixed rigidly to the table, loaded with sand and shaken at a range of frequencies and amplitudes, the displacement of a small block embedded in its surface, and of the car itself, being measured by two pens contacting a paper-covered rotating drum; a third mark recorded the beat of an electromagnet connected to a seconds pendulum. Because no devices for measuring either acceleration or velocity existed, Rogers deduced both of them by graphical differentiation with respect to time of the displacement traces at a few critical points. For example, at 20% water content he calculated that the maximum acceleration of the soil is around three times that of the car, and since force is proportional to acceleration, this explains the destructiveness of earthquakes in wet soils.

Starting with dry sand Rogers observed that the sand "moved perfectly with the car" but he then proceeded to increase the water content in a number of steps. At 12% (Fig. 16.2) the behaviour was similar to that of dry sand, but from there on the behaviour became more complex, and at 20% Rogers describes the sand as being ". . . very soft, almost semi-fluid." Figure 16.2 shows that at this high water content, for low frequencies the sand oscillates with greater amplitude than the car, an amplitude which decreases as frequency increases, reducing almost to zero at the higher frequencies. But at 15% water content the sand and car moves almost together at low frequencies, and amplitude increases with frequency. Thus, a 5% difference in water content results in remarkable behavioural changes, which Roger explains by the sand becoming more "cohesive".

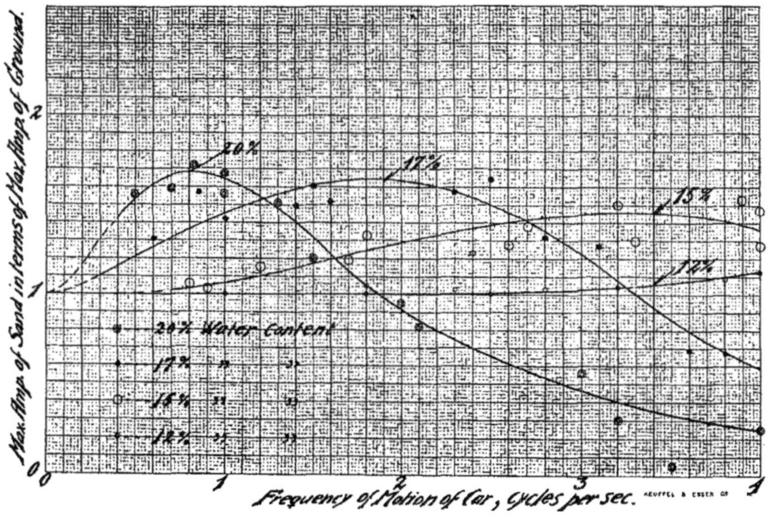

Fig. 16.2 Graphical presentation of Rogers' results

In another series of experiments Rogers tried to replicate a layered foundation with upper and lower layers of sand containing 10 and 20% water, respectively, separated by a piece of oil-cloth, observing that the upper layer moved as though it was floating on a semi-fluid mass. His conclusion after many such experiments was that such a situation led to an increase in the destructive effects of the shaking motion. However, he was aware that the small size of his car would affect replication of the real situation, and questioned whether what we now call "resonance" would occur between the semi-fluid mass of soil and the motion being impressed upon it by the car.

An attempt was made by H.F. Reid (1906) to provide theoretical support to Rogers' observations but the classical analytical methods available to him were not able to do so.

16.2.2 Further Studies on Sand, by L.S. Jacobsen

It is highly probable that the renewal of interest in earthquake engineering, and in the construction of shaking tables, was caused by the 1923 Tokyo earthquake, which, like that in California in 1906, caused great damage and loss of life.

The Stanford group was now led by L.S. Jacobsen (1930), whose 10×12 ft table (Fig. 16.3) was mounted on street-car wheels running on rails. Because the details of real earthquake motion was now better understood to be more than an initial impact, input by two separate devices were constructed, the first being impact at one end by a pendulum, followed by damped oscillation controlled by anchor springs at the other end. Alternatively, input could be produced by an ingeniously designed rotating wheel fixed to the table (Fig. 16.4) which had a tank eccentrically located

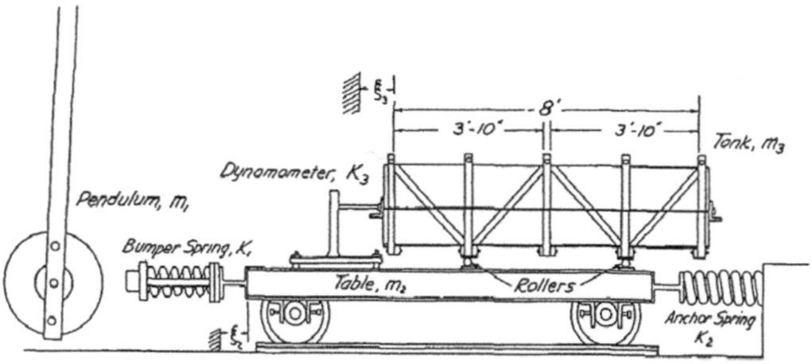

Fig. 16.3 Jacobsen's shaking table using pendulum input

to produce harmonic motion – requiring anchor springs at each end of the table. The ingenuity lay in the fact that the tank contained lead-shot, the amount of which was capable of variation with the wheel in motion, thus allowing amplitude and frequency to be controlled without interruption to the test programme. This was, in fact, the earliest form of shaking table control, and it was this second form of input which Jacobsen used in his continuation of Rogers' work on Monterey sand.

The dimensions of his sand box were $2.5 \times 3.5 \times 16$ ft and his method of recording motion was to embed a small anchor of sheet metal at appropriate levels in the sand, its movement being recorded by an attached loop of piano wire containing a recording pen (Fig. 16.5). The two shapes of plate used are shown in Fig. 16.6 (a) being for the free surface, and (b) between any two layers, possibly having different properties.

In his first series of tests he used only dry sand in a box reduced in length to 8 ft, with the sand level increasing by 12 in. layers from 12 to 36 in. His procedure was to increase the frequency and amplitude of vibration until the recorded motion of sand and table showed a difference, the results being as indicated in Fig. 16.7. It

Fig. 16.4 Jacobsen's unbalanced wheel input mounted on his table

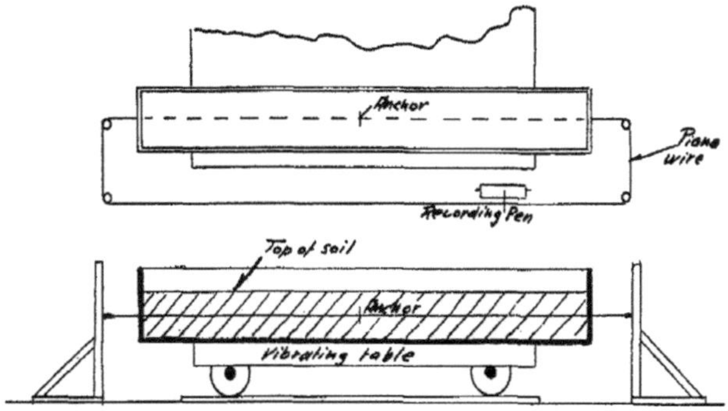

Fig. 16.5 Jacobsen's looped-wire method of measuring soil displacement

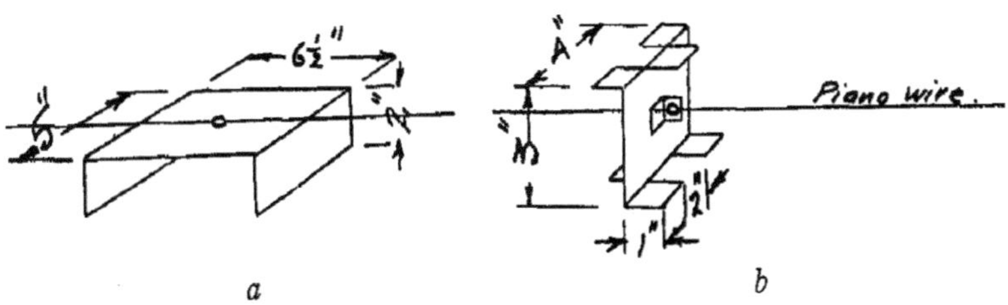

Fig. 16.6 The two types of anchor used for measuring soil displacement

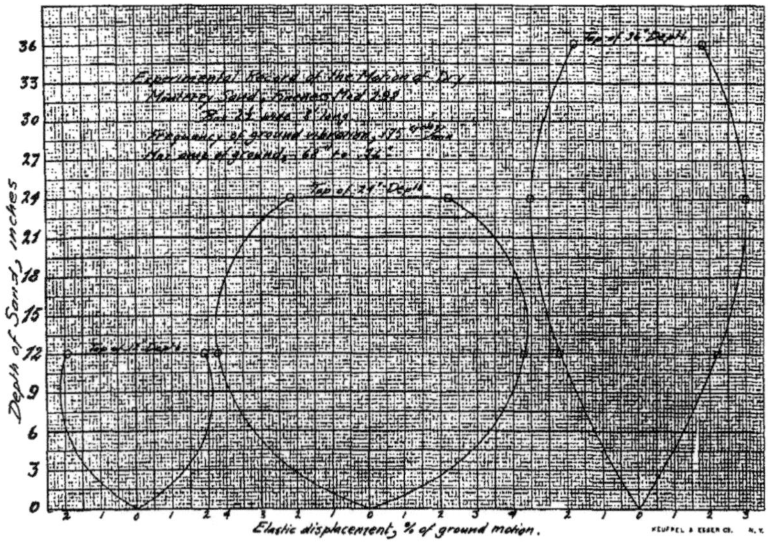

Fig. 16.7 Jacobsen's displacement measurements for one, two, and three layers of dry sand

surprised Jacobsen that for the 24 and 36 in. surface levels the maximum displacement occurred at the second level down, but he pointed out that he was attempting to measure hundredths of an inch and was unsure about his measuring system! In an attempt to remove "end-effects" he repeated the tests in the 16 ft long box and obtained similar results.

When he turned to wet sand, Jacobsen found the same inconsistencies as Rogers had done 20 years earlier, and commented that his results did agree qualitatively with those of Rogers. But at this point he stopped testing, making the very significant observation that experiments which had no theory to guide their direction would not lead to useful results.

Following this precept, he first repeated Reid's analysis (Reid, 1906) assuming that he was dealing with an elastic soil subject to simple harmonic ground vibration with no dissipative forces acting. He followed this by introducing viscous damping depending on velocity, which resulted in an expression for the displacement in terms of hyperbolic functions with complex arguments. His aim was to succeed, where Reid had failed, in producing a satisfactory explanation of Rogers' results, and this he did by finding the equivalent elastic $(Gg/\mu)^{1/2}$ and viscous (gc/μ) properties of the sand which Rogers had used – G,g,c and μ representing modulus of rigidity, gravitational acceleration, viscous coefficient, and density, respectively. His calculated values for the viscous function for 20, 17 and 15% water content were 4.7, 12 and 23 respectively, which accorded, in sequence, with the actual observations. For the equivalent elastic constant, his own value and that of Rogers differed by about 12%.

16.2.3 Seismic Study of Earth Dams by Mononobe et al.

The state-of-the-art in soil mechanics in the mid-1930s is exemplified by this research (Mononobe et al., 1936), again prompted by the Tokyo earthquake of 1923 which caused cracks in two Japanese earth dams. Accepting that available methods in soil mechanics did not enable them to calculate the free vibration of actual earth dams, they simplified the problem to that of shear vibration of a uniform elastic body of appropriate shape, which allowed them to calculate its first natural frequency knowing only the density and modulus of rigidity.

No details are given of the shaking table used, but the size of the model and the results presented indicate that it was small and capable of applying only harmonic motion. The first model, made of agar–agar – a gelatinous substance made from seaweed – was 20.1 cm high with top and bottom widths of 19.5 and 98 cm, respectively, its density and modulus of rigidity being deduced from the period of free vibration. A second model of similar size was constructed from sandy clay with a specific gravity of 2.45 and water content of 45.3%; because its fundamental period could not be measured directly, it was obtained from the measured deformation of the crest Although the results from the first model supported the theory used, those from the second did not, due, it was said, to its small size, but more probably to the linear elastic theory which they had used.

To conclude their study, Monobe et al. measured the modulus of rigidity of four existing Japanese earth dams (including the two damaged in the 1923 earthquake), ranging in height from 30 to 37 m, by measuring the vibrations through them caused by dropping a heavy weight on the crest. Their conclusion was that the low rigidity value would have allowed them to resonate with a seismic motion having a period of around 0.6 s.

16.2.4 Studies on Rockfill Dams by Clough and Pirtz

This study (Clough and Pirtz, 1958) is included in the section on soil dynamics because its principal purpose was to evaluate the relative merits of vertical and sloping clay cores in rockfill dams. The uni-directional table constructed by Clough and Pirtz in the mid-1950s (Fig. 16.8) was similar to that of Jacobsen, with motion being caused by a pendulum striking one end through a buffer spring, the other end being restricted by an anchor spring of suitable stiffness. The major difference from Jacobsen was that the box carrying the model was supported by 4 corner steel bars having flexibility in the vibration direction but great stiffness in all others.

The two models (Fig. 16.9) were based on existing dams with a modification to one of them in the interest of comparability between the two types of core. The size of the table (1×7×2.3 ft) dictated that the 300 ft high dams must be scaled at 150:1, which meant a model height of 2 ft and a base length of about 7 ft. Modelling of such dams requires compromises to be made, and here crushed quartzite was used throughout. As a consequence, in order to attempt to replicate the graded rockfill

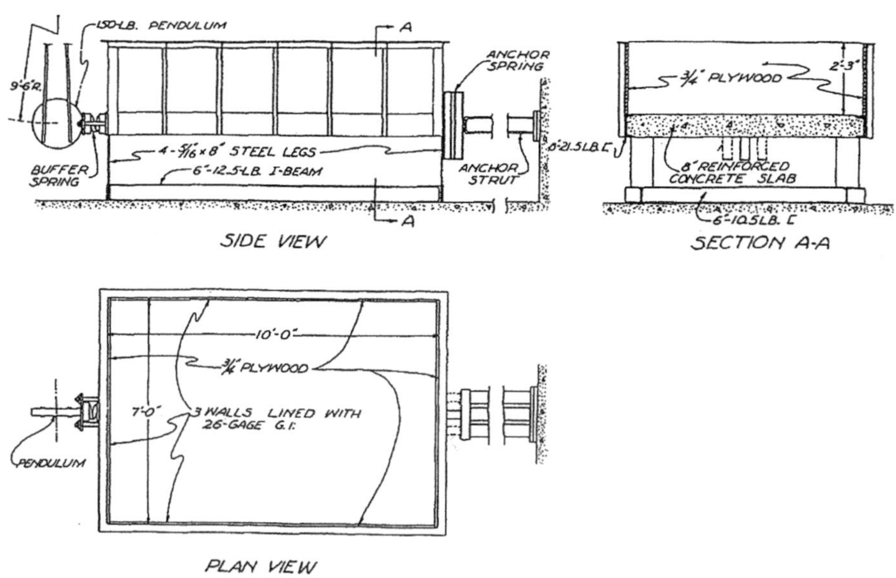

Fig. 16.8 The Clough and Pirtz shaking table

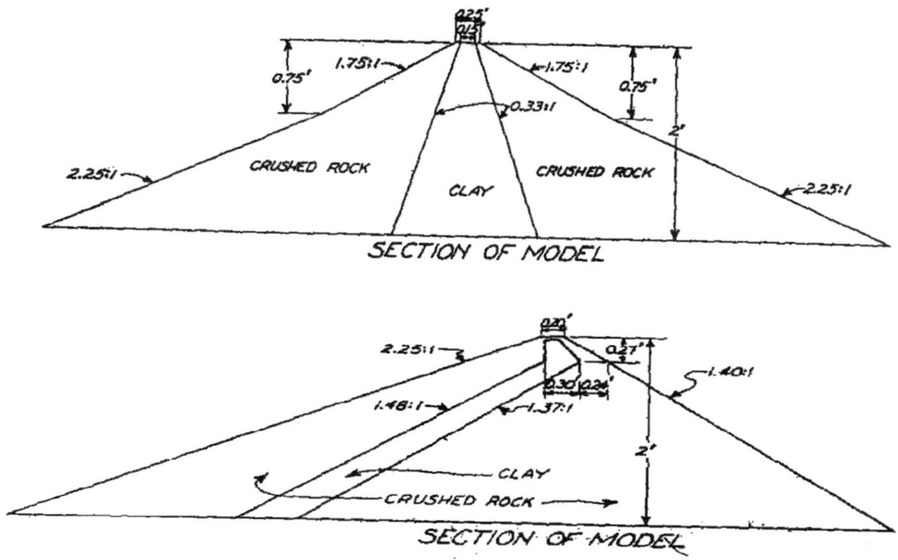

Fig. 16.9 The vertical- and sloping-core rockfill dam models tested by Clough and Pirtz

used as a filter next to the clay core in the real dams, the thickness of the model core – made of kaolin clay- was increased by half the equivalent model thickness of the filter layers. Overall, it proved possible to meet the similitude requirements with satisfactory accuracy save those of the modulus of rigidity and the specific gravity of the reservoir liquid. In both cases the effect was deemed to be of a conservative nature. The existence of the reservoir in such a small tank raised the usual problem of "end-effects", to be countered here by a blanket of rubber tubes against the upstream wall of the tank.

The testing programme was extensive, with 8 models being tested, 7 of which concerned the sloping core; 3 water-levels were used, empty, 4/10 full, and full, with observations of displacement and acceleration being measured by LVDTs and accelerometers at the base and at the crest, and recorded on an accelerograph. A profilometer mounted on a rail above the model was used to measure shape. An example of displacement results for the top and base of the sloping core model is shown in Fig. 16.10, the upper and lower curves being displayed in a more diagrammatic form in its middle section.

Of great interest to this discussion of the practical value of the results obtained, is the fidelity with which the shaking table input could approximate to real earthquake input, and it is fortunate here that the concept of acceleration spectra had been recently introduced in analytical studies in earthquake engineering, and was used by the authors. Figure 16.11 summarises their results. From the measured input to the table, an idealization shown at the upper right-hand side was made, with its spectral acceleration being computed for relevant natural periods at model scale and shown alongside corresponding curves for the N-S component of the 1940 El Centro event

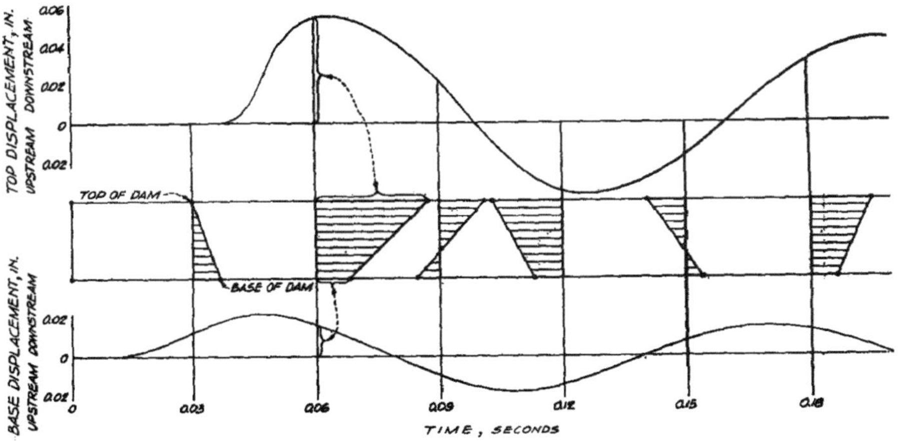

Fig. 16.10 Displacement with time of the crest and base of the sloping-core model; reservoir 0.4 full

for 2 and 20% critical damping. The first point of interest is that the two upper curves have reasonable agreement up to the first (and predominant) period of the model, but subsequently there is resonance with the low frequency harmonic motion of the table, and the curves diverge. But looking at the two lower curves, both computed for 20% damping – which is more relevant to rock-fill dam material – the coincidence is very close, allowing the conclusion to be made that the input produced by the table can indeed simulate the El Centro earthquake for this type of model.

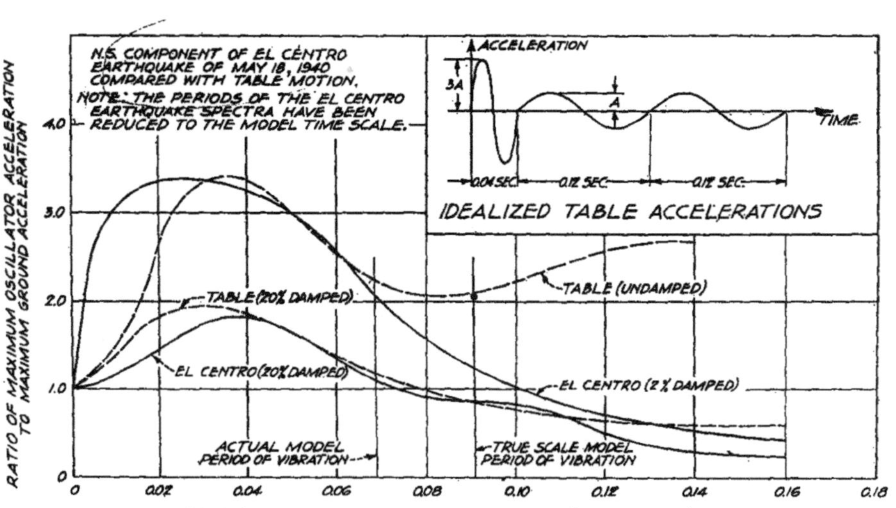

Fig. 16.11 Clough and Pirtz shaking table: comparison of acceleration spectra for El Centro earthquake and the shaking table input

As a postscript to this very significant and innovatory research, it is noted that the authors unwisely concluded that rockfill dams were inherently safe against earthquakes because of their flexible structure and the fact that no catastrophic slippages occurred in the models when they were subjected to a ground acceleration exceeding that of gravity. Contributors to the discussion pointed out that rockfill dams were not usually built on a finite rigid base, that vertical components of acceleration were important, that only one recorded earthquake input had been used, and that such a sweeping conclusion could not be made from tests on a small-scale model.

16.3 Fluid – Structure Interaction

16.3.1 Studies Using Jacobsen's Table at Stanford University

The possibility of earthquake-induced dynamic pressure on dams had by 1930 caused concern amongst engineers concerned with their safety, prompting Hoskins and Jacobsen (Hoskins and Jacobsen, 1934) to begin experiments using the latter's table for input by the pendulum device shown in Fig. 16.3. For some reason not stated, the experiments were stopped, and it may reasonably be conjectured that they were started again in response to Westergaard's seminal publication in 1931 (Westergaard, 1931). But that theoretical work was restricted to the two-dimensional problem of a vertical dam face vibrating with simple harmonic motion in an infinitely-long reservoir. For any relevant shaking table tests it was of course necessary to deal with a relatively small volume of water confined in a vessel small enough to be vibrated with available equipment. This caused Jacobsen to introduce his own two-dimensional theory defining the behaviour of water in a box of *finite* length. Westergaard's solution was dealt with as a limiting case. Although he included compressibility, as Westergaard had done, his calculation became much simplified if it was neglected; he correctly argued that it was not of any sensible importance!

The classical hydrodynamic theory was used by both men, but Jacobsen included only the first derivatives of the two displacement components, which made his solution only valid for very small motions, the practical consequence of which was that from his shaking table experiments he could only use the motion induced by the pendulum immediately after impact. Even though he was not able to study steady harmonic vibration, as we shall see his experimental results were of considerable value.

Jacobsen's shaking table arrangement (Fig. 16.3) for these tests consisted of a wooden box lined with sheet metal, 8 ft long, 2 ft high and 1.5 ft wide. It was carried on the table by rollers, but was actually attached to it through one of three dynamometers, which measured the force produced by the box and its contents for recording by a cine camera. The three dynamometers were described as being "rigid", "medium", and "flexible", giving 10,500, 6,400 and 2,206 pounds per inch deflection, respectively. His analysis indicated to him, as it did to Westergaard, that

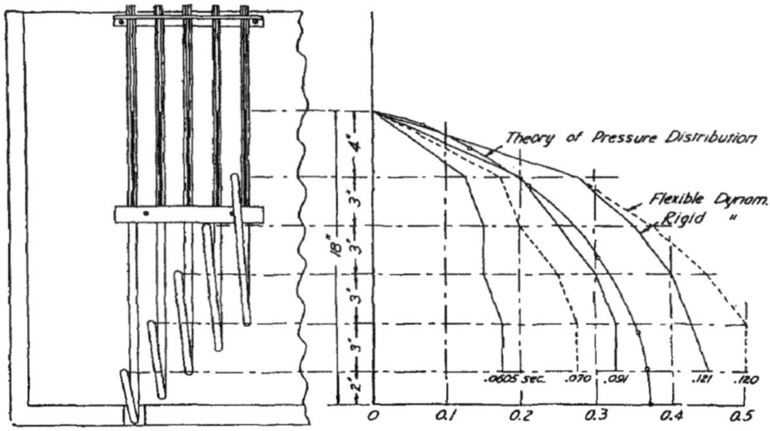

Fig. 16.12 Jacobsen's piezometer measurements and corresponding theoretical values of pressure on the wall of the tank

the overall effect was equivalent to a segment of the water taking part in the motion, and this he set out to measure – its total mass and its pressure distribution on the end of the tank, where his theory showed the approximately circular shape as indicated in Fig. 16.12, not essentially different from Westergaard's. The experimental approach to this achievement used different water levels in the range 6–18 in. in 3 in. increments, different magnitudes of pendulum impact, and the tank divided by a central partition so that hydrodynamic forces on 4 surfaces was produced. In parallel, the water in the tank was replaced by a range of rigid masses from zero to 450 lbs, so that comparison of the two sets would indicate the equivalent mass of water affecting the motion of the tank. For its distribution on one face, 5 piezometers were attached to the face at different levels (Fig. 16.12).

Only a summary of the results obtained by Jacobsen can be given here, noting first that his ratio of experimental to theoretical "added-mass" taken over all tests was 0.772, but if he ignored those for 6 and 9 in. water levels – where he suspected the accuracy of his measurements, being very small differences between large values, it rose to 0.843. He considered this to be acceptable because the theoretical values assumed a completely rigid box. For the distribution of pressure on the end of the tank, Fig. 16.12 reproduces piezometer values recorded by a cine camera operating at 128 exposures per second for the 0–0.5 s period, after which the theory became invalid. It will be seen that the distributions closely follow the theoretical predictions of both Westergaard and Jacobsen himself.

Sometime later, in the late 1940s, Jacobsen returned to studies on structure/fluid interaction, with J.S. Ayres as a colleague, this time with a study of the vibration of water in fixed-base cylindrical tanks (Jacobsen and Ayre, 1951), with tops either open or closed. Figure 16.13 shows four such tanks, having diameters of 6, 12, 23 and 47 in., fixed to the Jacobsen table and using the pendulum to produce the required input. His purpose was first to study gravitational surface waves, which he did by photographically recording the profile in a card inserted across the diameter

Fig. 16.13 The four cylindrical tanks on Jacobsen's shaking table

of the tank (Fig. 16.13), and second to measure – using dynamometers – an equivalent mass and overturning moment of the vibrating fluid; this he did, as before, by replacing the water by a range of solid masses and comparing responses.

A matter of importance here, in its application particularly to oil and water storage tanks, was whether the tank – if covered – was completely full, or had a clearance between fluid surface and tank cover. It was shown that for an open tank the effective mass and overturning moment can be calculated (within 5%) from classical theory, but when the tank has a rigid cover, the absence of only a few percent of fluid make it effectively open.

In view of what he had said earlier about the need for theory to guide experiments, in this case Jacobsen had no need to produce his own, because this had been done for him by Poisson in 1828 and Lord Rayleigh in 1876.

16.3.2 A Study of Elevated Water Tanks by A.C. Ruge

This work (Ruge, 1938) was prompted by the large number of failures of elevated water tanks in the Long Beach, California earthquake of March 1933, and was aimed at showing the inadequacy of the current design method, which simply took a proportion of the total mass as acting statically at the centre of mass in the horizontal direction. Ruge's shaking table was initially a simple arrangement of a light table suspended by a number of piano wires and prevented from rotation by other wires fixed in the ground. Additional weights were attached so that any model would weigh less than 1/15 of the total weight. In his first series of tests it was actuated by compressed springs, but for his second series he developed an input device (Ruge,

1936) which was the antecedent of shaking tables as we know them today. The innovation was an analogue device which allowed the table to be driven by the first earthquake motion ever recorded – that at Long Beach, California.

In the first series of tower tests, a 1:46.5 scale model accurately represented the tank filled with water, but the tower structure was idealized by a simpler arrangement of equal stiffness, and all tests were accompanied on the table by a "dummy" system consisting of an inverted simple pendulum, with the bob having the same mass as the tank plus water. Because the response of the dummy to the simple harmonic input could be accurately predicted theoretically, it acted as a check on the fidelity of the input to the actual model. The input itself was either pull-back followed by free vibration with the table locked in position, or harmonic tests when the table springs were compressed and then released. All motions were recorded photographically using a lens and mirror arrangement devised by Ruge.

In the free vibration a sloshing mode in the tank of period 0.39 s was recorded with little general movement of the tower. This was preceded by a mode at 0.24 s of the whole tower, for which the corresponding "dummy" mode was at 0.27 s. When the water in the tank was replaced by a sawdust/sand mixture the first mode period agreed with the dummy mode at 0.27 s. From these and similar tests on many other models, Ruge concluded that the accepted method of designing elevated tanks led to large errors over the entire range of possible earthquake frequencies, and that the best way to obtain earthquake resistance was to provide the tower with greater elasticity by the introduction of "spring elements" with special damping devices, and he proceeded to investigate these ideas in a further period of testing using simple harmonic input.

The truly innovative nature of the shaking table (Ruge, 1936) which Ruge built in 1935 and was used in the second part of his work on elevated water tanks has been described in Severn (to be published) The important feature was that a displacement trace of each component of the 1933 Long Beach earthquake, obtained by double integration of the recorded acceleration, was cut into the periphery of a disc which in the tests was rotated by a motor. An "electric-eye" was caused to follow this indented periphery and to feed an electric signal into an oil-filled piston which drove the table in one horizontal direction, with an "error-drive" feedback to ensure that the electric eye was following the edge of the disc. The disc is to be seen in the bottom right-hand corner of Fig. 16.14, which also shows the 1:25 scale model used in these tests. Ruge indicated that he intended to develop the idea so that he could use all three of the displacement components simultaneously, but he never did so. Present day control engineers aver that he could not have done so with the facilities at his disposal.

The model shown represented a 60,000 gallon tank on a 100 ft tower which contained Ruge's idea of "spring elements" and for full compliance with scaling laws the tank fluid was required to be mercury, but this was too dangerous and water was used. Because of the difficulty of introducing damping into each of the 24 spring-elements, it was provided by a solid friction device attached directly between the tank and a support.

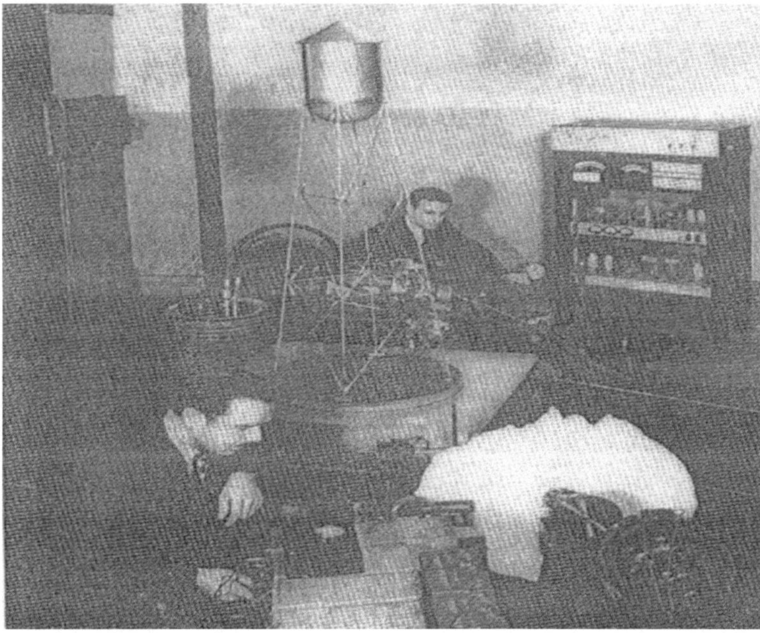

Fig. 16.14 Ruge's innovative shaking table showing the model elevated water tank and the indented disc representing ground displacement

Not wishing to base general conclusions on the one, and only, available record, Ruge modified its time factor, faster and slower and also the amplitude. To introduce some semblance of bi-directional input, he rotated the model through 45 degrees. Half-full tanks were also tested, but although he did produce design rules based on his research, Ruge recognized that the rigid foundation used in all tests prevented firm conclusions about the efficacy of his "spring elements", as did the fact that he discontinued the work due to lack of funding! It is of interest to note that D.S. Calder (1936) corroborated Ruge's work by full-scale measurement on similar water tanks.

16.4 Bridges

16.4.1 Dynamic Model Studies of the Ruck-A-Chucky Cable-Stayed Bridge

Important bridges for which seismic effects need to be considered present two difficulties for laboratory testing. First, their size means that models must be of so small a scale that important structural details are difficult to replicate; second, if the span is very large the time-lag of the earthquake reaching different parts ought to be considered. This second difficulty requires a number of different shaking tables

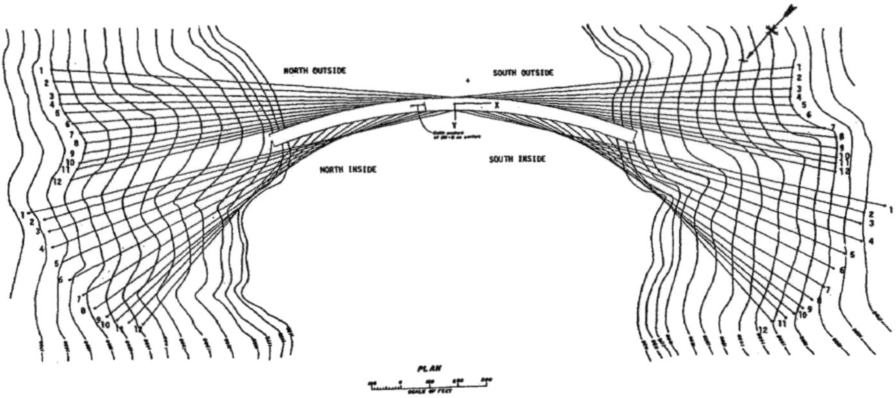

Fig. 16.15 The Ruck-A-Chucky bridge, California, USA

working together, but the control systems to do this effectively did not become available until c.1995 and so examples of their use are not included here.

The Ruck-A-Chucky bridge over the American river in N. California was of such an unusual design that shaking table tests were of great importance in corroborating analysis. Its box-girder deck of 1,300 ft span was curved in the horizontal plane (Fig. 16.15) with supporting cables attached directly to the valley walls, making the span to be modeled in excess of 3,000 ft. The facility available for this study by Godden and Asham (1978) was the 2×20 ft table built by Rae and Penzien (1973) at the University of California, Berkeley (Fig. 16.16) as capable of simultaneous input in the vertical and one of the two horizontal directions in the 0–20 Hz range. The tripartite plot of performance details in Fig. 16.16 shows that this shaking table had made great advances on anything previously available. Elastic analysis indicated that the frequency range covered the first 10 modes of vibration of a 1:100 scale model, but only 8 modes if the scale was 1:200. This latter scale was chosen, partly to avoid overhang of the model on the table, and partly because analysis indicated that the ninth and tenth modes added little to the response of this unique bridge.

For this highly redundant system, model construction involved initially supporting the deck on temporary piers, prestressing each individual cable to its dead-load value and then removing the piers. Forced vibration tests carried out by an eccentric-mass device measured modes which were close to those calculated. The earthquake input to the model was an artificially generated record having horizontal and vertical accelerations of 0.12 and 0.08 g, respectively. But as was usual at that time, not only was the input increased in stages to avoid damaging the model, but for correlation with calculations, the *measured table motions* were used, not the actual input. It would not be for another 20 years that fast adaptive control of input would become possible, thereby enabling the testpiece to receive the intended, uncorrupted, input increased in stages to avoid damaging the model, but for correlation with calculations, the *measured table motions* were used, not the actual input.

Fig. 16.16 The Rae and Penzien 2×20 ft shaking table at UCB

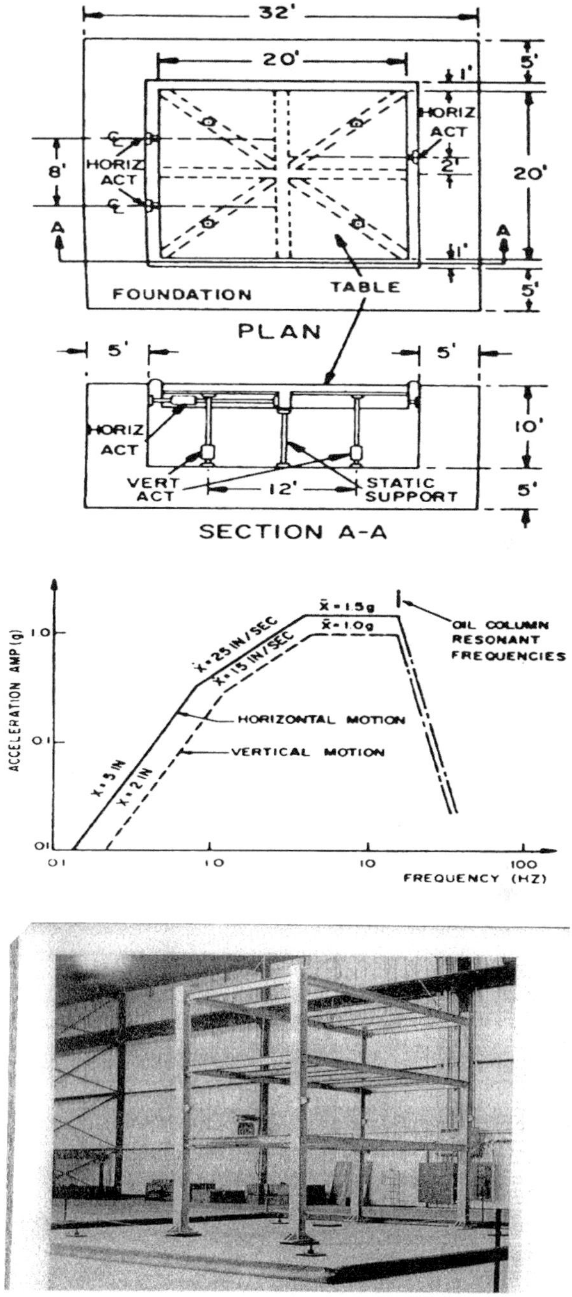

The extensive test programme consisted of input in the two horizontal directions separately, followed by vertical input combined with each of the horizontal components separately.

The overall conclusion for this unusual bridge was that the response to horizontal input was that of a fixed arch, with the cables being virtually inactive. But vertical input produced the predominant response which was in the vertical direction, that is, in the fundamental, and symmetrical mode. Combined vertical and horizontal input made little difference to this observation. Damping was found to be much lower than in conventional cable-stayed bridges.

16.4.2 Curved Highway Bridges

The UCB table was also used by Williams and Godden (1979) in an attempt to explain why curved highway bridges, particularly the 5/14 Freeway Interchange, suffered extensive damage in the February 1971 earthquake The fact that the two halves of the bridge were effectively uncoupled allowed only the eastern half to be modelled at 1/30 scale. But the usual difficulties of meeting the requirements of total similitude meant that the model (Fig. 16.17) was only *representative* of this type of structure, on which trends of behaviour could be studied, together with correlation with corresponding theoretical analysis. Accordingly, the input was generated from a filtered white noise. The test process itself consisted of excitation in the two horizontal directions separately, and also in one horizontal combined with the vertical direction. The input was increased incrementally in severity until damage

Fig. 16.17 A 1:30 model of a curved US highway bridge on the 2×20 ft UCB shaking table

first occurred at the expansion joints. Measured dynamic response, at 100 samples per second, was the displacement of the table itself coupled with displacements at the expansion joints and on both edges of the deck.

Because the tests were carried out in accord with theoretical analysis, many valuable conclusions could be made about the behaviour of this type of structure, but this did not extend as far as the effects of differential ground motion, and although it was stated that both linear and non-linear behaviour were to be studied, each model was only taken to the point of failure, indicating that control was not effective beyond that situation.

16.4.3 The International Guadiana Bridge

The actual earthquake which prompted this research was that at Fruili, Italy in May 1976 when a viaduct over the Tagliamento river having 28 spans each of 45 m was seriously damaged, and several smaller concrete bridges were also damaged. In the same period, the earthquakes at Tokachi-Oki (1968), San Fernando (1971) and Vranca (1977), also caused major damage to bridges.

This study, by Carvalho et al. (1978) is of particular interest because of its attempt to model foundation conditions which included sets of vertical and inclined piles driven down to bed-rock which at its deepest was 70 m below river level (Fig. 16.18). The shaking table available at LNEC Lisbon in the mid-1970s was small, and was similar to that shown in Fig. 16.20. It was able to input only sine-sweeps and shock tests in one direction, and the box, which carried the 1:100 perspex scale model of structure and piled foundation, was only 2 m in length and 0.7×.0.7 m in cross-section. The actual model was, in fact, only the left-hand half of that shown in Fig. 16.18. Due to the difficulty of scaling both the specific mass and the shear modulus, the shear-wave velocity of the soil was used as the scaling parameter. This

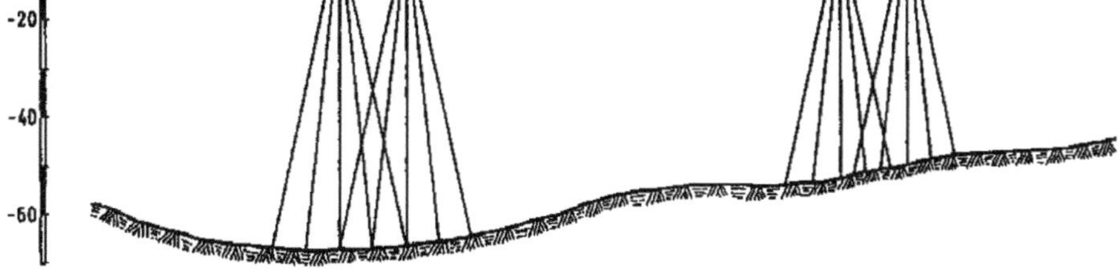

Fig. 16.18 Schematic elevation of the International Guadiana bridge

led to the soft superficial section being modeled by a mixture of sand and sawdust and the deeper, firmer, layer by very fine sand, with shear-wave velocities of 60 and 135 m/s, respectively, whereas the similitude values based on in situ measurements of 150 and 450 m/s should have been 62 and 185 m/s. In an attempt to overcome "end-effects" of the box, thick foam layers covered the ends.

When the box, without model, was filled (60 cm) with fine sand, frequency sweeps indicated a natural frequency around 50 Hz, with 12% viscous damping, but the results were less clear when the box had a 15 cm upper and 45 cm lower layer. With the perspex model in place, 60 cm of fine sand in the box, and the model oriented transversely to the direction of shaking table motion, natural frequencies were obtained by displacement and sudden release. The values 42, 47, and 58 Hz were obtained for torsion about the vertical axis, horizontal translation along the deck axis and horizontal translation transversely to the deck, vibration decay indicated 5% viscous damping. The corresponding calculated frequencies, assuming the piers to be built-in at the pile caps were 41, 44, and 55 Hz, respectively.

Several of the perspex piles were instrumented with strain gauges, which indicated that the main force in the pile was axial due to the overturning moment which the bridge structure induced in the pile-caps, but it was accepted that the simple nature of the modelled foundation, and the finite dimensions of the box, made this an unreliable conclusion. In fact, as a general result of this adventurous study, the authors accepted that their goals had only been partially fulfilled, but, as Jacobsen had done before them, they pointed to the imperative need for theoretical analysis and experiments to proceed in tandem; here it was theoretical analysis which was lacking!

16.4.4 Further Studies on Pile Foundations for Bridges

This study was made by Tatsuoka et al. (1978), citing earthquakes in Alaska (1964) and Niigata (1964) as the reasons. No details are given of the shaking table used, but Fig. 16.19 indicates that it was small in size, the model having dimensions of $12 \times 80 \times 50$ cm. Its input, in one horizontal direction only, could be either sinusoidal or random and was produced by a displacement-controlled dynamic actuator. It is noted that at this time there were several large multi-axis tables available in Japan which were built expressly for soil dynamics research as a result of the Niigata (1964) earthquake, such as the 15×15 m table at Tsukuba Science City, but they were owned either by large commercial groups, or by Government organizations, with model and testing costs being beyond the budgets of university researchers.

The series of tests, 14 in all, involved small groups of aluminium piles, varying in number between 4 and 9, placed on liquefaction-susceptible saturated loose sand layers with initial mean relative densities of 30% on average. The top of the pile group carried a large mass, and for about half the tests each pile was rigidly fixed to the base of the tank, otherwise they were free. Various types of response measuring devices were used as indicated in Fig. 16.19. The finite size of the tank meant that "end-effects" had to be considered, and this was achieved in 13 tests by hinging the

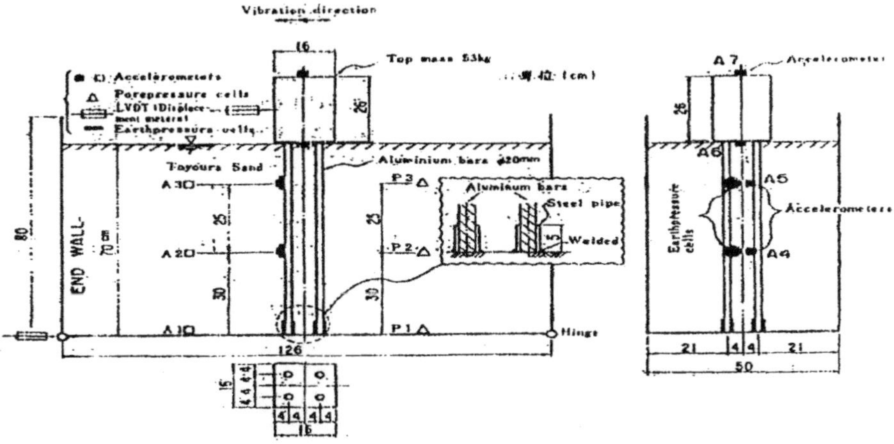

Fig. 16.19 Tatsuoka's shaking table arrangement for bridge piles in liquefying sand

base of the end walls and connecting the tops by wires. For this arrangement the sand layers did not move after liquefaction had occurred; but when the base of one end wall was made rigid the liquefied sand layers continued to move.

From the 14 tests made it is not easy to draw general conclusions because there is no obvious rationale to the choice of conditions; indeed, here is another example of ignoring what might be called the "Jacobsen hypothesis" (v.16.2.2) that for experiments to have a successful outcome they should be guided by theory.

16.4.5 Reinforced Concrete Bridge Columns

The importance of good detailing in RC columns was made evident by the 1971 San Fernando earthquake, particularly lapping starter bars and spacing of transverse bars in plastic hinge zones, and the anchorage of confining reinforcement. These issues were taken up by Priestley and Park (1979) who used a small uniaxial table to input the first 10 s of the N-S record of the 1940 El Centro event into a 1/6 scale model of an individual column (Fig. 16.20). Apart from measurements of the displacements of the centre of mass – which agreed well with computer analysis – no other experimental values appear to have been obtained.

16.5 General Structures

16.5.1 Nuclear Reactor Cores

This research was not caused by any particular earthquake, but by an urgent need of the construction industry. The world's first commercial nuclear power station

Fig. 16.20 The Priestley/Park test on a 1/6 scale model of a RC bridge pier

was built at Calder Hall in the UK during 1953–1956 and its design was used for the Tokai plant 70 miles north-east of Tokyo as a joint venture between British and Japanese engineers. Special studies were required for its design to withstand earthquakes and these were presented by Muto et al. (1962). There were no shaking tables in the UK at that time, and in Japan only those built before the second world war were available; one such is shown in Fig. 16.21. It was actuated by releasing a

Fig. 16.21 A model reactor core on a Japanese shaking table using compressed springs for input

series of compressed springs, so there would have been an initial impact followed by damped oscillation. It was capable of producing acceleration up to 2 g. The test shown is for a core consisting of 6 layers, each containing 450 graphite bricks and the containment vessel has been replaced by a braced framework for greater ease in reading the strain gauges attached to a selection of the bricks, and between the bricks and the framework.

The principal outcome of these tests was that they supported the very important design assumption that seismic loads acting on the graphite core would be carried by the restraint structure, with no transfer of shear between layers of bricks. It perhaps indicates the difficulties associated with shaking table testing at this time, that although all other or components in this nuclear power station were *designed* against seismic risk by a variety of analytical methods, no shaking table tests were carried out upon them.

16.5.2 Concrete Dam Studies at ISMES

In 1951 a large number of Italian companies came together to create, at Bergamo in N. Italy and with Guido Oberti as its first Director, a laboratory which soon earned a worldwide reputation for seismic model studies on a wide variety of structures, particularly concrete arch dams (Oberti, 1956). A principal facility for doing so was a shaking table which they themselves referred to as "...being of the Jacobsen type" (v.16.2.2). Figure 16.22 shows that the rigid frame table measuring 10×15 ft was suspended by wires from a very stiff gantry; the Jacobsen input arrangement of pendulum and reaction springs is shown on the left of Fig. 16.22, whilst the right-hand side shows a contra-rotating mass machine, capable of an input of 10T, situated for vertical input in the 2–25 Hz range. Not shown was a third input facility consisting of 4 electronically controlled electro-magnetic vibrators which could be used for multiple sinusoidal input in any direction.

Figure 16.22 shows the test on a concrete arch dam model, a form of construction in which a number of Italian engineers – notably Carlo Semenza – became

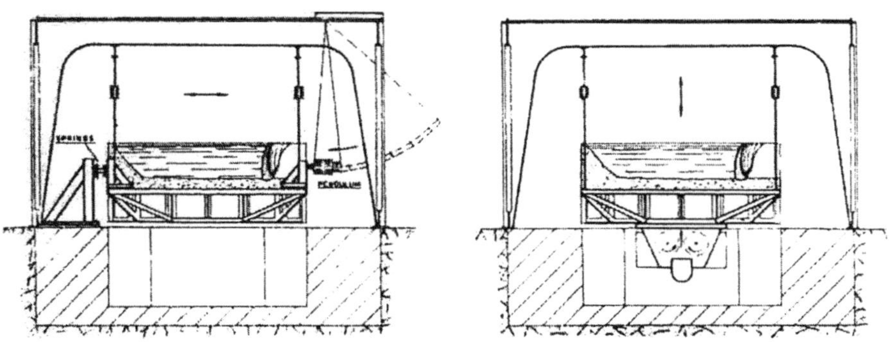

Fig. 16.22 The "Jacobsen-type" shaking table built at ISMES, Italy

internationally famous, resulting in contracts with many countries including Japan, Mexico, Yugoslavia and the United States, using models constructed from litharge and plaster of Paris in various combinations to produce E-values in the range 3,000–10,000 kg/cm^2, and compressive strength in the range 2–11 kg/cm^2. The scale of the models varied between 1:75 and 1:100, allowing the reservoir length to be three times the dam height, thus satisfying the Westergaard criterion required for removing "end-effects".

Being a commercial organization, ISMES published only scant details of its test procedures and results.

16.5.3 Tall Buildings

The device used by Ruge (v.16.3.2) to input a real earthquake signal into his model did not become more common due to the cumbersome nature of his analogue device, and also his arrangement for ensuring the accuracy of his input – what we now refer to as "control" – left a lot to be desired. The second world war, and its aftermath, then intervened to inhibit progress in shaking table design, and at leading laboratories, such as ISMES, the introduction by Biot in 1943 of his earthquake spectra concept, perhaps diverted attention somewhat from the need to develop the input of recorded earthquakes. But the war had a beneficial effect in producing major advances in electronic methods of measurement and in the control of all types of mechanical systems for military purposes, which within a few years had been adapted for use in shaking tables.

A further stimulating factor was the development in the construction industry of tall, and irregular-shaped buildings, bringing with them the need to consider multi-axis motion, particularly torsion, and their high centre of mass in inducing overturning moments in the shaking table. A first attempt at solution (Sesan et al., 1969) appears to have been made at Jassy in Romania (c.1960) using a hydraulic principle indicated in Fig. 16.23. A fixed base carrying the table contains an array of 16 water jets, each in a sealed compartment, which support the table itself by pressure from a header tank. When a tall building on the table is vibrated in one or two horizontal directions by oil-filled actuators, the overturning moments are corrected (within limits) by changes in the water pressure in the compartments.

A more versatile approach, particularly for tall buildings, was that by Penzien et al. (1967) at University College, Berkeley in 1967. This team explored the possibility of constructing a 100$\times$100 ft welded steel table which could vibrate a large structure in all three translational degrees of freedom, with control to zero of the three rotations. Unfortunately, control engineering would not be in a position to achieve this for another 20 years, causing UCB to build a 20$\times$20 ft table with periodic and random input capabilities in the vertical and one horizontal direction simultaneously. Figure 16.16 shows a full-scale RC 3-storey frame on this UCB table.

At the beginning of this chapter it was recorded that the study of sand deposits in California was a major, but unfinished, use of shaking tables. At UCB at this time it was proposed to return to this topic by mounting a smaller, uni-axial, table

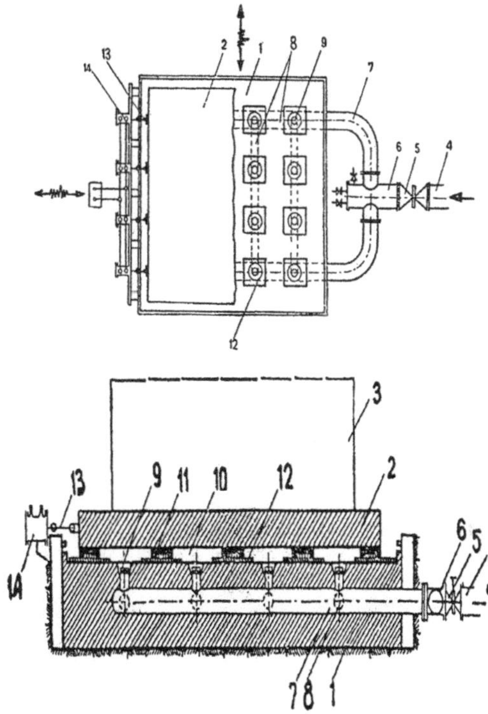

Fig. 16.23 The Jassy hydraulically-supported shaking table

containing the sand model on the existing bi-axial larger one, so that the sand received input in all three translational directions simultaneously. It is not known whether this idea was actually implemented.

16.5.4 Infrastructure – Fitness-for-Purpose

One of the important uses of shaking tables is in “fitness-for-purpose” tests, meaning that the manufactured article should not fail to perform its function as a result of an earthquake. Water and gas pipelines are good examples, as are power station machinery and transmission systems, but their complexity means that such facilities must be tested at full-scale. Figure 16.24 shows a transmission line circuit-breaker (Fischer and Daube, 1976) on the UCB shaking table in 1974. Because its centre-of-mass is high above the surface of the table any input motion will induce large overturning moments on the table itself and it is necessary to counteract these by the table control system, either to zero – as in the UCB in 1974 – or to a specified amount which might be required by the Certifying Authority.

The type of circuit-breaker shown in Fig. 16.24 had been designed for an equivalent static load of 0.2 g, and was severely damaged in the 1971 San Fernando earthquake, probably caused by quasi-resonance at about 2.5 Hz, which was

Fig. 16.24 A one-column circuit breaker on the UCB table

confirmed by theoretical analysis. A new design requirement was that it should survive at least a 0.5 g earthquake, so that in the actual tests the El Centro record multiplied by 1.5 was used. In addition, a sine-beat motion with an increasing amplitude was used to produce 1.0 g. The shaking table testing also allowed the addition of add-on damping devices to be assessed.

Acknowledgements The author is grateful to his many colleagues around the world who have given him the benefit of their knowledge in the task of collecting material for this chapter. Also, to the libraries of the University of Bristol and the Institutions of Civil Engineers for tracing copies of early publications.

I cannot claim that this chapter is in any way a complete record of the assistance which shaking tables have given towards advancing knowledge in earthquake engineering, and I shall be grateful if serious omissions are pointed out to me.

References

Calder DS (1936) Observed vibration of steel water towers. Bull Seis Soc Am 26:69–81

Carvalho EC, Ravara A, Duarte RT (1978) Analytical and model studies for the international Guadiana bridge. In: Proceedings of the 6th European Conference on Earthquake Engineering, Dubrovnik, pp 227–234

Clough RW, Pirtz D (1958) Earthquake Resistance of rock fill dams. Trans ASCE 123:792–816

Fischer EG, Daube WM (1976) Combined analysis and test of earthquake resistant current breakers. Earthquake Eng Struct Dyn 4:231–244

Godden WG, Asham M (December 1978) Dynamic model studies of Ruck-A-Chucky Bridge. J Struct Div ASCE ST12, 104:1827–1845

Hoskins LM, Jacobsen LS (January 1934) Water pressure in a tank caused by a simulated earthquake. Bull Seis Soc Am 24(1):1–22

Jacobsen LS (September 1930) Motion of soils subjected to simple harmonic motion vibration. Bull Seis Soc Am 20:160–195

Jacobsen LS, Ayre RS (October 1951) Hydrodynamic experiments with rigid cylindrical tanks subjected to transient motions. Bull Seis Soc Am 41:313–346

Mononobe N, Takata A, Matamura M (1936) Seismic stability of the earth dam. In: Proceedings of the 2nd congress on large dams, Washington, DC, Ques. vii, pp 435–442

Muir-Wood R (September 1988) Robert Mallet and John Milne – earthquakes incorporated in Victorian Britain. J Earthquake Eng Struct Dyn 7(1):107–142

Muto K, Bailey RW, Mitchell KJ (1962) Special requirements for the design of nuclear power stations to withstand earthquakes. In: Proceedings of the Institutions of Mechanical Engineering, Nominated lecture to the Nuclear Power Group, London, 10 October, pp 1–46

Oberti G (1956) Development of aseimic design and contribution in Italy by means of research on large model tests, 1st WCEE, Berkeley, CA, pp 25.1–25.12

Penzien J et al (1967) Feasibility study large scale earthquake simulator facility. Report EERC-67-1, University of California, Berkeley, September

Priestley MJN, Park R (1979) Seimic resistance of reinforced bridge columns. A conference organized by the US Applied T, 29–31 January, pp 253–283

Rae D, Penzien J (1973) Dynamic response of a 20×20 ft shaking table. In: Proceedings of the 5th WCEE, Rome, June 1973, vol 2, pp 1725–1752

Reid HF (1906) The theory of Mr. Rogers' experiments. The Californian earthquake of April 1906, Report of the State Investigation Commission, vol 2, Part I, p 49

Rogers FJ (1906) Experiments with a shaking machine. The Californian earthquake of April 1906. Report of the State Investigation Commission, vol 1, Part II, pp 326–325

Ruge AC (July 1936) A machine for reproducing earthquake motion from a shadowgraph of the earthquake. Bull Seis Soc Am 26(3):201–205

Ruge AC (1938) Earthquake resistance of elevated water tanks. Trans ASCE 103:889–949

Sesan A et al (1969) A new shaking table used to test structures at seismic actions. Bulletiniul Institutului Politehnic, Bucharest

Severn RT (Accepted for publication) The development of shaking tables – a historical note. J Earthquake Eng Struct Dyn

Taksuoka F, Tokida K, Yoshida S, Morayama I (1978) Shake table tests on dynamic behaviour of pile foundation model in liquefying sand layers. In: Proceedings of the 5th Japanese Earthquake Engineering Symposium

Westergaard HM (November 1931) Water pressure on dams during earthquakes. Proc. ASCE, 57:1303–1318

Williams D, Godden W (1979) Seismic response of long curved bridge structures: experimental model studies. Earthquake Eng Struc Dyn 7:107–128

Chapter 17
Development, Production and Implementation of Low Cost Rubber Bearings

Mihail Garevski

Abstract The investigations and the results discussed in this chapter are related to development, production and implementation of low cost rubber isolators. In addition to production of isolators, one of the main objectives of the project was replacement of the old bearings of the Pestalozzi school building with new ones. The initial dimensions of the prototype bearing were obtained through analysis of the dynamic response of the school building. The characteristics of the bearings had to satisfy two conditions; (1) the maximum lateral displacements of the new bearings had to be less than 20 cm in order not to exceed the seismic gap; (2) the efficiency of the new rubber bearings regarding reduction of seismic forces had to be better or identical to that of the old bearings. Once the initial dimensions and characteristics of the new bearings were adopted, there started the production of the bearings of smaller proportions. The tests performed on the bearings enabled: (1) definition of errors made during the trial production; (2) selection of an appropriate compound for preparation of rubber to provide damping of around 10% and (3) proving of the product quality. A large number of small bearings were produced and tested until the tests showed that the bearings possessed the necessary damping and the required quality. Then, the production of the prototype bearings for the Pestalozzi school building started. After adopting the production process, 54 large bearings were produced and used to replace the old bearings of the Pestalozzi school building. The successful accomplishment of the project was a good promotion of base isolation in the Balkan.

M. Garevski (✉)
Institute of Earthquake Engineering and Engineering Seismology, Ss. Cyril and Methodius University, 1000 Skopje, Republic of Macedonia
e-mail: garevski@pluto.iziis.ukim.edu.mk

M. Garevski, A. Ansal (eds.), *Earthquake Engineering in Europe*, Geotechnical, Geological, and Earthquake Engineering 17, DOI 10.1007/978-90-481-9544-2_17,

17.1 Introduction

Placement of engineering structures on rubber bearings for protection against earthquakes is being frequently practiced in developed countries today. The popularity of modern bearings for base isolation of structures has been increased particularly after the 1995 Kobe earthquake (Fujita, 1995, 1998). During this earthquake, a good behaviour of base isolated structures was observed (Fujita, 1999, 2001). The structure and the equipment of the base isolated buildings in Kobe did not suffer any damage during the said earthquake. Unlike these buildings, similar buildings constructed in the conventional way (with a fixed base) that were located close to the isolated structures suffered huge losses. Because of providing greater safety of structures and equipment, this technique is applied in structures of vital importance as are crisis centres, fire fighting centres, city halls, hospitals and schools. This technique can equally be used also for old and new structures. The city halls in San Francisco, Oakland and Los Angeles representing historic buildings are base isolated for the purpose of enabling their immediate functioning even after the strongest earthquakes and protecting their historic values. Until recently, this type of seismic protection of structures has been extremely costly so that only countries like the USA, Japan, New Zealand, Italy and other countries of powerful economy could afford incorporation of bearings since, for some of the buildings, the cost of installation of isolators amounted to millions of dollars.

In the developing countries, this technique has rarely been used due to non-existence of domestic production of bearings and high cost of the bearings produced in the developed countries. In some of these countries, as is Indonesia (Taniwangsa et al., 1995; Taniwangsa, 2002), Iran and Algeria, there have been some attempts to popularize this technique through development of low-cost bearings and their installation in demonstration structures, but no attempt for production has been made and hence there hasn't been any mass application of such bearings. A greater success in application of base isolation (with isolation of a large number of buildings) was achieved in Armenia where, in addition to placement of isolators in a building, their production was also adopted (Melkumyan et al., 1993, 1997).

The Balkan peninsula represents a region exposed to high risk pertaining to occurrence of catastrophic earthquakes.

In Macedonia, Albania and in the remaining Balkan countries, seismic design of fixed base structures is frequently practiced. Although the seismic regulations referring to fixed base structures provide protection of human lives, damage to the structures due to catastrophic earthquakes could be very severe.

In 1963, the primary school "Njegosh" in Skopje was completely ruined under the effect of the catastrophic earthquake. In 1965, a new building, i.e., the "Pestalozzi" school building was built in the same place. The Pestalozzi school building was placed on 54 rubber isolators which was the first application of rubber isolators worldwide (Naeim and Kelly, 1999). The construction of the school and the installation of the bearings was financed by the Swiss government. Since then, this technique has been applied neither in R. Macedonia nor in the remaining Balkan countries. The high cost of isolators and the insufficient knowledge of this technique pose a problem in adapting isolation to developing countries (Garevski, 2005).

For the purpose of increasing the interest in application of base isolation in R. Macedonia and the remaining countries in the Balkan, the Institute of Earthquake Engineering and Engineering Seismology (IZIIS) proposed a project on development of low-cost rubber bearings. After adopting the production technology, 54 bearings were produced and installed in the Pestalozzi school building to replace the old bearings. The project was financed by NATO through the Science for Peace programme. In addition to IZIIS, the Civil Engineering Faculty of Tirana and the Earthquake Engineering Research Centre from Berkeley, California also took part in the project (Garevski et al., 2008).

This chapter starts with a brief information on the Pestalozzi school building. The seismic input necessary for the dynamic analysis of the school building and the design of the new bearings are presented as well. Through this analysis, the characteristics of the new bearings and their proportions were defined for the purpose of trial production of large bearings.

The second part of the paper deals with the production, the difficulties faced during the production as well as the tests aimed at finding the most appropriate rubber compound (high damping rubber) and assuring quality of production.

Then, the dismantling of the old bearings and mantling of the newly designed ones is described.

Finally, corresponding conclusions on the project benefits are drawn.

17.2 Design of New Rubber Bearings for the Pestalozzi School Building

To establish the necessary production of seismic rubber isolators, i.e., high damping rubber isolators with steel layers, certain analytical and experimental investigations were carried out.

The analytical investigations were aimed at obtaining the proportions of the prototype bearings and the series of bearings to be produced later. They were related to a particular case, i.e., dynamic (seismic) nonlinear analysis and design of a new system for isolation of the Pestalozzi school building.

17.2.1 Description of Structural System of Base Isolated Part of Pestalozzi School

The building, i.e., the J.H. Pestalozzi primary school consists of several units of different number of stories and outline. The base isolation system was applied only for the main school building, while the other units were founded in the conventional way.

The main school building consists of a ground floor and two stories. The proportions of the structure at plan are 11.42/61.20 m, while its height is 10 m. It represents a reinforced concrete box system. The bearing wall system is composed of shear walls with a thickness of 0.18 m. The floor slab is 0.20 m thick. The building has a strip foundation forming a beam grid, which is sufficiently rigid to sustain all the effects from the upper structure without heavier deformations. There is a seismic gap with a width of 0.2 m along the edges of the building,. The gap is covered with reinforced concrete plates.

The structure was base isolated by rubber bearings inserted between the foundation and the structure itself, Fig. 17.1. A total of 54 bearings made of natural

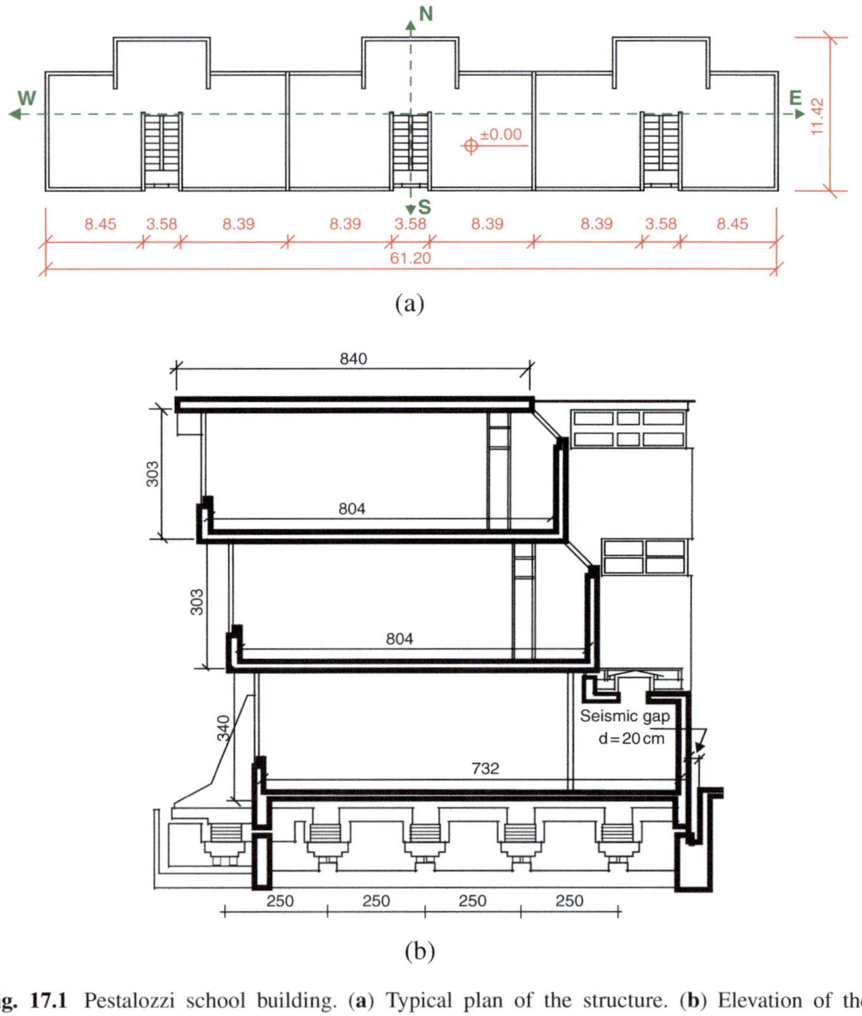

Fig. 17.1 Pestalozzi school building. (**a**) Typical plan of the structure. (**b**) Elevation of the structure

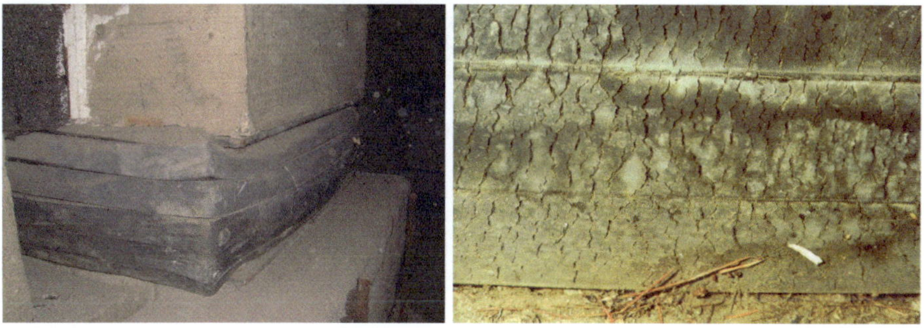

Fig. 17.2 Smashed and cracked old bearings

rubber, with a square shape, side of 0.7 m and height of 0.35 m were incorporated. Each bearing transfers a vertical force amounting to approximately 470 kN. The rubber bearings were produced by the Swiss firm HUBER SUHNER from Zurich (Staudacher, 1982, 1985). Unlike the presently used 2D rubber isolators that reduce the seismic forces only in two lateral directions, the objective of the design engineer of the school building was to install a 3D isolation system. Therefore, the isolators don't contain reinforcing steel plates. They are fabricated by gluing together seven rubber layers in order to obtain low rigidity in the vertical direction. In this way, a low vertical stiffness ($T_{1v} \approx 3T_{1\,h}$) was obtained (Garevski et al., 1998, 2000, 2002). However, due to the low vertical stiffness, they have to have large proportions in order to be able to sustain vertical forces due to dead weight and additional vertical forces caused by earthquakes. Due to the large dimensions, the first two fundamental modes of vibration are below 2 s ($T_{1\,N\text{-}S}$=1.32 s i $T_{1E\text{-}W}$=1.14 s) so that the effect of base isolation is not so high. In addition to the negative effect of the low vertical stiffness of the bearings on the possible rocking motion, there is another disadvantage of these bearings, which is the extensive bulging of the rubber bearings (subsidence of the school building) due to dead load effects, Fig. 17.2. The subsidence of the school building and the small cracks in the bearings due to ageing of the rubber were the main reasons for replacement of these isolators (Garevski and Kelly, 2001).

17.2.2 Computation of Stiffness Characteristics of New Bearings

The new bearings had to satisfy the following conditions: (1) the lateral displacement of the bearings under the maximum capable earthquake had to be less than 20 cm; (2) the new isolators had to enable the same or better reduction of seismic forces than the old isolators.

Dynamic analysis of the school building was carried out for two levels of earthquake excitation (the design and the maximum capable earthquake). With these analyses, the stiffness of the new bearings was defined, satisfying the main design conditions.

– The horizontal stiffness of the new isolation system was selected following the subsequently presented procedure:

- definition of stiffness of the bearings;
- definition of the seismic input for the corresponding fundamental period of the structure;
- computation of maximal lateral displacements due to the effect of the maximal capable earthquake;
- performing new analysis using the corrected lateral stiffness values of the bearings.

To satisfy these conditions, the procedure was conducted for three models with different stiffness characteristics of the bearings.

17.2.3 Initial Characteristics of Isolators

The first step in these analyses was definition of the initial striffness characteristics of the new rubber bearings. The natural rubber planned to be used for production of the new bearings had to be high damping rubber with an average effective damping of β_{eff}=0.10.

The effective damping, β_{eff}, of an isolator unit is calculated for each cycle of loading by using the following formula:

$$\beta_{eff} = \frac{2}{\pi} \cdot \left[\frac{E_{loop}}{K_{eff}\left(\left|\Delta^{+}\right| + \left|\Delta^{-}\right|\right)^{2}} \right] \tag{1}$$

where,

E_{loop} – area of a hysteretic cycle
Δ^{+} and Δ^{-} – positive and negative displacement amplitude.

The effective stiffness of an isolator unit, K_{eff}, is calculated for each cycle of loading by using the following formula:

$$K_{eff} = \frac{F^{+} - F^{-}}{\Delta^{+} - \Delta^{-}} \tag{2}$$

where F^{+} and F^{-} are the positive and negative forces at Δ^{+} and Δ^{-}, respectively.

To satisfy the design parameters, the initial calculations showed that the new period of the structure should be within the range of $T_1 = 1.30 \div 1.60$ s. The total effective stiffness of the bearings was computed on the basis of the assumed fundamental period of the structure and it ranged between K_{eff}=(40,789÷61,787) kN/m. The minimal effective stiffness of a single bearing was $K_{effmin} = 755$ kN/m, whereas the maximal one was $K_{effmax} = 1{,}144$ kN/m. To select the stiffness characteristics of the new bearings, dynamic analyses were carried out on base isolated models with dominant periods of T_1=1.32, T_2=1.46 and T_3=1.61 s.

The behaviour of the bearings was modelled by using a bilinear diagram. This diagram was defined by three parameters: initial elastic stiffness K_1, post-elastic stiffness K_2 and yield force F_Y. Using Eq. (1), the stiffness characteristics K_1 and K_2, as well as the yield force F_Y were calculated for $\beta = 0.10$ and K_{eff}=755–1,144 kN/m (Table 17.1).

Table 17.1 Parameters of the bilinear models

Period	K_1 [kN/m]	K_2 [kN/m]	FY [kN]
$T_1 = 1.32$ s	4,210.42	935.65	44.83
$T_1 = 1.46$ s	3,441.00	746.50	36.64
$T_1 = 1.61$ s	2,829.80	628.85	30.12

This initial analysis enabled definition of the characteristics of the final bearings that were to be produced. Such defined characteristics had to be verified by corresponding experimental tests. In cases of considerable difference between the experimentally obtained behaviour of the produced bearings and the assumed behaviour, the analyses had to be repeated.

17.2.4 Definition of Seismic Input

The dynamic analysis of the structure was carried out for two earthquake levels:

- Design earthquake with 10% probability that the earthquake excitation level will be exceeded in 50 years (return period of 475 years);
- The maximum capable earthquake: maximum level of earthquake excitation that can be expected in the region. It can be taken as an earthquake excitation level with the probability of 10% that it will be exceeded in a 100 year period (return period of 1,000 years).

For these defined seismic hazard levels, three pairs of corresponding time histories of acceleration were selected (load cases). The time histories had to include magnitude, distance from the fault and source mechanism consistent with those that control the design or the maximum earthquake. For each pair of horizontal earthquake components, the square root of the sum of squares (SRSS) for a spectrum with 5% damping had to be established from the scaled horizontal components. The horizontal components were scaled in such a way that the mean value of the

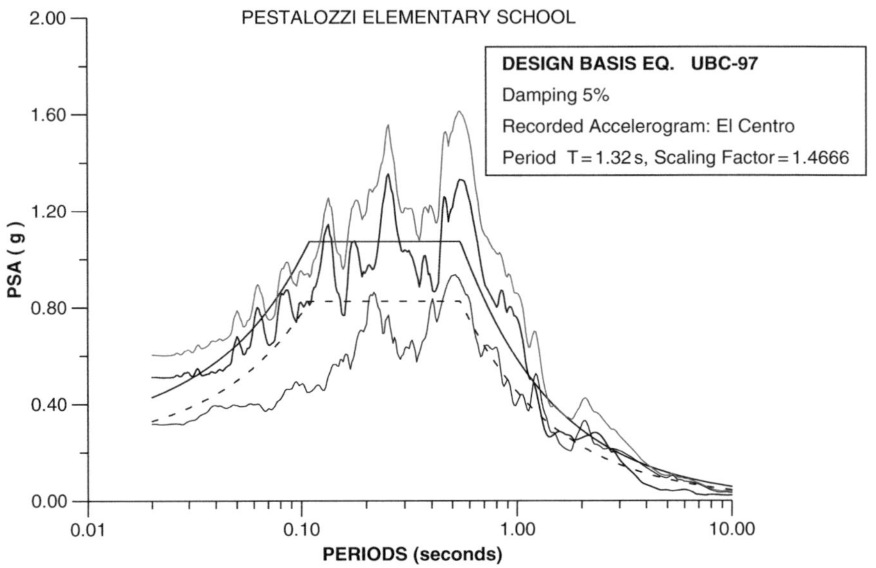

Fig. 17.3 Imperial valley EQ, El centro record, 1940-May-18, design spectrum, T = 1.32 s

SRSS spectrum did not drop below 1.3 times the spectrum for the design and maximal earthquake and not more than 10% for periods within the range of $0.50\times T_D$ to $1.25\times T_M$.

Figure 17.3 shows the selected pairs of accelerations for the design spectrum of the Imperial Valley EQ, El Centro record, 1940-May-18 and period of $T_1 = 1.32$ s.

17.2.5 Dynamic Response of the Structure

For each model ($T_1 = 1.32$, 1.46 and 1.61 s), 3D time history analyses were performed under two levels of earthquake excitation: design (return period of 475 years) and maximum capable earthquake (return period of 1,000 years). For each finite element model, three real earthquake records were selected with corresponding amplitude scaling factor. The scaling factors depended on the fundamental period of the structure and the defined design spectrum.

All dynamic analyses were performed for linear behaviour of the upper structure, while local nonlinearity was considered regarding the bilinear behavior of the isolators.

The results from the performed dynamic analysis were presented only for two selected nodes: JOZ1000 (the isolated level) and JOZ1019 (the top floor) (Fig. 17.4).

The lateral displacement of the isolation system was checked for the effect of the maximum capable earthquake. The maximum lateral displacements of the bearings

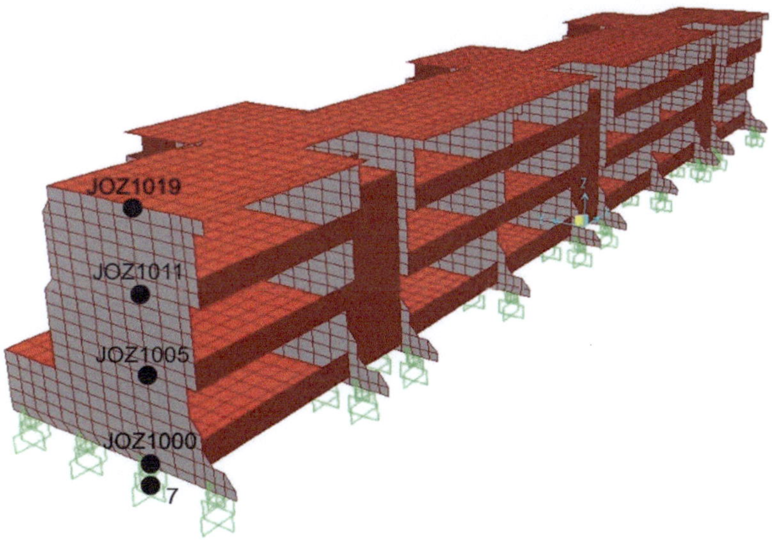

Fig. 17.4 Finite element model with the selected nodes

in function of the fundamental period of the structure are presented in Fig. 17.5. Each value represents an absolute maximal lateral displacement obtained for three load cases with pairs of actual earthquake excitations.

The bearing capacity of the superstructure was controlled under the effect of the design earthquake. The floor accelerations of the mathematical model with the old isolators and the model with the new isolators were also compared. The maximal

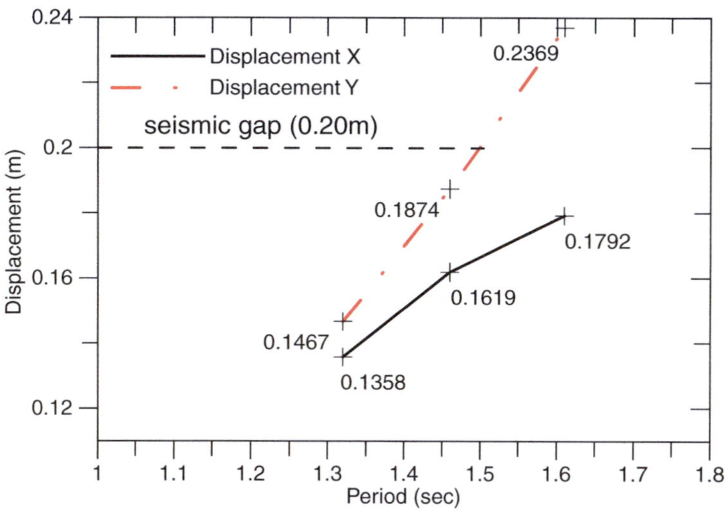

Fig. 17.5 Lateral deformation of the bearing under the effect of the maximum capable earthquake

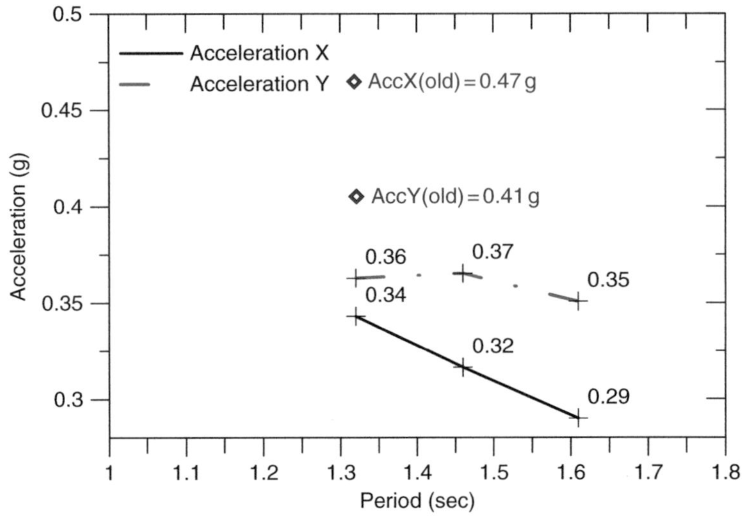

Fig. 17.6 Accelerations at the top floor under the effect of the maximum capable earthquake

horizontal accelerations in both orthogonal directions at the top floor under the effect of the three excitations (load cases) are given in Fig. 17.6.

From the performed analyses, it can be said that the second isolation system with period $T_1 = 1.46$ s represents the optimal solution for both design parameters. All further detailed analyses were carried out for this model. For the model with the first fundamental period of vibration of 1.46 s, time history analyses were performed by using the following real earthquake records:

- Imperial Valley EQ, El Centro record, 1940-May-18 (ElCentro);
- Montenegro EQ, Herceg Novi record, 1979-April-15 (HercegNovi);
- Kern Country EQ, Taft-Lincoln School Tunnel record, 1952-July-21 (Taft).

The horizontal displacement in y-direction (N–S), at the isolated level, was the greatest under the effect of the Hertseg Novi earthquake, amounting to $\Delta y = 18.5$ cm. The horizontal displacement in x-direction (E–W) at the isolated level was the greatest under the Taft earthquake, amounting to $\Delta x = 16.2$ cm. These displacement values didn't exceed the existing seismic gap of 20 cm.

The accelerations at the top floor, in longitudinal x-direction (E–W), were the highest under the effect of the design Taft earthquake (Accx = 2.2 m/s^2). The design El Centro earthquake produced the highest acceleration response in y-direction (N–S), at the top floor (Accy = 2.5 m/s^2).

From the performed dynamic analyses of the structure with the old isolation system (Garevski), it was obtained that the maximal acceleration at the top floor in transverse "y" direction was Accy = 4.7 m/s^2, whereas in the longitudinal "x" direction, it was Accx = 4.1 m/s^2. The graph presented in Fig. 17.6 shows that

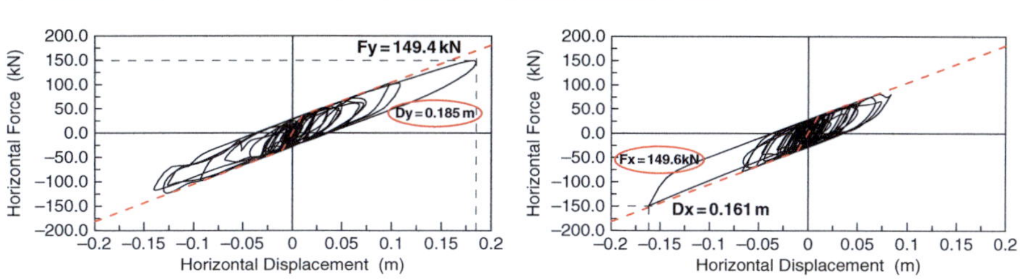

(a) (b)

Fig. 17.7 Force-displacement relationship of the isolatior in “x” and “y” direction (T_1=1.46 s). (**a**) Montenegro EQ, Hertseg Novi record, 1979-April-15. (**b**) Kern Country EQ, Traft-Lincoln School Tunnel record, 1952-July-21

the maximum acceleration at the top floor of the model with a vibration period T_1=1.46 s is far below the acceleration values obtained for the model with the old isolation system.

The horizontal force – displacement loops for the new isolator in transverse “y”(N–S) and longitudinal “x”(E–W) direction are presented in Fig. 17.7. The maximum base shear force in the bearings does not exceed 150 kN.

17.3 Production of Rubber Bearings

As mentioned, one of the main goals of the project was adoption of production of high damping rubber bearings with steel shims.

Until the beginning of this project, there wasn’t any experience in R. Macedonia, Albania and in the Balkan in general in production of rubber bearings – isolators. Trial production of rubber bearings with different dimensions and later, mass production, was carried out in a small workshoop (MAK-OZ) in Kavadartsi.

The trial and the mass production of bearings could not start without performance of a series of tests on small and large prototype specimens. Until the beginning of the project, IZIIS didn’t have a corresponding equipment for testing of bearings. Therefore, within the frames of the project, the existing single component small dynamic machine was, first of all, modified into a two-componental one. To test the prototype bearings, a big biaxial machine was designed and constructed. The main characteristics of the used machines will be given in the subsequent text.

To save materials, the production started first with scaled (small size) bearings. Later, during the tests, certain errors in the production were detected. These were corrected and the bearings were subjected to dynamic tests, which proved that the bearings satisfied the necessary requirements. Then, the trial production of the large bearings started. In the end, after eliminating the individual defects in the production of the large bearings and proving their quality through tests on the big testing machine, the mass production of 54 big bearings began.

This chapter deals with description of some phases of the production as well as some of the errors made during the trial production. The most important tests that contributed to the definition of the characteristics of the rubber and the bearings are also given.

17.3.1 Small Bearings

Our consideration to start first with production and testing of small bearings proved to be justified. Due to the errors made in the beginning of the production and the need for testing of bearings made of different compounds, a large number of bearings were produced and tested. If these experiments hadn't been done and the production and testing of large rubber bearings had started directly, a much greater quantity of rubber and remaining material would have been spent.

For the rectangular isolators (200/200/50 mm), steel sheet metal plates with thickness of 2 mm and proportions of 190/190 mm were prepared. The raw rubber was precisely assembled in the mold with a certain thickness, i.e., precisely tailored to be proportioned 195/195 mm at plan. The raw rubber was previously rolled to acquire rubber sheets with the required thickness. It can be said that rolling represents one of the key factors for precise assembling of the raw rubber in the mold. The insufficient preciseness during assembling of the rubber may lead to layers of different heights and inclination of the steel plates after curing.

Prior to the very process of vulcanization, the steel areas underwent three additional treatments. The first phase consisted of cleaning of the sandblasted areas with medical gasoline. This helped remove all the dirt occurring from the moment of sandblasting to the moment of vulcanization. The second phase involved coating of the areas with an appropriate adhesive. This coating protects the steel areas against any external effects, excluding mechanical ones. At the same time, this coating plays the role of a base for the second coat. Such coated plates are kept in premises with normal humidity and average temperature of 20–25°C. After 2 h, the third phase takes place, i.e., treatment of the steel areas with the second coat. As in the case of the first coat, after the second coating, the plates are also left to dry at room temperature for about an hour.

Upon completion of the preparation works, the tailored rubber and the treated steel plates are inserted in a mold. The mold with the inserted layers of rubber and steel plates is placed in a press to perform vulcanization (Fig. 17.8a).

Once the vulcanization is over, the element is dismantled from the mold and left to cool at room temperature for 24 h (Fig. 17.8b).

In addition to the bearings with rectangular form, rubber bearings with a cyllindrical form were produced. The external diameter of the cyllindrical bearings was 150 mm, whereas the total height was 100 mm. The rubber bearings were produced with inner steel plates with a thickness of 2 mm. The end plates in these elements had a thickness of 10 mm.

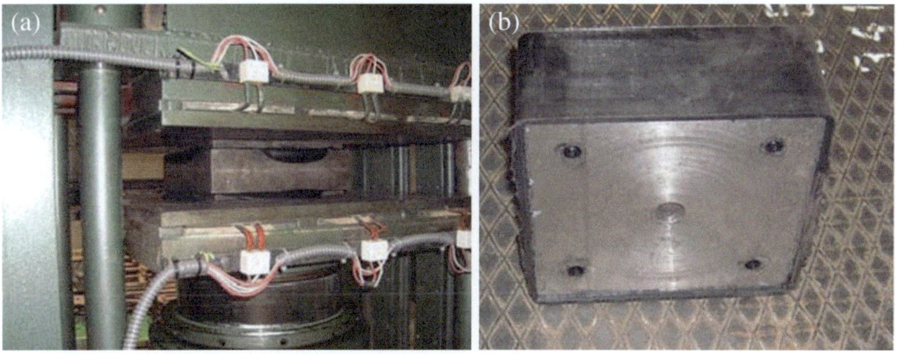

Fig. 17.8 (**a**) Vulcanization of isolator. (**b**) View of isolator

17.3.2 Testing and Selection of High Damping Rubber

To perform testing of the scaled rubber isolators in the IZIIS' laboratory, the existing single-component dynamic machine was modified into a two-component one.

The specimens were excited in two orthogonal directions. One direction of excitation was the dynamic one (the lateral one), whereas the other was the static (the vertical) one. The machine has the possibility of generating dynamic excitation of up to 10 Hz. The stroke of the dynamic actuator was ±7.5 cm, while the maximal dynamic force was 100 kN. The second vertical direction of excitation was with a maximal stroke of 15 cm and maximal compressive force of 200 kN. The specimens were tested in pairs. The same machine was used for two main tests of isolators (vertical and biaxial test). To test the quality of the produced rubber bearings, they were tested according to a prescribed testing procedure involving a series of horizontal and vertical tests.

To produce the seismic isolators, it was necessary to select a high damping raw rubber compound (8–10% on the average). Natural rubber with such characteristics could not be found as a product neither in Macedonia nor in the closer region. A number of rubber making companies were therefore engaged in development of a high damping rubber compound.

None had experience in developing such compounds but they expressed their willingness to participate in the project. Different natural rubber compounds were developed and supplied by different copanies (Avtoguma – Macedonia, Zebra – Bulgaria and Gomline – Slovenia). The specimens were exposed to horizontal excitation of different frequency. Figure 17.9 shows testing of small square bearings. The tests proved that all the compounds satisfied the elongation condition at failure and that almost all the compounds possessed an acceptable modulus of elasticity.

Most of the compounds showed an insufficient damping or the damping drastically deteriorated with the increase of the excitation frequency. The damping of the

Fig. 17.9 Horizontal test on scaled isolators

rubber was defined through biaxial tests on scaled seismic isolators (200/200/50 mm and Ø=15 cm).

The compound developed by Avtoguma (Fig. 17.10) possessed very low damping, i.e., under excitation of 0.2 Hz, the equivalent damping ratio was ζ_{eq} =3.51%, whereas under excitation of 0.4 Hz, it amounted to 0.71%.

Gomline developed eight compounds. Most of them exhibited behaviour similar to the behaviour of the Avtoguma compound. Figure 17.11a shows the response to a horizontal test under dynamic excitation of 0.2 Hz (ζ_{eq} = 5.16%) and 0.4 Hz (ζ_{eq} = 2.83%). The better modification of this compound is presented in Fig. 17.11b. A damping ratio of 7.5% was achieved under excitation of 0.2 Hz and 4.86% under excitation of 0.4 Hz. Although the damping was increased, it was not sufficient enough to classify the rubber as a high damping one.

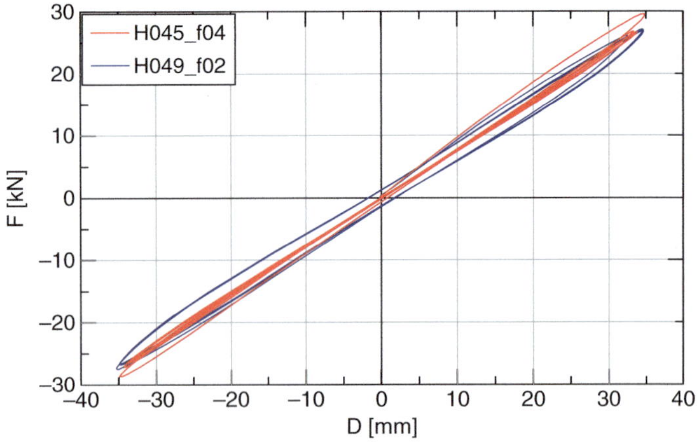

Fig. 17.10 Test H049 and H045, Avtoguma – Macedonia

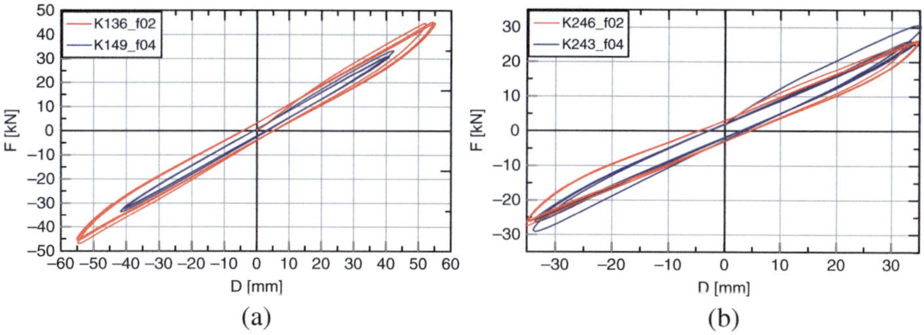

Fig. 17.11 **(a)** Test K136 and K149, Gomline. **(b)** Test K243 and K246, Gomline

A far more stable behaviour under the effect of high frequencies was exhibited by the compounds developed by Zebra (Fig. 17.12). Damping under frequencies of 0.2 Hz was 9.8%, while under 0.4 Hz, it was 7.0%.

The Zebra factory made one more improvement of the compound (Test K563). Figure 17.13. shows test loops from a lateral test on a bearing (Ø = 15 cm) with increased excitation amplitude and unchanged frequency. The computed damping coefficient under shear strain of the bearing of $\gamma \approx 100\%$ and under excitation with a frequency of $f = 0.3$ Hz amounts to $\zeta_{eq} = 12.5\%$ of the critical damping, whereas under excitation of $f = 0.2$ Hz, it is higher, amounting to $\zeta_{eq} = 13.5\%$. These tests also confirmed the characteristics of the rubber showing stable behaviour under different excitation conditions.

Given that the main requirements regarding obtaining of damping of more than 10% were satisfied, this compound was used for the production of all the bearings to be installed in the Pestalozzi school building.

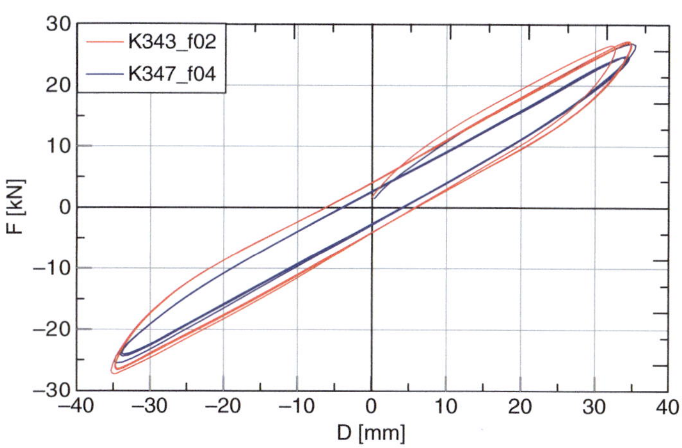

Fig. 17.12 Test K343 and K347, Zebra, Bulgaria

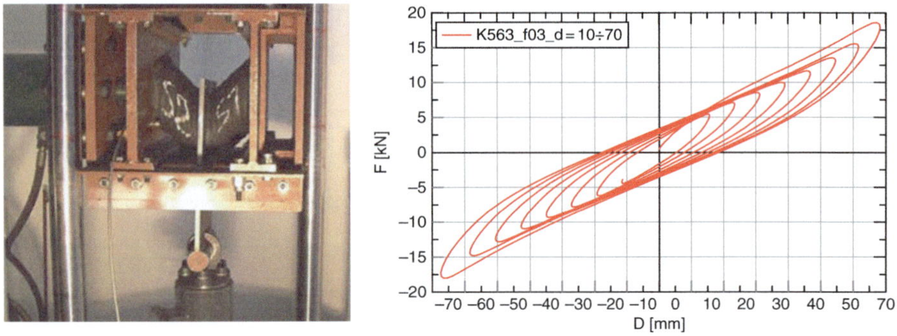

Fig. 17.13 Test on a pair of cylindrical bearings and associated force – displacement loop

The shear modulus of rubber materials can easily be measured and it depends, only slightly, on the dimensions of the specimen. There is a difference between a static and dynamic (frequency-dependent) shear modulus. The shear modulus is calculated by using the following expression:

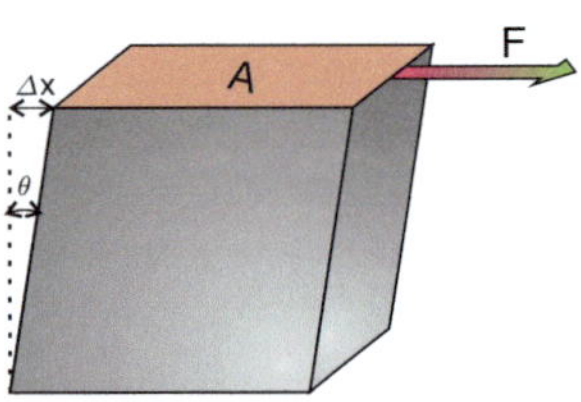

$$G = \frac{\sigma}{\tau} = \frac{\frac{F}{A}}{\frac{\Delta x}{h}} = \frac{F \cdot h}{\Delta x \cdot A} \qquad (3)$$

During the dynamic test with a frequency of 0.2 Hz, the shear modulus ranged between 1,240 kPa – the first part of the hysteretic loop, 800 kPa – the second part of the hysteretic loop and 1,200 kPa – the last part of the hysteretic loop. While under frequency of 0.4 Hz, the modulus was 1,080 kPa in the initial part and almost constant between the first and the third part amounting to 760 kPa, with a slight increase to 990 kPa in the third part.

Figure 17.14a shows one of the histories of vertical excitation composed of two individual different displacement histories. The first part is defined by two parameters: time and position, whereas the second part of the excitation represents harmonic excitation, which is also defined by two parameters: amplitude and frequency. The result of this tests is presented in the form of force – displacement diagram in Fig. 17.14b.

In addition to the presented tests, these bearings were also loaded with excitations with different velocities and amplitudes. The force-deflection relationship of the HDR isolator depended on the rate of loading and additional tests were performed under a frequency equal to the first fundamental frequency of the isolated structure.

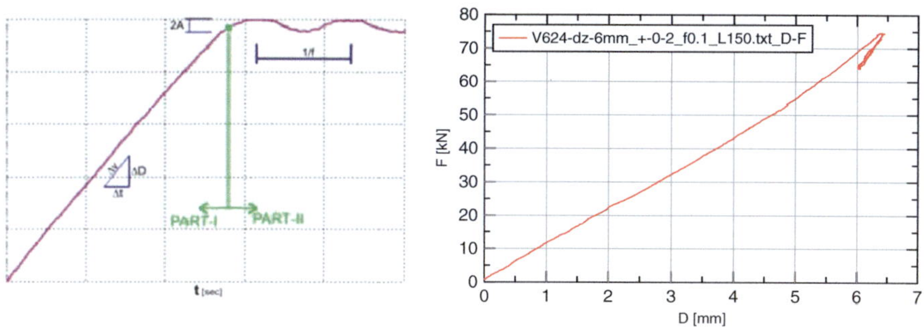

(a) (b)

Fig. 17.14 Vertical tests on cyllindrical bearings: (**a**) History of vertical excitation (**b**) Force-displacement diagram

The objective of these tests was to define the excitation effect upon the response of the bearings to be used for the Pestalozzi school.

As mentioned, the initial analyses were done by use of a bilinear link element. Upon completion of the tests, a finite element analysis was performed for the school building, whereat for the isolated element, the nonlinear loop obtained from the tests was used.

17.3.3 Large Bearings

Once the quality of the produced scaled bearings was verified through the dynamic tests performed on the small dynamic testing machine, the production of the large prototype bearings started (Fig. 17.15). Generally, the procedure for the production of the large bearings was completely the same as that for the small bearings.

Since the procedure for production of the small and the large bearings was the same, it was expected that there would be no greater problems with the production

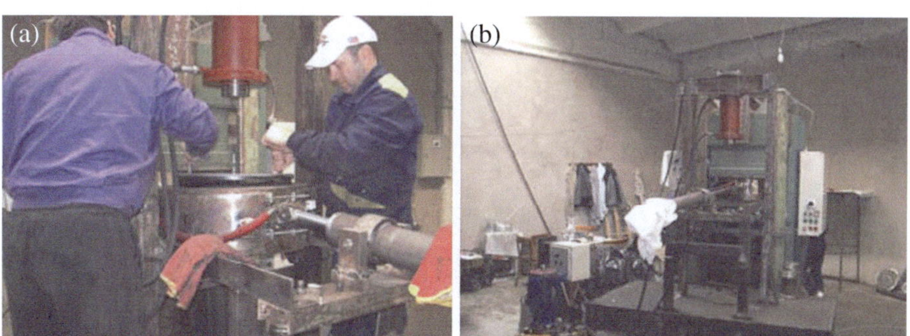

Fig. 17.15 Production of prototype large bearing: (**a**) assemblage of the shim plates and rubber sheets in the mold (**b**) curing process

of the large bearings and that we would be able to start with the mass production of these bearings at once. However, the tests on the first two prototype large bearings carried out on the big two componental machine showed that the production did not achieve the required quality since the bearings were damaged under the effect of horizontal force prior to achieving horizontal displacement of 20 cm. After getting insight into the individual defects in the production and their elimination, the mass production of the large bearings started.

17.3.4 Testing of Large Prototype Bearings

Mass production of the bearings was possible only after proving their quality through experimental tests. Their testing in horizontal and vertical direction was carried out on the big two-componental machine that was designed and constructed specially for testing of the prototype bearings.

The full scale tests on the isolators were carried out by use of the big two-componental testing machine "SBP 2007" (Fig. 17.16). The main purpose of this machine is testing of rubber bearings with maximum proportions of b/d/h = 700/800/400 mm or Ø = 700 mm.

The main characteristics of the machine are:

- Maximum vertical force: 3,000 kN
- Maximum horizontal force: 1,000 kN
- Maximal piston stroke: horizontal ±1,000 mm; vertical 500 mm

Fig. 17.16 Big two-componental testing machine

Horizontal tests were performed on two full-size specimens, at a constant room temperature of 22°C and at a low strain rate such that all the dynamic effects were negligible (f = 0.03 Hz at displacement of ± 200 mm).

Each bearing was subjected to three loading–unloading cycles up to maximum shear displacement of 20 mm. After completion of the third cycle of unloading, the displacement amplitude was increased to 35 mm and the bearing was again subjected to three full cycles. The amplitude was increased further to 100 mm and finally, after other three loading-unloading cycles of 150 mm, the lateral displacement of the specimen was 200 mm. The lateral response of the bearings is shown as the applied horizontal force versus the lateral displacement (Fig. 17.17a). In all the shear tests, the vertical load was constant and equal to 500 kN.

Analyzing the bearing behavior in lateral direction (Fig. 17.17b), it could be concluded that this rubber was quite different from the rubber used for the old bearings of the Pestalozzi school building. Three parts were distinguished in the obtained force – displacement loop. The first part went up to 18% of the lateral deformation and was characterized by the greatest stiffness of $K_1 \approx 2{,}000$ kN/m. In the second, post-elastic part, the stiffness deteriorated and remained constant until lateral deformation of about (75÷80)%. The average stiffness in this part was $K_2 \approx 900$ kN/m. Then, in the third part, the rubber was strengthened and the stiffness was permanently increased until failure of the specimen. The final average stiffness was $K_3 \approx 2{,}300$ kN/m. The referent stiffness of the isolation system taken in our computations was the stiffness corresponding to the second region.

All the large bearings that were previously subjected to lateral tests were also tested under the effect of vertical load. These tests were carried out by using a pre-defined procedure. This procedure involved loading of the specimen up to a certain threshold around which three sinusoidal cycles were performed. After this, the specimen was unloaded. The stiffness characteristics in vertical direction were almost identical and very close to the designed stiffness. Figure 17.18 shows the

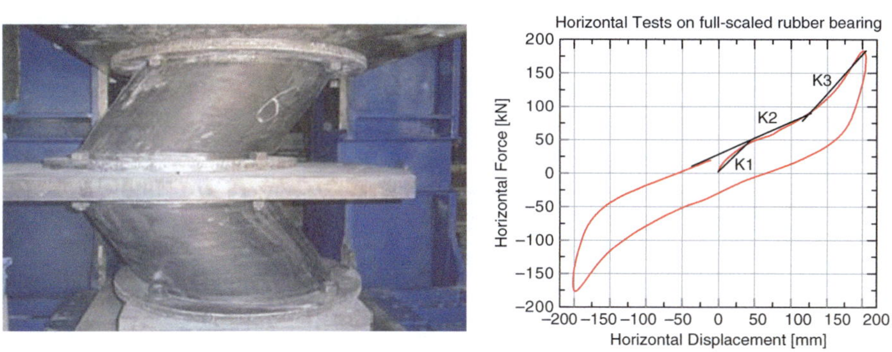

(a) (b)

Fig. 17.17 Lateral test on the new bearings and associated hysteretic loop (vertical force of 500 kN and maximum lateral displacement of 20 cm)

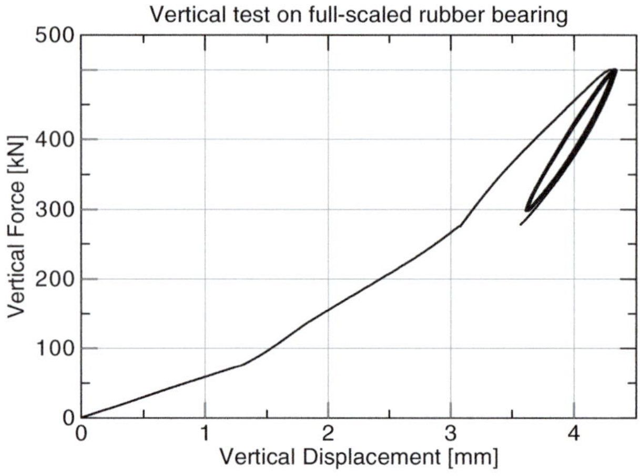

Fig. 17.18 Vertical test on large bearings with inner steel shims

results from the vertical test of specimens with seven layers of inner plates. Under a serviceability load of 500 kN, the isolator had a vertical stiffness of 280,000 kN/m.

17.3.5 Final Proportions and Mass Production of the Bearings

After all the analytical investigations and performed detailed tests on the small square and cyllindrical specimens in horizontal and vertical direction, the final dimensions of the new bearings for the Pestalozzi school building were obtained. In the course of the calculation of the dimensions of the large bearings, there were limiations from the aspect of the capacity of the vulcanisation press. The maximum diameter of the bearings that could be produced with the available press was Ø45 cm. So, the resulting final proportions of the bearings were: diameter of Ø45 cm and total height of rubber t = 16.8 cm.

Figure 17.19 shows section of the bearing and a photo of a produced bearing.

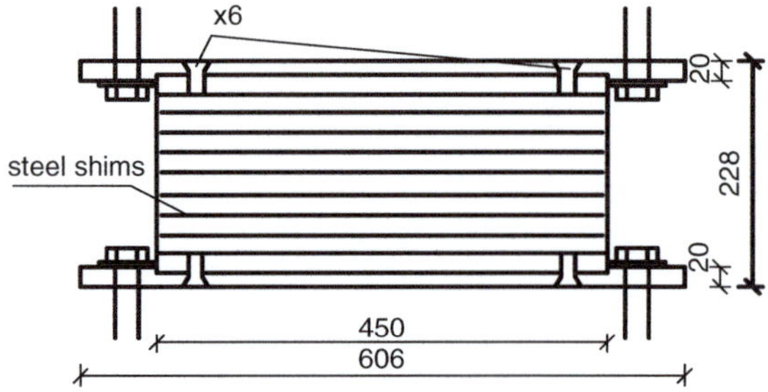

Fig. 17.19 Section and photo of the bearing

17.3.6 Problems in Production of Bearings

Since there wasn't any experience with production of bearings in the region, it was normally expected that certain errors would occur in the trial production. The main problems that occurred during the trial production of the bearings are summarized in the subsequent text.

One of the problems in the trial production was the inability to achieve equal distance between the inner steel shims (Fig. 17.20). This was due to the change of the rubber volume during vulcanization. It is known that the rubber prior and after vulcanization doesn't have the same bulk density. In the course of curing, the rubber filled all the cavities in the mold and tended to get out from it. The steel plates were not sufficiently strong to resist such motion of the rubber, which altogether led to a change of their position and form.

After the performed multiple changes, it was concluded that this phenomenon could be reduced to a negligible level only by precise assembling of the rubber. By means of calculation, the quantity of rubber to be placed in each individual layer was defined. The amount of inserted rubber mainly depended on the density of the raw and vulcanized rubber. In our case, we had a number of types of rubber of different bulk density wherefore we had to calculate precisely the amount of raw rubber of each type to be incorporated.

Another problem was the bond between the rubber and the steel area. A strong bond between the rubber and the steel areas is achieved provided that the steel areas are mechanically and chemically treated and the coating procedure is appropriately carried out. Figure 17.21a, b shows an inappropriate bond due to non-observation of this rule.

For high quality production, proper selection of optimal temperature, vulcanization time and corresponding pressure are also important parameters. The producer of the raw rubber supplied the vulcanization curve for the corresponding raw rubber compound. It must be mentioned however that the producer obtained this vulcanization curve for small and thin specimens. In our case, we had to vulcanize a big amount of rubber, whereat the proportions of the small specimens were 200/200/50, Ø150 and those of the large ones were Ø450/168 mm. For such specimens, it was impossible to stick to the vulcanization time prescribed by the producer. The main task in this phase was to vulcanize the rubber inside the specimen (which was at a distance of 170 mm from the heater) and at the same time, not to over-vulcanize,

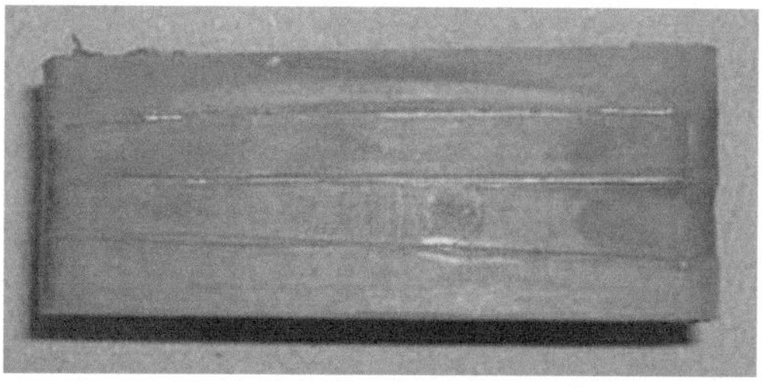

Fig. 17.20 Non-maintenance of inter distance of the steel plates

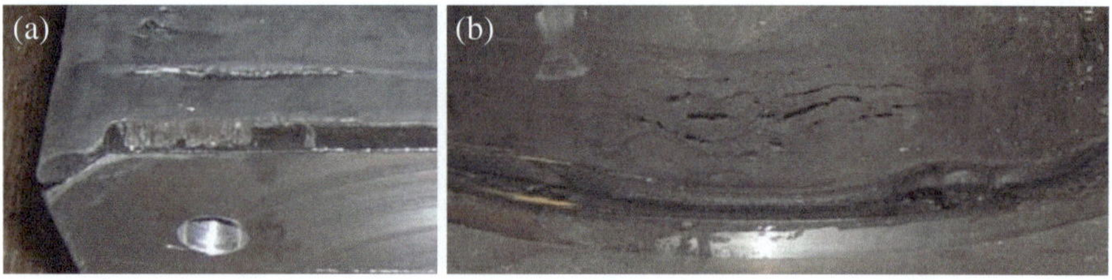

Fig. 17.21 Errors made in trial production. (**a**) Improper treatment of steel surfaces. (**b**) Effect of temperature and vulcanization time (separation of the steel from the rubber)

i.e., "burn" the rubber on the external side (which was at a distance of only 25 mm from the heater), Fig. 17.21b.

The control of the rubber from the outside, along the entire perimeter of the bearing, was easy since the rubber was visible, whereas the control of the vulcanization of the rubber inside the bearing was not possible. Therefore, selection of the optimal time and temperature of vulcanization for the large bearings was made by using special specimens of large bearings. In these specimens, the steel plates and the rubber layers were not glued so that, after the vulcanization, it was possible to see whether the rubber was well vulcanized over the entire inner rubber surface.

Regarding the small isolators (200/200/50), the vulcanization time was defined to be 50 min in a mold previously heated to 120°C. The temperature at the level of the compressed plate was defined to be 150°C. In these bearings, the heat was applied only on the upper and the lower side. In the course of the vulcanization, the rubber is constantly, from the upper and the lower side, exposed to external pressure of 100 bars.

The large isolators Ø450/168 mm were vulcanized for 150 min. The heat was applied over two independent surfaces (Fig. 17.22). A temperature of 137°C was continuously applied on the upper and the lower side, whereas along the mold perimeter, a temperature of 110°C was applied.

In addition to being exposed to temperature, the mold was also exposed to the inner pressure caused by the expansion of the rubber. It was designed by taking into

Fig. 17.22 Vulcanization of large isolators

account these pressures. The rubber was kept in the mold by means of a hydraulic press. The specimen was placed in such a way that it could be inserted and taken from the mold on one side only. During the vulcanization, it was necessary to block that side. The blocking of the specimen in the mold was done from the upper and the lower side, by continuous application of pressure of 300 bars.

17.4 Replacement of Bearings

In the last phase of the proposed project, the old bearings of the Pestalozzi school building were replaced by new ones. The behaviour of the old bearings in lateral direction proved to be quite well since failure occurred at lateral displacement of 25 cm. Lateral tests on the old bearings were carried out twice. The first tests were done in 2,000 when a bearing was taken out and tested to define its behaviour (Garevski and Kelly, 2001). 8 years later, such tests were performed upon replacement of the old bearings. In both tests, the lateral displacement at failure of the bearings was 25 cm. The failure mechanism of the bearings was also the same (Fig. 17.23a, b).

The need for the replacement arose from the manifested lateral deformations of the bearings causing permanent tensile stresses. These deformations also led to unwanted subsidence of the school building. Due to high tensile stresses and ageing of the rubber, small cracks were observed in the bearings, as it was stated in the very beginning of the paper.

The dismantling of the old bearings and the mantling of the new ones was carried out by the same firm (MAK-OZ) that produced the bearings. The entire process of replacement of the old bearings (54) by the new ones was performed by only two workers within 27 days. The local lifting of the 20,540 kN structure for the purpose of placement of the new bearings was done by means of manual hydraulic presses. It should be noted that the concrete blocks were lifted by use of the same presses (43 years old) that were used for the installation of the first bearings in the Pestalozzi school building.

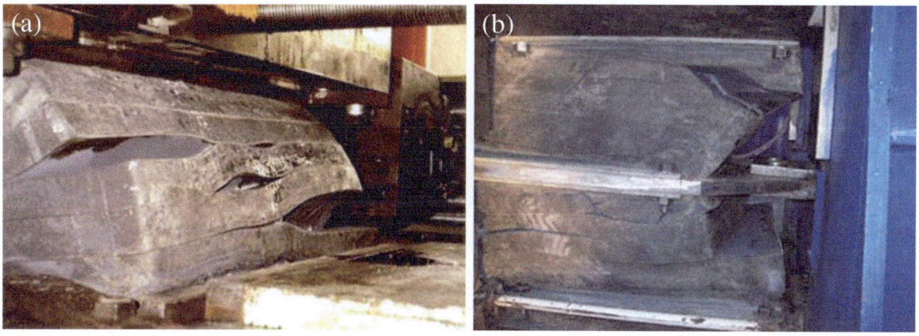

Fig. 17.23 Lateral tests on an old bearing. (**a**) Test done in 2000. (**b**) Test done in 2008

The procedure for replacement of the bearings was as follows:

- Insertion of two temporary supports to the left and the right side of the bearing;
- Removal of the steel bars from the concrete supports;
- Elevation of the concrete block below the bearings by means of 4 presses (two presses on each side);
- Removal of the concrete beams below the concrete block;
- Lowering of the concrete block and the bearing;
- Replacement of the old bearing by a new bearing;
- Insertion of additional steel washers to compensate for the difference in height of the bearings;
- Fixation of the concrete block below the bearing.

As presented in the graph obtained from the vertical tests performed on the isolators, the vertical stiffness of the new isolators was much higher than that of the old ones. This meant that a greater force was necessary to reach the same displacement in the new bearings. In this case, a considerably greater vertical force (from 1,000 to 1,500 kN) was used to fix the new bearings. Hence, it was clear that the use of the old hydraulic presses to replace the bearings was not sufficient enough. For that purpose, additional hydraulic presses were procured. Their dimensions were selected such that they could easily be placed in the free space around the existing bearings (Fig. 17.24).

After adopting the procedure, the bearings were replaced. This was done with a simple equipment and in a very short time, Fig. 17.25.

Before and after the removal of each bearing, the height of the old bearing was measured. The measurements showed that the height of each bearing was different and ranged between 24 and 26 cm. The average height of a loaded (old) bearing was 25.2 cm, while the average height (immediately after unloading) of an unloaded bearing was 30.2 cm.

Therefore, it was necessary to use an enormous number of washers of different thickness for precise leveling of the bearings. This was necessary because even very

Fig. 17.24 Placement of the hydraulic presses and removal of the central bearing

Fig. 17.25 Installed new and old bearing

small changes of vertical displacement could have led to great differences in vertical stress of the bearings. Using the inventive solution of the design engineer regarding the replacement of the bearings, it was easy to equally load all the new bearings.

The new bearings were fixed to the structure using the fixation points of the previous ones. End plates that transfer the effect from the structure to the isolator were specially constructed for that purpose (Fig. 17.25).

Figure 17.25 shows a mounted new and old bearing in the Pestalozzi school building.

17.5 Conclusions

Although very complex and ambitious, all the goals of the project were achieved.

In cooperation with IZIIS, several Macedonian companies produced the equipment necessary for the successful accomplishment of the project. The already existing single-component dynamic testing machine was modified into two-component. A vertical component of 200 kN, a new controlling device and software for control of the machine were added. A large two-component machine that was necessary for the testing of the prototype bearings of the Pestalozzi school building was also produced.

Through the project, the technique of static and dynamic tests of rubber bearings was completely adopted, representing the basis for scientific investigations and development associated with rubber isolators. The adoption of testing procedures for these bearings is of a great importance for their production since, through these tests, the quality of the bearings can be proved.

Through the horizontal tests on small and large bearings, all the errors leading to premature tearing of the rubber were detected. The bearing capacity of the connection between the steel plates and the rubber was defined as well. By preparation of a series of small bearings made of different rubber compounds and their testing, a high damping rubber compound was obtained. The tests proved that the rubber used for the production of the new bearings of the Pestalozzi school building had constant damping in a large range of frequencies that can be excited under earthquake effect.

At the end of the project, the production of rubber isolators was completely adopted. A small company for production of rubber products "MAK-OZ" from Macedonia that had no previous experience in production of bearings succeeded in mastering the production of rubber isolators.

With the accomplishment of the project, not only the process of production of rubber bearings was completely mastered, but a mass production of 54 high damping bearings was carried out. The produced bearings were implemented in the "Pestalozzi" school building, i.e., they were installed as a replacement for the 44 year old rubber bearings.

At the end of this project, it was shown that the production of rubber bearings was not complex and that it was possible to be carried out in developing countries of none or little experience in production of rubber bearings. It was shown that the mounting of the bearings can be simple and cheap. Starting with these conclusions, the construction companies from the Balkan region are realistically expected to show an increased interest in application of these technique in reduction of seismic risk.

Acknowledgements The author wishes to express his gratitude to NATO for the complete financing of these research investigations through the SfP 978028 project. The author is also indebted to the Swiss Agency for Development and Cooperation that financed the mounting and dismantling of the bearings. The author also wishes to extend his gratitude to Prof. J. Kelly (Project Director on the USA part) for the continuous assistance in all the project phases.

Gratitude is also accorded to I. Gjorgjiev – doctoral student who also participated in all the stages of the project.

Thanks are also extended to the producers of the bearings for the successful accomplishment of the project.

References

Fujita T (1995) Behaviour of base isolated buildings in the 1995 Great Hanshin earthquake and overview of recent activities on seismic isolation in Japan. In: Proceedings of the international post-SmiRT conference seminar on seismic isolation, passive energy dissipation and control of vibrations of structures, Santiago, Chile, 20–39

Fujita T (1998) Seismic isolation of civil buildings in Japan. Prog Struct Eng Mater 1(3):295–300

Fujita T (1999) Demonstration of effectiveness of seismic isolation in the Hanshin-Awaji earthquake and progress of applications of base-isolated buildings. Report on 1995 Kobe Earthquake by INCEDE, ERC and KOBnet. IIS, University of Tokyo – Voluntary information network for earthquake disaster mitigation, Serial Number 15, pp 197–216

Fujita T (2001) Progress of application and development for seismic isolation of civil and industrial structures in Japan. In: Proceedings of the 7th international seminar on seismic isolation, passive energy dissipation and active control of vibrations of structures, Assissi, Italy

Garevski M (2002) Evaluation of the functioning of rubber isolators of the elementary school "Pestalozzi" under strong earthquake effects. Comput Struct Dyn, Balkema Publishers, 85–93

Garevski M (2005) Application and development of passive seismic hardware in Macedonia. In: Proceedings of the 9th world Seminar on seismic isolation, energy dissipation and active vibration control of structures, Kobe, Japan

Garevski M, Kelly J (2001) Experimental tests for determination of the effect of natural ageing of the 31 year old rubber bearings. In: Proceedings of the 8th East Asia-Pacific conference on structural engineering and construction, 5–7 December 2001, Nanyang Technological University, Singapore

Garevski M, Kelly J, Bojadziev M (1998) Experimental dynamic testing of the first structure in the world isolated by rubber bearings. In: Proceedings of the 11th European conference on earthquake engineering, Paris

Garevski M, Kelly J, Gjorgjiev I (2008) Development of low-cost rubber bearings for seismic safety of structures in Macedonia and Balkan. Final report of SfP project 978028

Garevski M, Kelly J, Zisi N (2000) Analysis of 3D vibrations of the base isolated school building "Pestalozzi" by analytical and experimental approach. In: Proceedings of the 12th WCEE, Auckland, New Zealand

Melkumyan MG et al (1993) Development and research of laminated rubber bearings for seismic isolation of buildings in Armenia. In: Proceedings of the international conference continental collision zone earthquakes and earthquake hazard prediction, Yerevan-Sevan, Armenia, pp 228–233

Melkumyan MG et al (1997) Design, manufacturing, testing and first application of seismic isolation bearings in Armenia. In: Proceedings of the 4th international conference on civil engineering, Tehran, I.R. Iran, vol 1, pp 306–312

Naeim F, Kelly J (1999) Design of seismic isolated structures, John Wiley, New York, NY

Staudacher K (1982) The Swiss full base isolated system (3-D) for extreme earthquake safety of structures, Convention of the Structural Engineers Association of California, Sacramento, CA, USA

Staudacher K (1985) Protection of structures in extreme earthquakes: full base isolation (3-D) by the Swiss Seismafloat System. Nuclear Engineering Design, Vol 84, Issue 3, 1 Feb 1985, pp. 343–357

Taniwangsa W (2002) Design considerations for a base-isolated demonstration building. Earthquake Spectra 18(4):761–776

Taniwangsa W, Peter W, Kelly JM (1995) Natural rubber isolation systems for earthquake protection of low-cost buildings, Preport No. UCB/EERC-95/12, Earthquake Engineering Research Center, University of California, Berkeley, CA

Part VI
Managing Risk in Seismic Regions

Chapter 18
Investing Today for a Safer Future: How the Hyogo Framework for Action can Contribute to Reducing Deaths During Earthquakes

Sálvano Briceño

Abstract This article provides a brief review of the progress made in the 5 years since implementation of the Hyogo Framework for Action. It also explains the importance of the Hyogo Framework as the appropriate policy guidance tool for reducing risk and vulnerability to natural hazards or disaster risk reduction (DRR). It is based on the premise that natural hazards are increasingly threatening the development gains made in many countries. It is only by addressing all of the five priority areas identified in the Hyogo Framework that a particular government or community can effectively reduce its risk to natural hazards and thus reduce mortality and other negative impacts from these hazards; hence, the relevance of understanding the Hyogo Framework's various components and approaches (multi-hazard, multi-sectoral, multi-disciplinary, multi-stakeholder) as well as its implications with regard to other policy areas, in particular adapting to climate change, achieving the Millennium Development Goals (MDGs) and ensuring sustainable development. The article further explains how the use of the Hyogo Framework for adaptation to climate change also contributes to reducing risk from geological hazards such as earthquakes and volcanic eruptions. Finally, some specific recommendations are made to enhance the implementation of the Hyogo Framework in its second 5 years of application (2010–2015).

S. Briceño (✉)
UNISDR, Palais des Nations, Geneva, Switzerland
e-mail: briceno@un.org

The Hyogo Framework for Action (2005–2015): Building the Resilience of Nations and Communities to Disasters, available at http://www.unisdr.org/eng/hfa/docs/Hyogo-framework-for-action-english.pdf

M. Garevski, A. Ansal (eds.), *Earthquake Engineering in Europe*, Geotechnical, Geological, and Earthquake Engineering 17, DOI 10.1007/978-90-481-9544-2_18,

18.1 Introduction

In an unprecedented way, governments at the second World Conference on Disaster Reduction (Kobe, Hyogo, Japan, 18–22 January 2005)[1] agreed on a series of specific policies and measures to be taken to substantially reduce disaster losses by 2015 – in terms of loss of life and in the loss of social, economic and environmental assets of communities and countries. Governments adopted the Hyogo Framework for Action, which represented an historic engagement of governments and the international community to respond forcefully to increasing vulnerability to natural hazards around the world. Governments' collective decision to engage at this level was made all the more important by the tsunami that struck the Indian Ocean on 26 December 2004, just a few weeks prior to the Conference. While Governments' adoption of this set of specific recommendations has not yet translated into bold changes in policies and investments around the world, the process toward implementing such changes is irreversible.

A few pioneering countries have begun to review their policies and investments in order to make risk reduction an objective and risk management as a tool in implementation of the Hyogo Framework. Most notably, the Association of South East Asian Nations (ASEAN) members have committed to implementing the Hyogo Framework in a legal agreement that has recently been ratified by their respective parliaments.[2] A large majority of countries are implementing at least one of the Hyogo Framework's five priorities for action and in some cases, even several of its recommendations. However, very few countries have undertaken to implement the Hyogo Framework in its entirety, which is essential to the achievement of this important tool's objectives.

The key message that emerged from the Kobe-Hyogo Conference is that disaster risk reduction needs to be mainstreamed into development agendas. A key consideration in the Hyogo Framework is that "disaster risk reduction is a cross-cutting issue in the context of sustainable development and therefore an important element for the achievement of internationally agreed development goals, including those contained in the Millennium Declaration [2000]. In addition, every effort should be made to use humanitarian assistance in such a way that risks and future vulnerabilities will be lessened as much as possible". Reality shows, however, that disaster reduction is

---

[1] The first World Conference on the subject was held at Yokohama, Japan, 23–27 May 1994. At that time, the concept of "natural" disasters still prevailed among experts and thus was included in the title of the Conference. By the time the second conference was held in 2005, the adjective "natural" had been dropped in order to emphasize that disasters are caused mainly by social vulnerability and not only by natural hazards. The scope of the Hyogo Framework was defined in its preamble as encompassing disasters caused by hazards of natural origin and related environmental and technological hazards and risks. It thus reflects a holistic and multi-hazard approach to disaster risk management and the relationship between these hazards, which can have a significant impact on social, economic, cultural and environmental systems, as stressed in the Yokohama Strategy (section I, part B, letter I, p. 8).

[2] http://www.aseansec.org/PR-AADMER-EIF-End-2009.pdf

still perceived by many as only part of the narrower disaster management field and mainly as a humanitarian issue.

This article is aimed primarily at generating discussions during the 14th European Conference on Earthquake Engineering (ECEE-14, Ohrid, 30 August–3 September 2010); hence, the reference to earthquakes in the article's title and in some sections of this chapter. However, the topic of reducing risk from natural hazards is foremost a topic that requires a multi-hazard approach. The multi-hazard approach is explained below as is the special relevance of the Hyogo Framework in respect of seismic risks.

18.2 Brief Summary and Review of the Hyogo Framework

Following the 10-year review of the Yokohama Strategy,[3] 2 years of negotiations in preparation of the second World Conference on Disaster Reduction at Kobe, Hyogo, Japan and the deliberations at the Conference itself, governments adopted the Hyogo Framework, which provides an expected outcome as well as provides three strategic goals and five areas as priorities for action.

The expected outcome for the next 10 years; that is, up to 2015, is "the substantial reduction of disaster losses, in lives and in the social, economic and environmental assets of communities and countries". It was recognized that the achievement of this outcome will require the full commitment and involvement of all actors concerned, including governments, regional and international organizations, civil society including volunteers, the private sector and the scientific community; hence, the value of forums such as the 14th European Conference on Earthquake Engineering, among others in contributing to these discussions.

The Hyogo Framework for Action's strategic goals are:

(a) The more effective integration of disaster risk considerations into sustainable development policies, planning and programming at all levels, with special emphasis on disaster prevention, mitigation, preparedness and vulnerability reduction;
(b) The development and strengthening of institutions, mechanisms and capacities at all levels, in particular at the community level, that can systematically contribute to building resilience to hazards; and
(c) The systematic incorporation of risk reduction approaches into the design and implementation of emergency preparedness, response and recovery programmes as part of the reconstruction of affected communities.

---

[3]The Yokohama Strategy for a Safer World: Guidelines for Natural Disaster Prevention, Preparedness and Mitigation and its Plan of Action ("Yokohama Strategy") was adopted (1994) at the first World Conference on Natural Disaster Reduction (Yokohama, 23–27 May 1994), building on the mid-term review of the International Decade for Natural Disaster Reduction (1990–1999).

The five priorities for action are:

1. Ensure that disaster risk reduction is a national and a local priority with a strong institutional basis for implementation
2. Identify, assess and monitor disaster risks and enhance early warning systems
3. Use knowledge, innovation and education to build a culture of safety and resilience at all levels
4. Reduce the underlying risk factors
5. Strengthen disaster preparedness for effective response at all levels.

18.3 Priority One: Ensure that Disaster Risk Reduction Is a National and a Local Priority with a Strong Institutional Basis for Implementation

Developing institutional capacities to focus on risk and vulnerability reduction greatly facilitates more effective outcomes in reducing disasters. Just as public health policies that focus on prevention have been able to reduce society's risk from diseases and illnesses so too are policies that focus on risk reduction more capable of effectively reducing mortality and other negative impacts arising from natural hazards.

Paragraph 16 of the Hyogo Framework recognizes therefore that "countries that develop policy, legislative and institutional frameworks for disaster risk reduction and that are able to develop and track progress through specific and measurable indicators have greater capacity to manage risks and to achieve widespread consensus for, engagement in and compliance with disaster risk reduction measures across all sectors of society".

A number of political and economic leaders at all levels have misconstrued this priority, however, and in some cases reduced it to merely setting up new institutional mechanisms for disaster reduction without sufficient emphasis and dedication to making risk reduction a higher policy and political priority. It is clear that without the personal involvement and engagement of leaders at national and community levels and from both the public and private sectors, it will not be possible to change the attitudes and behaviour that this new focus requires.

The focus on reducing risk and vulnerability to natural hazards requires a paradigm shift. The traditional belief that disasters are "acts of God" can only be overcome with strong leadership engagement. People often relate to natural phenomena and hazards differently. Hazards are part of nature; they are not preventable but are manageable by citizens, communities and governments. It is important to understand that human, social, economic and ecological vulnerabilities turn hazards into disasters and are the main cause of disasters. Also, this vulnerability can be overcome by the joint efforts of governments, community organizations and citizens by including hazard risk management in land use, in environmental and development sector planning and by choosing the location and materials to build houses, schools and other critical infrastructure, in expanding urban areas.

This particular aspect of engagement of the highest levels of government needs to be emphasized by the mid-term review of the Hyogo Framework in 2010–2011 followed by appropriate awareness-raising campaigns that target political leaders. Such high-level awareness is beneficial and can contribute greatly to other priority processes such as adaptation to climate change, achievement of the MDGs and ensuring sustainable development.

The fact that financial savings can be realized by investing in disaster risk reduction, thereby avoiding or reducing the destruction of development gains, should be more than sufficient to convince top-level decision-makers, in particular in ministries of finance, international financial institutions and other multilateral and bilateral development investors.

18.4 Priority Two: Identify, Assess and Monitor Disaster Risks and Enhance Early Warning Systems

Paragraph 17 of the Hyogo Framework recognizes that "the starting point for reducing disaster risk and for promoting a culture of disaster resilience lies in the knowledge of the hazards and the physical, social, economic and environmental vulnerabilities to disasters that most societies face, and of the ways in which hazards and vulnerabilities are changing in the short and long term, followed by action taken on the basis of that knowledge".

The notion that risk assessment is primarily a technical exercise to understand natural hazards needs to be complemented by a wider and more intense understanding of social vulnerability. Understanding and assessing hazards is, of course, essential and knowledge is widely available on almost all hazards. However, when it comes to understanding and assessing vulnerability, which is composed of human, social, economic and ecological elements, it is a much less developed exercise. Vulnerability is the risk component to which changes can be made. It also requires the wider involvement of community and stakeholder representatives and not simply technical experts. In other words, to understand the vulnerability of a given city, town or neighborhood, there is no better contribution to the exercise than that of those who are most vulnerable.

Furthermore, assessing risk can and should become an exercise for every person and every family as citizens are often the first respondents in times of disaster and citizens are the ones who have a vested interest in understanding and managing their own vulnerability. Therefore, risk assessment should be promoted as an exercise that every person, family and community must participate in regularly with regard to their homes, offices, livelihoods, schools, etc. This risk assessment exercise has been referred to as "self-assessment" or "citizen-based vulnerability assessment."[4]

It would be helpful if the mid-term review of the Hyogo Framework raised awareness on this more public and general understanding of risk assessment, which

[4]See Ben Wisner and Peter Walker in Beyond Kobe (2005), pp 4–5, at http://www.unisdr.org/wcdr/thematic-sessions/Beyond-Kobe-may-2005.pdf

requires broader media and educational campaigns at all levels to be carried out in response to the Hyogo Framework's third priority area (see next item). In addition to the work being done by some local governments and inter-governmental institutions, a number of NGOs have advanced work on this issue at the community local levels. Specific information about their work is available on their respective websites.[5]

Further, as an essential component of sustainable development policy, disaster risk reduction provides an opportunity to address other related topics. "In education and public awareness, disaster risk reduction can be a window through which discussion of less dramatic but important development issues gain visibility – concerns such as urban sprawl, management of water resources, teacher and health provider pay and conditions of work, even very broad issues such as fair and free trade.[6]

An additional component of updating Priority Area Two during the mid-term review of the Hyogo Framework relates to early warning systems. Two things are required: (a) that a global early warning system is fully developed; one to which all countries contribute and from which all countries benefit; and (b) that the concept of early warning systems and related procedures are expanded to include the provision of warnings on the development of vulnerabilities linked to the occurence of hazards.. For a global early warning system to be developed, UNISDR,[7] in concert with other United Nations and non-UN partners, conducted a global survey of early warning systems in 2006 and issued a report, which still needs to be fully implemented.[8] On the second point, which speaks to the provision of warnings related to hazards' occurrence, further thinking is required as it is a new concept. The first issue of the Global Assessment Report on Disaster Risk Reduction (2009)[9] provides preliminary advice in this regard.

18.5 Priority Three: Use Knowledge, Innovation and Education to Build a Culture of Safety and Resilience at All Levels

Paragraph 18 of the Hyogo Framework for Action states that "disasters can be substantially reduced if people are well informed and motivated towards a culture of disaster prevention and resilience, which in turn requires the collection, compilation

[5]A non-exhaustive list includes the ProVention Consortium, International Federation of Red Cross and Red Crescent societies, ActionAid International, OXFAM, CARE International, Save the Children, Asian Disaster Preparedness Center, Asian Disaster Reduction Center, All India Disaster Mitigation Institute, Bangladesh Disaster Preparedness Centre, National Society for Earthquake Technology-Nepal.

[6]Ben Wisner, Let the children teach us (2006), p 6.

[7]UNISDR is the secretariat of the International Strategy for Disaster Reduction (ISDR).

[8]http://www.unisdr.org/ppew/info-resources/ewc3/Global-Survey-of-Early-Warning-Systems.pdf

[9]Global assessment report on disaster risk reduction 2009: risk and poverty in a changing climate, invest today for a safer tomorrow. Full text at http://www.preventionweb.net/english/hyogo/gar/report/index.php?id=9413

and dissemination of relevant knowledge and information on hazards, vulnerabilities and capacities". Reflections in the previous item are also relevant to the third priority of the Hyogo Framework since understanding risk and vulnerability are mainly research and educational tasks.

The third priority area focuses on knowledge, innovation and education, which are broad areas of focus and which require massive investments by all sectors at all levels. The particular role and capacities of the private sector, at all levels, are areas for further enhancement in the second period of implementation of the Hyogo Framework up to 2015. Reducing risk and vulnerability to natural hazards is particularly relevant for business continuity and job security.

At the government and public policy levels, investing in education is perhaps the most relevant action for reducing risk. Investment in education must address two different but complementary objectives; namely, introducing disaster risk reduction into school curricula and ensuring the safety of schools and other educational facilities.

On the subject of integrating risk reduction into school curricula, Ben Wisner expresses it clearly in his report "Let the Children Teach Us" (2006): "The Hyogo Framework, building on numerous studies and reports, articulates a worldwide consensus that disaster risk reduction is an integral part of sustainable human development, not a side issue of limited, merely specialist/technical interest or concern. In the words of the title of an excellent short video/DVD released of ISDR on the anniversary of the Hyogo Framework, disaster risk reduction is "Everybody's Business." If this is so, and disaster risk reduction is an essential part of sustainable development, then "education for sustainable development" should logically include risk reduction in the curriculum".[10]

On school safety, the role and responsibilities of ministries of education and international organizations such as UNESCO,[11] UNICEF[12] and others, must be expressed more clearly in education policies. Despite holding a side event and specific discussions on disaster risk reduction at the 48th International Conference on Education (Geneva, 25–28 November 2008), the issue still does not feature prominently in the programmes of organizations involved in education policy. By extension, ministries of education do not always feel responsible for the safety of schools, which is perceived by many in the education sector as the responsibility of other branches of government (ministries of construction, public works or disaster management). There is more to say about this topic in the next priority area of action.

[10] Ben Wisner, Let the children teach us (2006), p 8.

[11] UNESCO is the UN Education, Science and Culture Organization.

[12] UNICEF is the UN's Children Fund.

18.6 Priority Four: Reduce the Underlying Risk Factors

Paragraph 19 of the Hyogo Framework recommends that "disaster risks related to changing social, economic, environmental conditions and land use, and the impact of hazards associated with geological events, weather, water, climate variability and climate change, are addressed in sector development planning and programmes as well as in post-disaster situations".

This priority area is arguably the one that requires further development, primarily because it groups so many different issues together, whereas each of these would benefit from more focused treatment. This priority area includes everything from introducing disaster risk reduction into the policies and management and recovery efforts of the environment, health, agriculture, tourism and other sectors, to ensuring building code enforcement and enhancing insurance and instituting other financial and risk sharing mechanisms.

With regard to environmental policy, a specific recommendation that was not developed well enough in the Hyogo Framework was the concept that disaster risk reduction is an essential service provided by ecosystems. In point of fact, all ecosystems perform a buffer function to natural phenomena and such function need to be recognized through environmental policies and legislation in order to protect ecosystems, thereby reducing risk and vulnerability to natural hazards. It would contribute, therefore, to reducing ecological vulnerability, which is a significant and increasingly key factor in the intensity and frequency of disasters.

Recognizing disaster risk reduction as an ecosystem service would have the additional value of immediately converting ministries of environment and other governmental and non-governmental organizations working on environmental matters into disaster risk reduction actors and promoters. This would have the effect of raising people's awareness and collective consciousness, which is urgently required to ensure a paradigm shift in this field.

Another key issue under this priority area is that of building codes. The introduction of a multi-hazard approach to developing and implementing building codes requires greater recognition as a technical issue in that it is frequently linked to specific fields of expertise; for example, to seismic codes, fire codes or wind codes. Greater interaction between the fields of earthquake engineering, wind engineering and fire management, among others, and among policy makers and the media, would be beneficial. A consolidated and integrated approach to building codes still needs further development and wider promotion.

Furthermore, in the same way that people are increasingly aware of factors that affect their health, it is important to develop awareness-raising programmes for people to learn to look at factors that affect the "health" of their houses, their offices and the children's schools and to develop systematic approaches to examining vulnerability trends with regard to the location of these structures, the manner in which they are built and the materials used to construct them.

These are just a few of the issues under priority four that would benefit from further study during the mid-term review of the Hyogo Framework.

18.7 Priority Five: Strengthen Disaster Preparedness for Effective Response at All Levels

Paragraph 20 recognizes that "at times of disaster, impacts and losses can be substantially reduced if authorities, individuals and communities in hazard-prone areas are well prepared and ready to act and are equipped with the knowledge and capacities for effective disaster management".

The 2009 Global Assessment Report on Disaster Risk Reduction provides valuable information on the worldwide progress made in strengthening disaster preparedness, with many disaster management and humanitarian organizations contributing greatly to the progress achieved and the Red Cross and Red Crescent Societies and their International Federation having played a pioneering role. A direct consequence has been the significant reduction of the mortality rate associated with some types of hazards and a reduction in that rate for many communities, with the cases of Bangladesh and Cuba serving as noteworthy examples. However, many other communities around the world still need to enhance their disaster preparedness capacity. The anticipated impacts of climate change enable us to see the challenges ahead, requiring a more comprehensive response such as full implementation of the Hyogo Framework recommendations.

In this field of disaster preparedness and in addition to developing these programmes at the community level, the boldest action required is to expand disaster preparedness programmes to include more thorough risk and vulnerability assessments. Similar to early warning systems, many of the preparedness programmes still focus too heavily on hazard assessments without paying sufficient attention to identifying vulnerability and its possible reduction. Thus, vulnerability reduction must become a key component of disaster preparedness programmes at all levels with the appropriate degree of universally-accepted guidance. On this latter element, the role of entities such as the International Strategy for Disaster Reduction and the Hyogo Framework is crucial.

18.8 Review of Progress in Implementation of the Hyogo Framework and Current Status of Risk[13]

Implementing the five priorities for action of the Hyogo Framework will enable any country or community to significantly reduce its risk from natural hazards and reduce both the mortality caused by disasters and the social and economic impacts that ensue from disasters, which tend to affect great numbers of people.[14]

---

[13] A summary of findings from the review carried out by the Global Assessment Report on DRR is presented in Annex 1.

[14] A set of indicators was developed following the Kobe-Hyogo Conference in order to guide and assess progress in HFA implementation, These are contained in Annex 2.

While almost all governments have undertaken some type of action over the last 5 years in response to the recommendations arising from the Hyogo Framework, it is more difficult to assess whether and how many of these have addressed the entire set of recommendations and whether they have done so in a consistent and effective manner. Many governments have undertaken to develop policies, enact legislation and set up organizational arrangements to undertake disaster risk reduction and their initial steps have been encouraging.[15] Countries that have experienced large disasters have moved faster in this regard, while others still lag behind, but there are some notable exceptions in both cases.

The first global review of progress was carried out in 2007 in preparation for the first session of the Global Platform for Disaster Risk Reduction, held in Geneva (5–7 June 2007). At that time, it was too early to identify any significant trend in implementation of the Hyogo Framework and this first Global Platform served mainly to assess how governments were preparing to address the issue as well as their first steps in setting up national mechanisms to guide and support the process.

In 2008, a second set of reports was requested for the production of the first Global Assessment Report on Disaster Risk Reduction (GAR 2009).[16] A summary of findings is contained in Annex 1. The assessment served as the basis for discussions at the second session of the Global Platform for Disaster Risk Reduction, held in Geneva (16–19 June 2009) and was instrumental in devising the key recommendations flowing from that session.[17]

A new review of implementation progress will take place during 2010, to be completed in time for the next version of the Global Assessment Report on Disaster Risk Reduction (GAR 2011) and for the third session of the Global Platform for Disaster Risk Reduction in 2011. Governments will report directly through a Hyogo Framework – HFA Monitoring – facility, available on Preventionweb.net.[18] Additionally, international organizations, non-governmental organizations (NGOs) and other institutions will be expected to report on their progress in implementing the Hyogo Framework.

As part of the GAR 2011, an independent assessment will also be carried out in concert with various expert institutions around the world to identify the evolution and trends of vulnerability, hazards and overall risk. Results are expected to be ready some time in April 2011. The assessment of risk trends and of progress in the implementation of the Hyogo Framework will be major contributions made to the mid-term review of this tool as it reaches the 5th year of its validity. The mid-term review process will benefit from various consultations among expert institutions,

---

[15] More detailed information on each country's experience may be found at: http://www.preventionweb.net/english/hyogo/progress/?pid:3&pil:1

[16] Global assessment report on disaster risk reduction 2009: risk and poverty in a changing climate, invest today for a safer tomorrow. Full text at http://www.preventionweb.net/english/hyogo/gar/report/index.php?id=9413

[17] The full report of the second session of the Global platform for disaster risk reduction can be found at: http://www.preventionweb.net/english/professional/publications/v.php?id=11963

[18] http://www.preventionweb.net/english/hyogo/hfa-monitoring/

inter-agency and intergovernmental actors at the global, regional and national levels. These consultations are already taking place in various regions and will culminate in the third session of the Global Platform for Disaster Risk Reduction to be held in Geneva 8–13 May 2011.[19]

At the local level, many community-based organizations and local governments are taking bold steps to reduce risk and vulnerability, although their efforts remain largely scattered and are of varying degrees of effectiveness, particularly with regard to the increasing vulnerability of communities to natural hazards.

At the international level, several important initiatives have been developed, however. The International Strategy for Disaster Reduction that was launched at the United Nations in 2000 has been working steadily with governments and other partners in advocating, developing knowledge, and monitoring, assessing, reporting and guiding progress in the implementation of the Hyogo Framework for Action and in disaster risk reduction. To date, its efforts have been only cautiously supported by donor agencies, although there are some notable exceptions, but overall, the Organization's impact remains less significant than originally anticipated.

An increasing number of international organizations, such as the World Bank, UNDP,[20] UNICEF, WFP,[21] WMO[22] and others, are developing programmes that focus on risk reduction. These programmes remain much less important than the investments these organizations have made in their disaster response preparedness efforts. This is because there is still a tendency for some organizations to confuse programmes that deal with either emergency response or conflict. Both of these require capacities and knowledge that are distinct from reducing risk from, and vulnerability to natural hazards, for which knowledge is quite developed and widely available.

> Words and plans, demonstration projects and pilots, must be differentiated from serious national efforts with money and political will behind them. It is one thing for a ministry of education to say it will introduce knowledge about natural hazards into school curricula. It is another for that to happen in a few pilot schools. It is yet another for that curriculum to be used widely, routinely and creatively.[23]

Five years later, the world continues to witness ever increasing impacts and losses due to human and social vulnerability to natural hazards. While the number of geological and hydro-meteorological hazards has not increased on average, the numbers of victims and people and communities affected by these hazards have grown enormously due primarily to three factors:

---

[19]See paragraph 20 of the Chair's summary of the second session of the global platform for DRR at http://www.preventionweb.net/globalplatform/2009/background/documents/GP09-Chair's-Summary.pdf

[20]UNDP is the United Nations Development Programme.

[21]WFP is the World Food Programme of the UN system.

[22]WMO is the World Meteorological Organization of the UN system.

[23]Ben Wisner, "Let the children teach us", p 7, UNISDR, 2006.

(a) the increase in urban density, mainly in coastal areas, with increasing numbers of migrants moving to urban areas, amplified by a lack of adequate policies that integrate risk reduction with increasing poverty and economic vulnerability in many developing countries[24];
(b) the increasing deterioration of natural ecosystems,[25] all of which provide an essential service as buffers to natural hazards as well as to support the livelihoods of communities and nations; and
(c) the increasing impact of climate change, which is changing weather patterns, increasing the number and intensity of some hazards and forcing communities to move inland because of rising sea levels along many coastal areas around the world, thus increasing the already worrying urban migration trend.[26]

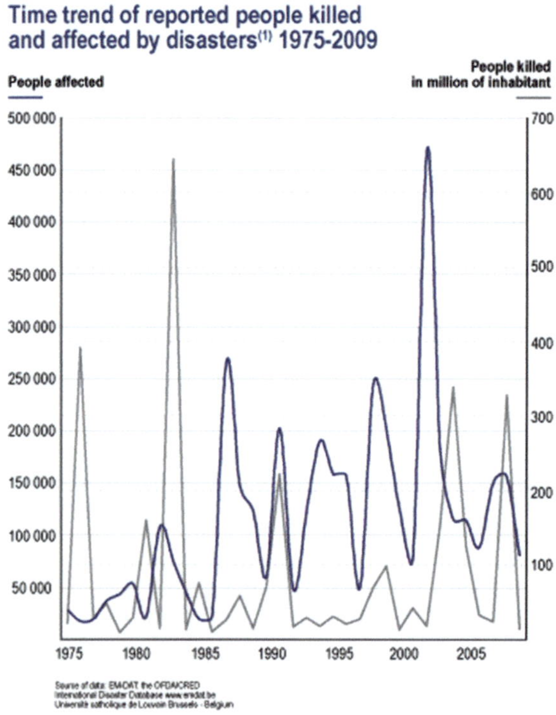

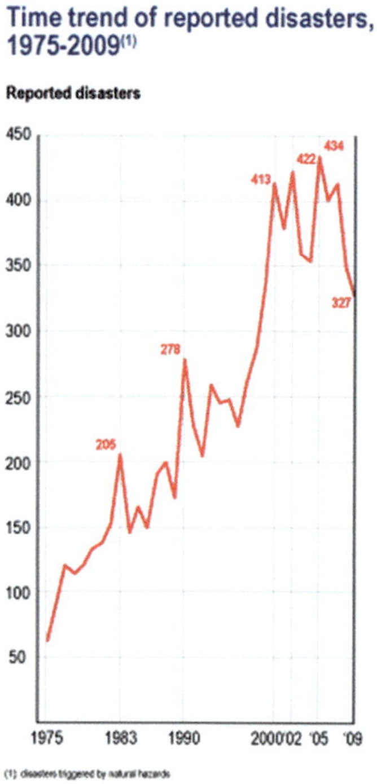

---

[24] With notable exceptions of countries like China, India and Brazil, which, by the sheer size of their populations, distort the average figures for the world, since numbers are still increasing in many other countries.

[25] Although apparently redundant, the term "natural" is utilized here to differentiate these ecosystems from those that have been created or enhanced by human intervention and which tend to be in a better state.

[26] Among many other sources, RTBF (Belgian TV) produced evidence in a video aired on 6 January 2010 of such impacts, called "De plein fouet, le climat vu du Sud" of the series called "Planète en question".

The combination of increasing human and social vulnerability with the changes in the climate and ecosystems are expected to result in more severe and frequent disasters in the coming years.

18.9 Can Reducing Risk from Earthquakes be Included in Adaptation to Climate Change?

This is one of the most frequent questions asked when referring to the Hyogo Framework for Action and climate change. Also, whether any agreement on climate change that includes disaster risk reduction can also help to reduce risk from earthquakes and other geological or non-climate related hazards. If these hazards are not related to climate change, then the agreements on climate change should not be applicable. However, since an essential principle of the Hyogo Framework is for risk reduction policies to take a multi-hazard approach, and assuming that this tool will be promoted further to guide work on disaster risk reduction for climate change, it is possible that such policies can also contribute to reducing risk from other non-climate related hazards.

For example, if risk assessments are introduced as an essential tool in land-use planning, environmental planning and planning in a variety of sectors (health, agriculture, tourism, etc.), as guided by the Hyogo Framework for Action following this multi-hazard approach, it would then become routine to address all hazards through such assessments.

Further, educational, awareness-raising and community-led preparedness programmes should be developed using this multi-hazard approach. Moreover, if new institutional mechanisms are developed as the Hyogo Framework recommends – through a multi-sectoral, multi-disciplinary, multi-stakeholder and multi-hazard approach, such as the National Platforms for disaster risk reduction, and by means of new policies and legislation aimed at risk reduction – then it may be reasonably expected that such institutional mechanisms would address all natural hazards, whether climate and non-climate related.[27]

The multi-hazard approach is more relevant for some policies and measures than for others. Some particular measures should be hazard-specific in their design in

---

[27] The Hyogo Framework defines its scope as "encompassing disasters caused by hazards of natural origin and related environmental and technological hazards and risks. It thus reflects a holistic and multi-hazard approach to disaster risk management and the relationship between natural hazards and disasters, which can have a significant impact on social, economic, cultural and environmental systems, as stressed in the Yokohama Strategy (section I, part B, letter I, p. 8)". It further recommends that "an integrated, multi-hazard approach to disaster risk reduction should be factored into policies, planning and programming related to sustainable development, relief, rehabilitation, and recovery activities in post-disaster and post-conflict situations in disaster-prone countries" and it refers to a similar recommendation in the Johannesburg Plan of Implementation, World Summit on Sustainable Development, Johannesburg, South Africa, 26 August–4 September 2002, paragraphs 37, 65.

order to address specific challenges posed by each type of hazard (for example, forecasting and early warning systems must be adapted to each type of hazard while scientific research and preparedness programmes must also focus on specific hazards). However, other measures and policies that seek to reduce risk from, and vulnerability to all hazards, must have a stronger multi-hazard approach (through awareness-raising activities, basic education, institutional development, policies, legislation, organizational arrangements, risk assessments, etc.)

The mid-term review process of the Hyogo Framework, which is to be completed by the third session of the Global Platform for Disaster Risk Reduction (Geneva, 2011) will almost certainly identify climate change as a major area for updating this important tool as it had not been fully addressed at the time of its adoption. This will also be a key feature of the mid-term review process. Among other things, it will require a comprehensive review of decisions made during the climate change negotiations with a view to ensuring coherence, compatibility and complementarity between the two processes with a view to provide more effective guidance to governments and other organizations working on climate change and on disaster risk reduction issues.

UNISDR has proposed to the UNFCCC[28] process that efforts to reduce vulnerability and build resilience to extreme events should be made a priority in the immediate and short terms. Assigning these priorities would help to avoid humanitarian and economic losses in the short term, as well as secure development gains and provide a more sustainable basis for other adaptation actions over the long term. It would also capitalize on the knowledge and capacities that are currently available, especially in the fields of disaster risk reduction and risk management.[29]

In addition to identifying those elements of the climate change negotiations that either impact or could benefit from the Hyogo Framework, such as the recommendation to make disaster risk reduction the first line of defence in climate change adaptation, it will be important to also explain the value of such programmes in addressing the risks related with non-climate related hazards such as earthquakes and other natural hazards.

The most glaring case is that of Haiti. Although struck by a strong earthquake on 12 January 2010, the country continues to suffer from the impacts of hurricanes and heavy rains that occur each year with greater frequency and intensity. The vulnerabilities that made the January 2010 earthquake so devastating are the same human, social, economic and ecological vulnerabilities that transform hurricanes and heavy rains into large-scale disasters. These same hurricanes, which result in either no victims or very small losses of life in neighboring countries (Dominican Republic, Cuba, Jamaica, Bahamas, Cayman Islands and the United States), can cause large

---

[28]UNFCCC is the UN Framework Convention on Climate Change.

[29]Joint submission by the Inter-agency Standing Committee (IASC) and ISDR to the UNFCCC Ad-hoc Working Group on Long Term Cooperative Action, available at http://www.unisdr.org/eng/risk-reduction/climate-change/docs/IASC-ISDR_paper_cc_and_DDR.pdf

numbers of victims and great destruction in Haiti. It would be unimaginable that the new set of building codes and construction practices to be developed for Haiti would not take into account these natural hazards, whether climate or non-climate related.

With regard to climate change in particular, the Hyogo Framework requires updating to reflect the advances made in climate change negotiations. At the 13th session of the Conference of the Parties to the UNFCCC in Bali, December 2007, disaster risk reduction was recognized by the negotiators as an essential component for climate change adaptation, ultimately becoming part of the Bali Action Plan.[30]

The Bali Action Plan provides the roadmap toward a new international climate change agreement to be concluded by 2009 (now expected for 2010), and that will ultimately lead to a post-2012 international agreement on climate change. In paragraph 1c, the Bali Action Plan highlights the significance of disaster risk reduction as part of enhanced action on climate change adaptation. Among other things, it calls for the consideration of: "Risk management and risk reduction strategies, including risk sharing and transfer mechanisms such as insurance; and disaster reduction strategies and means to address loss and damage associated with climate change impacts in developing countries that are particularly vulnerable to the adverse effects of climate change".[31]

Despite the lack of a final agreement being reached by Parties at the 15th session of the UNFCCC Conference in Copenhagen in December 2009, delegates took note of the recommendations of the Ad-hoc Working Group on Long-term Cooperative Action on the Bali Action Plan, suggesting that negotiations will continue throughout 2010 and hopefully be agreed upon at the 16th session of the COP planned for Cancun, Mexico (29 November–10 December 2010).

In the text agreed upon at Copenhagen, those elements addressing disaster risk reduction were expanded to include a reference to the Hyogo Framework, which makes this important tool all the more meaningful. Moreover, such references were not placed in brackets, boding well for a future agreement to be reached on this subject in Mexico on this topic. If a final agreement is adopted in Mexico in late-2010 that includes disaster risk reduction, reducing risk from natural hazards could become a required element for adaptation to climate change, in turn making it a legal obligation for governments as well as a strong incentive for other relevant organizations in various sectors at all levels around the world. Further, if the Hyogo Framework is used to guide these efforts, risk reduction policies and measures have the potential to address all hazards since the multi-hazard approach is an essential component of this guiding document.

---

[30]More details at http://www.unisdr.org/eng/risk-reduction/climate-change/docs/Climate-Change-DRR.pdf

[31]The full text of Bali Action Plan may be found at: http://unfccc.int/resource/docs/2007/cop13/eng/06a01.pdf#page=3

Governments that are Parties to the UNFCCC and to any expected new agreement will have to develop more intensive disaster risk reduction policies and measures. As a result, those countries that are prone to geological hazards (non-climate hazards related such as earthquakes, tsunamis and volcanic eruptions with their associated landslides), will also benefit indirectly from a legally binding commitment as these policies will need to be multi-hazard oriented following the guidance provided in the Hyogo Framework.

18.10 Challenges Ahead

A cursory reflection on the first 5 years of the Hyogo Framework reveals that larger and more intensive efforts are needed if a paradigm shift (that is, a change in mentalities, attitudes and behaviour) is to be achieved that will reduce both risk from, and vulnerability to, natural hazards. With the knowledge that is available and required to understand both hazards and vulnerability, the efforts that governments, communities and individuals need to make go beyond simple awareness-raising to consciousness-raising.[32]

Disasters provide a level of visibility that serves as an important attraction for leaders of some nations to expend greater efforts and resources on preparing for the disaster itself rather than investing in reducing vulnerability to disasters. Also, coupled with the short-term life span of governments between elections and traditional cultural beliefs that disasters are somehow "acts of God", allows some politicians to blame the disasters on God. This does not help people to understand that much more can and should be done to avoid or reduce deaths and the negative impacts of natural hazards.

For all of the above, it is urgent to invest in raising awareness of the fact that disasters result more from the existing vulnerability than from the hazards themselves. Part of this awareness or consciousness-raising campaign is the need to move beyond the term "natural" disasters, which tends to promote the belief that disasters, if natural, are thus inevitable.

The thousands of experts and representatives of governments and organizations from around the world who discussed the Hyogo Framework and who managed to avoid the use of the term "natural" disasters serve as a good example that must be encouraged and replicated (see footnote 3 above).

---

[32]Ben Wisner: The late Brazilian adult education pioneer, Paulo Freire, thought of education as a collective study of reality and problem solving, and made policy recommendations using a Portuguese term roughly translatable as "consciousness- raising". In relation to disaster risk, "consciousness" is a useful term, going far deeper into the root causes of vulnerability than does the common expression "risk awareness". For example, in Turkey between 1995 and 2003, a series of deadly earthquakes was met with a crescendo of public outcry and a slowly deepening public understanding of what had to be demanded of the construction industry (Mitchell and Page 2005). This, too, is education.

Thus, a broader and more aggressive campaign will be needed to ensure that the objectives set by the Hyogo Framework can be met by 2015. In this regard, campaigns to raise awareness on the importance of prevention and risk reduction can benefit from the experiences gained by the health sector in raising awareness on the importance of preventive measures for reducing the risk of illness. In this regard, natural hazards could be compared to nature's viruses for which individuals and society at large must learn to develop risk reduction approaches and behaviours.

Just as people can reduce the risk of falling ill by eating healthy food, exercising, having regular check-ups from their doctors and maintaining other good lifestyle habits, people should develop similar good practices with regard to natural hazards that can affect their communities or the places where they travel as well as the vulnerabilities that create or increase their risk. This means ensuring that their homes, offices and schools are built in less riskier areas and with the appropriate materials and design standards necessary to withstand the impacts of hazards (earthquakes, storms, fires and others) and knowing what to do in case these hazards strike.

And just as there are vaccines and treatments to protect people from individual viruses (such as for cardiovascular illnesses, malaria, tuberculosis, AIDS[33] or simple colds and heartburn), similarly, specific measures are required, either by individuals (looking after their homes or workplaces), or by governments, to reduce the risks from each type of natural hazard, by undertaking risk assessments, developing specific early warning systems, enforcing appropriate building codes and facilitating disaster preparedness programmes in communities.

Just as health insurance provides funds to people to treat them in times of disease, illness and emergency, so too are governments able to respond quickly and effectively in the provision of emergency relief and recovery in times of disaster when they and other organizations allocate such funds for disaster response, whether through disaster management or humanitarian funding.

However, it is not enough to manage disaster risk by simply putting aside funds in order to respond when they are required, yet this is what is being done in many places. Greater attention and resources are still being dedicated to preparing for disaster response rather than for reducing the risk. Or, as one expression puts it, "maintaining the ambulances at the bottom of the cliff rather than investing in building fences at the top." Disaster preparedness is to building resilience what health insurance is to developing a healthy lifestyle, a key requirement albeit an insufficient one.

To conclude, the purpose of this article was to raise only a few issues related to the implementation of the Hyogo Framework. Much more work is still required to address the broad spectrum of recommendations and gaps in the Hyogo Framework. Furthermore, as vulnerabilities in urban areas continue to increase

---

[33] AIDS is the acquired immunodeficiency syndrome.

exponentially through population migrations arising from economic, environmental, conflict and other reasons, and new vulnerabilities arise that are prompted by climate change, further research will be necessary. The scope of such a large task will inevitably extend beyond the time frame established for the Hyogo Framework.

The mid-term review of the Hyogo Framework must serve, therefore, not only to guide more intense action in the period 2011–2015 but at the same time, to lay the ground for a new and wider framework, expected to be implemented more intensively in the period post-2015, as it is links policies and measures more closely to climate change adaptation.

Like the Yokohama Strategy and Plan of Action of 1994, the expected outcome, goals and priority areas for action of the Hyogo Framework will remain valid over the long term if not forever. If the Hyogo Framework can serve to advance the process of understanding disasters as essentially a human creation and that these same disasters can also be reduced through human action, it will have served its purpose well. With such an understanding, it will be much easier for us to address the issues at stake by instilling effective attitudes and behaviour in people, families, communities, nations and international organizations, enabling them all to play their respective critical roles, hopefully cooperatively, by working together and with the greatest possible respect for Nature – the basis for all sustainable development.

Annex 1: Review of Progress in the Implementation of the Hyogo Framework for Action (HFA), Chapter Five of the Global Assessment Report on Disaster Risk Reduction 2009: Risk and Poverty in a Changing Climate, Invest Today for a Safer Tomorrow[34]

Summary of Findings

1. *Areas of HFA progress reported*: Significant progress has been made in strengthening capacities, institutional systems and legislation to address deficiencies in disaster preparedness and response. Good progress is also being made in the identification, assessment and monitoring of disaster risks and in the enhancement of early warning systems. However, little progress is being made in the use of knowledge, innovation and education and in particular in the mainstreaming of disaster risk reduction into economic, social, urban, rural, environmental and infrastructure planning.

[34] As mentioned in the article, a new review of progress will take place during 2010 to be concluded in 2011 in time for the next version of the Global Assessment Report on DRR (GAR2011).

2. *Progress by income and regional classification*: High-income countries have achieved greater progress across all HFA Priorities for Action than middle- and low-income countries. However, while disaster risk reduction considerations are well integrated into different sectors, many countries lack a holistic policy and strategic framework for addressing disaster risk. Some least developed countries report major gaps in institutional, technical, human and financial capacities, which limit their ability to address the HFA. While many low- and middle-income countries have made good progress in developing national policies, legislation and institutional systems, they are challenged by the issue of mainstreaming disaster risk reduction into sectoral and local development.
3. *Challenges reported*: Specific challenges were highlighted by the review, including an ad hoc and dispersed approach to hazard monitoring and risk identification that does not facilitate comprehensive multi-hazard risk assessments; difficulties faced by national disaster risk reduction organizations in engaging development sectors; and a lack of accountability and enforcement in implementation. At the same time, however, the review highlights innovations in disaster risk reduction governance, showing that some of these challenges can be addressed.
4. *Climate change and disaster risk reduction*: Adaptation to climate change faces many of the same challenges as disaster risk reduction. In addition, implementation is still incipient and its policy and planning frameworks are rarely integrated with those for disaster risk reduction.
5. *Poverty reduction and underlying risk drivers*: Many poverty reduction strategies have potential to address the underlying risk drivers and do recognize disaster impacts as a contributing factor to poverty. However, the disaster risk reduction components in such strategies are often limited to preparedness and response aspects. In many countries, poverty reduction and disaster risk reduction are not strongly integrated in terms of policy and planning.

Annex 2: The Hyogo Framework for Action: Strategic Goals, Priorities for Action, Core Indicators and Levels of Progress

Five Priorities for Action and 22 Core Indicators

HFA Priority for Action 1: Ensure that disaster risk reduction is a national and local priority with a strong institutional basis for implementation.

Core Indicator 1: National policy and legal framework for disaster risk reduction exists with decentralized responsibilities and capacities at all levels.

Core Indicator 2: Dedicated and adequate resources are available to implement disaster risk reduction plans and activities at all administrative levels.

Core Indicator 3: Community participation and decentralization are ensured through the delegation of authority and resources to local levels.

Core Indicator 4: A national multi-sectoral platform for disaster risk reduction is functioning.

HFA Priority for Action 2: Identify, assess and monitor disaster risks and enhance early warning.

Core Indicator 1: National and local risk assessments based on hazard data and vulnerability information are available and include risk assessments for key sectors.
Core Indicator 2: Systems are in place to monitor, archive and disseminate data on key hazards and vulnerabilities.
Core Indicator 3: Early warning systems are in place for all major hazards, with outreach to communities.
Core Indicator 4: National and local risk assessments take account of regional/trans-boundary risks, with a view to regional cooperation on risk reduction.

HFA Priority for Action 3: Use knowledge, innovation and education to build a culture of safety and resilience at all levels.

Core Indicator 1: Relevant information on disasters is available and accessible at all levels, to all stakeholders (through networks, development of information sharing systems, etc.)
Core Indicator 2: School curricula, education material and relevant training include disaster risk reduction and recovery concepts and practices.
Core Indicator 3: Research methods and tools for multi-risk assessments and cost–benefit analysis are developed and strengthened.
Core Indicator 4: Countrywide public awareness strategy exists to stimulate a culture of disaster resilience, with outreach to urban and rural communities.

HFA Priority for Action 4: Reduce the underlying risk factors.

Core Indicator 1: Disaster risk reduction is an integral objective of environment related policies and plans, including for land use, natural resource management and adaptation to climate change.
Core Indicator 2: Social development policies and plans are being implemented to reduce the vulnerability of populations most at risk.
Core Indicator 3: Economic and productive sectoral policies and plans have been implemented to reduce the vulnerability of economic activities.
Core Indicator 4: Planning and management of human settlements incorporate disaster risk reduction elements, including enforcement of building codes.
Core Indicator 5: Disaster risk reduction measures are integrated into post-disaster recovery and rehabilitation processes.

Core Indicator 6: Procedures are in place to assess the disaster risk impacts of major development projects, especially infrastructure.

HFA Priority for Action 5: Strengthen disaster preparedness for effective response at all levels.

Core Indicator 1: Strong policy, technical and institutional capacities and mechanisms for disaster risk management, with a disaster risk reduction perspective are in place.
Core Indicator 2: Disaster preparedness plans and contingency plans are in place at all administrative levels, and regular training drills and rehearsals are held to test and develop disaster response programmes.
Core Indicator 3: Financial reserves and contingency mechanisms are in place to support effective response and recovery when required.
Core Indicator 4: Procedures are in place to exchange relevant information during hazard events and disasters, and to undertake post-event reviews.

The specific levels of progress of countries and regions in each priority area can be seen at: http://www.preventionweb.net/english/hyogo/gar/report/documents/GAR_Chapter_5_2009_eng.pdf

Chapter 19
Emergency and Post-emergency Management of the Abruzzi Earthquake

Mauro Dolce

Abstract The activities initiated after the Mw 6.3 Abruzzi Earthquake of April 6, 2010 can be distinguished in three phases: emergency, post-emergency and reconstruction. Although they are not exactly sequential, because of some overlapping between them, it is useful to use such a scheme, also in relation to the objective to be reached at the end of each of them. The management of the emergency and post-emergency phases was relied upon a governmental commissioner, that, until January 31, 2010, was the chief of the Civil Protection Department. From February 1, 2010, a new commissioner was enforced, with the full responsibility of the reconstruction phase. In this paper the main problems and solutions dealt with in the three phases are described, with a main concern for engineering aspects.

19.1 Introduction

On April 6, 2009, a Mw 6.3 earthquake struck the town of L'Aquila and the surrounding areas in Central Italy, causing more than 300 casualties. The shake in the epicentral area was characterized by a maximum Intensity IX–X in the MCS scale and a maximum peak ground acceleration as high as 0.66 g. The mainshock was preceded by a swarm which started at the end of 2008; the largest earthquake of this swarm had magnitude 4.0 and occurred on March 30, 2009 (Chiarabba et al., 2009). The mainshock was followed by thousands of aftershocks. Within the first 3 weeks, over 4,000 aftershocks occurred, 7 of which with magnitude greater than Mw 5.0.

The mainshock was caused by a ~N135°-striking, ~50°-dipping, 12–19 km-long, normal fault located at a depth between 1–3 and 12–14 km (Chiarabba et al., 2009). Satellite interferometry analyses highlighted a maximum coseismic subsidence of about 20–30 cm perfectly fitting with the active depocenter of the L'Aquila

M. Dolce (✉)
Seismic Risk Office, Italian Department of Civil Protection, 00189 Rome, Italy
e-mail: mauro.dolce@protezionecivile.it

M. Garevski, A. Ansal (eds.), *Earthquake Engineering in Europe*, Geotechnical, Geological, and Earthquake Engineering 17, DOI 10.1007/978-90-481-9544-2_19,

Quaternary basin. Surface coseismic ruptures reactivated minor faults just in correspondence with the slope bounding the basin. Surface ruptures were observed in particular along the Paganica fault, whereas other normal faults known in the area were not activated (e.g. the Mt. Pettino fault; EMERGEO Working Group, 2009).

The 2009 L'Aquila earthquake was the strongest in the Central Apennines since the 1915 Avezzano event (Mw 7.0), and struck a region affected by severe seismicity with magnitudes up to 7 (CPTI Working Group, 2004). This region is undergoing SW-NE active extension and has been deeply investigated from a seismotectonic point of view. Several maps of active faults have been compiled through the years for this region (among the others, see Boncio et al., 2009; Vezzani and Ghisetti, 1998); the mapped faults are mostly SW-dipping and display normal kinematics. Accordingly, all the seismogenic faults proposed in works preceding the L'Aquila earthquake dip to SW and display normal kinematics, whereas there are discrepancies on their location and length (e.g. Akinci et al., 2009; DISS Working Group, 2009; Pace et al., 2006).

The aftershock sequence developed not only on the seismogenic fault which caused the mainshock, but it also activated the deeper part of a second fault plane parallel to the first one and located more to the NNW, in correspondence with the Campotosto Lake. In this area, fault activity has been thoroughly investigated and there is good accordance in interpreting as active the N150°-striking, 60–70°-dipping normal fault which bounds the western slope of the Mt. Gorzano, in particular its southernmost 10 km (Galadini and Galli, 2003). Accordingly, the seismogenic faults available in the literature have been modeled extending the Gorzano fault down to the hypocentral depths, which in this region can reach $\sim$15 km.

The effects of the earthquake were devastating, due to the density of population in the stricken area. As a matter of fact the epicenter was just few kilometers far from the centre of L'Aquila, a city with about 75,000 inhabitants and important masterpieces in the historical centre. Many vulnerable buildings were severely damaged and collapsed, with a tremendous impact on the cultural heritage. Moreover, L'Aquila, being the capital of the Abruzzi Region, also hosts all the main governmental offices of the Region, so that the main public activities were interrupted or spoiled by the collapse of many public buildings.

Soon after the earthquake the Italian National Civil Protection Service was activated, with all its components, to face up the emergency phase, aimed at rescuing as many people as possible and providing a safe shelter to all people. After some days the problems related to the second phase, post emergency, were started to be considered, with the aim to re-establish acceptable life conditions, until the third long phase, the reconstruction phase, will be completed. The three phases are not perfectly sequential, as actually there are some overlapping, in order to meet immediately the needs arising at the end of one phase with the first results or products prepared for the subsequent phase.

The management of the emergency and post-emergency phases was relied upon a governmental commissioner, that, until January 31, 2010, was the chief of the Civil

Protection Department. From February 1, 2010, a new commissioner was enforced, with the full responsibility of the reconstruction phase.

In this chapter the main problems and solutions dealt with in the three phases are described, with a main concern for engineering aspects.

19.2 Emergency Management

Immediately after the earthquake, the first picture of the possible consequences was obtained from the data provided by the National Earthquake Centre, the seismic surveillance system for Civil Protection managed by the National Institute of Geophysics and Volcanology. It collects data from a national network of about 300 seismometric stations all over the Italian territory transmitting data in real time. After few minutes the epicentral coordinates and the Richter magnitude were estimated and communicated to the Civil Protection Department.

In half an hour, the simulation scenario prepared by DPC with the internally developed software SIGE shown in Fig. 19.1, made it immediately apparent the serious consequences of the earthquake in terms of number of persons involved in building collapses (between 200 and 2,200), homeless people (between 8,700 and 54,000), collapsed or unusable buildings (between 4,000 and 24,000). These estimates resulted to be quite accurate, as the real figures turned out to be quite close to the predicted ones. Actually the upper bound values provided the more realistic estimates, because of an initial underestimate of the local magnitude.

The estimate of the possible effects was consolidated by the first data collected in the first hour from the national strong motion network (RAN), owned and directly managed by DPC. Today RAN counts 422 stations and provide a dense station coverage for all high seismic hazard areas. Eighty-five stations still have analog instruments, that are however scheduled for replacement with as many digital accelerometers within 2010. In L'Aquila there is an array of six stations within few kilometers from the epicenter, while 64 stations, up to 273 km distance, recorded the event. Few minutes after the quake, the ground motion parameters from the near source RAN station and full waveforms from farther stations started to arrive to the DPC data centre. The high values of the epicentral peak ground accelerations, from 0.37 g up to 0.66 g, further substantiated the results of the damage scenario. In Fig. 19.2 there are listed the ten records of the stations with the highest PGA values during the mainshock.

In the meanwhile the first pieces of information from site were collected in the operational room of DPC, confirming the dramatic consequences of the quake, although the information was initially quite confused. Less than a hour after the mainshock, teams of experts moved towards the epicentral area to carry out a first survey of the damage distribution (QUEST Group, 2009). Their work was preliminarily aimed at defining the localities with middle-to-high level of damage. At the end of the 7th of April the macroseismic field of the most damaged area was already

PRESIDENZA DEL CONSIGLIO DEI MINISTRI
Dipartimento della protezione civile
Ufficio rischio sismico

RAPPORTO E.S.
Rapporto Emergenza Sismica

3 *Scala di emergenza sismica*

Evento sismico

Comune epicentrale L'Aquila (L'Aquila)

Data	06-04-09	Longitudine	13.33	Profondità	4.9
Ora	03.32	Latitudine	42.33	Magnitudo Ml	5.8

Stime complessive (Attenzione: scenario calcolato su una profondità media ipocentrale di 10 km)

Probabili persone coinvolte in crolli (min-max)	200	2200
Probabili persone senza tetto (min-max)	8700	54000
Probabili abitazioni crollate o inagibili (min-max)	6700	38000
Probabili abitazioni danneggiate (min-max)	61000	207000
Probabile Intensità (MCS) massima raggiunta		VIII-IX
Probabili COM/Prefetture da allertare		16 (+ 46 con ES = 0)
Comuni afferenti ai COM		284 (+ 244)

Fig. 19.1 Summary report of the SIGE damage scenario software, as it appeared in the first simulation carried out soon after the 06.04.2009, 3.32 a.m. earthquake

UFFICIO VALUTAZIONE PREVENZIONE E MITIGAZIONE DEL RISCHIO SISMICO
SERVIZIO MONITORAGGIO DEL TERRITORIO E GESTIONE BANCHE DATI

RAN - RETE ACCELEROMETRICA NAZIONALE

TERREMOTO DEL 6 APRILE 2009 - ore 01:32 (UTC) - AQUILANO - M = 5.8

N.	Codice record	Codice stazione acc.	Località	Provincia	Regione	Lat N	Long E	PGA (cm/s^2)	Distanza epicentrale (Km)
1	GX066	aqv	L'Aquila - V. Aterno - Centro Valle	L'Aquila	ABRUZZO	42,377	13,344	662,599	4,80
2	FA030	aqg	L'Aquila - V. Aterno - Colle Grilli	L'Aquila	ABRUZZO	42,373	13,337	504,921	4,30
3	CU104	aqa	L'Aquila - V. Aterno - F. Aterno	L'Aquila	ABRUZZO	42,376	13,339	478,000	5,80
4	AM043	aqk	Aquil PARK ing.	L'Aquila	ABRUZZO	42,345	13,401	366,285	5,60
5	EF021	gsa	GRAN SASSO (Assergi)	L'Aquila	ABRUZZO	42,421	13,519	148,862	18,00
6	TK033	cln	CELANO	L'Aquila	ABRUZZO	42,085	13,521	89,381	31,60
7	BI106	avz	AVEZZANO	L'Aquila	ABRUZZO	42,027	13,426	67,687	34,90
8	CR008	orc	ORTUCCHIO	L'Aquila	ABRUZZO	41,954	13,642	64,399	49,40
9	BY048	mtr	MONTEREALE	L'Aquila	ABRUZZO	42,524	13,245	62,233	22,40
10	CR003	sul	SULMONA	L'Aquila	ABRUZZO	42,089	13,934	33,656	56,50

Fig. 19.2 List of the ten strong motion stations that recorded the highest PGA values during the 6th April, 2009 event

depicted; the assessment was later completed and therefore extended over a much larger area, as described ahead.

The National Civil Protection Service (SNPC) was immediately activated, its mandate being the safeguarding of human life and health, goods, national heritage, human settlements and the environment from all natural or man-made disasters. Italy's SNPC was formally established in 1992 through the National Protection Act 225. The Act establishes DPC as the key coordinating body in the event of a national emergency. The DPC is located within the Prime Minister's office, which affords it close contact with the Prime Minister and gives it effective authority over all government functions during the response to a major disaster. All the ministries, regions, provinces and municipalities are components of SNPC and cooperate in case of a national emergency. In particular the operative structures of the Ministry of Interior, i.e. National Fire-fighters Corps, Police, Prefecture, and of the Ministry of Defense, i.e. Army, Navy, Air Force and Carabinieri, contribute to SNPC actions, as well as the State Forest Corps and the Financial Police. Companies of road and railway transportation, electricity, telecommunication are also part of the system. Another important strength of the system is the link to the academic community through the centers of competence, which enables timely translation of up to date scientific knowledge into operability.

While the described first assessment activities were under way, the following actions were undertaken in order to activate the rescue and population assistance organization of SNPC.

- 4.15 a.m.: Meeting of the Crisis Unit at the headquarter of DPC in Rome, with the chief of DPC and the board of directors.
- 4.40 a.m.: The first teams of DPC leaves from Rome to L'Aquila, one for the organization on site, the other for the macroseismic survey.
- 4.40 a.m.: The Operational Committee, involving all the components of SNPC, is convened in Rome.
- 9:00 a.m.: The Direction of Command and Control (DiComaC) is established in L'Aquila.

The choice of the headquarter of Civil Protection in L'Aquila was of particular importance. The availability of the great school barracks of the Financial Police, with about 60 buildings, none of them damaged in their structural parts, and only slightly damaged in the non structural elements, was fundamental for the best management of the emergency and post-emergency phases. The DiComaC was established in the great gymnasium of the school, the same that, after transferring the DiComaC in a close building, hosted the G8 meetings in July.

In the following days and weeks eight Mixed Operative Centers were established in the territory stricken by the earthquake for local assistance and coordination of the operations. They were mainly installed in undamaged schools, and other public buildings and sport facilities (Fig. 19.3).

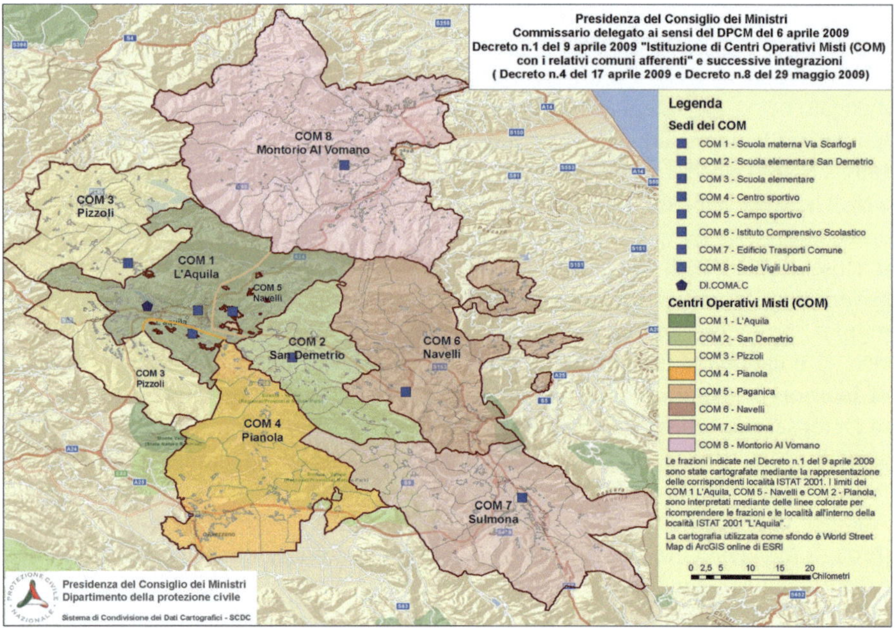

Fig. 19.3 Map of the mixed operative centers

19.2.1 Search and Rescue and Assistance to the Population

The search and rescue operations started few hours after the event and lasted 1 week. At the end of this period 300 corps were retrieved and more than 100 alive people were extracted from the ruins of collapsed buildings.

Table 19.1 reports the number of operators of the main operational components of SNPC (firefigthers, army, police, redcross, volunteers), as well as the number of regions and autonomous provinces present in the first 24 and 48 h. The total number of operators was about 8,000 the first day, 10,000 the second day, then about 12,000 in the following days. This figure was kept practically constant for many months. The offer of help for the emergency operations from foreign countries was declined by the Italian Government, having judged that the capability of SNPC to manage the emergency was adequate to the needs created by the event.

The assistance to the population was activated since the first moments, at the same time as the search and rescue operations. It was necessary to provide suitable shelter for homeless people, that was estimated of the order of some tens of thousands since the beginning. The figures drawn from the simulation scenario, however, were further increased by the ordinance of evacuation for all the citizens, promulgated by the mayor of L'Aquila, soon after the quake. Therefore, even people living in undamaged houses needed a shelter.

Two solutions were immediately conceived and activated: tents for people that did not want to go far away from their home, hotels for the others. Actually, the

Table 19.1 Main operational components of SNPC and relevant number of operators

	24 h	48 h
Firefighters	2,010	2,400
Army	1,520	1,650
Police	1,500	2,000
Red Cross	800	800
Volunteers	2,000	4,300
	8 regions+1 Aut. Province	14 regions+1 Aut. Province

touristic activity in the Adriatic sea coast provides a large number of hotels available and capable to lodge temporarily some tens of thousands of people.

Many camps of tents were immediately set up in L'Aquila and in the nearby municipalities since the first days (Fig. 19.4). After 2 days, 30 camps were set up with about 3,000 tents, 24 field kitchen and 13 advanced medical posts. The maximum numbers of tents and camps during the emergency phase were, respectively, 5,957 and 171. Tents for maximum eight people were used. The maximum reached number of people in tents and in hotels was, respectively, 39,193 and 33,964, and the maximum number of assisted people was 67,459. These relative maxima were reached in different moments: mid-April in tents and mid-May for hotels and end of April for the totally assisted population.

The tents and the relevant camps were so organized as to assure the maximum possible comfort to the population, compatibly with the arrangement in tents. Electricity, heating and air conditioning (during summer) in tents, running water in toilets and shower.

As can be seen in the diagrams of Fig. 19.5, the number of people in tents, after reaching its maximum value at mid-April, progressively reduced until the end of August, when the number of people in tents became of the order of 15,000, then rapidly decreasing to some hundreds by the end of October, when the temperature started to create very difficult conditions. The reduction of people arranged

Fig. 19.4 Aerial view of two tent camps

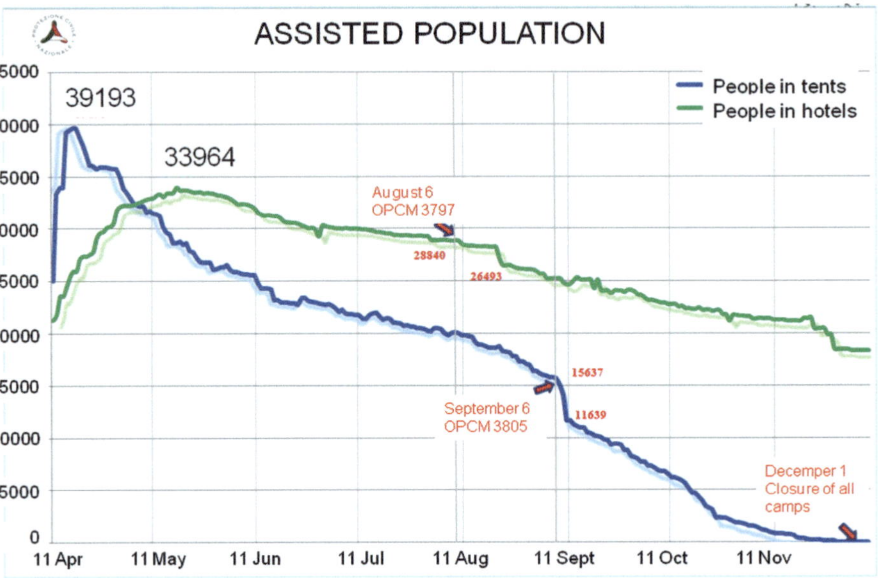

Fig. 19.5 Number of people assisted in tents and hotels

in tents was obtained by (i) providing the temporary long term housing solutions (C.A.S.E. and M.A.P., see later), (ii) arranging about 1,500 people in purposely arranged rooms in one barrack and in the School of the Financial Police in L'Aquila, (iii) moving people from tents to hotels (in the intermediate period) (iv) providing a fixed amount of money per person in order to find autonomous housing arrangement.

The need to keep people wishing to stay close to their home in tents is strictly related to the general strategy for temporary long term housing that will be explained in the following paragraph, as well as to the favorable climatic conditions due to the date of the event and the possibility of keeping acceptable conditions in tents during Spring, Summer and the beginning of Autumn.

The assistance to the population was completed by the arrangement of field kitchen, mainly in or near the tent camps and of advanced medical posts, whose numbers reached 107 and 47 respectively, during the first 2 months after the event.

Psychological and sociological assistance was also provided to the population, especially in the camps. Sport trainers contributed to improve live conditions of the youngest population, especially during summer. A special program of artistic performances (music, theatre, cinema, ballet, etc.) called "Campi sonori" was purposely organized during the summer period, in various facilities, taking profit of the offer of several artists to provide their help to Abruzzi. It was greatly appreciated by the population as a contribution to the recover of normality.

19.2.2 Technical Activities

Several technical activities were started soon after the earthquake and were carried out in the following weeks and months. Most of them are the continuation of the activities aimed at assessing the effects of the earthquake that were started just after the event, some others are related to the management of the emergency and to support the post-emergency phase. Some of them are reviewed in this paragraph.

It must be emphasized that many of the technical activities have seen the strong cooperation of many scientific institutions, with hundreds of researchers fully involved in technical evaluations, always under the coordination of DPC, as envisaged in Dolce (2008). Certainly most of the data that have been drawn and partially elaborated during this phase will constitute a precious source for future in-depth scientific studies on seismology and earthquake engineering.

The collection and processing of data from the strong motion network (Fig. 19.6) continued for some months, also because of the several aftershocks and the addition of further 15 digital stations in the epicentral area. Seven aftershocks with Mw $\geq$ 5 occurred during the first week. Besides the main shock (see Fig. 19.7), the RAN stations totally recorded 78 foreshocks and aftershocks of Ml $\geq$ 3.5, during the period from January to December 2009.

More than 1,000 records of Mw $\geq$ 3.5 earthquakes have been produced by digital instruments and are now all available at the Civil Protection and at the ITACA Websites (www.protezionecivile.it, http://itaca.mi.ingv.it/ItacaNet/). They constitute more than 25% of the records of the RAN database. Moreover it is the first time that several near-fault records are available. Actually 48 three component observations of Mw $\geq$ 5 events, recorded at a distance less than 15 km from each of the major involved faults, provide a significant increase of near-field records available for the Italian territory.

Many more seismometers and accelerometers were installed by INGV and several other scientific institutions, producing aftershock records that were precious for the subsequent microzonation studies (see following paragraph) and for future scientific studies.

Many structures were also monitored with temporary local accelerometric network, within the activities of the Seismic Observatory of Structures, a permanent project of DPC that has now about 100 structures permanently monitored with 12–30 sensors systems. Unfortunately in Abruzzi and some other regions there were not yet monitored structures at the time of the quake, as the project, that started about 10 years ago, will be completed in the next 2 years.

The scope of the temporary monitoring systems was to keep some important buildings under control. Among them, eight buildings of the school of the Financial Police in Coppito, in the headquarters of Civil Protection, and some others hosting COM's, the local centres of activity coordination. The fast acquisition of the acceleration time histories at the stories of the building and their fast processing could give immediate information on the eventual increase of the damage state and, then, on the structural safety of the monitored buildings. One good example is the Town Hall of Pizzoli, where COM 3 was initially established, with the monitoring

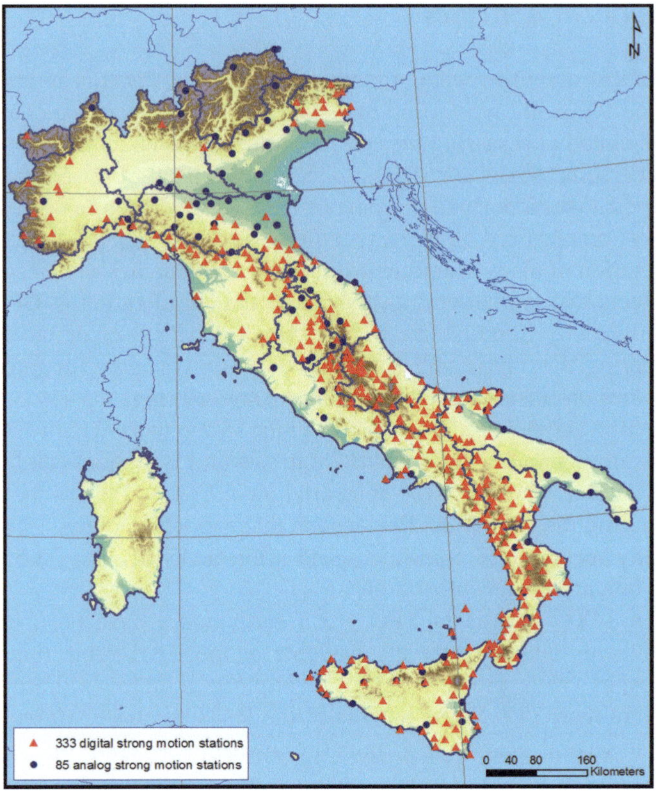

Fig. 19.6 Spatial distribution of the strong motion stations of RAN

system recording several events of magnitude greater than 5 in the first week after the mainshock.

It is the first earthquake in Italy whose emergency management could benefit of the rapidly available data of the accelerometric monitoring networks. This is obviously due to the digital instrumentation as well as to the telecommunication system advancements. Obviously all these data will be precious for future elaboration and improvement of the knowledge in seismology and earthquake engineering, but important perspectives are opened for the use of monitoring systems to support emergency and post-emergency activities, especially if telecommunications are further improved, both in velocity and in reliability, and tools now under development, such as early warning systems and shake maps, reach a sufficient level of maturity for civil protection purpose.

The Macroseismic surveys continued for some weeks after the event. At the end, 315 localities were monitored and the related macroseismic intensities assigned. The zone of maximum damage strikes NW-SE, in accordance with the direction of the seismogenic fault. The maximum intensity value Imax=IX–X (MCS intensity

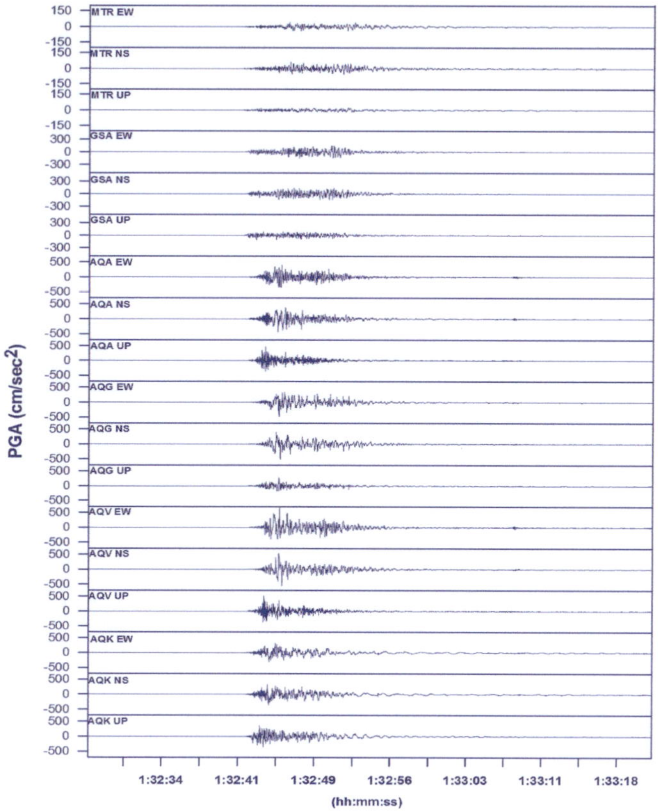

Fig. 19.7 Digital records of the April 6, 2010 quake, in the epicentral area

scale) was assigned to two villages affected by local amplification, whereas four localities suffered an intensity I_{MCS}=IX. In total, more than 130 localities, belonging to 57 municipalities, experienced intensity $I_{MCS} \geq$VI (Fig. 19.8). The intensity assignment to municipalities assumed a great importance, since an Ordinance of the Prime Minister limited the "damaged area" to the municipalities where $I_{MCS}\geq$VI had been felt. The "damaged area" could benefit of some economic provision (e.g.: tax exemption). This made the macroseismic assessment particularly critical and required a special care by the surveyors and the team. All the localities had to be surveyed many times by different surveyors in order to reach a consensus on the assignment.

Two important activities lasting many months were related to the assessment of the damage on the ground and on structures.

The damage and seismic usability assessment of public and private buildings is the most demanding technical activity in terms of needed resources. It also has a strong impact on the management of the emergency and on the reconstruction phase. The seismic usability assessment, in fact, aims to evaluate the safety conditions of

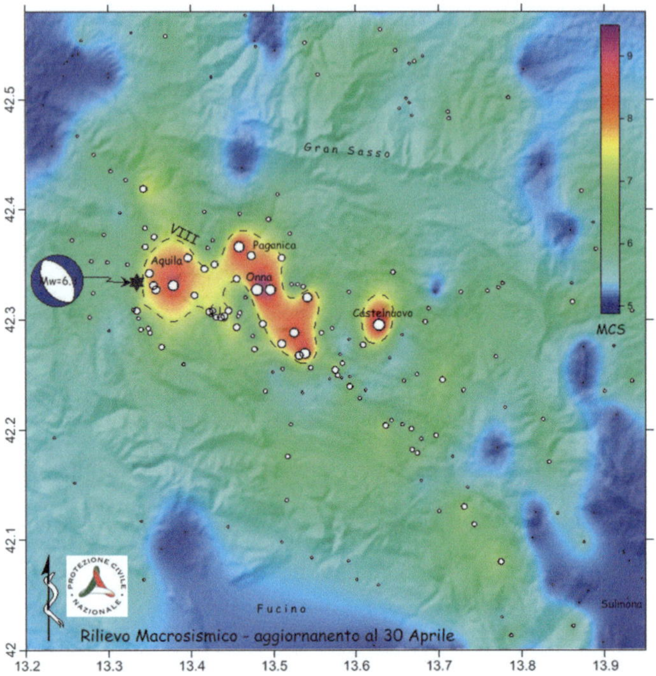

Fig. 19.8 Macroseismic field as resulting from the surveys

the buildings affected by earthquake damage, in order to enable people to return to their houses and social and economical activities to restart. Therefore they condition the possibility of recovering people back to their home, thus reducing the needs for medium and long term temporary housing. But the outcome of the survey is also the main parameter that govern the amount of contribution of the State for the repair and reconstruction of the single buildings.

Like for other past Italian earthquakes (Pollino 98, Molise 02), the damage and usability assessment survey has been coordinated by DPC, with a substantial support from Regions, Provinces, Municipalities, Firemen Corps, and research institutions, such as ReLUIS, EUCENTRE, CNR, and the national councils of professionals CNI, CNA, CNG, using a well established procedure. This is based on the inspection form and manual called AeDES, according to which buildings are classified into the following categories of usability:

A- Usable building. The building, even if slightly damaged, can keep on housing the functions to which it was dedicated, remaining the human life reasonably protected in case of an aftershock at least as strong as the one that motivated the inspections.

B- Building usable only after short term countermeasures. It is the case of a building with limited or no structural damage, but with severe non-structural damage. However, once countermeasures are taken the building can be used as classified under category A.

C- Partially usable building. It is the case of a building with limited or no structural damage, but with severe non-structural damage located in a part of the building. The possible partial or total collapse of the damaged part must not imply a risk for the usable part.
D- Building to be re-inspected. It is the case of atypical damage scenario or geological, geotechnical or other situations that require a specific, but still visual, investigation.
E- Unusable building, as a consequence of at least one of the following conditions: high structural risk, high non structural risk, high external risk or high geotechnical risk.
F- Unusable building for external risk only.

The form also includes general information on the geometry of the building and the categorization of the type of vertical and horizontal structures, as well as their damage level.

The teams were trained with specific on-site short courses held on a weekly basis, on Mondays. More than 5,000 inspectors (up to 500 daily), engineers, architects and undergraduates, operated as volunteers.

Two months after the earthquake, about 50,000 inspections had already been carried out, and they increased up to more than 72,000 by September 28.

The data entry requested the daily input of 1,000–1,800 forms in the first 5 weeks and was performed according to two different methodologies, a fast one, to transfer the main data day-by-day in a data base that enabled the automatic issue of summary reports published every day from April 18, and a longer one, to report all the data collected in the form. The fast data entry was carried out by a staff of several operators belonging to voluntary organizations, public administrations and the army. The complete data entry was carried out by an IT service company.

The potential scientific value of these data must be again underlined. Their analysis can contribute to the improvement of the vulnerability assessment of existing buildings, as in the past occurred since the 1980s in Italy (Braga et al., 1982).

The results of the damage and usability assessment of buildings are summarized in Fig. 19.9. As can be seen, more than 50% of the inspected buildings were in category A, i.e. immediately usable, while about ¼ of the total resulted

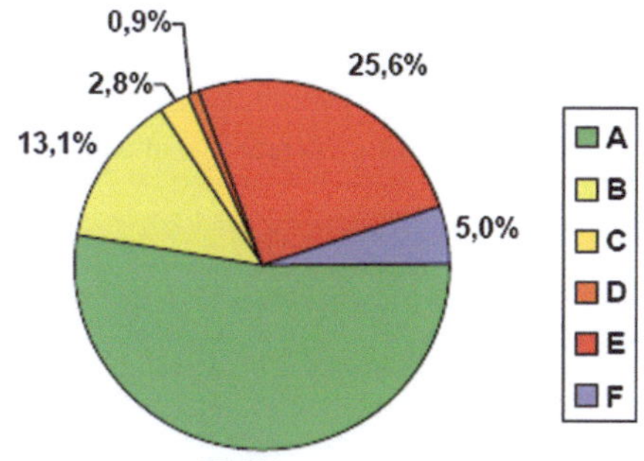

Fig. 19.9 Statistics of usability categories of the damage/usability assessment survey for all buildings

so severely damaged (or collapsed) that they cannot be used before important repair or reconstruction works are made. Intermediate situations (categories B and C), which need minor repair works, are relevant to 16% of the building population, while 5% of the buildings cannot be used because of external risks (Dolce et al., 2009) The statistics provided in Fig. 19.9 are relevant to all the types of buildings, independently of their use. However, since dwelling buildings are more than 95% of them, they represent well the situation of dwelling buildings. Comparing the different kinds of buildings, it can be seen that public buildings generally were characterized by a better seismic behavior. The most effective ones are barracks, almost 70% resulting usable (A), and only 6% totally unusable (E). Also quite good performances were statistically exhibited by buildings for production activities, counting almost 60% usable buildings (A) and only 15% totally unusable (E).

In Table 19.2 the percentage of usability categories are provided for the different structural types of dwelling buildings, namely masonry, reinforced concrete, mixed masonry-R/C structures. As can be seen, R/C structures exhibit a considerable better behavior than masonry structures, with more than 60% in category A vs. less than 50% and about 14% in category E vs. more than 30%.

The survey followed different priorities for public, private and commercial buildings. In order to speed up the recovery and to reduce the social hardship, priority to public buildings has been given, in particular hospitals, schools and headquarter buildings, as well as to commercial buildings.

Schools had priority in the damage and usability assessment (Di Ludovico et al., 2009) and most of them were immediately inspected in the first days after the earthquake. In the municipality of L'Aquila, 106 school buildings were inspected. 309 school buildings were also inspected in other 64 municipalities of the Province of L'Aquila. All the technical teams that inspected schools were composed of university professors or researchers, coordinated by ReLUIS (the Italian consortium of earthquake engineering laboratories).

After some days the situation was quite clear, as shown in Fig. 19.10: about 50% of the inspected schools suffered no or slight damage, so that they could be immediately used (category "A"), while about 30% suffered some damage (category "B" or "C"), so that they could not be immediately re-used, needing some repair works. The remaining 20% approximately could not be used due to heavy damage (category "E") or external risk (category "F"), and this situation was confirmed by further inspections.

Table 19.2 Percentage of usability categories for different structural types

	A (%)	B (%)	C (%)	D (%)	E (%)	F (%)
Masonry	48.7	10.7	2.6	1.2	30.5	6.3
Mixed structure	62.9	11.3	3.0	0.6	17.1	5.1
Reinforced concrete	61.6	19.4	2.3	1.1	13.5	2.1
Total	52.0	12.5	2.6	1.0	26.5	5.4

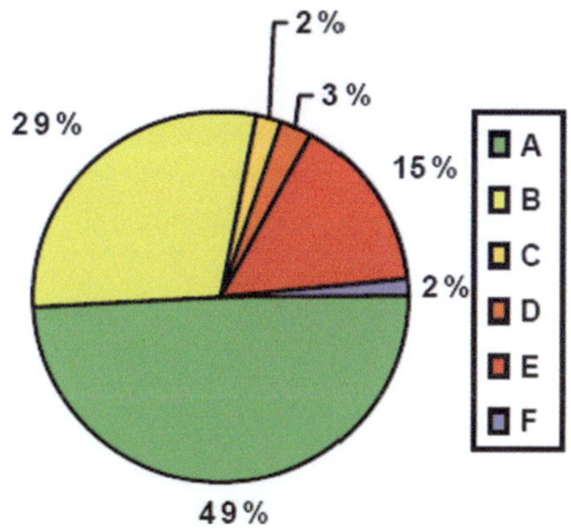

Fig. 19.10 Statistics of usability categories of the damage/usability assessment survey of schools

Cultural heritage deserved a special attention, due to the complexity of their behavior, also related to the many different structural configurations that characterize monumental buildings.

Different and more detailed inspection forms have been used for churches and ancient palaces. The description of their structural features and macro-elements as well as of their damage is much more accurate. Even in this case, the form and the procedures derive from the experience of recent past earthquakes and are now quite consolidated in Italy. Obviously, the time needed for an inspection to a monumental building is much longer than in case of an ordinary building.

Almost 1,800 inspections have been carried out, whose statistics are reported in Fig. 19.11. The much higher seismic vulnerability of monumental (Fig. 19.12) with respects to ordinary buildings is well represented in this diagram. More than 50% of them were strongly damaged or collapsed, and then totally unusable (E) while only 23% resulted to be in good conditions so as to be immediately reused (A).

Besides the inspections of buildings and other structures, so far described, many surveys, about 130, were carried out to verify geological and hidrogeological critical conditions, aimed at taking immediate decision on countermeasures to contrast dangerous situations. They were mainly related to landslides of various kind activated by the mainshock and the aftershocks.

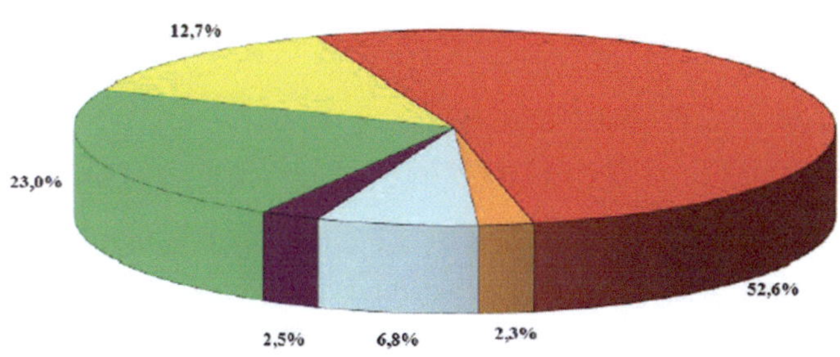

Fig. 19.11 Statistical distribution of usability categories of monumental buildings

Fig. 19.12 The partial collapses of S. Maria di Paganica and of the Abruzzi Museum in L'Aquila

19.3 Post-emergency

The post-emergency phase is characterized by all the actions that allow the population to restore normal life conditions. First of all, it is necessary to provide adequate temporary housing solutions, for people that will remain homeless for several years, according to the experience of past earthquakes. The second fundamental exigency is to provide full effectiveness of the main public offices and of the school system. This latter aspect was of great concern in order to allow children to restart and continue their regular educational activities, this being a primary concern of families, and thus avoid the depopulation of L'Aquila and of the epicentral area. The third exigency is the restart of the productive activities and an acceptable functioning of the different kinds of lifelines.

A peculiar concern is deserved by the cultural heritage, which was strongly affected by earthquake damage. It needed immediate safety measures to be undertaken, in order to avoid further damage and deterioration, both of their content and of their structural elements and, possibly, to get acceptable safety conditions for their use.

19.3.1 Long-term Temporary Housing

After the first arrangement in tents or in hotels, most of the latter ones far from the epicentral area, a solution conceived for a temporary but long term housing arrangements near the original towns and village must be in general found. In the emergency management of the Abruzzi Earthquake, given the favorable seasons following the earthquake (spring and summer) it was possible to avoid an intermediate arrangement, often based on caravans, to pass directly to a long term temporary solution.

Four different solutions have been conceived for the about 30,000 homeless people. One is based on the requisition of the unused apartment, the second is based on a monetary contribution, calibrated on the number of family components, to provide families with the money needed to find an autonomous lodging arrangement. These two solutions were conceived in order to limit the amount of new prefabricated buildings and avoid large land occupation with new constructions. However they soon appeared inadequate do the demand and the exigency of a considerable number of comfortable prefabricated houses appeared soon clear. Two different further solutions, which are adapted to the specific local characteristics of the territory, have been set up and realized: the projects C.A.S.E. and M.A.P.

The Project C.A.S.E. was conceived to realize durable and comfortable three-story buildings, with an underground parking, in a green environment, with a limited land consumption (see Figs. 19.13 and 19.14). For this reason, the Project C.A.S.E. was only adopted inside the municipality of L'Aquila, to solve more than 50% of the total temporary long-term housing demand. The Project M.A.P. was initially conceived to realize small independent houses, with a limited lifetime. This solution implies a large land occupation, and for this reason the Project M.A.P. was adopted to host people of small villages, outside the municipality of L'Aquila. The extension of the M.A.P. Project to some villages of the Municipality of L'Aquila, due to a specific request of the population, required a two-story multimodule solution, which is much less land-consuming.

Fig. 19.13 Aerial view of one settlement of the Project C.A.S.E. (Bazzano) on October 29, 2009. Twenty-one buildings are all completed and inhabited by about 1,600 people

Fig. 19.14 External view of a building in the settlement of Cese di Preturo

19.3.1.1 C.A.S.E.

The Project C.A.S.E. for about 15,000 people of the Municipality of L'Aquila is characterized by the most advanced solutions of the prefabricated building construction industry, in terms of seismic safety, comfort, sustainability and eco-compatibility. Obviously construction speed is a primary prerequisite (Calvi and Spaziante, 2009).

The idea of the Project C.A.S.E. was officially stated by the Decree n. 39/2009 (Decreto Legge 28 aprile 2009, n.39, 2009; Legge 24 giugno 2009, n.77, 2009). Once the population living in C.A.S.E. will return to their home, at the end of the reconstruction phase, these buildings could have alternative uses. One of them, which was envisaged since the beginning, could be as student residences. In fact about 15,000 students of the University of L'Aquila before the earthquake were not resident in L'Aquila and needed a housing solution.

The Project C.A.S.E. is a great challenge in relation to two main objectives.

The first objective consisted, fundamentally, of realizing about 4,500 apartments, in 6–9 months. This made it possible the strategy of keeping the population in tents and in hotels until Autumn, to progressively proceed to deliver the new apartments starting from September. During these months, starting from May, it was necessary to (i) find suitable areas for the new settlements, (ii) verify their compliance with the city urban planning, and the geological, geotechnical, hydrogeological and seismic requirements, (iii) design the new constructions, carry out the call for tenders and contract out the different works, (iv) construct the buildings and make the urbanization and the green arrangement works, (v) complete the apartments with all their furniture, (vi) assign the apartments to homeless families.

The second objective was that of adopting the highest standards with respect to seismic safety, comfort and environmental sustainability, the meaning of the acronym C.A.S.E. being Antiseismic Sustainable and Eco-compatible Complexes.

The technical solution to guarantee the maximum seismic safety is based on seismic isolation (Naeim and Kelly, 1999). Isolators of the pendulum type are arranged between the building and its foundation. The configuration adopted for C.A.S.E. is quite original and was studied to maximize the benefits related to design, constructive and functional exigencies. It is characterized by two identical reinforced concrete plates, whose size is 21 × 57× 0.50 m. The lower plate constitutes the foundation structure, while the upper plate, which is supported by 40 columns and 40 isolators on top of the columns, is the basement of the buildings. Between the two plates there are arranged 32–34 parking car places, as well as the ducts and the flexible connections for the equipments of the building.

Besides assuring the maximum seismic safety, the substantial reduction of the seismic forces makes the design of the three story buildings quite independent of any special geometrical and structural performance constraints. Therefore, only minimum general requirements related to functionality and sustainability had to be complied with by the architectural design. Actually 16 different building types have been adopted for the 185 buildings of the Project, thus producing an agreeable variety of the realized architectures. Fifty percent of them has timber structure, 30% reinforced concrete structure and 20% steel structure.

The Project C.A.S.E. has been realized in 19 areas, each area with 3–25 buildings, occupying 1,200,000 sqm of land. The 4,449 apartments have a total dwelling surface of 330,000 sqm, while in the 220,000 sqm of plates about 6,000 parking car places are arranged. About 600,000 sqm are devoted to green areas. More than 7300 pendulum isolators have been used, resulting in the hugest application of seismic isolation in the world. About 1,500 of them have been statically and dynamically tested in Laboratory, according to the Italian and European norms, and subjected to their maximum allowable cyclic displacement, i.e. 260 mm. In situ tests have been carried out, applying up to 200 mm slow cyclic displacement and up to 100 mm in dynamic conditions simulating the earthquake, using a purposely made testing system (Fig. 19.15). About 600 isolators have been tested through the in-situ test.

Considerable energy savings are obtained, due to the thermo-insulation characteristics of the building walls and to the considerable renewable energy production capability, given by 7,000 sqm of solar panels and 35,000 sqm of photovoltaic panels. The total cost of all the works is about 725 millions of euros.

The works started on June 8 and the first 400 apartments were delivered on September 29. On December 25 about 12,000 persons were lodged in 152 buildings of the Project C.A.S.E. On February 19 all the 185 buildings have been delivered to their inhabitants.

The selection of the families having right to get an apartment was carried during August and September, based on rigorous criteria related to the unusability of their original house and the family composition, as well as the location of their home with respect to one of the C.A.S.E. settlements.

Fig. 19.15 Test set-up for the on-site experimental check

19.3.1.2 M.A.P.

The solution adopted mainly out of L'Aquila, but also for some villages inside the municipality of L'Aquila, is called M.A.P., an acronym that means temporary housing modules. A total number of 3,535 M.A.P.'s has been realized, in 141 areas, 19 of which in the villages of the Municipality of L'Aquila, with a total of 1,273 modules, and 121 in other municipalities, with a total of 2,262 modules. The total number of people to be hosted in M.A.P.'s is about 8,500.

M.A.P.'s are timber houses with good comfort standards and size. They are usually single or two-family one-story houses arranged in small settlements near the original village (Fig. 19.16), for the modules out of L'Aquila, while the modules are arranged in two-story small buildings for M.A.P.'s (Fig. 19.17) in the municipality of L'Aquila.

Starting from the first call for tender on June 18, the first M.A.P.'s were delivered to the population on October 21 outside L'Aquila. Inside L'Aquila Municipality, the Project was started later and the first deliver was on December 30, 2010, apart from some M.A.P. settlements funded by donations, that were completed in August and September.

19.3.2 Schools

After the earthquake, schools remained closed in the damaged area. The regular restart of the scholar year on September 21 appeared immediately to be one of the

Fig. 19.16 One-story single-module M.A.P. out of L'Aquila

Fig. 19.17 Two-story multimodule M.A.P. inside L'Aquila municipality

strategic objective to achieve in order to enables many families to return from the coast hotels and houses. Therefore a big effort was made in order to reset the full capability with respect to the number of students before the event. Parents, however, demanded for safer schools, so that the need for a comprehensive program to guarantee seismically safe schools by the end of September was immediately apparent, also for slightly damaged schools. A specific strategy for the works needed to rehabilitate damaged schools and construct new temporary prefabricated scholar modules in place of the severely damaged school buildings was set up.

Fast design procedures, complying with the new Italian seismic regulations, were adopted for seismic upgrading interventions on damaged reinforced concrete schools, finalized at eliminating the typical critical structural weaknesses of reinforced concrete buildings (see Figs. 19.18, 19.19, and 19.20). They were aimed at strengthening external beam-column joints and avoiding the dangerous tilting and collapse of non structural masonry panels, according to the observation of the damage produced by the earthquake. All the temporary school modules have a seismic resistant steel structure and external prefabricated panels.

The program was set up in May 2009 and was practically completed by the end of September 2009. It consisted of works for repair and strengthening of 35 slightly or moderately damaged school buildings for about 7,000 students in L'Aquila and 24 out of L'Aquila, and the construction of 32 prefabricated temporary school buildings (MUSP) for 6,000 students, with high comfort standards, as well as a

Fig. 19.18 Strengthening of beam-column external joints using carbon fibers

Fig. 19.19 Improvement of connections of infilled masonry panels with the R/C structure, against tilting

Fig. 19.20 A strengthened school at the start of the scholar year

Fig. 19.21 The new prefabricated music conservatory of L'Aquila

new prefabricated music conservatory, with the most advanced architectural and technological features, especially in relation to the peculiar acoustic requirements (Fig. 19.21).

19.3.3 Monumental Buildings

The main objective for monumental buildings in the post-emergency phase was the execution of provisional works, in order to avoid further damage, assure a minimum seismic safety during future restoration works and, in some cases, allow their, at least partial, immediate use.

A general positive consideration on the provisional works carried out in L'Aquila by firefigthers, not only on monumental buildings, is that rational interventions have been realized, very often using tendons instead of shoring. Thus several advantages are obtained, the most important one being that streets and spaces in front of the building façades are free. An example of this kind of intervention is shown in Fig. 19.22. About 200 buildings, mainly churches, have been subjected to provisional works.

Within the project called "One church for Christmas", 61 churches have been made safe and available in short time for their use. Some of them have been subjected to final restoration and strengthening, some other to provisional works and still need further final restoration works.

It is worthwhile to mention the big effort made for the two most important churches in L'Aquila, that were severely damaged, with some partial collapse of their domes. Complex provisional works have been made for a partial re-use of them, as illustrated in Figs. 19.23, 19.24, and 19.25.

Fig. 19.22 Provisional works on a building of the historical centre of L'Aquila

Fig. 19.23 Provisional works in the church of Anime Sante in L'Aquila

Fig. 19.24 The collapse of the dome of the Basilica of Santa Maria di Collemaggio in L'Aquila

Fig. 19.25 Christmas mass in the Basilica of Santa Maria di Collemaggio in L'Aquila

19.3.4 Lifelines

Performance of lifelines in the L'Aquila area (i.e., road network, water distribution, gas distribution, power distribution, water distribution and treatment, telecommunications) may be considered generally good if compared to the extended losses in buildings. Nevertheless damage and service downtime, which required recovery and emergency management, occurred.

From the reported investigations it is concluded that the emergency management of the lifelines networks provided a rapid and resilient response to the earthquake (Dolce et al., 2009).

This is because: (1) the main damaged areas were evacuated after the earthquake and their access was prohibited; and (2) the emergency management was effective in limiting the downtime of essential services.

Damages to structures of wastewater management plants reduced the service level, but even in this case, the evacuation reduced the demand of about 40%.

Critical elements of the transportation network did not suffer any significant damage, experienced only in secondary branches of the network.

Reduction of the traffic flow capacity was mainly due to debris from collapsed/damaged structures adjacent to the road in urban areas and to rock falls and landslides in mountainous areas. From the seismic risk reduction point of view, it was concluded that components in facilities should be anchored and that the use of flexible connections should become a standard practice.

As for the emergency management, the Civil Protection effectively coordinated a rapid and effective response. Chief executive and administrators of lifelines networks participated to the strategic decision making process since the very beginning of the post-event emergency-management.

The cooperation with the Civil Protection was continuous during all the phases of the emergency management ensured via daily meetings.

19.4 Reconstruction Start

In order to speed up the return to the normality, the ordinances of the Prime Minister for the repair and the reconstruction of private damaged buildings were set up and promulgated few months after the earthquake, on June 6, 2009, for buildings with usability grade “A”, “B”, “C”, on July 9, 2009 for “E” buildings, and on November 12, 2009 for masonry building complex.

In the meanwhile some activities were initiated, related to public buildings and facilities. Among them, the one relevant to schools, which have been described in the paragraph on the post-emergency phase, was the most significant in this first period.

Actually many buildings, hosting important public offices, were located in the historical centre, in the so-called red zone, and were severely damaged or almost collapsed during the earthquake. Therefore they need a longer time for their reconstruction, also due to the difficulties of operating in the red zones.

Similar considerations hold for monumental buildings, especially churches, that are frequently located in the historical centre.

A specific concern is deserved by the complicated problems determined by the disruption and heavy damage of the historical centre of L’Aquila and of some minor towns around L’Aquila. The recent history of Italian earthquakes frequently proposed this kind of problems, but in a considerably smaller scale. L’Aquila, indeed, is one of the city with the largest historical centre and cultural heritage concentration in Italy, which, of course, imply a high seismic vulnerability.

19.4.1 Rules for Private Buildings

The reconstruction process for the Abruzzi Earthquake includes the repair and strengthening of damaged buildings and the reconstruction of the collapsed ones.

As far as private buildings are concerned, the State reimbursements are allowed not only for repair works but also for seismic upgrading. The amount of reimbursement for repair works is calibrated on the state of damage and on the assessed usability state. The reimbursement is then regulated as follows:

1. Repair with max reimbursement of 10,000 €/apartment plus 2,500 €/apartment, for common parts, for "A" usable buildings; also light strengthening is allowed within such limits;
2. Full reimbursement of repair works plus 150–250 €/sqm for local strengthening (of critical structural and non structural elements) for "B", "C" buildings and for "E" buildings with slight or no structural damage;
3. Full reimbursement of repair works plus 400–600 €/sqm for seismic retrofit of "E" buildings.

A comment is deserved by the introduction of the local strengthening interventions of point 2. This kind of intervention is expressly considered by the new Italian technical code (Ministero delle Infrastrutture, 2008), in order to strengthen single structural elements or portions of a structure whose strength or ductility is inadequate with respect to the rest of the structure, and can therefore produce brittle failures. This is typically the case of external beam-column joints of reinforced concrete framed structures or of the connections of orthogonal walls in masonry buildings. This kind of interventions does not require the analysis of the entire structure, as the seismic behavior is not substantially changed but only improved. What has to be evaluated is, therefore, only the local increase of strength or ductility of the failure mechanism, thus speeding up considerably the time needed for design. Specific guidelines for such kind of interventions, as well as guidelines for planning and executing experimental tests to evaluate soil and structural characteristics as well as for structural masonry complex, have been drafted and put at disposal of the designer through WEB (De Sortis et al., 2009; DPC-ReLUIS, 2009; DPC-ReLUIS et al., 2010a, b).

In case of collapsed buildings, reconstruction cost reimbursement for an equivalent value of the old building is fully provided.

A special concern is deserved by the complex old masonry buildings that characterize the historical centers, due to the technical difficulties in their seismic upgrading and in order to promote effective well coordinated interventions, which should not be limited to single portions of the building complex. With this aim, an increase of the contributions is provided, so that even those parts that are only slightly damaged can be subjected to upgrading interventions similar to those of the most damaged parts of the complex.

19.4.2 Historical Centers and Cultural Heritage

The generalized heavy damage required the historical centre of L'Aquila and of other towns, to be declared inaccessible "red zone" for safety reasons, soon after the quake. According to a consolidated Italian practice and tradition, the tendency is to repair as many buildings in the historical centers as possible, avoiding demolition and reconstruction. However the rules for the choice of how and what repair and strengthen must follow more general urban planning and architectural restoration criteria, besides structural engineering criteria. A reconstruction plan must be set up, with specific guidelines.

Additional difficulties are created by the presence of the debris and the relevant problems for their disposal, given the severe EU rules and the need to recover the historically and monumentally significant parts of the debris (carved stones, etc.) to be reused. For these reasons, at the time being, the reconstruction of the red zones in the historical centers has not started yet and the regulatory tools have just been set up.

The situation of the monumental buildings is strictly related to that of the historical centre, where most of them are located. Apart from the interventions on the slightly damaged churches for the Christmas program, very few monumental buildings have been repaired so far.

19.4.3 Seismic Microzonation

Seismic Microzonation is an important tool both for prevention and for reconstruction after an earthquake. Significant amplification effects were detected after the event as well as soil instability. Comprehensive seismic microzonation multidisciplinary studies were therefore required to provide tools for the reconstruction, useful for urban planning, emergency planning and seismic design.

The Civil Protection Department promoted these studies soon after the first emergency phase. The aim was to produce microzoning maps in the most damaged areas, namely those affected by $I_{MCS} \geq 7$. The best skills on the different subjects of these studies were convened in L'Aquila, coming from scientific institutions and universities, technical chambers and local administrations, so that a group of about 200 researchers and experts were involved and coordinated by DPC (Dolce and Naso 2009).

The activities were carried out in compliance with the Guidelines for Microzonation (Working Group MS, 2008), that were published just a few months before the event.

Three different levels of accuracy are considered in the Guidelines. The lowest level, level 1, and the most accurate level, level 3, have been considered for all the areas. Level 1 is just aimed at identifying different microzones, according to the following categories: (i) stable, (ii) stable with amplification, (iii) unstable (landslide, liquefaction, surface faulting, etc.). Level 3 quantify the amplification effects quite accurately, also using simulation analyses.

The activities were organized in 10 tasks:

Task 1: Finding, archiving and informatizing already available data in a WebGis.
Task 2: Defining geological-technical models of the subsoil and of the coseismic phenomena.
Task 3: Geotechnical characterization of soils, exploiting existing and new data.
Task 4: Geophysical characterization of soils (100 down-hole, seismic investigation, etc.)
Task 5: Instrumental analyses of the mainshock, aftershocks and microtremors, recorded by temporary accelerometric arrays and by portable tromometers.
Task 6: Definition of the reference earthquake for the numerical simulation, based on seismotectonics, historical seismicity and accelerometric data.
Task 7: Numerical simulations with 1D and 2D models.
Task 8: Damage and vulnerability analysis of buildings, only in some areas.
Task 9: Comparison of the microzoning study results with local urban planning and National seismic regulations.
Task 10: Reports, maps and data dissemination.

The study was started in June 2009 and completed by November 2009. Maps of levels 1 and 3, such as the two examples shown in Fig. 19.26, are now available for public bodies and professionals through the web site www.protezionecivile.it

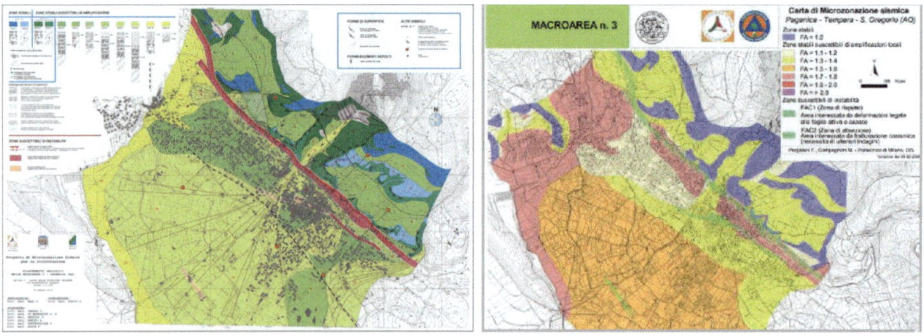

(a) Level 1 (b) Level 3

Fig. 19.26 Microzonation maps of the area of Paganica

19.5 Conclusion

The April 6, 2009 Earthquake of Abruzzi has been a big challenge for the response capability of the Italian National Civil Protection Service as a whole and for all their single components.

The emergency and post-emergency activities coordinated by the Civil Protection Department have been shortly described in this paper, paying greater attention to the technical aspects, for which the extended contribution of the scientific community has improved the quality of the intervention. Emergency and post-emergency objectives have been fully achieved. Short term shelters and long term temporary comfortable housing solutions have been timely provided to the population. The management of the two phases have been carefully coordinated, in order to avoid any waste of time in exploiting the outcomes of one phase for the achievement of the objectives of the subsequent phase.

Also the reconstruction process has been started a few months after the event, with the aim of speeding up the final recovery of normality for at least a part of the population. The rules for the state contributions and the technical criteria for the repair and strengthening interventions on single buildings have been progressively provided starting from less damaged to more damaged buildings. The most complex problems for the reconstruction are now relevant to the historical centers and to the cultural heritage, most of which are inside the great historical centre of L'Aquila and in the many other small historical centers severely damaged. The difficulties to be solved and overcome are many and multiform. Besides the well-known technical problems of the interventions on old masonry buildings, that must in any case comply with the good rules of the heritage restoration, there are the problems of the huge amount of debris, the overall big costs. Taking into account these well-known difficulties, the temporary housing solutions conceived and realized in the post-emergency phase are comfortable and durable, so that the population can immediately restart their normal life.

References

Akinci A, Galadini F, Pantosti D, Petersen M, Malagnini L, Perkins D (2009) Effect of time dependence on probabilistic seismic-hazard maps and deaggregation for the central apennines. Bull Seism Soc Am, Italy 99(2A):585–610

Boncio P, Tinari DP, Lavecchia G, Visini F, Milana G (2009) The instrumental seismicity of the Abruzzo region in Central Italy (1981–2003): seismotectonic implications. Ital J Geosci (Boll Soc Geol It) 128(2):367–380

Braga F, Dolce M, Liberatore D (1982) Southern Italy November 23, 1980 earthquake: a statistical study on damaged buildings and an ensuing review of the M.S.K.-76 scale. Publication CNR-PFG n.503, 1982, Roma

Calvi GM, Spaziante V (2009) La ricostruzione tra provvisorio e definitivo: il Progetto C.A.S.E. Progettazione Sismica, n.3, Pavia

Chiarabba C, Amato A, Anselmi M, Baccheschi P, Bianchi I, Cattaneo M et al (2009) The 2009 L'Aquila (central Italy) MW6.3 earthquake: main shock and aftershocks. Geophys Res Lett 36:L18308. doi:10.1029/2009GL039627

CPTI Working Group (2004) Catalogo Parametrico dei Terremoti Italiani, versione 2004 (CPTI04). INGV, Bologna. http://emidius.mi.ingv.it/CPTI/

DISS Working Group (2009) Database of individual seismogenic sources (DISS), Version 3.1.0: a compilation of potential sources for earthquakes larger than M 5.5 in Italy and surrounding areas. http://diss.rm.ingv.it/diss/, © INGV 2009 – Istituto Nazionale di Geofisica e Vulcanologia – All rights reserved

Decreto Legge 28 aprile 2009, n.39 (2009) Interventi urgenti in favore delle popolazioni colpite dagli eventi sismici della Regione Abruzzo nel mese di aprile 2009 e ulteriori interventi urgenti di protezione civile, Rome

De Sortis A, Di Pasquale G, Dolce M, Gregolo S, Papa S, Rettore GF (2009) Linee guida per la riduzione della vulnerabilità di elementi non strutturali, arredi e impianti. www.protezionecivile.it, www.reluis.it

Di Ludovico M, Di Pasquale G, Dolce M, Manfredi G, Moroni C, Prota A (2009) Seismic behaviour of schools after the earthquake of L'Aquila. Progettazione Sismica, n.3, Pavia

Dolce M (2008) Civil protection vs earthquake engineering and seismological research. In: Proceeding of the 14th world conference on earthquake engineering, Beijing, China, Keynote speech, October 2008

Dolce M, Di Pasquale G, Albanese V, Benetti D, Bramerini F, Coppari S et al (2009) Quick surveys: post-earthquake usability inspections. Progettazione Sismica, Pavia, n.3

Dolce M, Giovinazzi S, Iervolino I, Nigro E, Tang A (2009) Emergency management for lifelines and rapid response after L'Aquila earthquake. Progettazione Sismica, Pavia, n.3

Dolce M, Naso G (2009) The seismic microzonation for the reconstruction in L'Aquila territory: preliminary results. Progettazione Sismica, Pavia, n.3

DPC-ReLUIS (2009) Linee guida per la riparazione e il rafforzamento di elementi strutturali, tamponature e partizioni, www.protezionecivile.it, www.reluis.it

DPC-ReLUIS, Soprintendenza BAAS dell'Abruzzo, Struttura Tecnica di Missione (2010a) Linee guida per il rilievo, l'analisi ed il progetto di interventi di riparazione e rafforzamento/miglioramento di edifici in aggregato, www.protezionecivile.it, www.reluis.it

DPC-ReLUIS, Struttura Tecnica di missione, ALIG, ALGI (2010b) Modalità di indagine sulle strutture e sui terreni per i progetti di riparazione/miglioramento/ricostruzione di edifici inagibili, www.protezionecivile.it, www.reluis.it

EMERGEO Working Group (2009) Rilievi geologici nell'area epicentrale della sequenza sismica dell'Aquilano del 6 aprile 2009. Quaderni di Geofisica 70, ISSN 1590-2595

Galadini F, Galli P (2003) Paleoseismology of silent faults in the Central Apennines (Italy): the Mt. Vettore and Laga Mts. Faults. Ann Geophys 46(5):815–836

http://www.protezionecivile.it/cms/view.php?dir_pk=395&cms_pk=14798 (2008) Pianificazione dell'emergenza e scenari di danno, Rome

Legge 24 giugno 2009, n.77 (2009) Conversione del D.L. 28 aprile 2009, n.39, Rome

Ministero delle Infrastrutture, DM 14.01.08 (2008) Norme tecniche per le costruzioni, Rome

Naeim F, Kelly JM (1999) Design of seismic isolated structures. John Wiley & Sons Ltd, New York, NY

Pace B, Peruzza L, Lavecchia G, Boncio P (2006) Layered seismogenic source model and probabilistic seismic-hazard analyses in central Italy. Bull Seism Soc Am 96(1):107–132

QUEST Group (2009) Rapporto sugli effetti del terremoto aquilano del 6 aprile 2009. http://portale.ingv.it/real-time-monitoring/quest/aquilano-06-04-2009/

Vezzani L, Ghisetti F (1998) Carta Geologica dell'Abruzzo, 1:100,000 scale. S.E.L.CA. edn, Firenze

Working Group MS (2008) Guidelines for seismic microzonation. Civil Protection Department, Rome (in Italian)

Chapter 20
L'Aquila 6th April 2009 Earthquake: Emergency and Post-emergency Activities on Cultural Heritage Buildings

Claudio Modena, Filippo Casarin, Francesca da Porto, and Marco Munari

Abstract The earthquake that struck the Abruzzo region on 6th April 2009 at 3:32 a.m., had its epicentre in the capital of the region, L'Aquila, and seriously affected a wide area around the city, where many historic towns and villages are found. Due to the strategic importance of L'Aquila, a strong and organized civil protection action was necessary to face the emergency. In addition, the structural damage on historic buildings and centres was enormous. Therefore, it was necessary to carry out specific actions aimed at the safeguard of this heritage. These can be shortly listed as: set up of the organizational and decisional structure, damage surveys, temporary interventions to provide the minimum safety conditions, set up of a monitoring plan for some important monuments, set up of a methodology to intervene on complex and connected buildings in the historic centres, definition of adequate materials and techniques to intervene on the damaged buildings.

20.1 Introduction

The earthquake that struck the Abruzzo region of Italy the 6th of April 2009 at 3:32 a.m., affected a wide area among the cities of L'Aquila, Avezzano, Sulmona and Teramo. The ground morphology had an important role in the structural damage distribution and the most catastrophic effects were observed along the Aterno river valley, involving, besides L'Aquila, many historical centres like Paganica, Onna, Fossa, Sant'Eusanio Forconese, Villa Sant'Angelo and others.

The seriousness and the extension (Fig. 20.1) of damage on cultural heritage buildings were without precedent in the recent Italian seismic history, since an organized civil protection action had been developed. This was mainly due to the dimension and the strategic importance of L'Aquila, as capital of Abruzzo region.

C. Modena (✉)
Department of Structural and Transportation Engineering, University of Padova, 35131 Padova, Italy
e-mail: modena@dic.unipd.it

M. Garevski, A. Ansal (eds.), *Earthquake Engineering in Europe*, Geotechnical, Geological, and Earthquake Engineering 17, DOI 10.1007/978-90-481-9544-2_20,

Fig. 20.1 Santa Maria di Paganica's church in the historical centre of L'Aquila (*left*). Aerial view of the destroyed Onna village (*right*)

Indeed, the safeguards activities for the historical and architectural heritage in emergency conditions after the seismic event requested an exceptional effort. The aim was to reduce at minimum, but efficiently, the decisional "chain" that brings from damage surveys to execution of provisional safety measures, carried out on a series of different heritage structures, that goes from ruins, nowadays only historic testimonies, to survived parts of heavily damaged constructions, where testimonies of great artistic value are still present.

With reference to the organizational aspects, the choice to centralize those activities to one figure, the deputy Commissioner with delegation for the safeguard of cultural heritage, was crucial. This choice guaranteed homogeneity in the decisional response and allowed to concentrate the work in one working group, operating on a unique objective, relieving of it the other operational working teams (C.O.M.), which were already charged with all the other emergency problems.

The activities of Function 15 "Protection of Cultural Heritage" of the Civil Protection Department (DPC), under the direction of the delegate Commissioner, were carried out by the employers of the Cultural Heritage Ministry (MiBAC) with the help of the research centre of L'Aquila (CNR-ITC) and of a group of researchers from the University of Genova, Padova and Milano. The first objective of Function 15 was to elaborate a list, although provisory, of the protected architectural heritage to be urgently surveyed in the damaged area and to start the operation of filing the damage survey forms. This objective was crucial, as the Authority archives of L'Aquila were inside the damaged Spanish Fortress. The survey task teams were constituted by one or two representatives of the Cultural Heritage Authority (an architect and an historian or an art expert), a structural expert from the University, an officer from the Fire Brigade. One week after the earthquake, on 14th April 09, the on-site inspections started from churches located in the historic centre of L'Aquila and the surrounding areas, with the collaboration of the structural engineers from Universities participating in the ReLUIS network (Italian Laboratories University Network of Seismic Engineering).

During the filing works, the list of cultural heritage was constantly updated thanks to the notifications by municipalities, parishes and private landlords of protected constructions. The collected data were gradually recorded in a database implemented by the Ministry (MiBAC) with the support of the research centre of L'Aquila (CNR-ITC), who allowed fast access to the survey results, systematic organization of the subsequent on site inspections and the elaboration of some preliminary overall reflections on the activity progress and on the state of damage affecting the entire monumental heritage.

The creation of teams, made of people with different skills, allowed to fill a comprehensive survey form while giving first indications of the "emergency interventions" for the safety measures that were as much as possible respectful of the conservation principles, efficient from the structural point of view and feasible by operators, who initially were exclusively technicians of the Fire Brigade.

The Fire Brigade men had a decisive and irreplaceable role for their professional expertise, operational effectiveness and availability in the temporary safety measures. They were the only that could face extreme conditions (Fig. 20.5), which cannot be classified into standard and repeatable procedures, thanks to their background and to the legal provisions, in particular the possibility of acting in dispensation of the legislation on safety at work (Legislative Decree No. 81 dated 9 April 2008, on Health and Safety at Work).

20.2 Damage Surveys

For the damage survey of cultural heritage, dedicated survey forms were used, prepared by the Civil Protection (GLABEC – Working group for Cultural Heritage) respectively for churches (Model A-DC PCM-DPC MiBAC, 2006), also included in the (Guidelines for the evaluation and mitigation of seismic risk to architectural heritage, 2007), and for palaces (Models B-DP PCM-DPC MiBAC, 2006), approved with decree DPCM date 23 February 2006. The templates are constituted of different sections: general information; eventual presence of artworks, main dimensions, evaluation and economic quantification of structural and artistic damage; certification of fit for habitation; suggestions for temporary safety measures.

The structural damage survey is based on the identification of macro-elements that constitute a masonry building and on the evaluation of the level of activation of the kinematic mechanisms associated to the macro-element itself (Giuffré, 1991). In detail, the survey form for churches identifies 28 possible kinematic mechanisms, typically detectable in this building typology; the survey form for palaces identifies 22 possible kinematic mechanisms. These are subdivided into first-way mechanisms (out-of-plane) and second-way mechanisms (in-plane) (Giuffré and Carocci, 1999). In the case of churches, the mechanisms reported in the survey form were in good agreement with what was actually observed; this is due to the presence of typical elements such as façade, nave, transept, apse, vaults, etc.

The result was a form easy and quick to be filled on site, which provides a standardized evaluation on the level of building damage, substantially free of subjective evaluations by the compiler (Lagomarsino et al., 2001). The survey form used for the evaluation of the "palaces" has been used for the first time in Abruzzo region, unlike that of churches, which was already tested in previous earthquakes. The increased complexity of filling in this form for palaces is connected to the difficulty of bringing together all the non religious historic buildings into the same typology defined as "palaces". Indeed, in the case of complex structures, with different structural and geometric configuration and made of many additions, the identification of specific macro-elements is not unique for all cases.

20.2.1 Observed Damage

Almost 1,677 historical buildings were surveyed, from which 973 were churches, 649 were palaces and the remaining 55 correspond to other typologies of historical buildings, such as towers, fountains, etc. The survey of the damaged churches was finished about 6 months after the earthquake: 240 churches were certified fit for habitation and the most urgent and serious remaining cases were selected for the first safety measures performed by the Fire Brigade.

The systematic collection and data-processing of the damage survey forms will be followed by general considerations on the behaviour and specific vulnerabilities of the different building typologies. An initial analysis, carried out on a series of churches surveyed by the University of Padova during the first 3 months of activity, is given by (Costa, 2009) and (Modena and Binda, 2009). This analysis showed how the façade (Fig. 20.2 left) and the overhanging elements represented the main vulnerabilities of this building typology, according also to (Doglioni et al., 1994). A high level of damage associated with high levels of activation (greater than or equal to 3 on a scale from 1 to 5) was also obtained for mechanisms related to vaulted elements (Fig. 20.2 right), which typically have high intrinsic vulnerability, often worsened by the constructive technique, resorting on the use of sailor brick masonry.

The kinematic mechanisms approach adopted by the damage survey forms and also used in national codes for the seismic evaluation of existing historic buildings,

Fig. 20.2 Incipient out-of-plane overturning of the façade, church of Santa Gemma in Goriano Sicoli (*left*); collapsed vault, church of Santa Maria del Soccorso in L'Aquila (*right*)

is based on the monolithic behaviour of masonry structural elements. The damage survey in Abruzzo region showed that the monolithic behaviour in many cases did not develop because of the poor quality of masonry (Fig. 20.3). Indeed, the surveyed walls were often of considerable thickness, but made with irregular stones, irregularly arranged and assembled with a mortar of poor quality. An additional element of vulnerability was represented by the sequential and articulated constructive evolution of buildings, as often happens in historic centres.

Fig. 20.3 Outside view of Sant'Eusanio church in Sant'Eusanio Forconese (*left*) and inside view of San Michele Arcangelo in Celano (*right*)

Fig. 20.4 Inside view of San Marco's church (*left*) and of Beata Antonia's church (*right*) in the historical centre of L'Aquila: in the Beata Antonia's church the Fire Brigade provisional interventions repositioned the removed tie rods

A further observation and suggestion for investigations, to some extent already begun with the Umbria-Marche earthquake (Valluzzi, 2007), comes out from the visible and sometimes very serious effects of past structural interventions. The few cases listed below represent a non exhaustive list.

The effects of the wrong use of reinforced concrete are clearly evident (Figs. 20.4 left and 20.16 above), in particular the replacement of wooden trusses in the roofs, and the substitution of timber floors, with reinforced concrete trusses and gables or reinforced concrete slabs and hollow tiles.

Another intervention very often observed is the removal of the tie rods connecting orthogonal walls: emblematic examples are the Spanish Fortress and the Church of Beata Antonia (Fig. 20.4 right), both of great historical and artistic importance.

20.3 Emergency Interventions for Safety Measures

The design of temporary interventions for the safety of an historical building starts from the damage survey and from the identification of the activated collapse mechanisms. The filing in of the damage survey form is therefore the first tool to formulate an hypothesis of project for an intervention aimed to act against the specific occurring mechanism. The earthquake in the Abruzzo region was an opportunity to test on a large-scale the process of design and realization of interventions already drafted after the Umbria-Marche earthquake. This process involves three principal figures: an officer from the Cultural Heritage Authority, a structural engineer, and a team of operating Fire Brigade men. In few highly symbolic buildings, such as the Dome of the Anime Sante church (Fig. 20.5) and the roof of the Basilica of Collemaggio, particularly demanding works in terms of structural engineering and working conditions were carried out. Conversely, in the majority (hundreds) of badly damaged historic buildings, there was the need to quickly and effectively provide (in relation to structural safety, but also to the aftershocks occurring after the main shock), the "minimum survival conditions" of what was left after the mainshock.

In particular, on about 1,000 churches included in the MiBAC database, the churches certified fit to use after survey were approximately 25%, another 10% fell on other classes of use, which do not require provisional works, and about 65% needed an evaluation for safety measures. The standard procedure for their design and realization used to start by filling the damage survey form. Thereafter, the structural engineers could elaborate first project drafts, to be discussed in daily meetings at the operational core of the Fire Brigade (NCP, nucleus for the coordination of provisional interventions), with the engineers of the Nucleus and an officer of the Cultural Heritage Authority. In particular cases, the discussion might be followed by other on-site inspections. Following approval of the final project and opening of the site, subsequent inspections, aimed to determine the best solutions accounting for historical value and cost minimization, could also take place.

Fig. 20.5 Drum of the Anime Sante's church in the historical centre of L'Aquila (*left*); special groups of National Fire Brigade men at work on the Anime Sante's dome (*right*)

20.3.1 Type of Interventions

During the phase immediately following the 1997 Umbria-Marche earthquake, some types of provisional interventions and their design principles had been agreed. The main criteria were: avoiding the involvement of structures close to the building undergoing the interventions (e.g., propping of building façades made with elements acting on facing buildings); avoiding to occupy the roadways in order to allow accessibility after the earthquake; avoiding the use of interventions that could hinder the execution of subsequent works for the restoration of the structure. Other logistical considerations and operational criteria for selecting the provisional measures, already identified during the previous earthquake, focused on the choice of materials which are easy to be found and to be applied, in relation to the skills of the people involved in the interventions (VV.AA., 2007).

In designing the safety measures, besides the above mentioned logistical considerations, others of purely structural character were added concerning the static and dynamic behaviour of a building damaged by the action of the earthquake. These hints were the result of both experience gained in previous earthquakes and knowledge developed in university, through the research on earthquake engineering and specifically the behaviour of the historic buildings subjected to seismic action (Modena et al., 2008).

The most widespread provisional interventions are those referring to the so-called first-way mechanisms, i.e. out-of-plane overturning of walls. These mechanisms are characterized by a load multiplier value which activates the mechanism, and then determines the collapse of the element, which is smaller than that of second-way (in-plane) mechanisms. These latter rarely evolve until the total collapse of the element. The provisional interventions to hinder overturning mechanisms of façades and perimeter walls can be performed in two ways: through traditional propping

Fig. 20.6 Propping of the façade of Santa Giusta church in Bazzano (*left*). Polyester bands on the façade of San Giuseppe dei Minimi's church in the historical centre of L'Aquila (*right*)

systems using wooden poles (Fig. 20.6 left) or by tying with steel cables or polyester bands (Fig. 20.6 right). These two interventions correspond to different structural approaches: the first is intended to partially restore the stiffness of the structure; whereas in the second approach, bands (or ties) elastically connect the various stiff masonry blocks, defined by the activation of seismic damage, which form the kinematic chain (Bellizzi, 2000; Bellizzi et al., 2001; Dolce et al., 2002).

The preference given to the second method is determined by the considerations set out above. First, road practicability: interventions performed with traditional props imply that a large portion of ground surrounding the building is to be occupied, thus preventing the passage. In addition, the dense series of props need to be removed to carry out the final repair works. Interventions performed with bands or ties do not occupy the ground space and are minimal for the structures. In addition, their application and subsequent removal is faster. However, the use of polyester bands is based on the assumption that the façade is not disassembled, and forms a clearly defined rigid rotation mechanism. From a structural point of view, an intervention with bands or ties connects rigid blocks identified by cracks due to earthquake. The damaged structure has a lower overall stiffness than before the earthquake, and the exact positioning of the bands provides a great displacement capacity preventing collapses due to overturning.

The use of bands allows connecting perpendicular walls and thus the transfer of action from the out-of-plane loaded walls to the perpendicular walls, which act in their plan of higher stiffness. In case of high intensity aftershocks, the use of traditional props may cause pounding of the façade by the poles and thus local increase of structural damage (Calderini et al., 2004).

In most cases, the structures developed more than one collapse mechanism. The adopted design principles aimed at acting on a single mechanism, making different interventions, as much disconnected one from the other as possible, for each mechanism. This is due to the fact that "global" interventions can change the structural scheme in a non-predictable way, hence triggering different and unforeseen collapse mechanisms (Modena et al., 2000). As an example, two main mechanisms

Fig. 20.7 Intervention on Santa Margherita's church in the historical centre of L'Aquila

were identified in the church of Santa Margherita: transversal response of the nave and overturning of the apse. Two separate interventions were thus carried out. The side walls were connected by metallic wire ropes anchored to a steel counter beam placed in the upper part of the walls, in correspondence of the vault (Fig. 20.7), the apse was connected by metallic wire ropes to the walls which separate the nave from the lateral chapels.

When structural elements develop both in-plane shear mechanisms and out-of-plane overturning mechanisms, the monolithic nature of the overturning element is damaged and must be secured with an intervention, which is preliminary to tying for hindering the out-of-plane mechanism. This intervention can be made by wooden grids and bands (Fig. 20.8), which first are used to strengthen the wall in its plane, and then are connected to the orthogonal walls by tying, thus preventing the overturning.

The local collapses due to poor quality and irregularity of masonry suggested to intervene extensively by consolidating the non-collapsed portions of masonry with a surface rendering, based on natural hydraulic lime mortars with fast hardening and reduced shrinkage, which can be subsequently integrated into the restoration interventions.

The methodological approach used for the temporary safety measures adopted on churches could be then used on a larger scale for intervening on historic buildings and palaces. The greater number of the latter, compared to churches, suggested to operate with the more rapid and less invasive methods. Interventions tested on churches, with bands or tie rods associated with wooden grids, appeared to be the best suited because they do not occupy the building entrances and the roadways, but they also achieve a better response to additional seismic actions. In historic centres, where streets are usually narrow, shoring systems acting on facing buildings (Fig. 20.9), which cause interactions and pounding between different structures, were as much as possible avoided.

Fig. 20.8 Intervention during works at S. Domenico's church and concluded intervention at Evangelica's church in the historical centre of L'Aquila

Fig. 20.9 Three types of safety measures on façades in the historic centre of L'Aquila: wooden grids and ties, traditional props, shoring systems acting on facing buildings

20.4 Structural Monitoring of Damaged Heritage Buildings

The enormous structural damage on heritage buildings caused by the earthquake posed not only the problem of providing the minimum safety conditions by means of temporary interventions, but also the problem of future strengthening and retrofitting interventions, to be carried out on the buildings. This problem is dual: on one hand, there the need to define adequate materials and techniques to intervene on these buildings, and to define proper strategies when it has to be dealt with buildings of high historic and artistic values, but with extensive portions of collapsed vertical and horizontal structures. On the other hand, it is evident that many monuments in L'Aquila and in the surrounding areas will not undergo the final works before some years from the seismic sequence of 6th April 2009.

In this context, the use of structural monitoring has been proposed and agreed, with the various involved authorities (MiBAC, Civil Protection, Local Authorities) to control eventual damage progression, to understand the structural behaviour of buildings after the earthquake and the execution of temporary interventions, and to evaluate the structural modifications that will occur with the final interventions. Different strategies are being employed in smaller churches and buildings on a more diffused scale, and in more emblematic buildings. In the following, two cases of the latter type (St. Mark Church and the Spanish Fortress) are discussed.

20.4.1 St. Mark Church

The St. Mark church, located in L'Aquila city centre, was severely damaged by the 6th of April 2009 earthquake. The first construction of the church dates back to the end of the thirteenth century – beginning of the fourteenth century. The building was completely restructured around 1750, after the 1703 devastating earthquake. The two bell towers enclosed in the façade belong to that period (Casarin et al., 2010b).

The church reported severe damage in the apsidal and transept area, where the external walls manifested a visible outward overturning, involving the four pillars sustaining the dome. Also the transversal response of the church proved to be inadequate, since most of the vaults collapsed, such as a big portion of the external wall, at the clerestory level (Fig. 20.10 left). Severe damage was finally reported in the vaults of the apse, of the presbytery, in the triumphal arch (Fig. 20.10 right).

In this church, such as in many others of the area, the original roof had been replaced by a heavy and stiff roof made by precast reinforced concrete beams and hollow tiles, with an upper layer composed by a concrete slab.

Fig. 20.10 St. Mark church in L'Aquila: external (*left*) and internal (*right*) views

20.4.1.1 Provisional Strengthening Interventions and Monitoring

The initial provisional strengthening intervention was financed by the Veneto Region, which paid 24,000,000 € as a first instalment. Works started on the 4th of July and were completed in November 2009.

This first intervention aimed at counteracting the most critical collapse mechanisms, such as the apsidal and transept walls overturning. This intervention entailed the construction of a retaining scaffolding made of hollow pipe steel trusses, constituting a portal spanning across the church. Two lattice towers were built on the two sides of the church, subsequently connected at their top by an open web girder, in its turn connected to the masonry structures of the church.

Further interventions in the façade area were made to counteract the overturning of a significant portion of the right side wall external veneer. Wooden struts were employed to counteract this mechanism and to sustain the stone elements of the valuable church portal, disconnected by the seismic motion. The façade was hooped on two levels with steel cables, positioned to avoid the progression of shear damage. Finally, several openings were propped, and the façade bell-towers were hooped at the level of the belfry (Fig. 20.11).

In parallel to the execution of the interventions, the structure was controlled by means of an automated low-cost monitoring system. This system is continuously acquiring data, and stores hourly the readouts coming from 5 linear displacement transducers, positioned across the main cracks of both apse and transept, where the worst damage scenario is observed (Fig. 20.12). Data are correlated to the environmental parameters recorded by a temperature – relative humidity sensor.

A couple of acceleration sensors is located at the base of the structure, in order to record any seismic event, of even low-moderate energy, and other two sensors are positioned at the top of the North wing of the transept, to store the structural response (amplification) of the church.

The system is able to automatically store the data when the acceleration in one of the sensors exceeds a predefined threshold, both in time and frequency domain, and to periodically record the data (e.g. at a 12–24 h intervals), to carry out sequential dynamic identifications and measure eventual variations in the modal parameters, with the progression of the strengthening interventions.

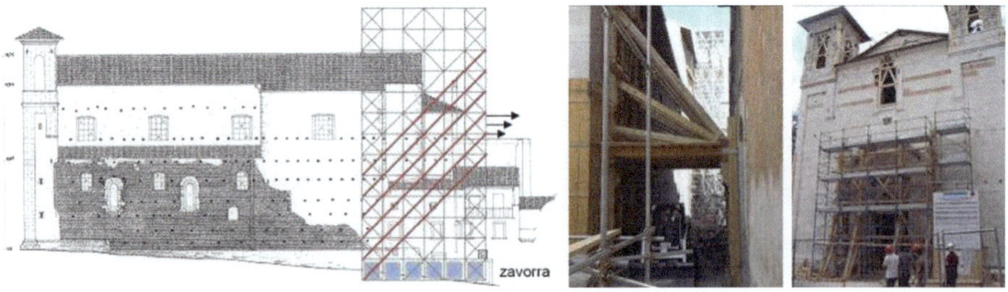

Fig. 20.11 St. Mark church: initial provisional strengthening interventions

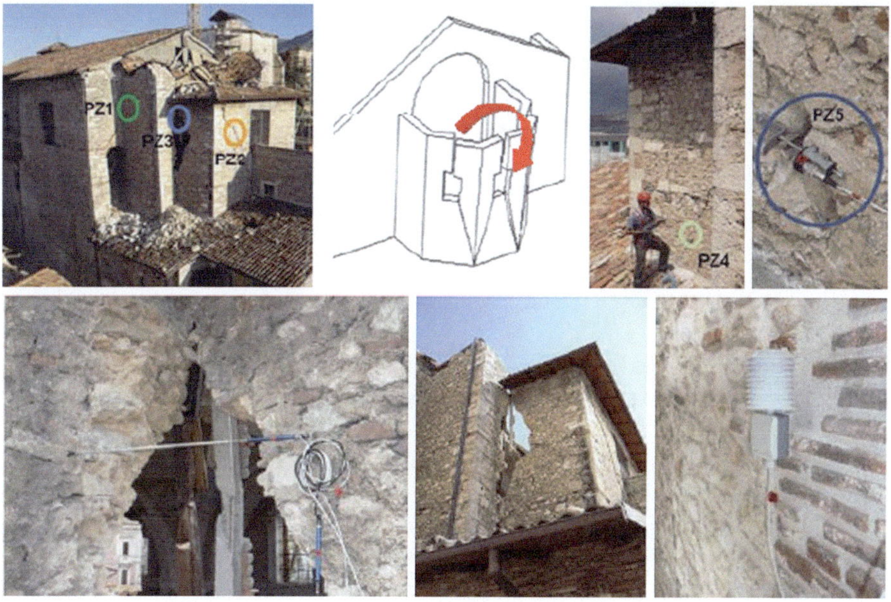

Fig. 20.12 View of the apsidal area of St. Mark church: displacement and temperature-relative humidity sensors positioning

The monitoring system was installed the 10th of August 2009. Displacement transducers data – up to January 2010 – are plotted in the graph of Fig. 20.13. In these first months, any worsening of the crack pattern can be excluded, confirming the effectiveness of the adopted provisional interventions. In this period, some minor seismic events (magnitude between 3.0 and 3.5) were recorded in the L'Aquila area, and some higher seismic events, despite of moderate magnitude (4.0–4.1), occurred in different districts (Ascoli Piceno, about 100 km far from L'Aquila). The

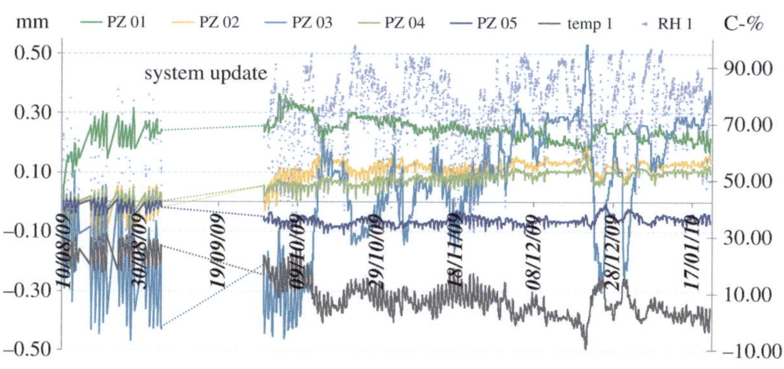

Fig. 20.13 St. Mark monitoring system: plots of the displacement sensors data vs. time

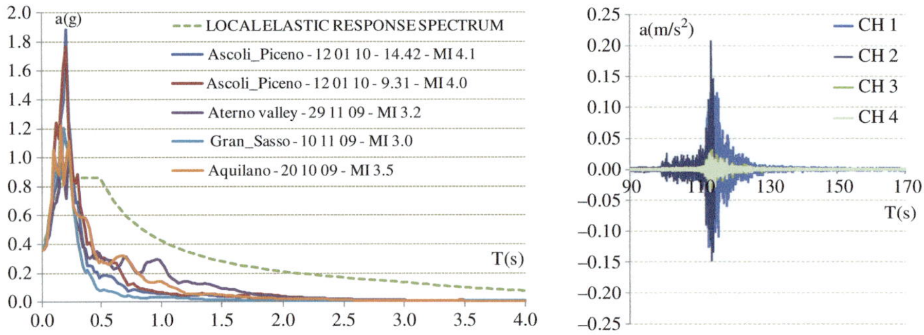

Fig. 20.14 Horizontal elastic response spectra (scaled to the local PGA value) calculated from the low magnitude recorded accelerograms (*left*); recorded MI 4.1 Ascoli Piceno event (*right*)

corresponding response spectra (Fig. 20.14), scaled to the local peak ground acceleration as prescribed by the Italian seismic code (DM 14/01/2008), indicate a local resonance with very high peaks around the frequency of 5 Hz (especially in the far source seismic events), in a band where several structural eigenvalues fall.

20.4.2 The Spanish Fortress

The Spanish Fortress of L'Aquila (Fig. 20.15) is one of the most impressive Renaissance castle in Central and Southern Italy. It was built starting from 1534, when L'Aquila had become the second most powerful city in the Kingdom of Naples, under the Spanish domination, to punish the citizens for their rebellion. The Fortress was never used in a battle, and, between 1949 and 1951, was restored and transformed into the National Museum of Abruzzo.

Fig. 20.15 Aerial view of the Spanish fortress of L'Aquila, before the earthquake

Fig. 20.16 *Above*: view of the L'Aquila castle (Spanish fortress) before and after the earthquake; *below*: severe damage recorded in the internal area of the S-East wing

The structure is made of four bastions connected through mighty walls, 60 m long, with a thickness of 30 m at the bottom and 5 m at top. All around the fortress was a ditch (never filled with water), 23 m wide and 14 m deep, aimed at defending the foundations from the enemy's artillery (Casarin et al., 2010b).

Following the 6th of April 2009 earthquake, the fortress was seriously damaged, especially in the upper floors. According to the damage survey form for palaces used in the inspections (Model B-DP PCM-DPC MiBAC, 2006), overturning and flexural mechanisms on the external walls, shear damage in the external and internal walls, damage to vaults and arches, local collapses of floors and vaults, corresponded to the worst activated mechanisms (Fig. 20.16). Damage was remarkable both for intensity and distribution, and was considered so serious to likely prejudice the overall stability of the building.

20.4.2.1 Provisional Strengthening Interventions and Monitoring

Provisional interventions were carried out on the S-East and S-West wings of the fortress, by groups of specialized Fire Brigades men. The structural stability was provided by connecting the internal and external façade by means of stainless steel cables, to avoid the incipient overturning mechanisms. In the S-East wing the roof was partially rebuilt using hollow section steel trusses and light wooden covering structures. In the S-West wing, steel frames were positioned in contrast to the external and internal façades before tensioning the cables (Fig. 20.17).

Between 17 and 19 December 2009 a dynamic monitoring system was installed in the fortress, following a first investigation campaign carried out in September,

Fig. 20.17 Provisional interventions: connection of the internal and external façades

including dynamic identification of the structure. The system complements a static monitoring system installed during the first months after the earthquake by the ISCR (National Conservation and Restoration Institute) of Rome, which controls the crack pattern and the environmental parameters. The dynamic system is composed by an acquisition unit connected to eight high sensitivity piezoelectric accelerometers. The central unit, located at the second floor of the fortress, in the S-East wing, is provided with a WiFi router for remote data transmission.

Two reference sensors are fixed at the base of the structure, in two orthogonal horizontal directions, to record the ground accelerations in both operational conditions and during seismic events. The positions of the acceleration sensors on the elevation of the S-East wing was based on the results of the previous dynamic identification. According to the predominant motion direction, sensors were fixed orthogonally to the internal and external façades (Fig. 20.18).

Dynamic data are collected both at fixed time intervals to allow sequential dynamic identification of the structure with different environmental conditions, and on a trigger basis, when the signal, on one of the acceleration channels, gets over a predefined threshold, on the time and/or frequency domain.

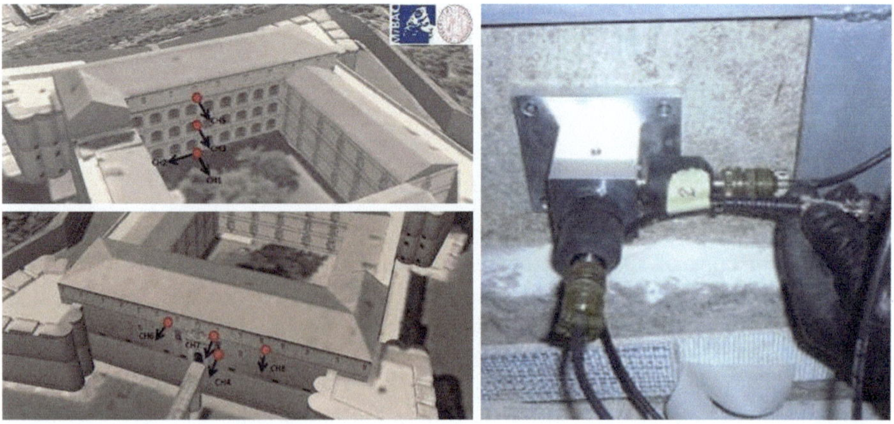

Fig. 20.18 L'Aquila castle: *left* – localization of the acceleration transducers; *right* – base acceleration transducers (CH1, 2) and acquisition unit

Fig. 20.19 Mode shapes emerged from the dynamic data acquisition

The frequencies and mode shapes emerged from the experimental investigation activities carried out before the installation of the monitoring systems (September 2009) are reported in Fig. 20.19. During the period of monitoring no significant seismic events were recorded (a power line suspension between 4th and 14th of January did not allow to record the 12th of January 2010 Ascoli Piceno events). Monitoring results show that the dynamic response of the monument vary with the environmental conditions. The natural frequencies tend to increase with the temperature (Casarin et al., 2010b), but fortunately a change in the structural response due to worsening of the damage state can be at present excluded.

20.5 Procedure for the Analysis of Aggregate Buildings

The Ordinance n. 3820 (12/11/09) requires that, to repair and reconstruct connected masonry buildings, the owners must constitute a consortium, representing at least 51% of the property. This ordinance reflects the state of knowledge in Italy where, since the OPCM 3431 (03/05/2005) and also in the current seismic code (D.M. 14/01/2008), it is asked to pay attention and try to analyze the behaviour of an entire aggregate, when intervening on a single portion or building. This is due to the fact that, in case of adjacent or connected buildings, the interactions between them play an important role in the definition of the structural and seismic behaviour of each of them. Therefore, to have access to the funding for reconstruction and avoid, as much as possible, erroneous interventions on single portions of building, the owners of aggregate buildings are asked to act together.

The operational tool to do so is contained in the Ordinance n. 3820 (as modified by the Ordinance 3832), which requires, as a first step, to identify the aggregate in a technical report, following the guidelines drafted by ReLUIS (March 2009). An aggregate is a complex of buildings, generally defined by roadways or open spaces. Very often an aggregate structurally describes what in town planning is called "block". The word "aggregate" does not implicitly identify the type of building connection. Indeed, buildings in an aggregate are very often autonomous under the constructive point of view. In addition, in many historic centres, covered passages or buttresses can connect adjacent aggregates, but when their dimensions are limited compared to the aggregate extension, they do not define a unique large aggregate. They can be simply accounted for into structural models as restraints or concentrated actions (Circolare 617, 2009).

Figure 20.20 (above) shows the properties included between the streets of San Francesco di Paola, degli Alemanni, Donadei, Piscignola and the Prefettura Square, whose owners founded a consortium with reference to the Ordinance n. 3820. In this area two aggregates can be identified, marked with green and red colours in Fig. 20.20 (below), and a large complex palace (grey colour).

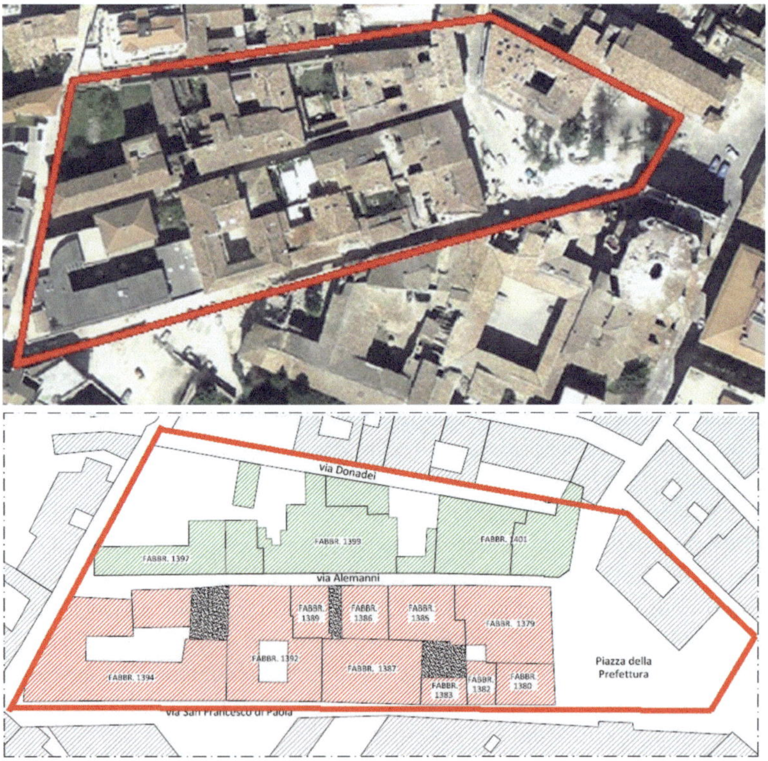

Fig. 20.20 Properties of the Prefettura Square Consortium and subdivision in aggregates

The analysis of the architectural and technological features that define the aggregate is based on studies related to materials and structures on one hand and, on the other, to its construction history. The different types of surveys to be carried out (to identify geometry, structural typologies, constructive techniques, damage) and the critical-historical analysis of the buildings (to define the site morphology, how the aggregate was created, how it evolved during time, how it interacts with the environment) are integrated with the aim of formulating a judgment on the structural efficiency and the seismic response of the buildings. This multi-level approach is being developed for years (Giuffrè, 1991, 1993; Binda et al., 1999).

The first steps to achieve this knowledge consist of on-site inspections, geometrical surveys, and examining the available cartography (Fig. 20.21) and historic documents. The various information that can be drawn concern the process of creation and transformation of the block (buildings, roadways and open spaces); the identification of the load bearing walls in relation to the sequence of construction and the degree of connection; the identification of original building portions and of those created to saturate open spaces; the analysis of openings to evaluate transformations during time and plausible load transfers today; the position of walls and floors to better understand the mechanism of aggregation and the observed damage. The results of this phase can be summarized in written reports, in plans and elevations (Fig. 20.22), where the information on connections and interactions between buildings are already evident (Fig. 20.23), and in first hypotheses on the construction phases of the aggregate (Fig. 20.24).

After this first study, the analysis is deepened by means of on-site inspections (ReLUIS, 2010a), aimed at cataloguing the various construction typologies, techniques and materials (Fig. 20.25), damage survey (Fig. 20.26), on-site tests for characterizing the materials. These studies allow to move from the overall analysis of the aggregate to the more detailed analysis of each single building in the block.

Fig. 20.21 Seismic microzonation of L'Aquila (available on www.protezionecivile.it, *left*); historic cartography in 1500, 1700, after the 1703 earthquake, and in 1858 (*right*)

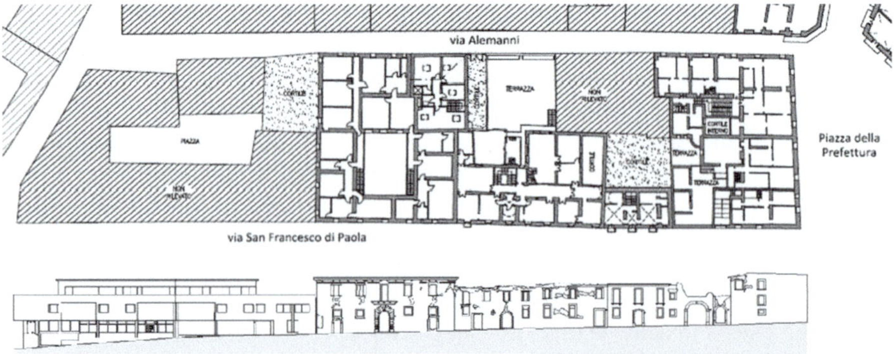

Fig. 20.22 Plan of level 1 and elevation along Via San Francesco di Paola

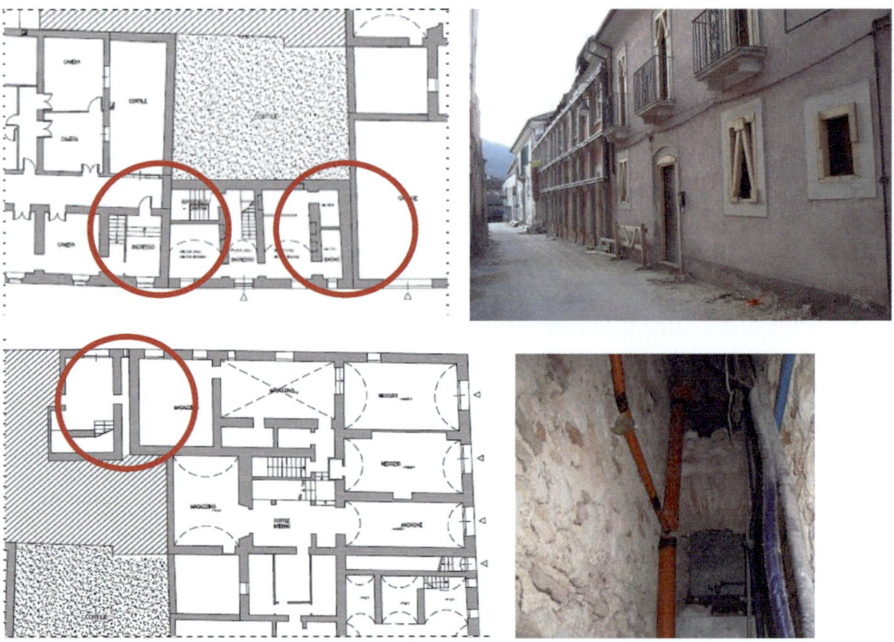

Fig. 20.23 Different types of connection between buildings

In this phase, the architectural, constructive and structural typologies are evidenced, together with the earthquake induced damage. On one hand, the general analysis carried out before helps the structural interpretations that are carried out in this second phase, on the other hand, the more detailed analysis that is carried out here can be used to update and correct some assumptions on the history and evolution of the aggregate, which are useful for the structural analysis of the same. The integration of the two level of analysis allows defining the units were reliable structural

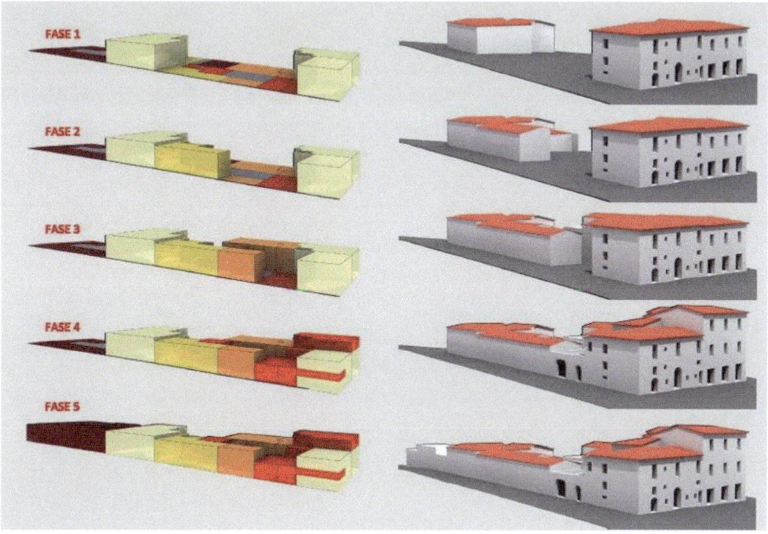

Fig. 20.24 Hypothesis of the phases of construction of the aggregate

Fig. 20.25 Vertical cross section of the building in Piazza della Prefettura (*left*); types of floors: stone vaults, steel beams and brick elements, vaults made of sailor brick masonry; types of masonry: irregular stone, stone with brick courses, brick masonry

Fig. 20.26 Damage state in the building in Piazza della Prefettura: elevations and pictures

analysis can be carried out, and helps in the choice and design of the strengthening interventions for repair and seismic improvement of the aggregate.

20.6 Selection of Techniques and Materials for the Interventions

The specific techniques to be adopted for repairing and strengthening the historic buildings affected by the 6th April 2009 earthquake are still being under definition, and will be reported in the ReLUIS guidelines (March 2010b). The outcomes of the analyses described in Section 20.5 is being used to define the main structural typologies and materials, and their state of damage, also with the purpose of identifying some typical interventions that can be applied. Conversely, the interventions on ordinary buildings have been mostly already carried out, and their exemplification and design procedure is contained in other guidelines (ReLUIS, August 2009). An overall discussion on structural interventions for historic masonry buildings can be found in (Modena et al. 2008, 2009).

In general, one of the first aspect to be taken into account when dealing with the seismic behaviour of existing masonry buildings, is the lack of good connections between structural elements, as evidenced by the type of damage observed after the Abruzzo earthquake too. Hence, it is necessary to improve connections between masonry walls, and between walls and floors and walls and roofs (Tomaževič and Weiss, 1994; Tomaževič and Lutman, 1996). This goal may be achieved inserting ties, confining rings, and tie-beams at the top of buildings (preferably in reinforced masonry or steel, also in r.c. but with restrictions). An effective connection between floors and walls is useful since it allows, above all, to apply a restraining action towards the walls' overturning. In the case of wooden floors, a satisfactory connection is provided by fasteners anchored on the external face of the walls.

Interventions aiming at enhancing the in-plane stiffness of existing floors must be carefully evaluated, since they change the redistribution of horizontal seismic action to the load-bearing walls, and may increase the seismic masses (Tomaževič et al., 1991). Providing a further layer of wooden planks is a limited, but useful, intervention (Parisi and Piazza, 2002). Some studies focused on double planking methods (Valluzzi et al., 2008), with different types of connectors (Modena et al., 2008). In addition, the use of metallic belts or FRP strips, or metallic tie-beams bracings, may improve not only the stiffening effect (Corradi et al., 2006), but also the wall to floor connection. In the case of traditional roofs, as already observed in Section 20.2.1, the application of an overlay of reinforced concrete or the entire substitution of the original wooden structure is in general dangerous. Conversely, a good effect can be obtained by bracing and anchoring the roof trusses to the supporting walls (Piazza and Candelpergher, 2001).

In the case of arches and vaults, which constitute a widespread type of horizontal diaphragm, a traditional strengthening method may be that of using tie-rods to compensate the thrust induced on the bearing walls (Oliveira and Lourenço, 2004). In addition, to absorb thrust of vaulted arches, the possibility of realizing buttresses or reinforced vertical diaphragms should be considered. Composite materials, such as FRP (Barbieri et al., 2002; Valluzzi et al., 2001; Valluzzi, 2008) or SGP/SRG (Borri et al., 2007) could be a suitable option in some cases.

Interventions, aimed at increasing the masonry strength, may be applied to re-establish the original mechanical properties of materials or to improve their performance. Techniques, employed with caution, should make use of materials with mechanical and chemical-physical properties similar to the original ones (Valluzzi, 2008). Some techniques of this type, with different specific aims, are the local rebuilding ("scuci-cuci"), the mortar bed-joint repointing, the structural repointing with bars inserted in the joints (D'Ayala, 1998; Valluzzi et al., 2005). The insertion of small-sized tie beams across the wall (Valluzzi et al., 2004), or the insertion of "diatoni" (masonry units disposed in a orthogonal direction with respect to the wall's plane), provide transversal connections between wall leaves.

The injection of grouting mixture through a regular pattern of drilled holes has been proposed since a long time, particularly for stone masonry walls (Vintzileou and Tassios, 1995). An extensive research has recently focused on the use of non cement-based mortar grouting, which is a more compatible material, to increase the strength of multi-leaf masonry walls (Valluzzi et al., 2004). A wide research, studying the dynamic behaviour of injections and transversal ties, is in progress (Aoki et al., 2010).

More specifically, considering that the damage observation after the seismic event indicated the scarce mechanical characteristics of the masonry types that can be found in the Abruzzo region, research for the definition of the mechanical

Fig. 20.27 Selected masonry wall specimens in Tempera (*left*) and in Onna (*right*)

properties of such walls, and the development of proper injection admixture, is being carried out. The experimental campaign consist in carrying out diagonal compression tests on 21 typologically representative masonry specimens, selected in several towns of the most affected area close to the city of L'Aquila (Fig. 20.27; Casarin et al., 2010a). The original constituents, especially the mortars, were fully characterized from the mechanic, petrographic, textural, mineralogical and chemical point of view, in order to choose the most suitable restoration products (Artioli et al., 2010).

20.7 Conclusions

The earthquake that hit the Abruzzo region on April 6th provided an opportunity to test a methodology for the damage survey of monuments and historic buildings. This method, already implemented during previous earthquakes, has largely proved its reliability. The damage and seismic vulnerability survey allowed to conduct a preparatory analysis to the design of provisional interventions. The study of provisional interventions made during previous earthquakes, showed how the adopted methodological choices and solutions are sometimes in contrast with any purpose and design criteria for safety measures. Thanks to the activity carried out at a larger scale during this earthquake, it was also possible to prepare guidelines for their design (NCP, 2009).

After this experience, the survey forms for damage assessment can be further refined, particularly in relation to the definition of synthetic damage index, the evaluation of economic damage, and the simplification of the palace survey form. The systematization of the observed damage will also hopefully bring new insight into the seismic behaviour of historic structures. In the case of temporary interventions, specific and systematic research activities are also necessary to optimize the measures for heritage buildings, through theoretical, numerical and experimental research (Liberatore et al., 2009).

The extent to which the city of L'Aquila and the surrounding towns have been hit by the earthquake, will require deep studies and a long time before all of the historic buildings will be repaired. In this contest, monitoring, which is being more and more considered as a key activity to increase knowledge on the structural behaviour of monuments, is being proposed as a tool for the post-emergency phase. In case of the seismic sequence of L'Aquila, monitoring can show the eventual progression of damage, can be a warning tool in case of sudden worsening of structural conditions, and allows minimizing and tailoring the repair works, when these will be carried out. The low cost distributed monitoring that is being implemented can also provide a tool to create a time-schedule for the most needed interventions.

The fact that buildings in historical centres are often organized into nuclei of buildings that interact in complex ways makes even more difficult the approach, the development of analysis and the interpretation of results. The general methodology that is going to be defined will systematize the analysis of buildings aggregates within city centres: the diagnosis based on the knowledge of the building's history, the detailed evaluation of the actual state of the structure and materials, the monitoring of the structural behaviour and the comprehension of each building's proper characteristics, can be transformed in indications necessary for restoration and seismic retrofitting.

Deep studies and analyses are needed to interpret correctly the peculiar damage mechanisms observed and the effect caused by past interventions, to select proper intervention techniques, and to adjust them or prove their compatibility to the local structural types, construction elements and materials. On one hand, attention has to be paid in avoiding unnecessary interventions, on the other hand, it will be necessary to determine some shared criteria to intervene on badly damaged buildings, where large portions of the original structural elements collapsed.

Acknowledgments The authors wish to thank all the researchers, collaborators, Ph.D. candidates and students for their significant help during the activities in L'Aquila. The activity was supported by ReLUIS.

References

Aoki T, Mazzon N, Valluzzi M, Casarin F, Modena C (2010) Dynamic identification and damage detection of multi-leaf stone masonry building by shaking table test. ASCE J Struct Eng 56

Artioli G, Casarin F, da Porto F, Mazzoli C, Secco M, Valluzzi MR (2010) Restoration of historic masonry structures damaged by the 2009 Abruzzo earthquake through cement- and polymer-free injection grouts. In: Proceedings of the 2nd Historic Mortars Conference (HCM 2010), Prague, Czech Republic

Barbieri A, Borri A, Corradi M, Di Tommaso A (2002) Dynamic behaviour of masonry vaults repaired with FRP: experimental analysis. In: Proceedings of the 6th International Masonry Conference, London

Bellizzi M (2000) Le opere provvisionali nellemergenza sismica. Civil Protection Agency – National Seismic Service, Roma

Bellizzi M, Colozza R, Dolce M (2001) Le opere provvisionali nellemergenza sismica. In: Proceedings of the 10th conference ANIDIS lingegneria sismica in Italia, Potenza-Matera, Italy

Binda L, Baronio G, Gambarotta L, Lagomarsino S, Modena C (1999) Masonry constructions in seismic areas of central Italy: a multi-level approach to conservation, In: Proceedings of the 8th North American masonry conference (8NAMC), Austin, USA, pp 44–55

Borri A, Casadei P, Castori G, Ebaugh S (2007) Research on composite strengthening of masonry arches. In: Proceedings of the 8th FRPRCS, Patras, Greece
Calderini C, Lagomarsino S, Podestà S, Lemme A (2004) La messa in sicurezza degli edifici monumentali. In: Proceedings of the 11th conference ANIDIS lingegneria sismica in Italia, Genova, Italy
Casarin F, Dalla Benetta M, da Porto F, Valluzzi MR, Modena C (2010a) Masonry panels strengthened by grout injections after the 2009 Abruzzo earthquake. In: Proceedings of the 14th European conference on earthquake engineering (14ECEE), Ohrid, Republic of Macedonia
Casarin F, da Porto F, Modena C (2010b) Structural monitoring of damaged cultural heritage buildings after the April 2009. Abruzzo earthquake. In: Proceedings of the 3rd international workshop on conservation of heritage structures using FRM and SHM, Canada
Circolare n. 617 02/02/2009. Istruzioni per l'applicazione delle "Norme tecniche per le costruzioni" di cui al DM 14/01/2008 – in Italian
Corradi M, Speranzini E, Borri A, Vignoli A (2006) In-plane shear reinforcement of wood beam floors with FRP. Elsevier Composites: Part B, 37:310–319
Costa CQM (2009) Seismic vulnerabilty of historical structures. Damage state of the Abruzzo churches, in the sequence of the 2009 earthquake. MSc thesis, University of Padova
D'Ayala D (1998) The use of bed joint reinforcement to improve the performance of historic masonry buildings. In: Proceedings of the 5th international masonry conference, London
D.M. (14/01/2008) Ministerial Decree: NTC 2008 – Norme tecniche per le costruzioni, Roma, Italy
Doglioni F, Moretti A, Petrini V (1994) Le chiese e il terremoto. LINT Ed, Trieste
Dolce M, Liberatore D, Moroni C, Perillo G, Spera G, Cacosso A (2002) OPUS: Manuale delle Opere Provvisionali Urgenti Post-Sisma (http://postterremoto.altervista.org/index.php)
Giuffré A (1991) Letture sulla meccanica delle murature storiche. Kappa Ed, Roma
Giuffré A (1993) Sicurezza e conservazione dei centri storici: Il caso Ortigia. Laterza Ed, Bari
Giuffré A, Carocci C (1999) Codice di pratica per la sicurezza e la conservazione del centro storico di Palermo. Laterza Ed, Bari
Guidelines for evaluation and mitigation of seismic risk to cultural heritage with reference to technical construction regulations (12/10/2007) Prime Minister Directive, Ministry for Cultural Heritage and Activities, Department of Civil Protection Agency
Lagomarsino S, Maggiolo L, Podestà S (2001) Vulnerabilità sismica delle chiese: proposta di una metodologia integrata per il rilievo, la prevenzione ed il rilievo del danno in emergenza. In: Proceedings of the 10th conference ANIDIS lingegneria sismica in Italia, Potenza-Matera, Italy
Liberatore D, Mattera M, Perillo G (2009) Opere provvisionali post-sisma per edifici in muratura. In: Proceedings of the 13th conference ANIDIS lingegneria sismica in Italia, Bologna, Italy
Model A-DC PCM-DPC MiBAC (2006) Scheda per il rilievo del danno ai beni culturali – Chiese, www.protezionecivile.it
Model B-DP PCM-DPC MiBAC (2006) – Scheda per il rilievo del danno ai beni culturali – Palazzi, www.protezionecivile.it
Modena C, Binda L (2009) Protection of the cultural heritage in the post-earthquake emergency. Progettazione Sismica 1(3):107–115
Modena C, Casarin F, da Porto F, Garbin E, Mazzon N, Munari M, Panizza M, Valluzzi MR (2009) Structural interventions on historical hasonry buildings: review of eurocode 8 provisions in the Light of the Italian experience. In: Proceedings of the workshop: quali prospettive per l'Eurocodice 8 alla luce delle esperienze Italiane, Napoli, Italy
Modena C, Pineschi F, Valluzzi MR (2000) Valutazione della vulnerabilità sismica di alcune classi di strutture esistenti. Sviluppo e valutazione di metodi di rinforzo, CNR – GNDT, Roma
Modena C, Valluzzi MR, da Porto F, Casarin F, Munari M, Mazzon N, Panizza M (2008) Assessment and improvement of the seismic safety of historic constructions: research and applications in Italy. In: Proceedings of the I congreso Iberoamericano sobre construcciones históricas y estructuras de mampostería, Bucaramanga, Colombia

NCP – Nucleo di Coordinamento per le Opere Provvisionali (2009) Vademecum STOP: Schede tecniche delle opere provvisionali per la messa in sicurezza post-sisma da parte dei Vigili del Fuoco – release 2.2

Oliveira DV, Lourenço PB (2004) Repair of stone masonry arch bridges. In: Proceedings of the arch bridges 04, Barcelona, Spain, pp 451–458

OPCM 3431 (03/05/2005) Ordinance of the Prime Minister. Further changes and upgrade to the OPCM 3274/2003

OPCM 3820 (12/11/2009) and OPCM (3832 22/12/2009) Ordinances of the prime minister: interventi urgenti diretti a fronteggiare gli eventi sismici verificatisi nella regione Abruzzo il giorno 6 Aprile 2009

Parisi MA, Piazza M (2002) Traditional timber joints in seismic areas: cyclic behaviour, numerical modelling, normative requirements. Eur Earthquake Eng 1:40–49

Piazza M, Candelpergher L (2001) Mechanics of traditional connections with metal devices in timber roof structures. In: Proceedings of the 7th international conference STREMAH, Bologna, Italy, pp 415–424

ReLUIS (2009) Linee guida per la riparazione e il rafforzamento di elementi strutturali, tamponature e partizioni. Guidelines drafted by ReLUIS, August 2009

ReLUIS (2010a) Modalità di indagine sulle strutture e sui terreni per i progetti di riparazione/miglioramento/ricostruzione di edifici inagibili. Guidelines drafted by ReLUIS, March 2010

ReLUIS (2010b) Linee guida per il rilievo, l'analisi ed il progetto di interventi di riparazione e rafforzamento/miglioramento di edifici in aggregato. First Draft. Guidelines drafted by ReLUIS, March 2010

Tomaževič M, Weiss P, Velechovsky T (1991) The influence of rigidity of floors on the seismic behaviour of old stone-masonry buildings. Eur Earthquake Eng 5(3):28–41

Tomaževič M, Lutman M (1996) Seismic behaviour of masonry walls: modelling of hysteretic rules. ASCE J Struct Eng 122(9):1048–1054

Tomaževič M, Weiss P (1994) Seismic behaviour of plain and reinforced masonry buildings. ASCE J Struct Eng 120(2):323–338

Valluzzi MR (2007) On the vulnerability of historical masonry structures: analysis and mitigation. RILEM Mater Struct 40:723–743

Valluzzi MR (2008) Strengthening of masonry structures with fibre reinforced plastics: from modern conception to historical building preservation. In: Proceedings of the 6th international conference on structural analysis of historical constructions – SAHC08, Bath

Valluzzi MR, Binda L, Modena C (2005) Mechanical behaviour of historic masonry structures strengthened by bed joints structural repointing. Elsevier Constr Building Mater 19:63–73

Valluzzi MR, da Porto F, Modena C (2004) Behavior and modeling of strengthened three-leaf stone masonry walls. RILEM Mater Struct 37(267):184–192

Valluzzi MR, Garbin E, Dalla Benetta M, Modena C (2008) Experimental assessment and modelling of in-plane behavior of timber floors. In: Proceedings of the 6th international conference on structural analysis of historical constructions – SAHC08, Bath

Valluzzi MR, Valdemarca M, Modena C (2001) Behaviour of brick masonry vaults strengthened by FRP laminates. ASCE Int J Compos Constr 5(3):163–169

Vintzileou E, Tassios TP (1995) Three leaf stone masonry strengthened by injecting cement grouts. ASCE J Struct Eng 121:848–856

VV.AA (2007) Beni Culturali in Umbria: dall'emergenza sismica alla ricostruzione. Commissario Delegato per i Beni Culturali, Ufficio del Vice Commissario, BetaGamma

Chapter 21
Rapid Earthquake Loss Assessment After Damaging Earthquakes

Mustafa Erdik, Karin Sesetyan, M. Betul Demircioglu, Ufuk Hancilar, and Can Zulfikar

Abstract This chapter summarizes the work done over last decades regarding the development of new approaches and setting up of new applications for earthquake rapid response systems that function to estimate earthquake losses in quasi real time after an earthquake. After a critical discussion of relevant earthquake loss estimation methodologies, the essential features and the characteristics of the available loss estimation software are summarized. Currently operating near real time loss estimation tools can be classified under two main categories depending on the size of area they cover: Global and Local Systems. For the global or regional near real time loss estimation systems: GDACS, WAPMERR, PAGER and NERIES-ELER methodologies are presented. Examples are provided for the local rapid earthquake loss estimation systems including: Taiwan Earthquake Rapid Reporting System, Real-time Earthquake Assessment Disaster System in Yokohama, Real Time Earthquake Disaster Mitigation System of the Tokyo Gas Co., and Istanbul Earthquake Rapid Response System.

21.1 Introduction

As illustrated in Fig. 21.1 (after Böse, 2006), management of earthquake risks is a process that involves pre-, co- and post-seismic phases. Earthquake Early Warning (EEW) systems are involved in the co-seismic phase. These involve the generation of real time ground motion estimation maps as products of real-time seismology and/or generation of alarm signals directly from on-line instrumental data. The Rapid Response Systems take part immediately after the earthquake and provide

M. Erdik (✉)
Kandilli Observatory and Earthquake Research Institute, Bogazici University, 34684 Istanbul, Turkey
e-mail: erdik@boun.edu.tr

M. Garevski, A. Ansal (eds.), *Earthquake Engineering in Europe*, Geotechnical, Geological, and Earthquake Engineering 17, DOI 10.1007/978-90-481-9544-2_21,

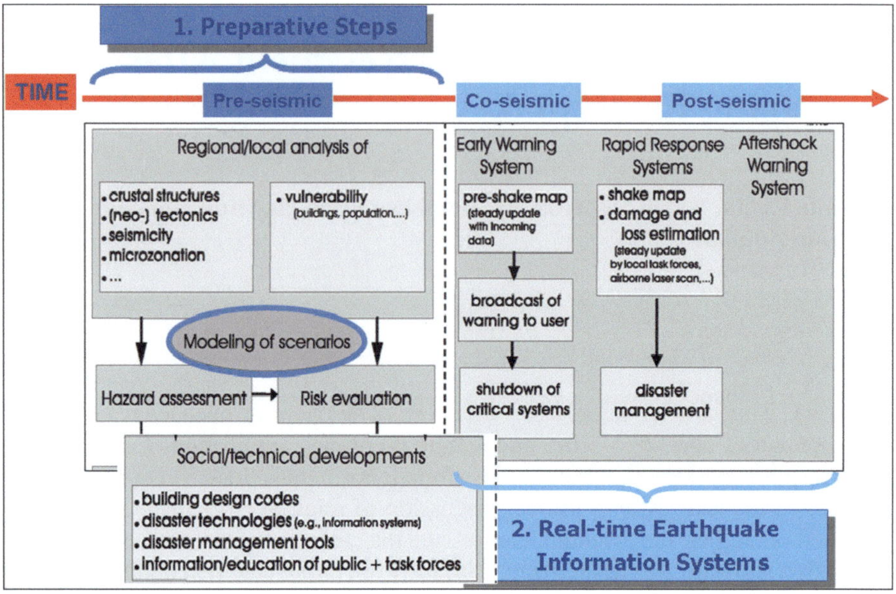

Fig. 21.1 Pre- co- and post-earthquake risk management activities (after Böse, 2006)

assessment of the distribution of ground shaking intensity (Shake Maps) or physical damage and casualties (Loss Maps). These maps can serve to direct the search and rescue teams to the areas most needed and assist civil protection authorities in the emergency action. This study will critically review the existing earthquake rapid response systems that serve to produce earthquake loss information (casualties and, building, lifeline and facility damages) immediately after an earthquake.

Potential impact of large earthquakes on urban societies can be reduced by timely and correct action after a disastrous earthquake. Modern technology permits measurements of strong ground shaking in near real-time for urban areas exposed to earthquake risk. The assessments of the distribution of strong ground motion, building damage and casualties can be made within few minutes after an earthquake. The ground motion measurement and data processing systems designed to provide this information are called Earthquake Rapid Response Systems.

The reduction of casualties in urban areas immediately following an earthquake can be improved if the location and severity of damages can be rapidly assessed by the information from Rapid Response Systems. Emergency management centers of both public and private sector with functions in the immediate post-earthquake period (i.e. SAR, fire and emergency medical deployments) can allocate and prioritize resources to minimize the loss of life. The emergency response capabilities can be significantly improved to reduce casualties and facilitate evacuations by permitting rapid and effective deployment of emergency operations. The Rapid Response data should possibly be linked with incident command and standard emergency management systems to increase effectiveness.

Ground motion data related with power transmission facilities, gas and oil lines and transportation systems (especially fast trains) allow for rapid assessment of possible damages to avoid secondary risks. Water, wastewater and gas utilities can locate the sites of possible leakage of hazardous materials and broken pipes. The prevention of gas-related damage in the event of an earthquake requires understanding of damage to pipeline networks and prompt shut-off of gas supply in regions of serious damage.

Available near real time loss estimation tools can be classified under two main categories depending on the size of area they cover: (1) Global Systems and (2) Local Systems.

For the global or regional near real time loss estimation efforts, Global Disaster Alert and Coordination System (GDACS, http://www.gdacs.org), World Agency of Planetary Monitoring Earthquake Risk Reduction (WAPMERR, http://www.wapmerr.org) and the Prompt Assessment of Global Earthquakes for Response (PAGER, http://earthquake.usgs.gov/eqcenter/pager/) system of USGS and NERIES-ELER (http://www.neries-eu.org) can be listed. For example, the PAGER system is designed to distribute three level alarms via internet similar to GDACS, in addition to information on the earthquake location, magnitude, depth, number of people exposed to varying levels of shaking, and region's fragility. This system uses the ground motion and intensity distribution maps that are produced by the USGS ShakeMap system (http://earthquake.usgs.gov/eqcenter/shakemap) as input. In addition to a three level approach for casualty and loss estimation, PAGER also estimates the number of fatalities based on empirical correlations between casualties and intensity. These systems are elaborated in detail in Section 21.4.

Several local systems capable of computing damage and casualties in near real time already exist in several cities of the world such as Yokohama, Tokyo, Istanbul, Taiwan, Bucharest and Naples (Erdik and Fahjan, 2006). For example, the Istanbul Rapid Response System (Erdik et al. 2003a) consists of 100 strong motion accelerometers distributed in the Metropolitan area of Istanbul. After triggered by an earthquake, each station processes the streaming strong motion to yield the spectral acceleration at specific periods and sends these parameters in the form of SMS messages to the main data center at through available GSM network services. A ground shaking and damage distribution map is then automatically generated based on the spectral acceleration data received, building inventory and the vulnerability relationships. Some of these local systems are covered in detail in Section 21.5.

21.2 Earthquake Loss Estimation

An extensive body of research, tools and applications exists that deals with all aspects of loss estimation methodologies. The components of rapid earthquake loss estimation will be addressed following the modular structure of the HAZUS (Whitman et al., 1997; Kircher et al., 2006; FEMA, 2003) methodology illustrated in Fig. 21.2.

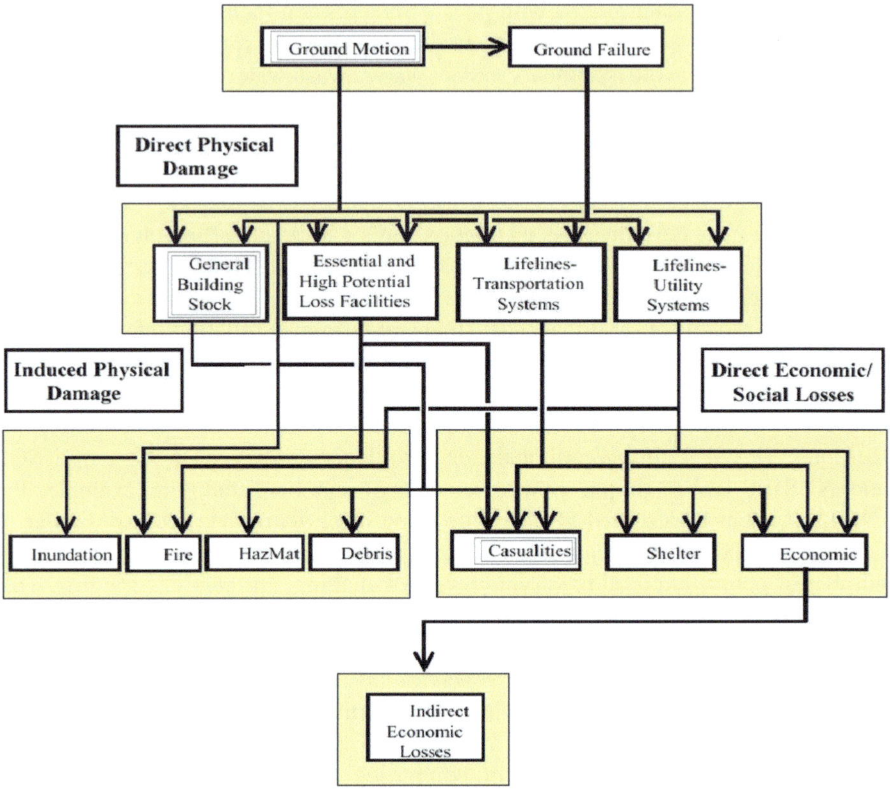

Fig. 21.2 Earthquake loss estimation structure (after HAZUS-MH, FEMA, 2003)

The important ingredients of this earthquake loss estimation flowchart in consideration of the "rapid" assessment of earthquake losses are Ground Motion, Direct Physical Damage to General Building Stock and Casualties as Direct Social Losses.

Almost all deterministic earthquake loss assessment schemes rely on the quantification of the earthquake shaking as intensity measure parameters in geographic gridded formats. The earthquake shaking can be determined theoretically for assumed (scenario) earthquake source parameters through ground motion prediction relationships GMPE's (i.e. attenuation relationships) or using a hybrid methodology that corrects the analytical data with empirical observations, after an earthquake. Either procedure yields the so-called, maps that display the spatial variation of the peak ground motion parameters or intensity measures. We owe this "ShakeMap" term to the USGS program that provides near-real-time maps of ground motion and shaking intensity following significant earthquakes in the United States as well as around the Globe (http://earthquake.usgs.gov/eqcenter/shakemap/). ShakeMap uses instrumental recordings of ground motions, kriging techniques and empirical ground motion functions to generate an approximately continuous representation of the shaking intensity shortly after the occurrence of an earthquake (Wald et al., 2005). The shape files that are generated via ShakeMap can be used as inputs for the casualty and damage assessment routines for rapid earthquake loss estimation. In USA, and increasingly in other countries, these maps are used for post-earthquake response and recovery, public and scientific information, as well as for loss assessment and disaster planning.

The earthquake loss estimations should consider the uncertainties in seismic hazard analyses, and in the vulnerability relationship. There exits considerable amount of epistemic uncertainty and aleatory variability in ShakeMaps. Accuracy of the ShakeMap is mainly related to two factors: (1) the proximity of a ground motion observation location, i.e. the density of the strong ground motion network in the affected area, and (2) the uncertainty of estimating ground motions from the GMPE, most notably, elevated uncertainty due to initial, and unconstrained source rupture geometry. The epistemic uncertainties become highest for larger magnitude events when rupture parameters are not yet well constrained (Wald et al., 2008). Aleatory uncertainties may be reduced if the bias correction with recorded amplitudes is performed directly on the ground surface rather than at bedrock level which the case in the current ShakeMap application.

The reliability of the vulnerability relationships is related to the conformity of the ground motion intensity measure with the earthquake performance (damage) of the building inventory. Estimates of human casualties are derived by uncertain relationships from already uncertain building loss estimates, so the uncertainties in these estimates are compounded (Coburn and Spence, 2002).

It is possible to examine the effect of cumulative uncertainties in loss estimates using discrete event simulation (or Monte-Carlo) techniques if the hazard and that the probability distribution of each of the constituent relationships is known. The general finding of the studies on the uncertainties in earthquake loss estimation is that the uncertainties are large and at least as equal to uncertainties in hazard analyses (Stafford et al., 2007).

21.3 Earthquake Loss Estimation Software Tools

In connection with the EU FP7 NERIES Project, Stafford et al. (2007) has reviewed Earthquake Loss Estimation software tools that have been developed worldwide over the past 10 years. The results are summarized in Table 21.1

For known inventories of buildings and under conditions where the earthquake hazard in terms of ground shaking distribution can be assessed rapidly after an earthquake, these tools can be adapted for rapid loss estimation. A brief description and references for the selected earthquake loss assessment software can be given as follows:

HAZUS: HAZUS-MH (NIBS and FEMA, 2003) is developed by the United States Federal Emergency Management Agency (FEMA) for the prediction and mitigation of losses due to earthquakes (HAZUS), hurricanes and floods

Table 21.1 List of software for earthquake loss estimation

Type	ELE software	Owner or developer
US	HAZUS-MH	FEMA/NIBS
US	EPEDAT	EQE International, Inc. for California Governor's Office of Emergency Services
US	REDARS	MCEER/FHWA
US	INLET	ImageCat, Inc. for RESCUE Project
Europe	SES 2002	Spanish Civil Protection
Europe	SIGE/ESPAS	Italian Civil Protection Department
Europe	KOERILOSS	KOERI for LESSLOSS Project (Bogazici Un.)
Europe	LNECLOSS	LNEC for LESSLOSS Project
Europe	SELENA	NORSAR
Europe	DBELA	ROSE School/EU-Centre
Europe	EQSIM	Karlsruhe University
Europe	ELER	E.U. F.P.6 Project NERIES
World	RADIUS	OYO Corporation for United Nations ISDR Secretariat
World	QLARM (QUAKELOSS)	ESRC/WAPMERR
World	NHEMATIS	Canadian Civil Protection
World	EQRM	Geoscience Australia

(Whitman et al., 1997; Kircher et al., 2006). The package is intended for US applications only and includes federally collected data as default. The inventory is classed based on 36 different types of building based on construction standards and material as well as size and building use. HAZUS-MH MR2 version, released in 2006, includes the capability for rapid post-event loss assessment.

EPEDAT: The EPEDAT (Early Post-Earthquake Damage Assessment Tool) is designed by EQE International, Inc. for post-earthquake loss estimation (Eguchi et al., 1997). The output encompasses damage (building and lifelines) and casualty for California based on county specific housing and demographic data. It is Windows-based and uses Modified Mercalli Intensity to quantify the hazard.

SIGE: SIGE, developed by Italian National Seismic Service of the Civil Protection Department, is used for rapid approximate estimate of the damage (Di Pasquale et al., 2004). The first update of the program (FACES) considers linear sources, directivity effects, and the influence of focal depth. The most recent modification of the codes has been implemented in a new model called ESPAS (Earthquake Scenario Probabilistic Assessment).

KOERILOSS: A scenario-based building loss and casualty estimation model developed by Bogazici University (Erdik and Aydinoglu, 2002; Erdik et al., 2003b; Erdik and Fahjan, 2006) for estimating earthquake losses in Istanbul, Izmir, Bishkek and Tashkent. Derivatives of the model were used in the

EU FP5 LessLoss project as well as for the assessment of scenario earthquake losses in Amman. The methodology considers both deterministic (scenario) and probabilistic forecasting approaches. The vulnerability calculations can be based on empirical results (EMS intensity-based) or on a response-spectrum-based method similar to HAZUS. It is used for rapid loss assessment in connection with the Istanbul Earthquake Rapid Response System, described in Section 21.5.3 of this chapter.

ESCENARI: ESCENARIS (Roca et al., 2006) is the software tool developed for Catalonia. The methodology relies on the use of scenario-based earthquake hazards and intensity-based empirical vulnerability functions of Giovinazzi (2005). The losses are based on the building stock and classes of social impact.

CAPRA: CAPRA (Central American Probabilistic Risk Assessment – www.ecapra.org) Project has developed a region-specific Earthquake Loss Estimation model using a Web 2.0 format. It is currently under construction (Anderson, 2008).

LNECLOSS: LNECLOSS is a software package developed by the Laboratorio Nacional de Engenharia Civil (LNEC) in Lisbon, Portugal (Sousa et al., 2004). LNECloss is an earthquake loss assessment tool, integrated on a Geographic Information System (GIS), which comprises modules to compute seismic scenario bedrock input, local soil effects, vulnerability and fragility analysis, human and economic losses. LNECloss was applied to Metropolitan Area of Lisbon (Zonno et al., 2009).

SELENA: SELENA (Seismic Loss Estimation Using a Logic Tree Approach) is a software package developed at NORSAR for earthquake building damage assessment (Molina and Lindholm, 2005). SELENA uses the capacity-spectrum method (HAZUS methodology, ATC-55-ATC, 2005) with a logic tree-based weighting of input parameters that reportedly allows for the computation of confidence intervals. GIS software can be utilized at multiple levels of resolution to display predicted losses graphically.

DBELA: DBELA (Displacement-Based Earthquake Loss Assessment) is an earthquake loss estimation tool currently being developed at the ROSE School/EU-Centre in Pavia (Crowley et al., 2004; Calvi et al., 2006; Bal et al., 2008a). The methodology is essentially based on comparison of the displacement capacity of the building stock (grouped by structural type and failure mechanism) and the imposed displacement demand from a given earthquake scenario. The methodology aims to allow a good correlation with damage, ease of calibration to varying building stock characteristics and systematic treatment of all sources of uncertainty. It takes into account the uncertainties associated through the process for demand and capacity. Applications of the methodology were carried out for loss assessment in the Marmara Region (Bommer et al., 2006).

EQSIM: EQSIM (EarthQuake damage SIMulation) is the rapid earthquake damage estimation component of the Disaster Management Tool (DMT) currently being developed at the University of Karlsruhe (Baur et al., 2001;

Markus et al., 2004). The loss estimation methodology is based on the adaptation capacity spectrum method used in HAZUS to reflect the European building practice. EQSIM has been used to assess earthquake losses in Bucharest on the basis of scenario earthquakes (Wenzel and Marmureanu, 2007).

QUAKELOSS: QUAKELOSS is a computer tool for estimating human loss and building damage due to Earthquakes developed by the staff of the Extreme Situations Research Center in Moscow. An earlier version of this program and data set is called EXTREMUM (Larionov et al., 2000). QUAKELOSS software is used by the World Agency of Planetary Monitoring and Earthquake Risk Reduction (WAPMERR) to provide near-real-time estimates of deaths and injuries caused by earthquakes anywhere in the world. The building inventory reportedly incorporates data from about two million settlements throughout the world.

NHEMATIS: NHEMATIS (Natural Hazards Electronic Map and Assessment Tools Information System) has been developed Emergency Preparedness Canada (Couture, 2002). It is a national-scale automated facility for the collection and analysis of natural hazard information combined with characterizations of population and infrastructure to allow analyses of risks. Similar to HAZUS, NHEMATIS integrates an expert system rule base, geographic information system (GIS), relational databases, and quantitative models to permit assessment of the hazard impact.

EQRM: EarthQuake Risk Management (EQRM), developed by Geoscience Australia, is an event-based tool for earthquake scenario ground motion and scenario loss modeling as well as probabilistic seismic hazard and risk modeling (Robinson et al., 2005, 2006). The risk assessment methodology is based on the HAZUS methodology with some modifications to adapt it to Australian conditions. It has the potential to be used with earthquake monitoring programs to provide automatic loss estimates.

OSRE: The Open Source Risk Engine (OSRE), developed in Kyoto University – Graduate School of Engineering, Department of Urban Management, is multi-hazards open-source software that can estimate the risk (damage) of a particular site (object) given a hazard and the vulnerability with their associate probability distributions (AGORA-Alliance for Open Risk Analysis, http://www.risk-agora.org). The catalogue vulnerability data for different facility classes was obtained from ATC-13.

ELER: The Joint Research Activity 3 (JRA3) of the EU Project NERIES (Erdik et al., 2008) has developed a methodology and software (Earthquake Loss Estimation Routine – ELER) for rapid estimation of earthquake damages and casualties throughout the Euro-Med Region. The ELER is designed as open source software to allow for community based maintenance and further development of the database and earthquake loss estimating procedures. The software provides for the estimation of losses in three levels of analysis. These levels of analysis are designed to commensurate with the quality of the available building inventory and demographic data. Detailed information on ELER is provided in Section 21.4.4 of this chapter.

MAEVIZ: MAEviz, developed in the Mid-America Earthquake Center in University of Illinois, integrates spatial information, data, and visual information to perform seismic risk assessment and analysis (http://mae.ce.uiuc.edu/software_and_tools/maeviz.html). It can perform earthquake risk assessment for buildings (structural and non-structural damage), bridges and gas networks with a built-in library of vulnerability relationships. In addition to applications in USA and important application of the software has been conducted for the Zeytinburnu District of Istanbul (Elnashai et al., 2007).

21.4 Global and Regional Earthquake Rapid Loss Assessment Systems

Available near real time loss estimation tools can be classified under two main categories depending on the size of area they cover: (1) Local Systems and (2) Global Systems. For the global or regional near real time loss estimation efforts, Global Disaster Alert and Coordination System (GDACS, http://www.gdacs.org), World Agency of Planetary Monitoring Earthquake Risk Reduction (WAPMERR, http://www.wapmerr.org) and the Prompt Assessment of Global Earthquakes for Response (PAGER, http://earthquake.usgs.gov/eqcenter/pager/) system of USGS and NERIES-ELER (http://www.neries-eu.org) can be listed. A description of the important rapid earthquake loss assessment systems with global or regional coverage will be provided in the following sub-sections.

21.4.1 PAGER – Prompt Assessment of Global Earthquakes for Response

The PAGER (Prompt Assessment of Global Earthquakes for Response) system operated by USGS (http://earthquake.usgs.gov/eqcenter/pager/) provides, within minutes of a significant earthquake, information on the earthquake location, magnitude, depth, the names of cities exposed to severe shaking, and estimates of the number of people exposed to varying levels of shaking and region's fragility. PAGER products are generated for all earthquakes of magnitude 5.5 and greater globally and for lower magnitudes of about 3.5–4.0 within the US. PAGER's results are posted on the USGS Earthquake Program Web site (http://earthquake.usgs.gov/), in form of a one-page report, and sent in near real-time to emergency responders, government agencies, and the media. In the hours following significant earthquakes, as more information becomes available, PAGER's content is modified both manually and automatically.

21.4.2 GDACS – The Global Disaster Alert and Coordination System

The Global Disaster Alert and Coordination System – GDACS (http://www.gdacs.org/) provides near realtime alerts about natural disasters around the world and tools to facilitate response. GDACS collects earthquake information every 5 min from: United States Geological Survey National Earthquake Information Center (NEIC), European-Mediterranean Seismological Centre (EMSC), GEOFON Program of the GFZ Potsdam and Japan Meteorological Agency (JMA). Using the reported earthquake parameters, a three level alert based on the LandScan population dataset and the population vulnerability (European Commission Humanitarian Aid Department Global Needs Assessment Indicator) in the region of interest. Currently, the evaluation of the potential humanitarian impact of earthquakes considers (1) earthquake magnitude (2) earthquake depth (3) population within 100 km of epicenter, and (4) national population vulnerability. The last two elements are automatically calculated by GIS based on the earthquake epicenter, the LandScan population dataset and ECHO's Global Needs Assessment indicator. The alerts are considered on the basis of the so-called alert score which combines the earthquake magnitude and depth, size of the exposed population and the country-specific vulnerability index. The alert score is transformed into three alert levels: red, orange and green.

21.4.3 WAPMERR – World Agency of Planetary Monitoring and Earthquake Risk Reduction

WAPMERR (http://www.wapmerr.org) provides loss estimates for earthquakes in global scale within less than 2 h of their occurrence. The alerts WAPMERR issues include number of fatalities and injured, as well as average damage to buildings in the affected settlements. This service is being carried out in partnership between WAPMERR and the Swiss Seismological Service with the support of the Swiss Agency for Development and Cooperation. QLARM (earthQuake Loss Assessment for Response and Mitigation – http://qlarm.ethz.ch), and outgrowth of the former QUAKELOSS software, is the computer tool used to estimate the building damage and casualties (Trendafiloski et al., 2009a). Loss estimations are done for the QLARM worldwide database constructed of (1) point city models for the cases where only summary data for the entire city are available; and, (2) discrete city models where data regarding city sub-divisions (districts) are available (Trendafiloski et al., 2009b).

21.4.4 NERIES Project – ELER: Earthquake Loss Estimation Routine for the Euro-Med Region

The Joint Research Activity JRA-3 of the EU Project NERIES aims at establishing rapid estimation of earthquake damages, casualties, shelters and food requirements throughout the Euro-Med Region. Within the scope of this activity, a rapid loss estimation tool is developed by researchers from KOERI, Imperial College, NORSAR and ETH-Zurich. The loss estimation is conducted under three levels of sophistication as elaborated in Fig. 21.3.

The ground motion estimation methodology is common in all levels of analysis. Based on the event parameters the distribution of PGA, PGV, SA ($T = 0.2$ s) and SA ($T = 1.0$ s) are estimated based on a choice of ground motion prediction models. Local site effects are incorporated either with the Borcherdt (1994) methodology or, if available, with the use of Vs30 based amplification functions within the ground motion prediction models. If strong ground motion recordings are also available, the prediction distributions are bias corrected using the peak values obtained from these recordings. EMS-98 Intensity distributions are obtained based on computed PGA and PGV values using the procedure proposed by Wald et al. (1999).

Both Level 0 (similar to PAGER system of USGS) and Level 1 analyses of ELER software are based on obtaining intensity distributions analytically and estimating total number of casualties either using regionally adjusted intensity-casualty or magnitude-casualty correlations (Level 0) or using regional building inventory databases (Level 1). Level 1 type analysis uses EMS98 based building vulnerability relationships (Lagomarsino and Giovinazzi, 2006) to estimate building damage and casualty distributions.

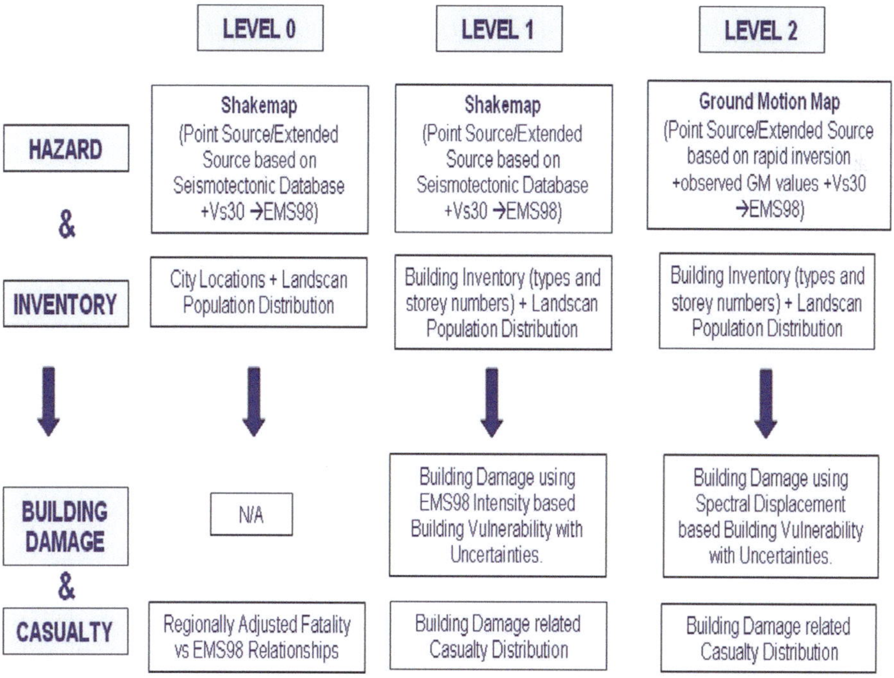

Fig. 21.3 The levels of analysis incorporated in the ELER software

Level 2 type analysis corresponds to the highest sophistication level in the loss estimation methodology developed. The building damage and casualty distributions are obtained using analytical vulnerability relationships and building damage related casualty vulnerability models, respectively.

21.4.4.1 Demographic and Building Inventory

For all levels of analysis the 30 arc s grid based LandScan population data (http://www.ornl.gov/sci/landscan/landscan2005/index.html) are used. For both the Level 1 and Level 2 analyses options exist for the use of local demographic data for casualty estimation.

ELER is structured in such a way that a building inventory can be classified in terms of any classification system as long as the empirical and/or mechanical vulnerabilities and performance curves associated with each building type is defined by the user. However, the HAZUS (FEMA, 2003), EMS-98 (Grünthal, 1998), and RISK-UE (2001–2004) building taxonomies are used as the default main classification systems in the development of ELER.

The regional scale building inventory used in Level 1 analysis corresponds to an approximated (proxy) European database consisting of the number of buildings and their geographic distribution. This approximated building database is obtained from CORINE Land Cover (European Environment Agency, 1999), LandScan population database (http://www.ornl.gov/sci/landscan/landscan2005/index.html) and Google Earth (http://earth.google.com) and is provided within ELER as the default data for Level 1 analysis. Following the determination governing land cover classes for each country, the basic methodology used in obtaining the country basis proxy distribution of the number of buildings (per unit area in each building class) is as follows (Demircioglu et al., 2009)

1. Select suitable sample areas from Google Earth for each Corine Land Cover class in all countries
2. Obtain the actual number of buildings in each sample area, automatically using image processing techniques
3. Approximate the total number of buildings in each country by spreading the sample area building counts to the country
4. Verify (and adjust) the number of buildings thus obtained by computing the population per building for each Corine Land Cover class, and by also checking with the actual number of buildings in a country if such information has been obtained from the corresponding country's statistical office.

The PAGER database (Jaiswal and Wald, 2010; Porter et al., 2008) provides the percentages of different construction types in all countries of the world for both urban and rural settlements and residential and non-residential occupancy types, making use of a HAZUS type classification. The corresponding European Building Taxonomy classes were identified and the associated percentages have been used to convert the grid based number of buildings to an inventory of differentiated structural types in each country. The grid based distribution of the number of buildings and population thus obtained is aggregated to 30 and 150 s arc grids to form the default data for Level 1 analysis.

21.4.4.2 Building Damage Assessment

Different vulnerability relationships and building damage assessment methodologies are used under the different levels of analysis.

The Level 0 analysis does not include any building damage assessment. The physical damage in cities and other populated areas can be roughly inferred through the intensities given by the Shakemaps.

For Level 1 damage assessment analysis, the intensity based empirical vulnerability relationships developed by Lagomarsino and Giovinazzi (2006) are used. ELER software allows for the incorporation of a regional variability factor in these relationships.

Level 2 analysis is essentially intended for earthquake risk assessment (building damage and consequential human casualties) in urban areas. As such, the building inventory data for the Level 2 analysis will consist of grid (geo-cell) based urban building (RISK-UE typology) and demographic inventories. The building damage assessment is based on the empirical vulnerability relationships developed by Lagomarsino and Giovinazzi (2006) and the analytical vulnerability relationships based on the Capacity Spectrum Method (so-called HAZUS methodology).

For the representation of seismic demand the 5%-damped elastic response spectrum provided EC8 Spectrum (Eurocode 8, CEN 2003) or IBC 2006 Spectrum (International Building Council, 2006) is used. For the estimation of the so-called "Performance Point", the intersection pint of the capacity and the demand curves, ELER uses the procedures based on: the Capacity Spectrum Method specified in ATC-40 (1996), its recently modified and improved version Modified Acceleration-Displacement Response Spectrum Method (FEMA-440) and the Coefficient Method originally incorporated in FEMA-356 (2000). ELER also incorporates another non-linear static procedure, the so-called "N2 – Reduction Factor Method" method (Fajfar, 2000) where the inelastic demand spectra is modified using ductility factor based reduction factors.

21.4.4.3 Casualty Assessment

For casualty Estimation in Level 0 analyses the empirical fatality rate correlation with magnitude (Samardjieva and Badal, 2002; Badal et al., 2005) and with EMS98 intensities are used. Casualties in Level 1 analysis is assessed on the basis of the simple correlations with fatalities and the number of buildings damaged beyond repair. The casualty estimation methodology of Coburn and Spence (2002) based on the number of buildings in D5 damage state of EMS98 is also coded in ELER. The estimation of casualties in Level 2 analysis is the one used in HAZUS based on the number of buildings of a given type at different damaged states and the associated casualty rates. The casualty rates corresponding to reinforced concrete and masonry structures given in HAZUS-MH (FEMA, 2003) are adopted in ELER.

21.5 Local (Country, City and Facility Specific) Earthquake Rapid Loss Assessment Systems

Several local systems (country-, city- or, facility-specific) capable of computing damage and casualties in near real time already exist in several regions of the world. For example the Taiwan Earthquake Rapid Reporting System, the Real-time Earthquake Assessment Disaster System in Yokohama (READY), The Real Time Earthquake Disaster Mitigation System of the Tokyo Gas Co. (SUPREME) and the Istanbul Earthquake Rapid Response System (IERRS) provide near-real time damage estimation after major earthquakes (Erdik and Fahjan, 2006). Almost all of these systems are based on the assessment of demand in real time from dense strong motion instrument arrays and the estimation of damage on the basis of known inventory of elements exposed to hazard and the related vulnerability relationships. After an earthquake the shaking and damage distribution maps are automatically generated on the basis of the ground motion intensity measure data received from the field stations, building inventory and the vulnerability relationships.

21.5.1 Earthquake Rapid Reporting System in Taiwan

Earthquake Rapid Reporting and Early Warning Systems in Taiwan, operated by Taiwan Central Weather Bureau, uses a real-time strong-motion accelerograph network that currently consists of 82 telemetered strong-motion stations distributed across Taiwan.

The rapid reporting system can offer information about 1 min after an earthquake occurrence, that includes earthquake location, magnitude and shaking maps (Tsai and Wu, 1997; Teng et al., 1997; Wu et al., 1998, 1999; Shin and Teng, 2001; Wu and Teng, 2002).

Central Weather Bureau of Taiwan operates two dense digital strong-motion networks: (1) The Taiwan Rapid Earthquake Information Release System (TREIRS), and (2) The Taiwan Strong Motion Instrumentation Program (TSMIP). The locations of these two networks are shown in Fig. 21.4.

TREIRS can obtain earthquake magnitude, epicenter location and focal depth within 90 s after occurrence of earthquakes. The TSMIP system consist of more than 650 stations spaced approximately every 5 km in populated areas in Taiwan. The Early Seismic Loss Estimation (ESLE) module has been developed and integrated with the application software "Taiwan Earthquake Loss Estimation System (TELES) provides decision support soon after occurrence of strong earthquakes for emergency providers" (Yeh et al., 2003). TELES software, essentially modeled after HAZUS, acts as a decision support tool in emergency responses.

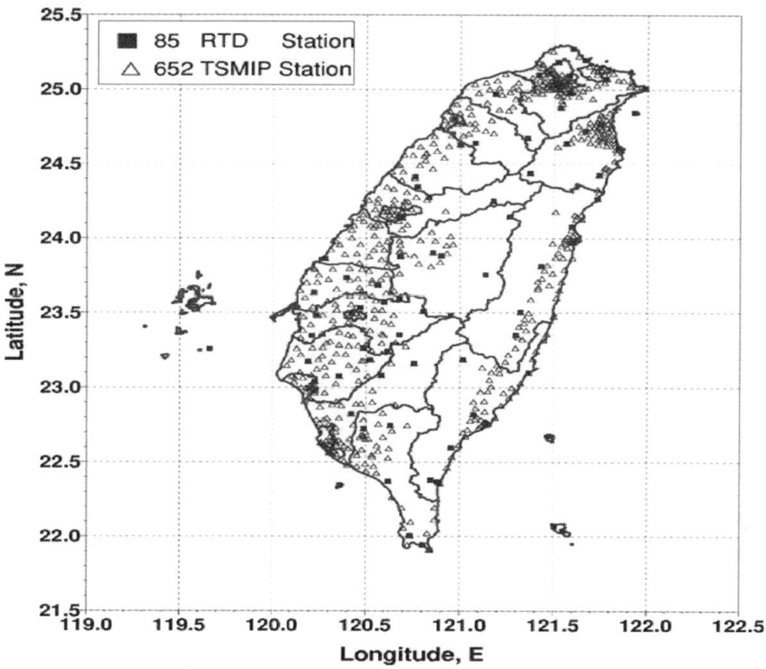

Fig. 21.4 Strong motion networks operated by Central Weather Bureau of Taiwan

The ESLE module is automatically triggered after receiving earthquake alerts. The estimated damages and casualties are then provided in the form of maps and tables automatically. Currently, the time span to complete the hazard analysis and damage assessment needs 3–5 min depending on the earthquake magnitude, epicenter location and focal depth.

21.5.2 USGS-Shake Cast

ShakeCast (http://earthquake.usgs.gov/shakecast/), short for ShakeMap Broadcast, is a fully automated system for delivering specific ShakeMap (http://earthquake.usgs.gov/shakemap/) products to critical users and for triggering established post-earthquake response protocols.

It was developed and is used primarily for emergency response, loss estimation, and public information. ShakeCast allows utilities, transportation agencies, businesses, and other large organizations to control and optimize the earthquake information they receive. With ShakeCast, they can automatically determine the shaking value at their facilities, set thresholds for notification of damage states for each facility, and then automatically notify (by pager, cell phone, or email) specified operators and inspectors within their organizations who are responsible for those particular facilities so they can set priorities for response (Lin and Wald, 2008).

ShakeCast software is built upon open-source tools, providing standard, freely available software for all users, encouraging user improvements, and simplifying interfacing with existing users' response tools.

21.5.3 Istanbul Earthquake Rapid Response System

To assist in the reduction of losses in a disastrous earthquake in Istanbul a dense strong motion network has been implemented. All together this network and its functions is called Istanbul Earthquake Rapid Response and Early Warning System (IERREWS). The system is designed and operated by Bogazici University with the logistical support of the Governorate of Istanbul, First Army Headquarters and Istanbul Metropolitan Municipality (Erdik et al., 2003a; Erdik and Fahjan, 2006). One hundred (100) of the strong motion recorders are stationed in dense settlements in the Metropolitan area of Istanbul in dial-up mode for Rapid Response information generation. Post-earthquake rapid response information is achieved through fast acquisition, analysis and elaboration of data obtained from the 100 stations of the IERREWS network (Fig. 21.5).

The Rapid Response part of the IERREWS System satisfies the COSMOS (The Consortium of Organizations for Strong-Motion Observation Systems) Urban Strong-Motion Reference Station Guidelines (www.cosmos-eq.org) for the location of instruments, instrument specifications and housing specifications. The relative instrument spacing is about 2–3 km which corresponds to about 3 wavelengths in firm ground conditions and more than 10 wavelengths for soft soils for horizontally

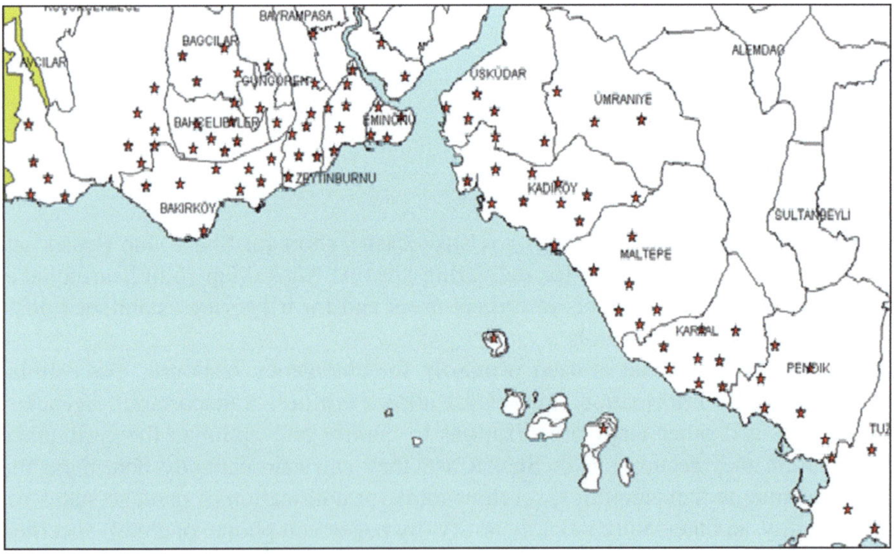

Fig. 21.5 Rapid response stations of IERREWS

propagating 1 s shear waves. For communication of data from the rapid response stations to the data processing center and for instrument monitoring a reliable and redundant GSM communication system (backed up by dedicated landlines and a microwave system) is used.

In normal times the rapid response stations are be interrogated (for health monitoring and instrument monitoring) on regular basis. Once triggered by an earthquake, each station process the streaming three-channel strong motion data to yield the spectral accelerations at specific periods, 12 Hz filtered PGA and PGV and transmits these parameters in the form of SMS messages at every 20 s directly to the main data center through the GSM communication system.

For the computation of the input ground motion parameters, spectral displacements obtained from the SMS messages sent from stations are interpolated to determine the spectral displacement values at the center of each geo-cell using two-dimensional splines. The earthquake demand at the center of each geo-cell is computed using these spectral displacements. For the generation of Rapid Response information (Loss Maps) specific software that follows the same methodology (based on spectral displacements) used in the development of the ELER tool (Erdik et al., 2008) is used.

The loss estimation relies on the building inventory database, fragility curves and the direct physical damage and casualty assessment techniques. The computations are conducted at the centers of a $0.01^{\circ}\times0.01^{\circ}$ grid system. The building inventories (in 24 groups) for each geocell together with their spectral displacement curves are incorporated in the software. The casualties are estimated on the basis of the number of occupancies and degree of damage suffered by buildings. The resulting rapid response (i.e. LossMap) information is communicated to the concerned emergency response centers (currently Istanbul Governorate, Istanbul Municipality and First Army Headquarters). The ShakeMap and LossMap estimated by the rapid response system after the March 12, 2008 M4.3 Earthquake is provided in Fig. 21.6. A simulated LossMap is illustrated in Fig. 21.7.

"Earthquake Disaster Information System for the Marmara Region, Turkey" (EDIM, http://www.cedim.de/EDIM.php) is an interdisciplinary cooperative project to improve the real-time information before, during and after an earthquake. EDIM consortium consists of Bogazici University in Istanbul, Karlsruhe University, GeoForschungsZentrum Potsdam, Humboldt-University, lat/lon GmbH, and DELPHI InformationsMusterManagement GmbH. EDIM aims to provide algorithms for the rapid detection of earthquakes, calculation of near-real-time shake- and loss-maps and to establish a dynamic geo-information infrastructure to link the various user groups and to provide information (Koehler et al., 2007). The Building Damage Estimation Service (http://www.edim-des.delphi-imm.de/) developed by Delphi GmBH is a web-based application for the assessment of structural damages in Istanbul and Marmara region for a given scenario earthquake. This web application uses the spectral capacity-based analytical vulnerability relations adopted from Erdik et al. (2003b). Another application called "SOSEWIN", based on the innovative technology of self-organizing networks, has been set up in the Atakoy region of Istanbul as a prototype (Picozzi et al., 2008).

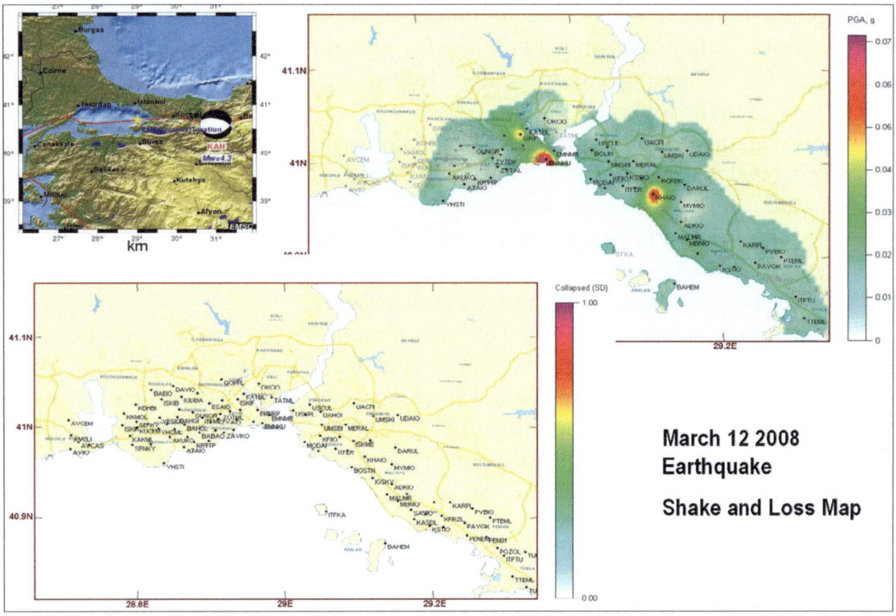

Fig. 21.6 Shake and LossMaps for March 12, 2008 M4.3 earthquake

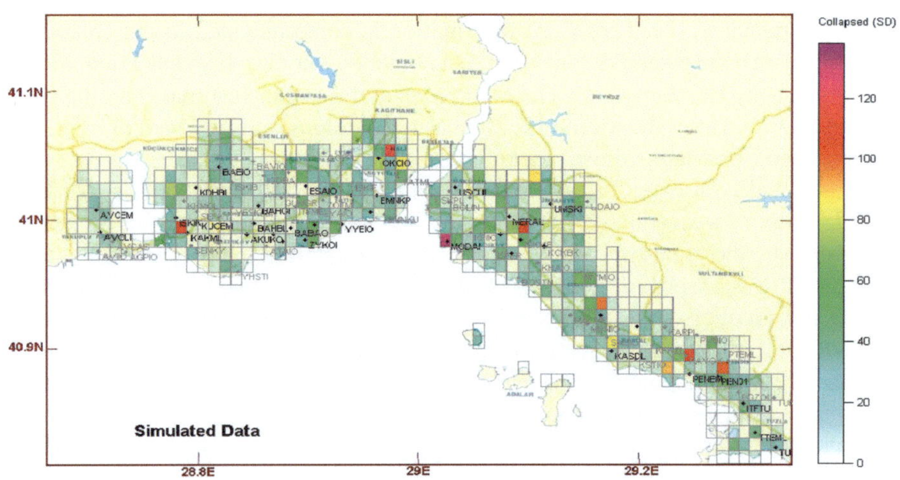

Fig. 21.7 A sample LossMap (Building damage distribution obtained from simulated data)

21.5.4 *Rapid Response and Disaster Management System in Yokohama, Japan*

In 1997 the city of Yokohama installed a dense strong-motion array for earthquake disaster management. The array (called, REal-time Assessment of earthquake Disaster in Yokohama-READY System) consists of 150 strong motion accelerographs at a spacing of about 2 km (Fig. 21.8). In addition borehole strong motion systems at installed at 9 different locations for liquefaction monitoring. It is currently used for strong motion monitoring, real-time seismic hazard and risk assessment and damage gathering systems (Midorikawa, 2005).

These stations are connected to three observation centers, the disaster preparedness office of the city hall, the fire department office of the city and Yokohama City University, by the high-speed and higher-priority telephone lines. At 18 stations, the backup communication system by satellite is available (Midorikawa, 2004).

When the accelerograph is triggered by an earthquake, the station computes ground-motion parameters such as the instrumental seismic intensity, peak amplitudes, predominant frequency, total power, duration and response spectral amplitudes. The seismic intensity data is conveyed to the city officials by the pager, and the intensity map of the city is drawn within a few minutes after the earthquake. The map is immediately open to the public through the Internet and local cable TV.

Rapid assessment of the damage to the timber houses is computed and mapped on the basis of their dynamic characteristics and the response spectrum of ground

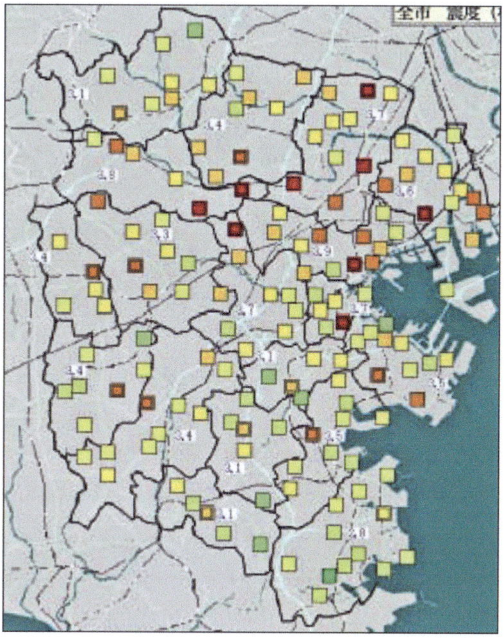

Fig. 21.8 Distribution of accelerographs

motion. The damage map is displayed with other information such as locations of hospitals, refuges and major roads for emergency transportation (Midorikawa, 2004).

21.5.5 Tokyo Gas – Supreme System

In order to avoid earthquake risks due to leakage of gas from breakage of buried pipes, Tokyo Gas Co. Ltd. has developed and put into use a real-time safety control system, SUPREME (former name SIGNAL, http://www.tokyo-gas.co.jp/signal/).

The system monitors the earthquake motion at 3,800 district regulators using spectrum intensity sensors, interprets the data, and assesses gas pipe damage in order to decide whether or not the gas supply should be interrupted (Yamazaki et al., 1995). Sensor locations are illustrated in Fig. 21.9. Spectrum intensity sensors computes the Housner Intensity (Housner, 1961) based on the integral of the 5% damped response spectra between the periods of 0.1 and 2.5 s.

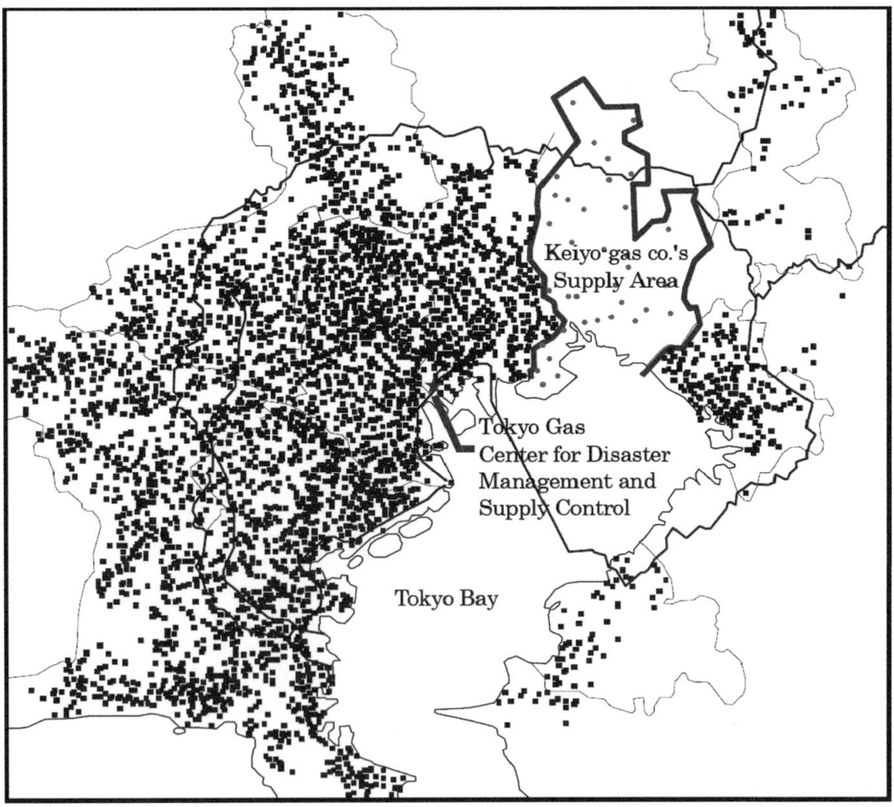

Fig. 21.9 Sensor locations of the SUPREME system (after Nakayama, 2004)

The communication between the sensors and headquarters where the shut-down decision is made by computer relies on two kinds of channels. Three hundred and thirty two stations are connected to the corporate wireless network and the remaining 3,500 stations rely on the ordinary telephone network. The gas supply at the district regulator is shut off when the SI value exceeds 30–40 cm/s.

21.6 Conclusions

Impact of large earthquakes in urban and critical facilities and infrastructure can be reduced by timely and correct action after a disastrous earthquake. Today's technology permits for the assessments of the distribution of strong ground motion and estimation of building damage and casualties within few minutes after an earthquake.

The reduction of casualties in urban areas immediately following an earthquake can be improved if the location and severity of damages can be rapidly assessed by the information from Rapid Response Systems. The emergency response capabilities can be significantly improved to reduce casualties and facilitate evacuations by permitting rapid, selective and effective deployment of emergency operations.

The ground motion measurement hardware, data transmission systems and the loss assessment methodologies and software needed for the implementation of such Earthquake Rapid Response Systems have reached to a degree of development that can ensure the feasible application of such systems and services throughout the world.

The reduction of the uncertainties inherent in the basic ingredients of earthquake loss assessment is an important issue that needs to be tackled in the future for viability and reliability of rapid loss assessments. We believe that the results of the relevant EU projects (such as NERIES and SAFER), as well as the recently started Global Earthquake Model (www.globalquakemodel.org) public-private partnership will provide the correct directions and developments in this regards.

References

Anderson E (2008) Central American Probabilistic Risk Assessment (CAPRA): objectives, applications and potential benefits of an open access architecture. Global Risk Forum, GRF Davos, Switzerland, August 2008

ATC-40, Applied Technology Council (1996) Seismic evaluation and retrofit of concrete buildings, vol 1. Applied Technology Council, Redwood City, CA

ATC-55, Applied Technology Council (2005) Improvement of nonlinear static seismic analysis procedures, FEMA 440, Redwood City, California, p 392

Badal J, Vazquez-Prada M, Gonzalez A (2005) Preliminary quantitative assessment of earthquake casualty and damage. Nat Hazards 34:353–374

Bal IE, Crowley H, Pinho R (2008a) Displacement-based earthquake loss assessment for an earthquake scenario in Istanbul. J Earthquake Eng 11(2):12–22

Baur M, Bayraktarli Y, Fiedrich F, Lungu D, Markus M (2001) EQSIM – a GIS-based damage estimation tool for Bucharest. Earthquake hazard and countermeasures for existing fragile buildings, Bucharest

Bommer J, Pinho R, Crowley H (2006) Using a displacement-based approach for earthquake loss estimation. In Wasti ST, Ozcebe G (ed) Advances in earthquake engineering for urban risk reduction. Springer, New York

Borcherdt RD (1994) Estimates of site-dependent response spectra for design (method and justification). Earthquake Spectra 10:617–654

Böse M (2006) Earthquake early warning for Istanbul using artificial neural networks. PhD thesis, University of Karlsruhe, Germany

Calvi GM, Pinho R, Magenes G, Bommer JJ, Restrepo-Vélez LF, Crowley H (2006) The development of seismic vulnerability assessment methodologies over the past 30 years. ISET J Earthquake Tech 43(4):75–104

Coburn A, Spence R (2002) Earthquake protection, 2nd edn. Wiley, Chichester

Corine Land Cover (1999) European Environment Agency, Brussels, Belgium

Couture R, Evans SG, Locat J (2002) Introduction. Nat Hazards 26(Special Issue):1–6

Crowley H, Pinho R, Bommer JJ (2004) A probabilistic displacement-based vulnerability assessment procedure for earthquake loss estimation. Bull Earthquake Eng 2(2):173–219

Demircioglu M, Erdik B, Hancilar M, Sesetyan U, Tuzun K, Yenidogan C, Zulfikar AC (2009) Technical manual – earthquake loss estimation routine ELER-v1.0. Department of Earthquake Engineering, Bogazici University, Istanbul

Di Pasquale G, Ferlito R, Orsini G, Papa F, Pizza AG, Van Dyck J, Veneziano D (2004) Seismic scenarios tools for emergency planning and management. In: Proceedings of the XXIX general assembly of the European Seismological Commission, Potsdam, Germany

EDIM, Earthquake Disaster Information System for the Marmara Region, Turkey, http://www.cedim.de/EDIM.php

Eguchi RT, Goltz JD, Seligson HA, Flores PJ, Blais NC, Heaton TH, Bortugno E (1997) Real-time loss estimation as an emergency response decision support system: the early post-earthquake damage assessment tool (EPEDAT). Earthquake Spectra 13(4):815–832

Elnashai AS, Hampton S, Karaman H, Lee JS, Mclaren T, Myers J, Navarro C, Sahin M, Spencer B, Tolbert N (2008) Overview and applications of Maeviz-Hazturk 2007. J Earthquake Eng 12(S2):100–108

EMS-98, European Seismic Commission Working Group on Macroseismic Scales. European Macroseismic Scale, Luxembourg http://www.gfz-potsdam.de/pb5/pb53/projekt/ems/eng/index_eng.html

Erdik M, Aydinoglu N (2002) Earthquake performance and vulnerability of buildings in Turkey: report prepared for World Bank Disaster Management Facility, Washington, DC

Erdik M, Aydinoglu N, Fahjan Y, Sesetyan K, Demircioglu M, Siyahi B, Durukal E, Ozbey C, Biro Y, Akman H, Yuzugullu O (2003b) Earthquake risk assessment for Istanbul metropolitan area. Earthquake Eng Eng Vib 2(1):1–25

Erdik M, Cagnan Z, Zulfikar C, Sesetyan K, Demircioglu MB, Durukal E, Kariptas C (2008) Development of rapid earthquake loss assessment methodologies for Euro-MED region. In: Proceedings of the 14th world conference on earthquake engineering, Paper ID: S04-004

Erdik M, Fahjan Y (2006) Damage scenarios and damage evaluation in 'assessing and managing earthquake risk.' In: Oliveira CS, Roca A, Goula X (eds), Assessing and managing earthquake risk. Springer, Netherlands, pp 213–237

Erdik M, Fahjan Y, Özel O, Alçık H, Mert A, Gül M (2003a) Istanbul earthquake rapid response and early warning system. Bull Earthquake Eng 1:157–163

Eurocode 8, CEN (2003) Eurocode 8: design of structures for earthquake resistance – Part 1: General rules, seismic actions and rules for buildings, prEN 1998-1, Doc CEN/TC250/SC8/N335, Comité Européen de Normalisation, Brussels, Belgium

Fajfar P (2000) A nonlinear analysis method for performance-based seismic design. Earthquake Spectra 16(3):573–592 Aug

FEMA-356 (2000) Prestandard and commentary for the seismic rehabilitation of building. Federal Emergency Management Agency and National Institute of Building Science, Washington, DC

FEMA (2003) HAZUS-MH technical manual. Federal Emergency Management Agency, Washington, DC

FEMA-440 (2005) Improvement of nonlinear static seismic analysis procedure. Federal emergency management agency, Washington, DC

FEMA, Federal Emergency Management Agency (2006) HAZUS-MH MR2 technical manual. Federal Emergency Management Agency, Washington, DC

Giovinazzi S (2005) Vulnerability assessment and the damage scenario in seismic risk analysis. PhD thesis, Department of Civil Engineering of the Technical University Carolo-Wilhelmina, Braunschweig and Department of Civil Engineering, University of Florence, Italy

Grünthal G (ed) (1998) European Macroseismic Scale 1998 (EMS-98). Cahiers du Centre Européen de Géodynamique et de Séismologie 15, Centre Européen de Géodynamique et de Séismologie, Luxembourg, 99pp

Housner GW (1961) Vibration of structures induced by seismic waves. Part I: Earthquakes. In: Harris CM, Crede CE (eds) Shock and vibration handbook. McGraw-Hill, New York, NY, pp 50-1–50-32

IBC International Building Code (2006) International Code Council, USA

Jaiswal KS, Wald DJ (2008) Creating a global building inventory for earthquake loss assessment and risk management. U.S.G.S. Open file report 2008-1160, Golden, 109 pp

Jaiswal KS, Wald DJ (2010) Development of a semi-empirical loss model within the USGS Prompt Assessment of Global Earthquakes for Response (PAGER) System. Proceedings of 9th US and 10th Canadian conference on earthquake engineering: reaching beyond borders, July 25–29, 2010 Toronto, ON, Canada

Kircher CA, Whitman RV, Holmes WT (2006) HAZUS earthquake loss estimation methods. Nat Hazards Rev 7(2):45–59

Koehler N, Wenzel F, Erdik M, Zschau J, Boese M (2007) An earthquake disaster information system for the Marmara region in Turkey (EDIM), Geophysical research abstracts, vol 9, 02006, 2007, SRef-ID: 1607-7962/gra/EGU2007-A-02006, European Geosciences Union 2007

Lagomarsino S, Giovinazzi S (2006) Macroseismic and mechanical models for the vulnerability and damage assessment of current buildings. Bull Earthquake Eng 4:415–443

Larionov V, Frolova N, Ugarov A (2000) Approaches to vulnerability evaluation and their application for operative forecast of earthquake consequences. In: Ragozin A (ed) All-Russian conference "Risk-2000". Ankil, Moscow, pp 132–135

Lin KW, Wald DJ (2008) ShakeCast manual. US Geological Survey Open-file report 2008-1158

Markus M, Fiedrich F, Leebmann J, Schweier C, Steinle E (2004) Concept for an integrated disaster management tool. In: Proceedings of the 13th world conference on earthquake engineering, Vancouver, BC

Midorikawa S (2004) Dense strong-motion array in Yokohama, Japan, and its use for disaster management. NATO meeting, Kusadasi, Turkey

Midorikawa S (2005) Dense strong-motion array in Yokohama, Japan, and its use for disaster management. In: Gulkan P, Anderson JG (eds) Directions in strong motion instrumentation. Springer, Netherlands, pp 197–208

Molina S, Lindholm C (2005) A logic tree extension of the capacity spectrum method developed to estimate seismic risk in Oslo Norway. J Earthquake Eng 9(6):877–897

Nakayama W, Shimizu Y, Koganemaru K (2004) Development of super dense realtime disaster mitigation system for urban gas supply network. J Jpn Assoc Earthquake Eng, Special Issue

Picozzi M, SAFER, EDIM Work Groups (2008) Seismological and early warning activities of the SOSEWIN. Geophysical research abstracts, vol 10, EGU2008-A-07001, 2008 SRef-ID: 1607-7962/gra/EGU2008-A-07001

Porter K, Jaiswal AK, Wald DJ, Greene M, Comartin C (2008) WHE-PAGER project: a new initiative in estimating global building inventory and its seismic vulnerability, 14WCEE, Beijing

RISK-UE (2004) The European risk-UE project: an advanced approach to earthquake risk scenarios. (2001–2004) www.risk-ue.net
Robinson D, Fulford G, Dhu T (2005) EQRM: geoscience Australia's earthquake risk model. Geoscience Australia record 2005/01, Canberra, Geoscience Australia, p 151
Robinson D, Fulford G, Dhu T (2006) EQRM: geoscience Australia's earthquake risk model: technical manual version 3.0, Book Bib ID 3794291, Geoscience Australia
Roca A, Goula X, Susagna T, Chavez J, Gonzalez M, Reinoso E (2006) A simplified method for vulnerability assessment of dwelling buildings and estimation of damage scenarios in Spain. Bull Earthquake Eng 4:2:141–158
Samardjieva E, Badal J (2002) Estimation of the expected number of casualties caused by strong earthquakes. Bull Seism Soc Am 92(6):2310–2322
Sousa ML, Campos Costa A, Carvalho A, Coelho E (2004) An automatic seismic scenario loss methodology integrated on a geographic information system. 13th world conference on earthquake engineering, Vancouver, BC, Canada, August 1–6, 2004, Paper No. 2526
Stafford PJ, Strasser FO, Bommer JJ (2007) Preliminary report on the evaluation of existing loss estimation methodologies. Report prepared for EU FP6 NERIES Project, Department of Civil & Environmental Engineering, Imperial College, London
Teng TL, Wu YM, Shin TC, Tsai YB, Lee WHK (1997) One minute after: strong motion map, effective epicenter, and effective magnitude. Bull Seism Soc Am 87:1209–1219
Trendafiloski G, Wyss M, Rosset Ph (2009b) Loss estimation module in the new generation software QLARM. In: Proceedings of the 2nd international workshop on disaster casualties, Cambridge. http://www.wapmerr.org/QLARM_Paper-Cambridge-def.pdf
Trendafiloski G, Wyss M, Rosset Ph, Marmureanu G (2009a) Constructing city models to estimate losses due to earthquakes worldwide: application to Bucharest, Romania. Earthquake Spectra 25(3):665–685 doi:10.1193/1.3159447
Tsai YB, Wu YM (1997) Quick determination of magnitude and intensity for seismic early warning. In: Proceedings of the 29th IASPEI meeting, Thessaloniki, Greece
Wald DJ, Earle PS, Allen TI, Jaiswal K, Porter K, Hearne M (2008) Development of the U.S. geological survey's pager system (Prompt assessment of global earthquakes for response). In: Proceedings of the 14th world conference on earthquake engineering, Beijing, China, October 12–17
Wald DJ, Quitoriano V, Heaton TH, Kanamori H (1999) Relationships between peak ground acceleration, peak ground velocity, and modified Mercalli intensity in California. Earthquake Spectra 15(3):557–564
Wald DJ, Worden BC, Quitoriano V, Pankow KL (2005) ShakeMap manual: technical manual, user's guide, and software guide. US geological survey techniques and methods, Book 12, Section A, Chap. 1. US Geological Survey, Reston, Virginia, p 132
Wenzel F, Marmuraenu G (2007) Rapid earthquake information for Bucharest. J Pure Appl Geophys 164(5):929–939 May
Whitman RV, Anagnos T, Kircher CA, Lagorio HJ, Lawson RS, Schneider P (1997) Development of a national earthquake loss estimation methodology. Earthquake Spectra 13(4): 643–661
Wu YM, Chung JK, Shin TC, Hsiao NC, Tsai YB, Lee WHK, Teng TL (1999) Development of an integrated seismic early warning system in Taiwan – case for the Hualien area earthquakes. TAO 10:719–736
Wu YM, Shin TC, Tsai YB (1998) Quick and reliable determination of magnitude for seismic early warning. Bull Seism Soc Am 88:1254–1259
Wu YM, Teng TL (2002) A virtual sub-network approach to earthquake early warning. Bull Seism Soc Am 92:2008–2018
Yamazaki F, Katayama T, Noda S, Yoshikawa Y, Ohtani Y (1995) Development of large-scale city-gas network alert system based on monitored earthquake ground motion. In: Proceedings of the JSCE, 525/I-33, pp 331–340 (in Japanese)

Yeh CH, Loh CH, Tsai KC (2003) Development of earthquake assessment methodology in NCREE. Joint NCREE/JRC workshop international collaboration on earthquake disaster mitigation research, Rep. No. NCREE-03-029, 83–92

Zonno G, Carvalho A, Franceschina G, Akinci A, Campos Costa A, Coelho E, Cultrera G, Pacor F, Pessina V, Cocco M (2009) Simulating earthquake scenarios in the European Project LESSLOSS: the case of Lisbon. In: Mendes-Victor LA, Oliveira CS, Azevedo J, Ribeiro A (eds) The 1755 Lisbon earthquake: revisited. Springer, Dordrecht

Chapter 22
Catastrophe Micro-Insurance for Those at the Bottom of the Pyramid: Bridging the Last Mile

Haresh C. Shah

Abstract Helping the poor in developing societies from the ravages of catastrophic events such as earthquakes should be one of the highest priorities of any society. It is always this sector of any society that takes a major portion of brunt and negative consequences. Providing them with a safety umbrella of decent and safe housing as well as economic relief should be on top of strategic risk reduction policy. In this paper, we discuss as to how those who are at the bottom of the economic ladder can be assisted by developing a safety economic net in the form of micro insurance. Such a product should be a sustainable, profitable and of true economic relevance to all the stake holders. The solution of micro insurance for the poor should help the insured, the insurance/reinsurance food chain and the governments. Any micro insurance solution that is based on pure altruistic and philanthropic considerations cannot be sustained as a viable solution in the long term. Two pilot projects currently under development by the authors are discussed in this paper. One micro insurance product is for the earthquake risk transfer for the poor in China based on two triggers. The second product is for the State of Gujarat in India where micro finance institutions and those who borrow from such institutions are both protected from earthquake risk. If we can truly implement some of these pilot projects on a wide basis, we can provide economic assistance through global insurance schemes to those at the bottom of the pyramid. This will be a much needed bridge to the last mile.

22.1 Introduction

It is widely known fact that those who are at the bottom of the economic ladder suffer the most in terms of loss of life, economic losses and social disruptions. These are the citizens who are at the Bottom of the Pyramid (BOP). Over and over again for the past century we have seen death and destruction, social and economic disruption

---

H.C. Shah (✉)
Department of Civil and Environmental Engineering, Stanford University, Stanford, CA 94305-4020, USA; RMS, Inc., Newark, CA 94560, USA; WSSI, Newark, CA, USA;
e-mail: Haresh.Shah@rms.com

M. Garevski, A. Ansal (eds.), *Earthquake Engineering in Europe*, Geotechnical, Geological, and Earthquake Engineering 17, DOI 10.1007/978-90-481-9544-2_22,

and just total devastation in poorer communities and poorer nations. Recent catastrophes of the past decade have amply demonstrated that on the global scale, poorer nations have taken the brunt of losses. The Asian Tsunami of 2004, the Kashmir earthquake of 2005, the Padang, Sumatra earthquake of 2009 and the most recent Haiti earthquake of 2010 have clearly demonstrated the high life, economic and social vulnerability of poor nations and of societies who are at the bottom of the economic pyramid.

The reasons for this disproportionate impact are obvious. These are the nations with limited resources, poor governance, competing demands on limited resources and to some degree a lack of awareness about strategic options for mitigating the impacts of catastrophes. With poverty comes the challenge of selecting between various risks and which ones to worry about and which ones to stop worrying about. It is often said that the appetite for risk (opposite of risk aversion) in developing countries and of those at the BOP are quite high. These individuals have so many things to worry about, including food, shelter, education for their children, day to day livelihood, etc that protecting themselves and their families from catastrophe risk is not high on their agenda. The perception is that unlike micro-finance where they get upfront financial assistance to lift themselves out of poverty into productive living, micro-insurance requires upfront investment (in terms of paying for catastrophe risk insurance premium) for an event that may or may not occur over many years. These are real issues and any solution that may help these sectors of societies must incorporate appropriate considerations for the solutions to be effective and widely used.

In this paper, we will explore some of these issues and propose that one of the most important strategies for helping those at the BOP is to develop a micro-insurance product. Such a product would put money in the hands of those who need it the most and who can put those funds to most appropriate use to recover from the catastrophic event rapidly. This strategy is not in any way contradictory to pre-event mitigation measures such as better construction standards and techniques. Those measures must be pursued and must be implemented whenever possible. Over the past five decades, we have seen great progress in improving the understanding of catastrophic events and measures for improving the vulnerability of housing and infra-structures. However, that still leaves the vulnerability of civil societies to such events at risk due to lack of resources available to poorer sections of the societies to rebuild their lives and livelihoods.

22.2 A Look at Vulnerabilities of Those at the BOP

Asia is the home to the largest number of "poor" people of the world. Even though the population at the bottom of the pyramid is most vulnerable to catastrophe events, and their needs for mitigating catastrophe risks are most urgent, the potential solutions are not obvious or available. Let us look at the demography of Asia as an example. Table 22.1 shows the 15 largest mega-urban cities of the world. 11 of these 15 cities are in Asia.

Table 22.1 Top 15 urban conglomerates (2009 data)

Sr. no.	Mega-urban city	Country	Continent	Population
1.	Tokyo	Japan	Asia	33,800,000
2.	Seoul	South Korea	Asia	23,900,000
3.	Mexico City	Mexico	North America	22,900,000
4.	Delhi	India	Asia	22,400,000
5.	Mumbai	India	Asia	22,300,000
6.	New York	USA	North America	21,900,000
7.	Sao Paulo	Brazil	South America	21,000,000
8.	Manila	Philippines	Asia	19,200,000
9.	Los Angeles	USA	North America	18,000,000
10.	Shanghai	China	Asia	17,900,000
11.	Osaka	Japan	Asia	16,700,000
12.	Kolkata	India	Asia	16,000,000
13.	Karachi	Pakistan	Asia	15,700,000
14.	Guangzhou	China	Asia	15,300,000
15.	Jakarta	Indonesia	Asia	15,100,000

(See Munich Re group; Megacities – Megarisks 2004)

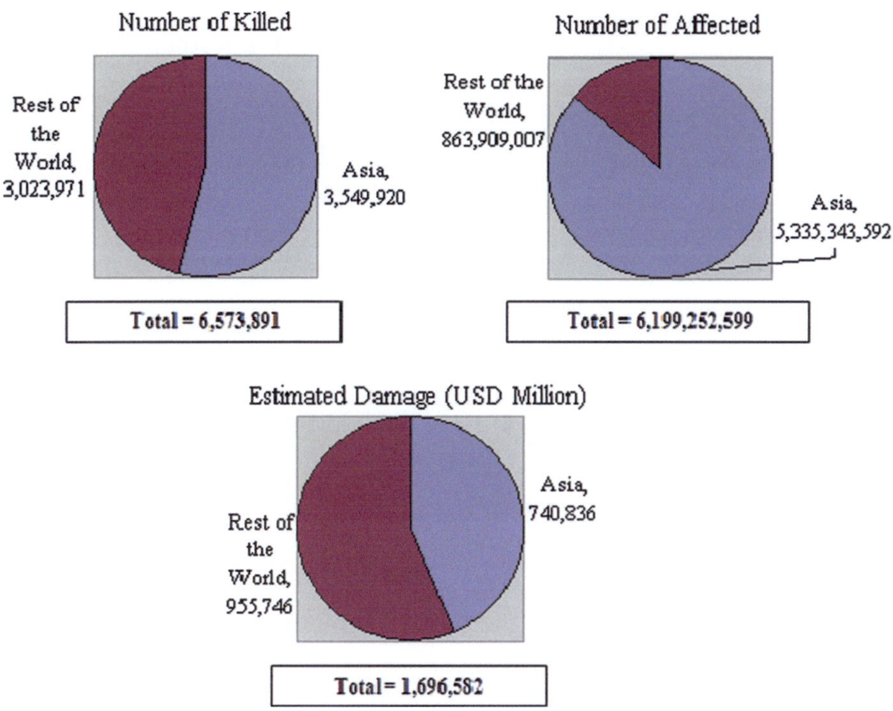

Fig. 22.1 All perils (1960–2008). Note: Perils include drought, earthquake (including Tsunami), epidemic, flood and storm. Source: EM-DAT: The OFDA/CRED International Disaster Database – www.emdat.be – Université Catholique de Louvain – Brussels – Belgium

If we look at the economic and life losses due to all perils (droughts, earthquakes, epidemics, floods and storms), it is clear that Asia dominates the global statistics. Figure 22.1 shows this data very clearly. The correlation between the economic development of these nations, their populations and the level of losses is obvious.

Figure 22.2 shows the same trends for earthquake related losses.

It has been observed that countries with low GDP per capita also have very low insurance penetration. Figure 22.3 shows this trend. This means that when a population is impacted by a catastrophe event, there are no resources to rebuild except to rely on international assistance. The poorer sections of the societies suffer the most under this state of affairs. Recent event of January 2010 in Haiti is a classic example. There have been numerous cases of farmers and urban poor committing suicides after major disasters because they cannot recover from economic ruin brought about by the catastrophe. In such cases, providing cash to these people could spell a difference between survival and economic ruin. Again, it is important to point out that post disaster assistance should not be a replacement for pre-disaster mitigation measures. It is only to make sure that civil societies made up of many at the BOP have the means to recover and rebuild after any catastrophe. It is important to mention at this time that with the massive urbanizations of developing countries, loss of life and

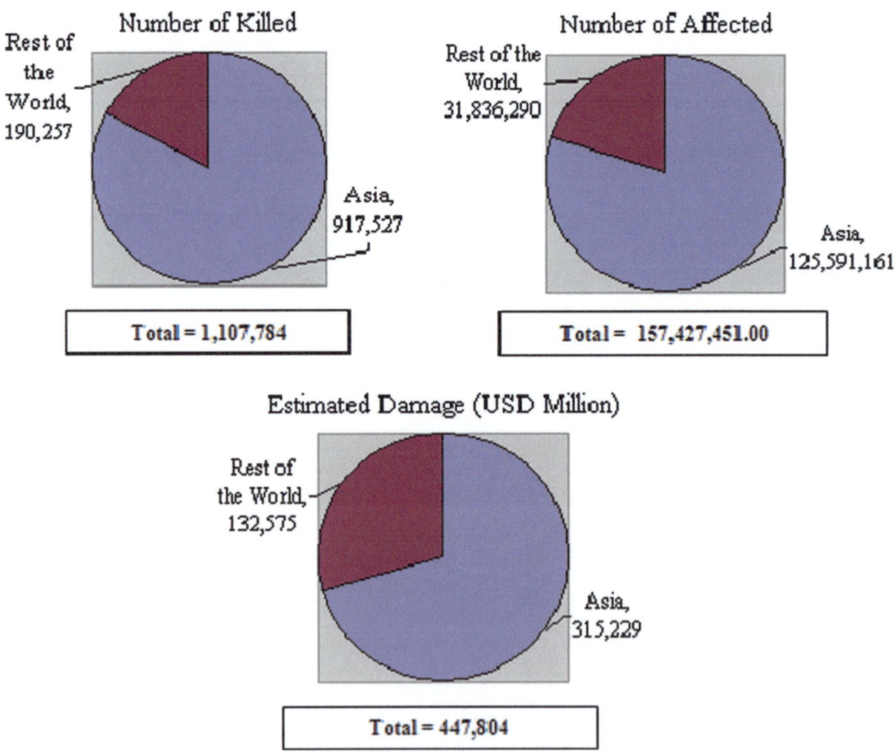

Fig. 22.2 Earthquake (1960–2008). Source: EM-DAT: The OFDA/CRED International Disaster Database – www.emdat.be – Université Catholique de Louvain – Brussels – Belgium

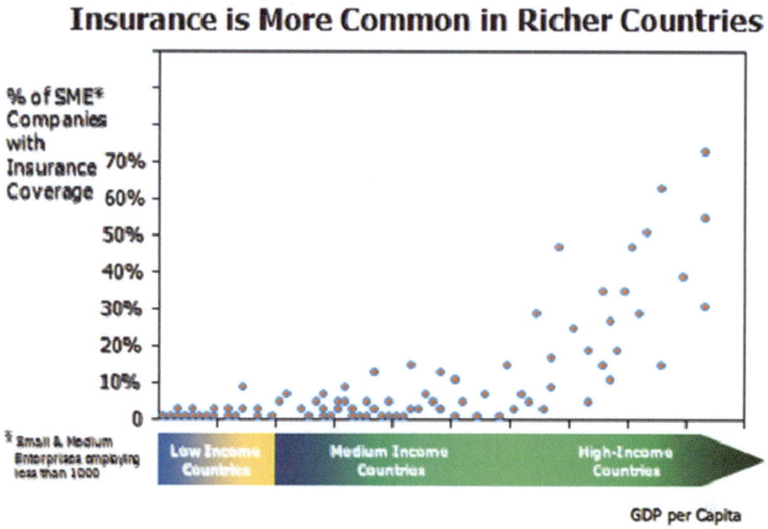

Fig. 22.3 Insurance is more common in richer countries

Table 22.2 Catastrophe insurance penetration in developing countries

Country	Catastrophe insurance protection
India	Under 0.5%
Philippines	Under 0.3%
Iran	Under 0.05%
Romania	Under 5%
Bulgaria	Under 3%
China	Under 0.5%
Turkey	Under 17%

(See Munich Re group Knowledge series, topics Geo 2004)

economic disruptions will increase. For catastrophes such as those brought about by climatic changes such as droughts, storms, floods, epidemics, it is expected that the frequency and severity of these disruptions will increase and those who are the people at the BOP will be the recipients of the negative consequences.

As an example, Table 22.2 provides insurance penetration of some of the developing countries. It is in these countries that a micro-insurance product may have the greatest applicability and impact. In the next section, we will describe two such pilot projects in India and China.

22.3 Micro-Insurance Pilot Projects

The biggest challenge in developing a viable and sustainable business model for micro-insurance is the cost of underwriting and the inability of those at the BOP to pay the premium. Another challenge and an obstacle is to reach millions of urban and rural poor who could benefit from a viable product. It must be kept in mind that any "solution" for micro-insurance that is based on the concept of charity or philanthropy is not sustainable. The only way to make this concept work on a sustained and affordable basis is to make sure that all the key players who participate in this venture must "win". The key players are the urban and the rural poor for whom micro-insurance product is designed, the government, the insurance and reinsurance industry and the civil society. Unless there is some incentive for all these players to adopt the solution of micro-insurance, it will not work. Also, it should be understood that there is no one solution to micro-insurance that will fit all situations. What may work in China may not be possible to implement in (say) India. With this in mind, we have developed two pilot projects; one for China and one for the State of Gujarat in India. Our goal in developing these two pilot projects is activation of cat micro insurance in sustainable and viable ways, and to make it part of the solution of the fundamental problem of poverty and lack of opportunities for more than 2 billion people worldwide. In that pursuit we will facilitate activation of compatible market opportunities for the global insurance industry. In the following pages, we will describe these two pilot projects.

22.3.1 Rural China Double Trigger Earthquake Micro-Insurance Program

During the past few years, as we have been developing sustainable and financially viable micro-insurance products, we came to realize that those who are at the Bottom of the Pyramid (BOP) did not trust insurance companies in general and they were of the opinion that when catastrophe hit them and when they needed the most help, insurance companies balked at providing the funds. On the other

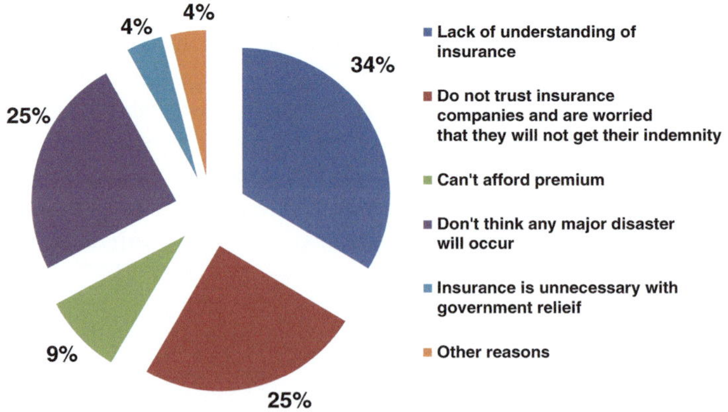

Fig. 22.4 The poor do not trust insurance. Courtesy: Institute of Integrated Risk Management of ADREM, Beijing Normal University 2008, "Summary of National Survey on Residential Catastrophe Insurance Reasons for Reluctance to Buy"

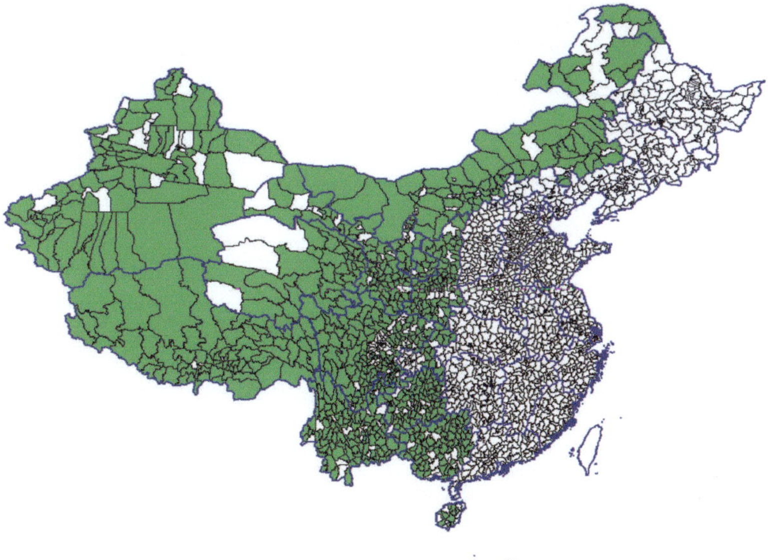

Fig. 22.5 Program coverage – rural/underdeveloped China

hand, insurance companies felt that the cost of insuring these BOP people was more than the premium they would get. In general the prevailing feeling was that "as a business proposition, risk transfer of poor citizens offered very little financial reward". This feeling is slowly changing but it still acts as a challenge for developing micro-insurance products. Figure 22.4 shows how poor people in China view micro-insurance as an alternative to managing catastrophe risk.

Learning from these observations, the China micro-insurance program developed a double trigger mechanism. The purpose of these two triggers is that a lower trigger when a moderate size earthquake occurs, a small amount of cash would be available to all those who experienced the event but did not really have catastrophic damage. This way, more frequent events would provide more frequent cash payout to insured. The higher trigger (bigger earthquake with severe damage) which would occur infrequently would provide the full amount of insured value. Figure 22.5 shows the region of China (which is predominantly rural and poor) where the designed micro-insurance program will be introduced. (See Stojanovski, P. et al.)

The micro-insurance program details are given below:

Non-indemnity cat coverage

- Limit of 16,000 RMB
- Step policy (minimal operational costs) qualified earthquakes in a single county
  - Full limit (16,000 RMB) for a collapsed or red tagged (uninhabitable) house
  - No payment for other levels of damage

Parametric non-cat coverage

- Fixed cash payout for occurrence of qualified earthquakes[a] in a single county of rural China
- Mag. 6.5–8,500 RMB
- Mag. 8 or higher 1,000 RMB

[a]Treatment of aftershocks needs to be detailed.

Exploring the feasibility of charging 10 RMB as the annual premium, the total country wide premium would be: 10 RMB × (222 million population/4 family members per family) = 550 million RMB.

The program expenses are:

Layer	Expense		Amt. (mil. RMB)
PI: Primary Ins.	Risk loaded AAL	193 + 0.2*344	262
AAL = 193 m	Fixed costs: 2.5% of collected premium	0.025*550	14
σ = 344 m	Loss adjustment: 5% of layer AAL	0.05*193	8
RI: Reinsurance	Risk loaded AAL	13.9 + 0.2*187	51
AAL = 13.9 m			
σ = 187 m	Brokerage: 25% of premium	0.25*51.3	13
GOV: Government	Risk loaded AAL	4.5 + 0.2*182	41
AAL = 4.5 m			
σ = 182 m	Brokerage: 11% of premium	0.11*40.9	4
All layers			393
Countrywide premium (100% participation)			550
Premium over expenses			157

Figure 22.6 provides the program structure. It can be seen that such a product can protect almost all the rural habitants of China from earthquake induced financial catastrophe. The benefits of such a program are:

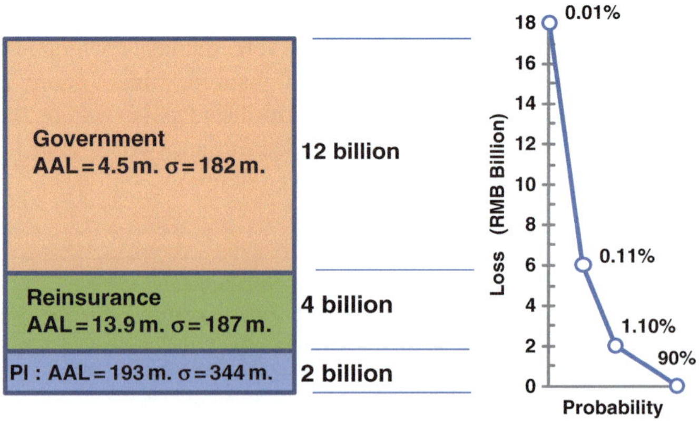

Fig. 22.6 Program structure

- Affordable for the farmers
- Parametric trigger provides payouts for smaller and more frequent events and addresses "insurance trust" issue
- Retains catastrophic non-indemnity "step policy" coverage of the original program
- Commercially viable and sustainable for all participants (primary, reinsurance, government)

Fronting operations and technology details need to be elaborated in more detail. From this example, it can be seen that an affordable product for those who are at the bottom of the pyramid can be developed which over a long period of time not only will result in providing profit potential to the insurers and reinsurers but will also improve the confidence of insured in the risk transfer mechanism. Finally, such a program could help governments in providing high liquidity for rapid reconstruction and renovation.

22.3.2 Micro-Insurance Product to Manage Earthquake Risk in Gujarat, India

Gujarat is prone to natural catastrophes – earthquakes, cyclones, floods, and drought. It is not unexpected to have two catastrophes hit the state in the same year. The economic consequences are significant. In the 2001 Bhuj earthquake more than 20,000 people were killed, more than 167,000 were injured, one million homes were destroyed, and economic loss reached USD 2 billion. Gujarat is considered as one of the more economically advanced Indian states. Yet, significant part of its population leaves below the poverty line. The following Fig. 22.7 portrays the percentage of the district population in Gujarat that falls below the Indian poverty line. In very few districts it is under 25%, majority of the districts are between 25 and 35%, some falling in the 35–45% range, and in extreme cases reaching 65% or more. The Figure below shows the percent of district populations below the poverty line in Gujarat. (See Johari, P. et al.)

Gujarat is one of the states in India that has a well developed network of financial institutions serving the low income people on commercial principles. The banking system is comprised of commercial banks, regional rural banks and cooperative banks. Also a significant number of micro finance institutions (MFIs) have mushroomed in the form of non-banking financial companies and self helping groups (SHG's), section 25 companies, societies, trusts, and other forms of MFI's. For this pilot project, we decided to insure the MFIs who are giving loans to those who are at the bottom of the pyramid rather than insuring individuals who have borrowed from the MFIs. This decision removes a major hurdle of reaching rural inhabitants and also the barrier that these poor people cannot afford to pay the needed small premium. By insuring the MFIs, we are developing a product that would payout to the MFIs the micro loans they have made and still put some cash in the hands

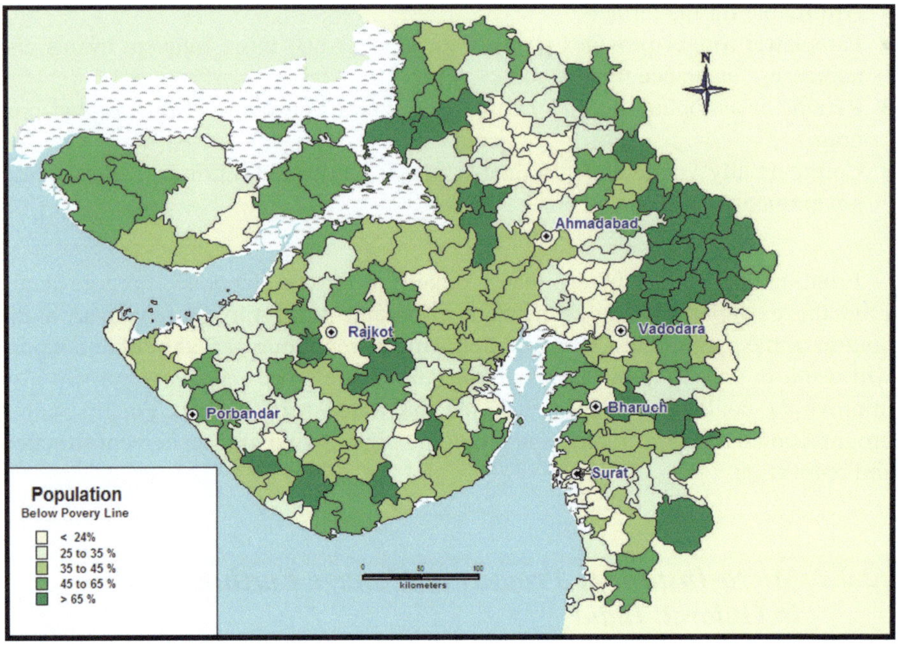

Fig. 22.7 Percentage of district population below poverty line

of those who have suffered due to an earthquake catastrophe. The loans taken out by these individual rural or urban poor will then be forgiven. As mentioned earlier, catastrophe events mostly affect those who are the poorest in any society and the scheme proposed in this pilot project will help mitigate that problem.

At the time of writing this paper, all the data collection for micro loans in Gujarat has not been complete. However based on what we have collected, the following hypothetical scenario is created. It is assumed that all the MFIs together have loaned the funds all over Gujarat.

- Total outstanding micro-loans 2,000,000,000 INR
- Number of outstanding loans (households served) 225,000
- Average (per loan) outstanding amount 8,888 INR
- Operating area: whole of Gujarat

We applied insurance industry standards for risk quantification and pricing, which would be necessary to ensure transparency in the risk transfer process within the insurance verticals.

Our assumption was that the premiums must be affordable to the poor, with stable subsidies if needed (Part of MFI's product offerings, Government, Diaspora and other "money pools", ...). The goal of this investigation is to determine how much coverage could be achieved with individual premiums as low as 7 INR. In this

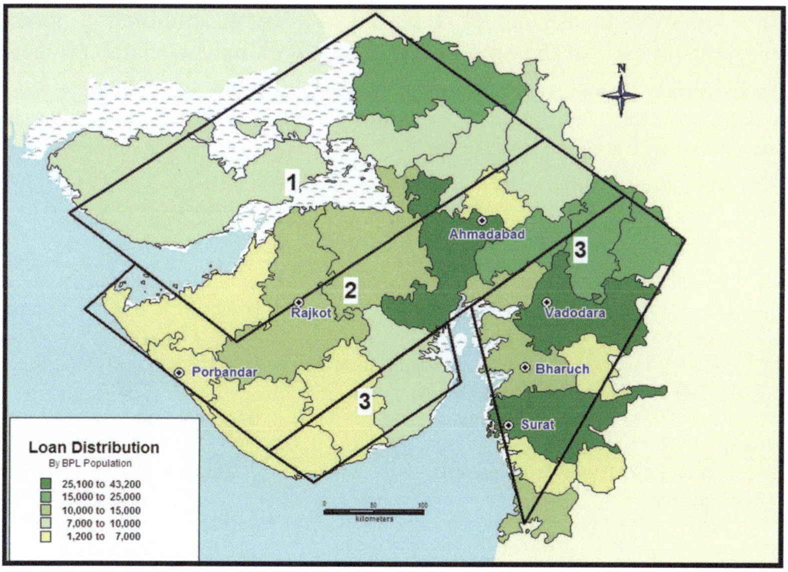

Fig. 22.8 Trigger location

case, the aggregate premium (assuming 100% participation) would be 225,000*7 = 1,575,000 INR. This premium will be examined against the risk cost and customary associated expenses in risk transfer transactions. To minimize the operational costs we have designed a parametric policy based on occurrence of qualifying earthquakes in Gujarat.

The policy pays out when a qualifying (trigger) earthquake of certain magnitude occurs in a predefined geographic area (box 1, box 2, or box 3) in Fig. 22.8 according to the rules defined below. In fact, we did not modify the payout rules as function of the bounding boxes. These boxes were introduced for the next more detailed investigation, in case it becomes necessary to vary the trigger and payout rules by box in order to minimize the basis risk.

Occurrence of a qualifying earthquake (trigger) would be certified by authorized and reputable public scientific agency (e.g. Indian Geological Survey). In this case the trigger is an earthquake of magnitude 5.25 or higher in box 1, box 2, or box 3.

We have designed the payout rules as follows:

For the MFI: Portion of the outstanding loans in the district where earthquake epicenter is, defined with the ratio of area with ground shaking ≥ 0.1 g (modeled) and the district area.

For the loan holders: Fixed amount to all loan holders in the district where the epicenter is, as function of the magnitude:

$$M > 7.25 \quad 1,000 \text{ INR}; \quad M > 6.75 \quad 750 \text{ INR}; \quad M > 6.25 \quad 500 \text{ INR}$$

RMS's India earthquake model was used to generate information about various size events in each of the boxes of Fig. 22.8, with the associated probabilities. Discussions of the model and results are beyond the scope of this paper. Based on the above payout rules and the total outstanding loans, the average annual losses and standard deviation for those losses are:

Layer	Expense		Amount (INR)
PI – Primary Ins.	Risk loaded AAL	476,000	1,198,000
AAL = 476,000 INR	Fixed costs: 2.5% of collected premium	+ 0.2*3,610,000 0.025*1,575,000	39,375
STD = 3,610,000 INR	Loss adjustment: 5% of layer AAL	0.05*476,000	23,800
Total expenses			1,261,175
Gujarat-wide premium (100% participation)		7*225,000	1,575,000
Premium over expenses			313,825

It can be seen from above simple calculations that even a 7 INR Premium per household per year (US 0.15 per year) would be able to help more than 4 million households, if such a product is made mandatory for all those who have taken out micro-loans. Such a product would not only protect the MFIs but will also relieve the poor from repaying the borrowed sum after the catastrophe event and will put some cash in their hands.

At the writing of this paper, a more realistic data base is being finalized for Gujarat and a product for multi-hazard catastrophe events is being designed. Such a product could include events such as earthquakes, droughts, floods, and wind. The purpose of demonstrating this product is to prove that by proper design of micro-insurance scheme, one could really assist a very large number of those who are at the bottom of the economic pyramid and help in restoring their lives and the vibrancy of the region to even pre-event level.

22.4 Concluding Remarks

Catastrophe events have devastated those who are already struggling with the normal everyday life sustenance. There are many technological solutions such as better construction on new dwellings or retro-fitting of existing dwellings. Globally there are many such schemes being promoted by governments, NGOs and academic institutions. These are truly valuable in minimizing the catastrophic loss of life and economy. Unfortunately, over the five decades during which time we have learned so much about the earthquakes and ways for designing and building dwellings and other structures, the devastations due to such events continue unabated. The micro-insurance products discussed in this paper may not save the lives of those who live in poorly constructed homes, but it will certainly help those who lived through the disaster towards putting their lives in order after the event. A true holistic solution would be a global thrust for better housing construction and a better way to help those who survived and who needs to deal with the economic calamity that follows for the poor of the world disaster after disaster. Those are the kinds of bridges we need to build for the last mile.

References

Johari P, Jindal A, Murthy B, Wagh S, Mohindra R, Dhore K, Satyanarayana P Multiperil pilot microinsurance program for Gujarat. In: 4th Microinsurance round table forum, Nanyang Technological University, Singapore, 8–10 Apr 2010

Munich Re Group; Knowledge Series, topics Geo 2004 Annual review: natural catastrophes 2004. Munchener Ruckversicherungs – Gesellschaft, awirtz@munichre.com, Order number 302-04321

Munich Re Group; Knowledge Series, Megacities-Megarisks 2004 Trends and challenges for insurance and risk management. Munchener Ruckversicherungs_Gessellschaft, awirtz@munichre.com, Order number 302-04271

Stojanovski P, Dong W, Wagh S, Mortgat C, Shah H; Rural China Double Trigger Earthquake Catastrophe Microinsurance Program. In: 4th Microinsurance round table forum, Nanyang Technological University, Singapore, 8–10 Apr 2010

Index

M. Garevski, A. Ansal (eds.), *Earthquake Engineering in Europe,* Geotechnical, Geological, and Earthquake Engineering 17, DOI 10.1007/978-90-481-9544-2,

MIX
Papier aus verantwortungsvollen Quellen
Paper from responsible sources
FSC® C105338

If you have any concerns about our products,
you can contact us on
ProductSafety@springernature.com

In case Publisher is established outside the EU,
the EU authorized representative is:
Springer Nature Customer Service Center GmbH
Europaplatz 3, 69115 Heidelberg, Germany

Printed by Libri Plureos GmbH
in Hamburg, Germany